BIOASSESSMENT AND MANAGEMENT OF NORTH AMERICAN FRESHWATER WETLANDS

BIOASSESSMENT AND MANAGEMENT OF NORTH AMERICAN FRESHWATER WETLANDS

Edited by

Russell B. Rader
Brigham Young University, Provo, Utah

Darold P. Batzer
University of Georgia, Athens, Georgia

Scott A. Wissinger
Allegheny College, Meadville, Pennsylvania

JOHN WILEY & SONS, INC.

New York • Chichester • Weinheim • Brisbane • Singapore • Toronto

This book is printed on acid-free paper. ♾

Published simultaneously in Canada.

Library of Congress Cataloging-in-Publication Data:

Rader, Russell Ben.
Bioassessment and management of North American freshwater wetlands / Edited by Russell B. Rader, Darold P. Batzer, and Scott A. Wissinger.
p. cm.
ISBN 0-471-35234-9 (cloth : alk. paper)
1. Wetland management—North America. 2. Environmental monitoring—North America. I. Batzer, Darold P. II. Title.

QH77.N56 R34 2001
333.91'8'097—dc21 2001022372

Printed in the United States of America.
10 9 8 7 6 5 4 3 2 1

Contents

Preface

The primary purpose of this book is to provide information on wetland bioassessment and management for resource managers. However, it could also be used as a textbook to supplement advanced undergraduate and graduate courses on ecology and conservation. There is no other book that reviews the application of bioassessment principles specific to wetlands. The management section emphasizes topics that are less known and often neglected by other textbooks on how to manage wetland resources.

Wetlands have recently emerged from an identity crisis. Since wetlands often alternate between dry and wet periods, they are neither truly aquatic nor truly terrestrial, but embody components of both. However, even though they share commonality and are often associated with other aquatic and terrestrial environments, they are clearly different from streams, rivers, deep lakes, and upland areas. In short, wetlands are a unique environment and should be recognized as a separate ecosystem on the terrestrial landscape.

As editors, we are particularly grateful to the individual authors who contributed their valuable time to make the production of this book possible. We also thank a number of colleagues who reviewed chapters. Their names are listed in the chapter acknowledgments with which they are affiliated.

It is our hope that this book will generate as much interest in managing, conserving, and protecting all wetlands as currently exists for some of our larger wetland habitats (e.g., the Everglades) and other freshwater environments (streams and rivers). For most states or provinces in North America, it is likely that a greater proportion of the land surface is covered with wetlands than any other freshwater environment.

Russell B. Rader
Department of Zoology
Brigham Young University

Darold P. Batzer
Department of Entomology
University of Georgia

Scott A. Wissinger
Departments of Biology and Environmental Science
Allegheny College

Contributors

RANDALL S. APFELBECK, Water Quality Specialist, Monitoring and Data Management Bureau, Montana Department of Environmental Quality, Helena, MT

DAROLD P. BATZER, Department of Entomology, University of Georgia, Athens, GA

BERND BLOSSEY, Biological Control of Non-Indigenous Plants Program, Department of Natural Resources, Cornell University, Ithaca, NY

JENNIFER L. BOGO, Environmental Science Department, Allegheny College, Meadville, PA

JULIA BOHNEN, University of Minnesota, Department of Horticultural Science, St. Paul, MN

STEPHEN C. BROWN, Manomet Center for Conservation Sciences, Manomet, MA

JOANNA BURGER, Division of Life Sciences, and Consortium for Risk Evaluation with Stakeholder Participation (CRESP), Environmental and Occupational Health Sciences Institute, Rutgers University, NJ

DAREN M. CARLISLE, Department of Fishery and Wildlife Biology, Colorado State University, Fort Collins, CO

KENNETH W. CUMMINS, California Cooperative Fishery Research Unit, Humboldt State University, Arcata, CA

THOMAS L. CRISMAN, Center for Wetlands, Department of Environmental Engineering Sciences, University of Florida, Gainesville, FL

SUSAN GALATOWITSCH, University of Minnesota, Department of Horticultural Science, St. Paul, MN

MARK C. GERNES, Minnesota Pollution Control Agency, St. Paul, MN

JEROEN GERRITSEN, Tetra Tech Incorporated, Owings Mills, MD

GORDON GOLDSBOROUGH, University Field Station (Delta Marsh) and Department of Botany, University of Manitoba, Winnipeg, Manitoba, Canada

CHARLES P. HAWKINS, Department of Fisheries and Wildlife, Utah State University, Logan, UT

SCOTT G. INGMIRE, Environmental Science Department, Allegheny College, Meadville, PA

JUDY C. HELGEN, Minnesota Pollution Control Agency, St. Paul, MN

MURRAY K. LAUBHAN, U.S. Geological Survey, Midcontinent Ecological Science Center, Fort Collins, CO

J. VAUN MCARTHUR, Savannah River Ecology Laboratory, Aiken, SC

RICHARD W. MERRITT, Department of Entomology, Michigan State University, East Lansing, MI

JOSEPH P. PRENGER, Wetland Biogeochemistry Laboratory, Soil and Water Science Department, University of Florida, Gainesville, FL

RUSSELL B. RADER, Department of Zoology, Brigham Young University, Provo, UT

STEVEN RATHBUN, Department of Statistics, University of Georgia, Athens, GA

VINCENT H. RESH, Department of Environmental Science, Policy and Management, Division of Insect Biology, University of California, Berkeley, CA

CURTIS J. RICHARDSON, Duke University Wetland Center, Nicholas School of the Environment, Durham, NC

JAMES E. ROELLE, U.S. Geological Survey, Midcontinent Ecological Science Center, Fort Collins, CO

DENNIS K. SHIOZAWA, Department of Zoology, Brigham Young University, Provo, UT

AARON S. SHURTLEFF, Department of Entomology, University of Georgia, Athens, GA

JOEL W. SNODGRASS, Department of Biological Sciences, Towson University, Towson, MD

R. JAN STEVENSON, Department of Zoology, Michigan State University, East Lansing, MI

JAN VYMAZAL, Ecology and Use of Wetlands, Praha Czech Republic and Duke University Wetland Center, Nicholas School of the Environment, Durham, NC

SCOTT A. WISSINGER, Departments of Biology and Environmental Sciences, Allegheny College, Meadville, PA, and Rocky Mountain Biological Laboratory, Crested Butte, CO

1 An Introduction to Wetland Bioassessment and Management

RUSSELL B. RADER

The ability of humans to manage the natural world depends on the wise application of ecological information. Individuals, populations, and communities are influenced by many interacting factors: physical and biological, genetic and environmental, strong and weak, positive and negative, direct and indirect, local and distant, current and historical. Because of the immense complexity of nature, we have a limited understanding of nature from autecological to ecosystem perspectives. However, potential threats to environmental health and integrity require efforts to find immediate solutions. The major themes of this book (wetland bioassessment and management) are particularly timely because of recent emphasis on the fact that wetlands are unique aquatic habitats ecologically separate from rivers, streams, and deep lakes (e.g., Batzer et al. 1999) that provide valuable goods and services for society (e.g., water purification, groundwater recharge, timber, fur, food).

WHAT IS A WETLAND?

Defining what constitutes a wetland is a complicated question because there are many different types of wetlands with different physical, chemical, and biological characteristics (marshes, bogs, swamps, beaver ponds, rock pools, pitcher plants, etc.). This variety makes it difficult to provide a definition that includes all wetlands while excluding other aquatic ecosystems. The U.S. Fish and Wildlife Service (Cowardin et al. 1979) offered the most widely accepted definition. They indicate that wetlands must have both of the following related characteristics: (1) saturated sediments or soils covered by shallow water for at least part of the growing season, and (2) hydrophytes as the dominant plant, at least periodically. Soils saturated by water eventually become *anaerobic* (low oxygen) or even *anoxic* (no oxygen), causing the accumulation of toxic compounds (e.g., methane, sulfur dioxide, some fatty acids)

Bioassessment and Management of North American Freshwater Wetlands, edited by Russell B. Rader, Darold P. Batzer, and Scott A. Wissinger.
0-471-35234-9

produced by bacterial respiration when oxygen is limiting. Emergent hydrophytes can "pump" oxygen from photosynthetic structures (leaves and stems) to the roots. This oxygen can detoxify noxious compounds, allowing hydrophytes to grow in conditions lethal to other plants. Therefore, plants adapted to anaerobiosis are an important component in defining wetlands (e.g., Tiner 1989, 1991).

Because the study of wetlands is a relatively new endeavor, there has been a recent profusion of terminology, which can often result in considerable confusion. For example, the original definition of wetlands by the U.S. Fish and Wildlife Service emphasized their shallow, often temporary nature and did not include lotic ecosystems (Shaw and Fredine 1956). However, when their objectives were to locate and enumerate "all continental aquatic environments" of the United States, including streams and rivers, the U.S. Fish and Wildlife Service (Cowardin et al. 1979) later decided to relabel running-water environments as *riverine wetlands*. Since then, other wetland definitions have also included running waters (e.g., Adamus 1983). Most recently, the U.S. Fish and Wildlife Service defined a *riverine wetland* as any wetland or deepwater habitat *contained within a stream channel* (Armantrout 1998). With the exception of beaver dams, water flowing through stream channels has always been a primary characteristic of lotic ecosystems. Expanding the term *wetlands* to include all aquatic environments of the continental United States (Cowardin et al. 1979), and relabeling streams and rivers as riverine wetlands not only confuses the distinction between different ecosystems, but fails to recognize the unique properties of wetlands as distinct aquatic environments. Lotic ecosystems are not wetlands. In most cases, stream and river channels are a useful means of separating lotic ecosystems from wetlands associated with running waters. Riverine wetlands are shallow, low-flow habitats adjacent to stream channels (floodplains) or located in river valleys (oxbows).

As with lotic ecosystems, deep lake basins are also functionally different from shallow wetlands. For example, temperature stratification with increasing water depth and the formation of zones (epilimnion, metalimnion, hypolimnion) with different physical-chemical properties is a common characteristic of lakes in the temperate zone (Wetzel 1983). Temperature stratification in wetlands would be only a temporary phenomenon. As with streams, wetlands associated with lakes are often located along the margin and are seasonally inundated and continuous with the deeper basins of lakes (e.g., Gathman et al. 1999). The boundary separating stream channels or lake basins from associated shallow wetland habitats will fluctuate through time and is often a subjective decision. For bioassessment objectives, the delineation of wetlands is less relevant than classifying the different types in a given area. Delineation provides a systematic procedure for determining wetland boundaries so that different investigators can approximate the same boundaries for similar types of wetlands. Of course, no boundaries actually exist. Wetlands (like other ecosystems) are continuous with other aquatic and terrestrial environments, but for legal purposes, the process has been standardized. Because like all environments, wetlands, are continuous with other ecosystems, some scientists have described them as a transition zone between terrestrial and aquatic environments, suggesting that they are some "halfway aquatic world" in transition toward the terrestrial environment (e.g., Smith 1980). There is little evidence to support the notion that wetlands are only intermediate stages in some short-lived transition between aquatic and terrestrial environments (hydrarchal succession). Most wetlands are an enduring part of the landscape (see Wissinger 1999).

How Are Wetlands Unique?

Wetlands are not unique in the sense that all wetland plants or animals occur only in wetlands and nowhere else. (In this sense, what ecosystems are unique?). However, many taxa are wetland specialists. For example, there are several invertebrates that are never or rarely encountered in other aquatic environments (streams, rivers, deep lakes), such as marsh beetles (Helodidae), minute bog beetles (Sphaeriidae), marsh flies (Sciomyzidae), water measurers (Hydrometridae), fairy shrimp (Anostraca), and tadpole shrimp (Notostraca) (Wissinger 1999). An unusual set of physical and chemical conditions, in addition to saturated soils and hydrophytes, also makes them a unique environment.

Most wetlands are typically *low-flow,* standing-water habitats that are poorly mixed. That is, flowing/turbulent water is necessary to mix and distribute chemicals and gases throughout the water column. A reduction or absence of flow in many wetlands results in steep chemical gradients (e.g., oxygen, temperature, pH, nutrients) that can fluctuate orders of magnitude over very short distances and on a 24-hour time scale (e.g., Rader and Richardson 1992). Although some wetlands are regularly mixed (e.g., prairie potholes) because they are poorly protected from the wind (e.g., Lovvorn et al. 1999), most wetland organisms are, none the less, adapted to wide chemical fluctuations over short spatial and temporal scales (Williams 1987).

Shallow water depth is also a distinguishing characteristic of wetlands that makes them prone to water-level fluctuations and periodic drying (e.g., Kusler et al. 1994). Many scientists accept the idea that wetlands are less than 2 to 3 m deep at their deepest point when water levels within the basin are at their lowest point (e.g., Mitsch and Gosselink 2000). Two to three meters is usually the maximum depth at which rooted hydrophytes can grow. Therefore, a body of water that is greater than 3 or 4 m deep at its lowest level may be considered a lake, not a wetland, even though the margins may be covered with hydrophytes.

Taken together, *anaerobiosis, hydrophytes, water-level fluctuations,* and *drying* can define wetlands. However, the distinction between wetlands and shallow lakes is tenuous. Are the patterns, processes, and functions of lentic systems less than 2 to 3 m deep at their lowest water level (wetlands) different from those of lentic systems 5 to 8 m deep at their lowest level (shallow lakes)? At least with respect to some processes (e.g., phosphorus cycling), wetlands and shallow lakes are similar (Moustafa 2000). Some authors in this book have included shallow lakes and reservoirs (greater than 2 to 3 m deep at their lowest water level) as part of their wetland bioassessment investigations (see Chapters 4 and 7). This should not be confusing as long as investigators describe the different types of water bodies included in their research.

DEFINITIONS

Science is necessarily filled with new terms, as new ideas often demand new terminology. In some cases the definition of old terms can be expanded to include recent concepts. Most biomonitoring and management terminology falls into the latter category. Although Karr and Chu (1999) provide an extensive glossary, a few definitions are necessary here to avoid confusion and refresh current understanding.

Ecosystem management is "driven by explicit goals, executed by policies, protocols, and practices, and made adaptable by monitoring and research based on our best understanding of the ecological interactions and processes necessary to sustain ecosystem composition, structure, and function" (Christensen et al. 1996). In short, we should manage to provide future generations with the opportunities and resources that we enjoy today; we should manage to ensure ecosystem sustainability. *Bioassessment* is the practice of using organisms to detect environmental health and integrity (Rosenberg and Resh 1993). *Health* and *integrity* refer to the proper functioning of ecosystems (e.g., Karr 1991). Although the term *function* is one of the most frequently used and confusing terms in ecology, in bioassessement it has two separate meanings: functions performed for society and ecosystem functions. Society functions include *goods* that have monetary value in the marketplace (food, medicine, construction materials, recreation) and *services* that are valuable but are rarely bought or sold (groundwater recharge, flood control, toxicant storage, nutrient removal, fish and wildlife habitat, soil maintenance, pest control). When many authors speak of "wetland functions" they are referring to services provided by wetlands to society, not ecosystem functions (e.g., Brinson 1993). *Ecosystem functions* refer to the processes that occur in all ecosystems (decomposition, community respiration, production, element cycling). Ecosystem goods and services depend on the proper functioning of these natural processes. Our ability to determine "proper" functioning of a wetland is based on comparing minimally impacted *reference sites* to *test sites* that are potentially impaired by human intervention. However, our ability to measure ecosystem functions is often limited and almost always costly. Also, Howarth (1991) suggested that the biological component of ecosystems should be more responsive to environmental stress than functional attributes.

RATIONALE FOR BIOASSESSMENT

Ecosystem health and integrity depend on the interaction between living organisms and ecosystem functions (Schulze and Mooney 1994). Ecosystem functions (e.g. decomposition and element cycling) depend on the biota (e.g., bacteria, fungi, and invertebrates) and the biota (e.g., algae and macrophytes) depend on ecosystem functions (decomposition and element cycling). Although it is not always clear how many species can be lost before ecosystem functions begin to fail (e.g., Ehrlich and Walker 1998), it is clear that at some point, the number of species present have an important impact on maintaining ecosystem functions (e.g., Covich et al. 1999). The role of biomonitoring in ecosystem management is to provide assessment prior to the initiation of new management practices, to determine the impact of established practices, or to track the impact of pollutants and other types of human intervention.

Alternative practices for determining ecosystem health (e.g., chemical sampling) may provide valuable information but are not adequate without associated biological data (Hart 1994), especially in wetlands. As mentioned, water chemistry in wetlands (oxygen, pH, nutrients) is extremely variable over small spatial and temporal scales. Numerous samples at appropriate microspatial scales, using 24-hour profiles, on at least a seasonal basis are necessary just to approximate chemical patterns (Rader and Richardson 1992). If we rely on chemical sampling to determine wetland degradation, we would have to separate chemical trends caused by human intervention from a natural signal that shows vast spatial and temporal fluctuation. Wetland organisms, however, provide a temporally integrated assessment of environmental conditions that more precisely represents ecosystem function and significant environmental change (Hart 1994).

WHICH ORGANISMS ARE THE BEST INDICATORS OF DEGRADATION?

All organisms or groups of organisms (e.g., algae versus invertebrates) do not respond the same to environmental stress. Depending on various disturbance characteristics (intensity, frequency, etc.), each group may respond at different rates and provide different information. For example, if the objectives are to detect potential environmental change attributed to increases in phosphorus or nitrogen, microbes or plants may best signal early changes and potential degradation (see Chapters 6 and 12). Changes in the abundance of small, rapidly reproducing species with wide dispersal capabilities are among the earliest responses to stress (Schindler 1988). Longer-lived organisms that are slower to recover (macrophytes, some invertebrates, and many fish) may indicate the impact of pulsed stressors that occur only periodically. However, if the objectives are to determine the impact of an exotic invasion, interacting species (prey and competitors) will be the best indicators. Clearly, it will be necessary to examine several groups in order to indicate overall environmental health and integrity given the potential for multiple, difficult to measure, non-point-source pollutants (see Chapters 7, 8, and 11).

LEGAL MANDATE

The tradition of land management began with the Organic Act of 1897. Since that time, several laws and statutes form the legislative mandate for current land management in the United States:

- *Organic Act of 1897:* specified the purposes (e.g., timber harvest and water supply) for which forest reserves could be established and managed

- *Multiple-Use Yield Act of 1960:* directed that the national forests be managed for multiple uses, including recreation, wildlife, and fish while providing a sustained yield of products and services (e.g., timber and water)
- *Clean Water Amendments Act of 1972:* established a policy to restore and maintain the chemical, physical, and biological integrity of the nation's waters
- *Endangered Species Act of 1973:* set the policy for conserving species and habitat of fish, wildlife, and plants that are in danger of or threatened by extinction
- *The Forest and Rangelands Renewable Resources Planning Act of 1974:* required a national assessment every 10 years to identify acceptable, deteriorating, or seriously impaired natural resources

This legislation shows a trend of increasing awareness of the relationship between healthy natural ecosystems and their ability to provide goods and services for humans. The most important legislation for wetlands are Sections 404 and 104 (b)(3) of the Clean Water Act (1972 and 1974). Section 404 regulates the dredging and filling of "waters of the United States". Also, Section 104 of the Clean Water Act specifies that water quality standards need to be developed to protect the beneficial uses of lakes, rivers, and streams. Wetlands are also included. Furthermore, many threatened and endangered mammals, birds, reptiles, and fish depend on wetlands for all or part of their life cycle (Niering 1988). Protecting and maintaining wetlands is an important part of maintaining overall ecosystem integrity (e.g., forest health) and achieving legally mandated management objectives.

BOOK ORGANIZATION AND OBJECTIVES

This book is separated into two sections: bioassessment and wetland management. The bioassessment section begins with Chapter 2, in which general principles and guidelines for establishing a biomonitoring program are described. This chapter is also designed to place the subsequent chapters on bioassessment into a larger context, pointing the reader to specific topics in each chapter. In general, the next three chapters cover statistical issues related to sampling numerous sites across the landscape (Chapter 3), the application of multivariate procedures (Chapter 4), and invertebrate functional groups (Chapter 5) to wetland bioassessment. Chapters 6 to 8 are case studies detailing the results of wetland bioassessment using various organismal groups in three different regions of the United States. One of the intriguing themes connecting these three case studies is how they integrate the two primary approaches to bioassessment, multimetric indices and multivariate statistics. In Chapters 9 to 11 the relationship between bioassessment and wetland restoration is discussed, beginning in Chapter 9 with a detailed explanation of how to restore wetland plant communities. In the remaining four chapters in Part 1 the authors describe how to use and sample bacteria (Chapter 12), algae (Chapter 13), macrophytes (Chapter 14), and invertebrates (Chapter 15) in wetland bioassessment.

When people hear the words, *wetland management* they often think of manipulating wetlands to maintain or maximize waterfowl production. Although waterfowl are certainly an important component of wetlands, other management topics are of equal importance. Numerous publications emphasize waterfowl management, including recent reviews by Baldassarre and Bolen (1994) and Weller (1999). Similarly, amphibians in wetland environments have been reviewed (Olson et al. 1997). Therefore, we have chosen to concentrate our efforts on other important but less known topics. Chapters 16 to 18 provide an extensive summary and review of how to manage fish (Chapter 16), waterbirds (Chapter 17), and mosquitoes (Chapter 18), in wetlands. Chapter 19 includes the first review of the effects of timber harvest on wetland ecosystems, with information on how to assess and manage wetlands to avoid the adverse impact of cutting practices. Finally, a comprehensive discussion of the impacts of an exotic invader, the macrophyte purple loosestrife, is given in Chapter 20. These chapters summarize information previously neglected in wetland management.

More and more, resource managers must attempt to strike a balance between resource exploitation to meet immediate human needs and resource protection to meet future human needs. Science has the responsibility of providing information to assist in finding this important balance.

REFERENCES

Adamus, P. R. 1983. A method for wetland functional assessment, Vol. I. Report FHWA-IP-82-23. U.S. Department of Commerce, National Technical Information Service, Washington, DC.

Armantrout, N. B. 1998. Glossary of aquatic inventory terminology. American Fisheries Society, Bethesda, MD.

Baldassarre, G. A. and E. Bolen. 1994. Waterfowl ecology and management. Wiley & Sons, Inc., New York.

Batzer, D. P., R. B. Rader, and S. A. Wissinger. 1999. Invertebrates in freshwater wetlands of North America: ecology and management. Wiley, New York.

Brinson, M. 1993. A hydrogeomorphic classification for wetlands. Wetlands Research Program Report TR-WRPDE-4. U.S. Army Corps of Engineers, Waterways Experiment Station, Vickburg, MS.

Christensen, N. L. and 12 others. 1996. The report of the Ecological Society of America committee on the scientific basis for ecosystem management. Ecological Applications 6:665–691.

Covich, A.P., M.A. Palmer, and T.A. Crowl. 1999. The role of benthic invertebrate species in freshwater ecosystems. Bioscience 49:119–127.

Cowardin, L. M., V. Carter, F. C. Golet, and E. T. LaRoe. 1979. Classification of wetlands and deepwater habitats of the United States. Report FWS/OBS-79-/31. U.S. Fish and Wildlife Service, Washington, DC.

Ehrlich, P. R., and B. Walker. 1998. Rivets and redundancy. Bioscience 48:387.

Gathman, J. P., T. M. Burton, and B. J. Armitage. 1999. Coastal wetlands of the upper Great Lakes: distribution of invertebrate communities in response to environmental variation. Pages 949–994 *in* D. P. Batzer, R. B. Rader, and S. A. Wissinger (eds.), Invertebrates in freshwater wetlands of North America: ecology and management. Wiley, New York.

Hart, D. D. 1994. Building a stronger partnership between ecological research and biological monitoring. Journal of North American Benthological Society 13:110–116.

Howarth, R. W. 1991. Comparative responses of aquatic ecosystems to toxic chemical stress. Pages 169–195 *in* J. Cole, G. Lovett, and S. Findlay (eds.), Comparative analysis of ecosystems: patterns, mechanisms, and theories. Springer-Verlag, New York.

Karr, J. R. 1991. Biological integrity: a long-neglected aspect of water resource management. Ecological Applications 1:66–84.

Karr, J. R., and E. W. Chu. 1999. Restoring life in running waters: better biological monitoring. Island Press, Washington, DC.

Kusler, J. A., W. J. Mitsch, and J. S. Larson. 1994. Wetlands. Scientific American, January: 64–70.

Lovvorn, J. R., W. M. Wollheim, and E. A. Hart. 1999. High plains wetlands of southeast Wyoming: salinity, vegetation, and invertebrate communities. Pages 603–634 *in* D. P. Batzer, R. B. Rader, and S. A. Wissinger (eds.), Invertebrates in freshwater wetlands of North America: ecology and management. Wiley, New York.

Mitsch, W. J. and J. G. Gosselink. 2000. Wetlands, 3rd ed. Wiley, New York.

Moustafa, M. Z. 2000. Do wetlands behave like shallow lakes in terms of phosphorus dynamics? Journal of the American Water Resources Association 36:43–54.

Niering, W. A. 1988. Endangered, threatened and rare wetland plants and animals of the continental United States. Pages 227–237 *in,* D. D. Hook et al. (eds.), The ecology and management of wetlands, Vol. I. Timer Press Publications, Portland, OR.

Olson, D. H., W. P. Leonard, and R. B. Bury. 1997. Sampling amphibians in lentic habitats: methods and approaches for the Pacific Northwest. Society for Northwestern Vertebrate Biology, Olympia, WA.

Rader, R. B., and C. J. Richardson. 1992. The effects of nutrient enrichment on macroinvertebrates and algae in the Everglades: a review. Wetlands 12: 34–41.

Rosenberg, D. M., and V. H. Resh. 1993. Introduction to freshwater biomonitoring and benthic macroinvertebrates. Pages 1–9 *in* D. M. Rosenberg and V. H. Resh (eds.), Freshwater biomonitoring and benthic macroinvertebrates. Chapman & Hall, New York.

Schindler, D. W. 1988. Experimental studies of chemical stressors on whole lake ecosystems. Verhandlungen der Internationalen Vereinigung fuer Limnologie 23:11–41.

Schulze, E. D., and H. A. Mooney. 1994. Biodiversity and ecosystem function. Springer-Verlag, New York.

Shaw, S. P., and C. G. Fredine. 1956. Wetlands of the United States. Circular 39. U.S. Fish and Wildlife Service, Washington, DC.

Smith, R. L. 1980. Ecology and field biology, 3rd ed. Harper & Rowe, New York.

Tiner, R. W. 1989. Wetland boundary delineation. Pages 232–248 *in* R. P. Brooks, F. J. Brenner, and R. W. Tiners (eds.), Wetlands ecology and conservation: emphasis in Pennsylvania. Pennsylvania Academy of Science, Philadelphia.

Tiner, R. W. 1991. The concept of a hydrophyte for wetland identifications. Bioscience 41:236–247.

Weller, M. W. 1999. Wetland birds: habitat resources and conservation implications. Cambridge University Press, New York.

Wetzel, R. G. 1983. Limnology. Saunders Publications, Philadelphia.

Williams, D. D. 1987. The ecology of temporary waters. Timber Press, Portland, OR.

Wissinger, S. A. 1999. Ecology of wetland invertebrates: synthesis and applications for conservation and management. Pages 1013–1042 *in* D. P. Batzer, R. B. Rader, and S. A. Wissinger (eds.), Invertebrates in freshwater wetlands of North America: ecology and management. Wiley, New York.

PART 1
Bioassessment in Freshwater Wetlands

2 General Principles of Establishing a Bioassessment Program

RUSSELL B. RADER and DENNIS K. SHIOZAWA

This chapter begins with a discussion of the importance of wetland classification as a means of reducing natural variation and refining comparisons between reference conditions and test sites. A three-tiered, spatially hierarchical, a priori classification involving subecoregions, watersheds, and sites nested in watersheds could stratify the landscape, ensuring that the full range of reference conditions or community types are identified. Procedures and criteria for identifying minimally impaired reference conditions at the watershed scale are also discussed. Additional emphasis is placed on detecting early signs of degradation and examining the trade-off between methods designed to reduce sampling variability and reductions in information needed to detect degradation. These decisions involve replication, qualitative versus quantitative samples, composite samples, habitat-specific sampling, when to sample, the number of taxa, and taxonomic resolution.

The idea that living organisms (microbes, algae, macrophytes, invertebrates, and vertebrates) are the best means to monitor environmental change has a strong intuitive appeal. However, the prevalence of bioassessment as a management tool has fluctuated over the past several years, in part because of the need to resolve questions regarding the application of bioassessment principles (sampling design, metrics, etc.). Recent success in the application of bioassessment in lotic environments suggests that bioassessment should also be a valuable tool when applied to wetlands.

Most questions regarding bioassessment involve what, how, when, and where to sample to gather useful information. The objective of this chapter is to address these questions and discuss general principles of how to establish a wetlands bioassessment program. However, this chapter also functions as a road map placing the following chapters in the bioassessment section of this book into a broader context, discussing how they fit into the overall theme of how to develop a bioassessment program.

Bioassessment and Management of North American Freshwater Wetlands, edited by Russell B. Rader, Darold P. Batzer, and Scott A. Wissinger.
0-471-35234-9

OBJECTIVES OF A BIOASSESSMENT PROGRAM

First, and one of the most important considerations when establishing a bioassessment program, is to determine the objectives (Green 1979). Bioassessment objectives fall within two general but not exclusive categories: (1) surveys of wetlands over a large geographical area, and (2) testing to determine the site-specific effects of a disturbance, recovery from a disturbance, or the condition of created or restored wetlands. Assessing the condition of numerous wetlands (objective 1) usually involves sampling a wide variety of wetland types over a large area, such as a national forest, the state of Wyoming, a fifth-order watershed, or a subecoregion. This type of assessment is used to detect potential changes in wetland condition attributed to a variety of ill-defined, non-point-source stressors or disturbances (e.g., acid deposition, sedimentation, eutrophication). Common questions addressed by the first objective are: (1) what percentage of the wetlands in this area are impaired or in poor condition?, (2) which wetlands should be set aside or preserved?, and (3) what areas/wetlands need restoration? The second objective pertains to the disturbance, recovery, restoration, or creation of specific sites. Questions commonly addressed by the second objective are: 1) what is the condition of this site?, 2) has this site recovered now that the disturbance has stopped?, 3) does this created or restored wetland function properly? These two general objectives are related and will often be part of the same bioassessment program. The first objective can serve as a screening process designed to locate specific sites pursuant to the second objective. That is, numerous sites can be screened to locate both minimally impacted sites and potentially impaired sites that may require further investigation, restoration, or protection. Methods used to accomplish the first objective may differ from the methods used in more detailed site-specific analyses.

SITE SELECTION, WETLAND CLASSIFICATION, AND DEFINING REFERENCE CONDITIONS

Natural populations and measures of community structure (e.g., richness and taxonomic composition) spatially and temporally vary in response to numerous biotic (predation, competition) and abiotic factors (temperature, hydroperiod). Separating natural variation from human impairment is critical to interpreting bioassessment results correctly. Objectives 1 and 2 both involve making comparisons between minimally impaired reference sites and test sites of potentially impaired but unknown condition. Ideally, reference sites would be identical to test sites except for human intervention. Since identical sites are highly unlikely, it is important to classify wetlands into groups with similar biological properties (e.g., community structure) and to define the range of variation associated with each class. At that point we can ask if potentially degraded test sites fall outside the range of variation established for reference sites of the same class. Without determining the full range of variation within a class, we might falsely conclude that a site is impaired if we have failed to identify it as a natural component of the variation within a class.

Can Wetland Communities Be Grouped into Distinct Classes?

All community classifications assume some level of predictability between community structure and controlling factors such as hydrology. Predictability suggests that deterministic forces underlie differences in community structure. Deterministic forces produce a set of predictable occurrences, whereas stochastic processes produce a range of random possibilities. All forces that influence community structure (e.g., hydroperiod, predation) have both deterministic and stochastic components. If deterministic influence is minimal, community structure will be driven by stochastic events, community composition will vary unpredictably, and classification would not be possible (White and Walker 1997). Although controversial, recent evidence suggests that the same type of communities can be identified predictably from year to year, suggesting that deterministic forces are sufficiently prevalent to warrant community classification (Wilson and Whittaker 1995, Wilson et al. 1996).

Classification also depends on whether geographical variation among communities distributed across the landscape is discrete or continuous (Keddy 1993). That is, the pattern of different sites or community types distributed along environmental gradients will influence our ability to distinguish different classes. Figure 2.1 shows two hypothetical continuous distributions of community structure: uniform versus aggregated. Given this particular aggregated-continuous distribution (Fig. 2.1a), samples collected along the entire gradient would reveal five classes with distinct differences. However, if the underlying pattern is uniform-continuous (Fig. 2.1b), then no matter how many sites are sampled, the number of classes identified will be arbitrary. The question of whether communities are discrete or continuous is also controversial and remains unresolved (Wilson 1991, Keddy 1993). However, exploratory statistical techniques may help to identify classes even when the distribution of sites/community types is uniformly continuous.

Assigning sites to specific classes can be done objectively by cluster analysis, whereas the number of classes may or may not be a subjective decision made by the investigator, depending on the underlying pattern of community types (Fig. 2.1). If communities form a discrete pattern across environmental gradients, then a hierarchical cluster analysis (e.g., Norris and Georges 1993) can objectively identify the number of classes. (In nonhierarchical cluster analysis the investigator specifies the number of classes). If the pattern is uniformly continuous, the investigator will have to specify the number of classes regardless of whether or not the clustering program is hierarchical. If many small classes are chosen, potentially degraded test sites might share similar characteristics with several different reference groups, which may obscure decisions concerning impairment. A few large classes, however, might mask potential degradation because of large variation within a class—it would be difficult to separate natural variability from changes caused by human intervention. Rerunning separate nonhierarchical cluster analyses while specifying a different number of classes for each run may help to identify the number of classes that group the greatest number of sites at an acceptable level of confidence while maintaining a relatively small level of variation within a class. However, if the distribution of sites on the landscape is uniform-continuous, some sites will always be difficult to place in a distinct group.

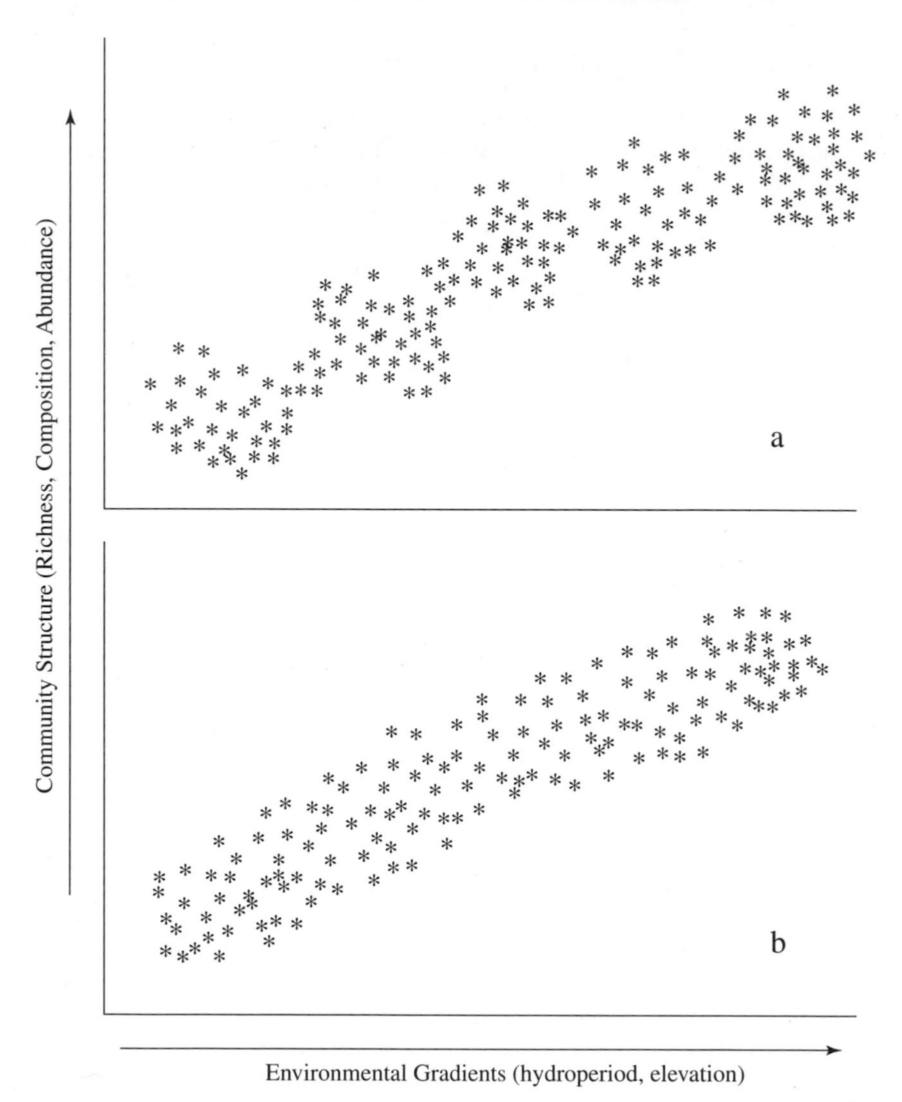

FIGURE 2.1. Conceptual representation of community types distributed along environmental gradients. Communities can be distributed in a continuous but discrete pattern (*a*) or in a more continuous-uniform pattern (*b*).

Classification Procedures

We can establish reference groups prior to (a priori) or following (a posteriori) sampling. The primary reason for a priori classification is to provide a rationale for specifying the number of reference classes initially and to ensure that the full range of biological communities within and among classes is identified. A priori classification usually involves classifying wetlands based on factors known or suspected to influence community structure but not subject to human intervention (e.g., elevation, latitude, longitude, aspect, soil type, geology, geomorphology, slope). Common terms for wetlands (marshes, bogs, swamps, pocosins) or the origins of wetlands are of little use in classification if they are not founded on the ecological forces that produce different communities. A review of the literature (Table 2.1) showing the factors used in wetland classifications, although not exhaustive, indicates that each factor is site-specific rather than a large-scale variable (watershed or regional environmental trait). Recent classifications in other environments (streams and lakes) are hierarchical and often span several spatial scales (region, watershed, segment, reach, habitat unit; e.g., Frissell et al. 1986). There is no spatially hierarchical classification for wetlands where smaller geographical units are nested within larger units. The classification by the U.S. Fish and Wildlife Service (Cowardin et al. 1979) is hierarchical in that some site-specific attributes (e.g., vegetation) are used in the classification prior to others. Since these classifications are site-specific they do not include factors that might operate at larger/regional scales (variations in rainfall, evapotranspiration, etc). Large-scale landscape variables could affect site-specific factors that control differences in community structure (hydroperiod and water chemistry). A spatially hierarchical classification is discussed in the section on "A Priori Classification."

It is also interesting to note that most factors are indirect measures of important ecological forces (e.g., hydroperiod) which are more easily measured than the actual force itself (Table 2.1). For example, using observations on the geomorphological setting of different sites to estimate differences in hydroperiod is far easier than recording water levels over the course of several years. These surrogates are useful when they represent the underlying reasons (ultimate causes) for differences in hydrology, water chemistry, and so on (proximate causes).

The *hydrogeomorphic model* (HGM) is a recent development that shows promise as a site-specific tool for classifying wetlands into ecologically meaningful groups (Brinson 1993, Shaffer et al. 1999). It is a site-specific tool because it is applied to individual sites as opposed to watersheds or regions. It is based primarily on the geomorphic setting of a wetland which is used to infer hydrology. HGM is a surrogate for actually measuring (1) water sources (surface flow, stream overflow, groundwater) and (2) the primary direction of water movements (e.g., vertical versus horizontal; Hruby 1999), both of which are difficult to measure but have an important impact on water chemistry and hydroperiod. Water chemistry and hydroperiod have a strong direct impact on wetland community structure (e.g., Mitsch and Gosselink 2000). For example, depressional wetlands with abundant vertical, groundwater inflows should have a longer hydroperiod and greater concentrations of nutrients than depressional wetlands with horizontal, surface water inflows (Brinson 1993). HGM is one of several potentially important site-specific factors that should be valuable in classifying wetlands into ecologically different groups.

A priori Classification. Local or site-specific community composition results from an interplay of local and regional factors (Ricklefs and Schluter 1993, Hildrew and Giller 1994). Large-scale or regional characteristics can be used initially to stratify the landscape to estimate the full range of variation among sites. Large-scale factors

TABLE 2.1. Ecological Basis for Various Wetland Classification Schemes

Reference	Scope[a]	Water Hydrology	Vegetation	Chemistry	Geology and Soils
Brinson 1993	Nationwide (all wetlands)	Geomorphology, water sources, water flows	—	—	—
Cowardin et al. 1979	Nationwide (all wetlands)	Hydroperiod (ephemeral, semipermanent, permanent)[b]	Forested, scrub, shrub, emergent, submergent	Salinity[b]	Mineral vs. organic
Driscoll et al. 1984	Nationwide (all wetlands)	—	Forested, scrub, shrub, emergent, submergent	—	—
Golet and Larson 1974	Northeast United States	Depth (deep vs. shallow) Hydroperiod (ephemeral, semipermanent, permanent)[b]	Wooded, scrubs, shrubs, meadows, emergent, submergent	Salinity[b]	—
Goodwin and Niering 1975	Inland wetlands of the United States	Depth (deep vs.shallow) Hydroperiod (ephemeral, semipermanent, permanent)[b]	Wooded scrubs, meadows	Salinity[b]	—
Gosselink and Turner 1978	Nationwide (nonforested wetlands)	Water sources, water flows, flood periodicity	—	—	—

Hills et al. 1994	European riverine wetlands	—	Functional groups	—	—
Millar 1976	Prairie potholes (Manitoba and Saskatchewan)	Hydroperiod (ephemeral, semipermanent, permanent)[b]	Vegetation in various zones	—	—
Shaw and Fredine 1956	Nationwide (all wetlands)	Depth (deep vs. shallow) Hydroperiod (ephemeral, semipermanent, permanent)[b]	Wooded, scrub, shrub, emergent	Salinity[b]	—
Stewart and Kantrud 1971	Prairie potholes (Iowa, Dakotas, Minnesota)	Depth (deep vs. shallow) Hydroperiod (ephemeral, semipermanent, permanent[b]	Vegetation in the deepest zone	—	—
Windell et al. 1986	Rocky Mountain wetlands	Depth (deep vs. shallow) Hydroperiod (ephemeral, semipermanent, permanent)[b]	Forested, scrub, shrub, emergent, submergent	Salinity, minerotrophic vs. ombrotrophic[b]	Mineral vs. organic

[a]Geographical extent of the classification and type of wetlands being classified.

[b]Inferred from vegetation.

(e.g., aspect, annual precipitation/evapotranspiration) may also predict community composition as surrogates for much more time intensive and costly site-specific factors (measuring annual hydroperiod) if the two forces are strongly correlated. However, large-scale factors also average over important environmental heterogeneity at local scales (e.g., site-specific geomorphological setting). Recent data from stream analyses indicated that landscape/large-scale factors had a weak correspondence with invertebrate community classifications (Hawkins and Vinson 2000). Furthermore, it was found that "if used alone landscape classifications lead to imprecise predictions of the expected biota at a site and, hence, a high likelihood of not detecting impairment" (Hawkins et al. 2000). Since it is unlikely that large-scale factors can override factors that operate at smaller scales (Allen and Hoekstra 1992), both site-specific and regional factors should help to distinguish classes in wetlands. If for no other reason, a priori landscape classifications are useful because they ensure sampling the full range of variation among sites in a given area.

Although the eventual goal of classification is to determine the condition of specific sites, the watershed may be the best scale for documenting disturbances and defining reference conditions. The watershed unit often used by management is defined as third- or fourth-order basins (Strahler 1957) that are 50 to 500 km^2 and easily distinguished on U.S. Geological Survey topographic maps at a ratio of 1:24,000 (Mount Hood Watershed Analysis Team 1994). (One kilometer on the map is equal to 24,000 km on the land surface). Although the decision to select a certain-sized basin to represent "the watershed" is somewhat arbitrary, it nonetheless provides a useful standard for management decisions. This is a good scale for capturing information on ecosystem processes (e.g., water yield) and human disturbances (e.g., road density).

At a larger scale, Omernik (1987) and Bailey et al. (1994) have also provided a geographical framework for partitioning or stratifying the landscape into areas (ecoregions) with similar broad environmental attributes (topography, climate, geology, vegetation, and land use) that may affect watershed and site-specific characters, such as nutrient levels and rates of drying. Omernik (1987) divided the entire continental United States into 76 ecoregions at a scale of 1:7,500,000. This resolution is not sufficiently detailed for most management purposes, including bioassessment. Therefore, recent efforts have sought to delineate subecoregions at a finer scale (1:250,000; Bailey et al. 1994, Omernik 1995). Although both state and federal agencies are accustomed to making management decisions at the watershed scale, Omernik (1995) makes the compelling argument that hydrologic units (watersheds) do not necessarily correspond to ecological units with similar environmental attributes useful in classifying different community types. A potential compromise that maximizes the advantages of both watershed and ecoregion concepts seems obvious. Reference conditions could initially be defined for watersheds, and as subecoregions are delineated, they should be defined for watersheds nested within subecoregions. This effort would create a three-tiered spatially hierarchical classification that would include the impact of landscape variables as summarized by the ecoregion concept (level 1) on watersheds nested in regions (level 2) and specific sites nested in watersheds (level 3). Since various factors have already been identified for classification at the landscape (ecoregions, Omernik 1987) and site-specific scales (e.g., Table 2.1), Table 2.2 lists some watershed factors potentially useful in wetland classification. Future research might identify the variables best correlated with differences in wetland community structure at all three scales. Such factors could be used to pair unknown test sites with their appropriate reference class in order to determine degradation (see the section "Reference Conditions and Site-Specific Evaluations").

Anderson and Vondracek (1999) found that ecoregions accounted for significant variation in wetland insect communities in the Prairie Pothole region of the United States. In streams, Hughes et al. (1990) found that differences in fish assemblages also corresponded to ecoregion boundaries. Similarly, Corkum (1990, 1991) found that differences between biomes accounted for the greatest amount of variation in the taxonomic composition of stream invertebrates over a broad geographical range. However, there were relatively low levels of similarity among rivers within a biome (Corkum 1991). Harding et al. (1997) examined invertebrate communities in 100 headwater streams in New Zealand. They concluded that their data "provided qualified support for the use of ecoregions as a basis for stream classification." Ideally, we would hope that differences among regions would be coupled with high similarity of communities within regions. This may depend on the taxonomic resolution. Coarser

TABLE 2.2. Watershed Variables Potentially Important in a Hierarchical Classification of Wetlands[a]

Watershed Variable	Description
Area	Total area upslope from the watershed outlet
Latitude	Latitude measured at the watershed outlet
Elevation	Average of highest and lowest elevations within a watershed
Relief	Difference between the highest and lowest elevations within a watershed
Slope	Mean percentage rise in elevation
Aspect	Mean direction of basin clockwise or counterclockwise from North
Vegetation	Dominant physiognomic type (as opposed to a single species descriptor)
Soils	Dominant type
Wetland number	Total number of wetlands per watershed
Wetland area	Total area of wetland sites within a watershed divided by watershed area.
Wetland connectance	Average distance among wetland sites within a watershed
Wetland clustering	Coefficient of variation of distance among wetland sites

[a]Individual sites could be nested within watersheds which could be nested within subecoregions.

levels of taxonomic resolution may supersede ecoregion boundaries. Chironomidae have a worldwide distribution, whereas species within Chironomidae are usually confined to a smaller range. If families are used to detect degradation, ecoregions may not help to distinguish different classes. However, ecoregions may distinguish different communities based on genus or species designations. At the coarsest taxonomic level, plants and animals, all communities (aquatic and terrestrial) would fall into the same class. We only poorly understand the impact of taxonomic resolution on our ability to classify natural communities. However, Hawkins and Vinson (2000) found that the strongest landscape-level stream classifications were based on genus designations, suggesting that the finest taxonomic resolution provided the best information for classification.

A posteriori Classification. A posteriori classification involves sampling individual sites and is usually based on the identification of classes using taxonomic lists or community composition (see Wright 1995). A posteriori classification makes no presampling assumptions about the number of potential classes or about the importance of various factors (e.g., large scale/ecoregions) in determining different community types (e.g., Norris 1995). Such assumptions can be tenuous since science has only a rudimentary understanding of how some forces determine differences in wetland community structure. This type of classification, however, makes no initial attempt to partition or stratify the variation among sites. That is, samples are collected from numerous sites in minimally impacted areas and multivariate techniques (e.g., clustering) are used to identify different classes based on biological similarity (e.g., Wright 1995; Chapter 4, this volume). A posteriori classification assumes that the number of sites sampled is sufficiently large to document the full range of community types in a specified area. Even if the number of sampled sites is large, there is still a risk of it being a biased subset. A priori classification could ensure that the full range of sites would be sampled and all community types identified. Recent research has used a combination of a priori and a posteriori techniques to classify wetlands (Chapters 7 and 8). Regional and watershed characteristics could be used initially to stratify the landscape and ensure sampling the full range of variation, whereas a clustering technique could be used to identify different biological classes after samples have been collected. In Chapter 3, Rathbun and Gerritson discuss sampling designs and randomization procedures applicable to site selection for both objectives 1 and 2, including how to select sites randomly for sampling after the landscape has been stratified (a priori) into classes but when the actual location of individual wetlands within strata is unknown. Also, Hawkins and Carlisle (Chapter 4) describe classification using cluster analysis.

Procedures for Identifying Reference Watersheds

Since most information on human impacts is summarized at the watershed scale, it is useful to define reference conditions at this scale (Hughes 1995). This is based on the relatively secure assumption that sites nested within a healthy watershed will also be healthy. Some basic steps may help to frame this process.

1. Stratify the landscape into ecologically similar groups using landscape variables (e.g., watershed classes and subecoregions).
2. Choose sections of land (e.g., watersheds) within each strata for site-specific sampling. The rationale for the first two steps was described above.
3. Find the general location of wetland sites within watersheds. The National Wetlands Inventory (NWI) has developed maps showing the location of wetlands based on high-altitude aerial photography (Cowardin et al. 1979). Consequently, they do not include smaller wetlands or wetlands hidden by surrounding vegetation.
4. Incorporate the location of landscape strata (subecoregions), watersheds, and wetland sites into a geographic information system (GIS) database. If subecoregions (e.g., Omernik 1987) have already been determined, there is no reason to re-create large-scale landscape strata using various GIS overlays (geology, soils, precipitation, etc.). GIS is a valuable tool because it can show the spatial relationship between the location of wetlands and potential causes of degradation (roads, timber harvest, mining, grazing, oil production, etc.).
5. Developing criteria for defining reference conditions for sites nested in watersheds. The following is a short list of potential criteria:
   a. Site-specific:
      (1) Presence of exotic invaders
      (2) Poor water quality (odors, oil slicks, etc.)
      (3) Signs of altered hydrology (ditches, berms, pipes, etc.)
      (4) Signs of altered riparian vegetation (young, uniform age, lacking complexity)
      (5) Signs of extensive human or livestock activity
   b. Watershed:
      (1) Watershed area affected by human disturbance.
      (2) Proximity of disturbances to wetlands or clusters of wetlands within watersheds.
      (3) Geomorphological position of disturbances relative to wetlands (upslope disturbances should have a greater impact than downslope disturbances).

Low-flying aerial photography and field reconnaissance are very valuable at identifying unmapped wetlands and locating recent or unmapped disturbances. Historical information (e.g., old surveys, old aerial photos, paleolimnology) and local experts (state and federal biologists) are important sources of information. However, historical information should be viewed with caution because of unmeasured changes in the environment and complex responses to past environmental change (time lags and nonlinear responses). Just because present conditions are different from past states does not necessarily indicate degradation.

What Is a Minimally Disturbed Watershed? How many species of exotic invaders should identify the threshold for violating minimally disturbed criteria or reference conditions? What nutrient concentrations, road density, and grazing intensity are permissible before reference criteria are violated? One cannot collect samples and allow the organisms to answer these questions. In order to avoid circular arguments in defining reference conditions, potential reference sites need to be identified prior to sample collection. When we must defend our choice of reference sites, it should be more than simply suggesting that they represent reference conditions because they differ from apparent degraded sites (e.g., have more species). A justification independent of and prior to sample collection that is related to watershed conditions will avoid circularity in defining reference conditions.

Because of the urgent need to protect our natural resources, perhaps the best way to proceed is to create standards by committee. That is, assemble experts and public representatives to define reference watersheds collectively. Because of the inescapable variation in nature, we should avoid standards that are represented by a single value and focus on a range of practical importance. This forces us to define the range of road density, grazing intensity, and so on, that constitutes minimally disturbed conditions. Proposed, standards, and guidelines could then be used as hypotheses for scientific verification and testing.

How Many Reference Sites Should Be Sampled? Enough minimally impacted sites need to be sampled to document the full range of community classes and the full range of variation within a class. Ideally, we could answer this question using *power analysis* (e.g., Krebs 1989). A power analysis is based on four interrelated concepts: (1) the amount of variation among sites within a class, (2) the "meaningful' difference between reference and test sites, (3) type I error (α value), and (4) type II error (β value). The larger the variation among sites within a reference class, the greater the number of replicates needed to detect differences. Similarly, more replicates are required to detect small compared to large differences. Consequently, if the variation among reference sites is large, we probably will not be able to detect small differences or early signs of degradation—hence, the need for classification.

At the beginning of a bioassessment program, we seldom know the variation among sites. One common suggestion is to conduct a pilot study prior to executing a bioassessment program (e.g., Norris et al. 1996). However, it is difficult to know how many sites is necessary for a pilot study in order to calculate the number of sites needed to define reference conditions for bioassessment. Occasionally, existing data can be found to estimate the variance among sites, but this is rare, especially in wetlands. However, if the variance among reference sites can be estimated at the beginning of a bioassessment program, several formulas are available to calculate the number of replicates needed to detect a "meaningful" difference (see Norris et al. 1996). The correct formula for a specific application depends on a variety of considerations that is best decided in consultation with a statistician.

Even when we can estimate the variance, there is still the problem of deciding what constitutes a meaningful difference. How dissimilar should the test site be to the reference condition to warrant a degraded designation? Is 35% dissimilarity (reference versus test) an appropriate threshold to indicate impairment? This question indicates the need to establish what degree of difference will have biological or ecological importance. What change in population abundances can occur or how many species can be lost before an ecosystem fails to function properly? These are complicated questions of current intensive research that may defy an easy consensus (e.g., McCann 2000). It is important to note that our inferences of what constitutes significance are based on our ecological understanding (Yoccoz 1991). In a very real sense, a difference is deemed significant because we define it as such. This observation is not unique to statistical inference (traditional tests of difference, multivariate techniques, etc.). Any analysis (e.g., multimetric graphs) relies on our scientific understanding of what constitutes a meaningful difference. Since information is lacking, such thresholds are determined arbitrarily and can be increased or decreased to tighten or loosen our definition of impairment (e.g., Hannaford and Resh 1995). For example, some investigators have suggested decreasing the threshold for concluding impairment from 85% to 65% similarity with reference sites (e.g., Resh and Jackson 1993). This has the effect of reducing the probability of identifying sites as impaired when they are not. However, it also increases the probability of failing to detect impairment when it actually exists, which therefore decreases the probability of detecting marginally impaired sites.

Potential difficulties in determining among site variability prior to sampling and in defining ecologically meaningful differences may force us to take a conservative approach. Perhaps the best way to proceed is to relax, not tighten, our criteria for committing a type I error (inferring a difference where none exists). We can do this by setting $\alpha = 0.1$ or 0.2, or by increasing, not decreasing, the threshold for identifying impairment (Hughes 1995). Sites may be designated as impaired if they are less than 70 to 80% similar to reference sites. This means that we will create a greater chance of inferring that a management activity is damaging when it really isn't. Which means that we will tend to err on the side of the resource; we will determine that more test sites are dissimilar to the reference condition and therefore warrant further investigation.

An example from streams indicates an upper limit on the number of sites sampled beyond which sampling additional sites does not increase our knowledge of the variation in reference conditions. Based on benthic macroinvertebrate samples from 234 stream sites distributed around the Great Lakes, Reynoldson and Rosenberg (1996) identified four or five biological classes and then determined the number of sites needed to adequately represent each class. They found that a minimum of 10 sites per class with a preferred target between 35 and 50 sites were necessary to distinguish different classes. More than 50 sites per class tended to blur the distinction among classes, making it difficult to assign sites to a distinct class. That is, the underlying pattern of community types was more uniform-continuous (Fig. 2.1b). However, selecting fewer than 35 sites per class resulted in high variation within a class, thus obscuring the ability of the model to detect impaired sites. In a similar analysis, Yoder and Rankin (1995) found that 35 to 40 reference streams per class were sufficient to distinguish classes and determine variation among sites. Assuming that the underlying pattern of community types is often uniform-continuous, we need to sample the number of sites that will balance the trade-off between a lack of distinct community classes versus a high degree of variation within classes. This may be an iterative process of collecting samples, running a cluster analyses and, if necessary, collecting more samples.

Over How Large an Area Do Reference Conditions Extend? Obviously, reference wetlands sampled in Florida would have no application to fens and bogs of the Rocky Mountains. It is important to remember that all else being equal, near sites are more similar than distant sites. Current environmental conditions, climatic gradients, disturbance history, and the biogeography of species all produce a general expectation termed the *distance decay of similarity,* which states that the similarity between any two sites will increase as the distance between them decreases (e.g., White and Walker 1997). But how close is "near" or how far is "distant"? This depends on the size of the area over which current environmental conditions, climatic gradients, disturbance history, and biogeography remain relatively uniform. Subecoregions (Bailey et al. 1994, Omernik 1987) include much of this information (boundaries of uniform climate) and therefore may help to define the size of the area over which reference conditions are applicable. That is, test sites compared to reference sites from the same class in the same subecoregion may have the best chance of eliminating errors, such as failing to detect degradation because of high variation among sites. The importance of the ecoregion concept in bioassessment is certainly promising, but remains relatively untested (Resh et al. 1995), especially for wetlands. For example, it is not clear if wetland communities with similar watershed and site-specific characteristics (e.g., hydroperiods) are more similar within versus among subecoregions.

Reference Conditions and Site-Specific Evaluations

The distance decay of similarity implies that a few minimally affected reference sites within the same vicinity as test sites may be sufficiently close to warrant comparison without sampling numerous sites over a large geographical area (e.g., Hannaford and Resh 1995). The cost of sampling a large number of wetlands to define regional reference conditions may not be necessary if physically similar sites (size, hydrology, elevation, etc.) exist in close proximity to the test site (Chapters 10 and 11). However, despite the distance decay of similarity, a test site (e.g., restored wetland) with a community different from a few reference wetlands might be misidentified as impaired if the full range of community types (variance among reference sites) has not been adequately identified. The decision to rely on a few nearby wetlands (if they exist) to define the reference condition, as opposed to sampling numerous reference sites and developing regional reference conditions, depends on the investigators' confidence that the few nearby reference sites have sufficiently similar physical-chemical attributes to the test site to warrant biological comparison to determine potential degradation. For this reason, describing the physical-chemical attributes of reference wetlands is important regardless of whether a few or numerous minimally impaired sites are used to define the reference condition. It is necessary to select physical-chemical variables least affected by human intervention (e.g., elevation). If a few local reference sites cannot be found, sampling numerous minimally impacted sites to define regional reference conditions can be a valuable approach (Wright 1995, Reynoldson et al. 1997; Chapter 4, this volume). This approach might be strengthened if landscape stratification were used to a priori classify reference wetlands to ensure sampling the full range of variation (see the section "A Priori Classification").

Classification Exceptions

Some wetlands seem to defy classification. The Everglades, the Okeefenokee Swamp, and the Great Dismal Swamp are examples of large wetlands that potentially warrant individual consideration. How do we identify reference conditions in such wetlands? It may be that despite their unique physical attributes the communities in these large wetlands are similar to other communities in smaller wetlands in the same area (e.g., subecoregion) in which case the condition of these wetlands could be compared to that of other reference wetlands in the same area. However, if their communities are unique, these wetlands could be compared to themselves. The wetland could be stratified into different habitat types composed of different community types (e.g., open-water sloughs versus sawgrass prairie in the Everglades). If present, minimally disturbed reference sites could be compared to potentially degraded sites within the same habitat type (i.e., minimally impacted sloughs versus potentially degraded sloughs).

REPLICATION

A replicate is often defined as the smallest experimental unit to which a treatment is applied (e.g., Krebs 1989). This definition does little more than defer the understanding of replication from one term (replicate) to another (experimental unit). Problems associated with replication have been described as an "insidious beast" (Heffner et al. 1996). I will not attempt to review replication and all of its ramifications (see Hurlbert 1984, Hawkins 1986). However, I will emphasize a few points that are particularly important to bioassessment.

Unfortunately, the term's *replication* and *sample size* are often used interchangeably. Replication is different from sample size depending on the scope or spatial domain of the study, which depends on the objectives (Heffner et al. 1996). The basic unit, or *replicate,* will change with changes in scale. The scope of the study refers to the spatial extent or area over which conclusions are drawn. For example, if the scope of the study is to assess the condition of wetlands in the state of Utah by taking dip-net samples of invertebrates, a replicate would be an individual wetland site and the sample size would be the number of dip-net samples collected at each site. If, however, the scope of the study is a single uniform habitat type (e.g., open-water slough) within a single wetland where this habitat type was represented by more than one unit, separate units would be replicates and the sample size would be the number of dip-net samples collected per unit. If the scope of the study were a single habitat unit (e.g., a single slough), the number of samples collected within the slough would be equivalent to the number of replicates; sample size and replication would be the same. Clearly, if the objective of the study dictates sampling a single habitat type within a single wetland, the results cannot be applied to other examples of the same habitat type within the same wetland, nor can they be applied to the same habitat type in other wetlands. There is no replication to allow such extrapolation. Hannaford and Resh (1995) compared a single restored site and a single unrestored site to a single upstream control site and used macroinvertebrate samples within sites as replicates. This is a pseudoreplicated design (Hurlbert 1984) that can be used to indicate differences among sites but not the causes of differences (pollution) because of unidentified, unmeasured factors that may also vary among sites (e.g., Beyers 1998). Consequently, Hannaford and Resh (1995) collected abundant physical/habitat data to strengthen their conclusion that pollution was the cause for their statistical differences.

SEVERAL SEPARATE SAMPLES VERSUS A SINGLE COMPOSITE SAMPLE

From a statistical perspective, the question of creating a composite sample or maintaining separate samples is unambiguous. When individual sites are the basic replicate, separate samples collected within sites do not contribute to our ability to detect differences between or among sites (e.g., reference sites versus test sites; e.g., Hurlbert 1984, Beyers 1998), they only give an estimate of the mean of the particular replicate or site. However, maintaining the identity of samples does allow us to estimate sample precision or within-site variability, which can be used for quality control. If a site appears to be an outlier (low species richness in a minimally disturbed watershed), perhaps it is because one or more of the samples were not collected properly (human error). Also, when pseudoreplication is unavoidable and a single reference site is compared to a single test site (e.g., Hannaford and Resh 1995), samples are the replicates and they must be kept separate if statistical inferences are necessary. In addition, Parsons and Norris (1996) found that a single composite sample from a site diluted important habitat specific information. The effects of stressors on a sensitive habitat-type might be diluted by unaffected habitats (see the section "Where to Sample"). Compared to separate samples, composite samples, from within a specific habitat type (e.g., within macrophyte beds) can help to reduce time and costs while retaining habitat-specific information.

DATA ANALYSIS

Prior to data analysis, the decisions concerning the type of data and when, where, and how to collect them are the same for each analysis scheme. Traditional parametric tests of significance [e.g., analysis of variance (ANOVA)] have limited application to most bioassessment programs. As mentioned, unreplicated designs may detect differences, but inferences concerning causes are tenuous (e.g., Hurlbert 1984). The avoidance of pseudoreplication is the primary reason for collecting samples at replicate reference sites to define regional reference conditions. However, traditional statistics also do not apply when comparing multiple reference sites to a single test site. Within-site variability of a single test site cannot be compared to between-site variability among multiple reference sites. Consequently, most bioassessment programs have currently settled on two types of data analysis: multivariate procedures (e.g., Wright et al. 2000) and multimetric graphical techniques (Karr and Chu 1999). The primary difference is that multivariate techniques include the fundamental tenet of all statistics, that there is a certain probability that any set of measurements will vary by chance alone. Since we cannot measure reference sites or test sites perfectly, what is the probability that they differ? Multivariate procedures allow us to approximate such probabilities, graphical techniques are more subjective. However, if multimetric indices include some estimate of variation (e.g., confidence intervals, or box-whisker plots), they too could estimate more objectively the probability that reference sites differ from test sites (e.g., Maxted et al. 2000). The advantages and disadvantages of both techniques have been compared (Resh et al. 1995, Reynoldson et al. 1997). In Chapter 4, Hawkins and Carlisle provide a lucid explanation of the multivariate technique, including an example of how it can be applied to wetland environments; Stevenson in Chapter 6, Apfelbeck in Chapter 7, and Helgen and Gernes in Chapter 8, describe multimetric procedures and show how both multivariate and multimetric analysis can be used to identify impaired conditions. Although some optimal technique that combines the strengths of both analyses may eventually be derived (Reynoldson et al. 1997), detecting degradation will still depend on how the data were collected and on our ability to maximize replication and minimize variation while maintaining sufficient information to detect ecologically meaningful differences.

TYPES OF DATA AND BIOLOGICAL METRICS

Although any type of biological information can be used to assess wetland health (genetic diversity, individual growth, population age structure), most bioassessment programs currently rely on five basic levels of data that differ with respect to information content: (1) richness (qualitative samples), (2) taxonomic composition (qualitative samples), (3) relative abundance (qualitative samples), (4) actual population abundance (quantitative samples), and (5) production (quantitative samples).

Richness is the number of taxa present at a site without regard to which taxa are present. Two or more sites may have identical richness but different community compositions. The identification of taxa present in a community (composition) can be compared using a variety of similarity indices (e.g., Krebs 1989). The similarity among reference sites and sites of unknown condition in the same class can be used to estimate degradation (Chapter 4). Estimates of relative abundance (e.g., percent representation of various taxa) can rank taxa into qualitative categories. Although it is not possible to determine numerical differences between categories, it is possible to identify abundant, common, and rare taxa, indicating qualitative differences between test and reference sites. Richness, composition, and relative abundance can be estimated using presence–absence data that require only qualitative samples. Quantitative estimates of actual population abundance numerically define a taxon using some measure of central tendency (e.g., average biomass per unit area) and the spread around the central parameter (standard deviation). Numerical differences between actual population estimates can be calculated. Production, the rate that biomass is fixed over a set period of time, is measured by a time series of population abundance estimates (density and biomass), often over the complete life cycle of a species. For example, if a species requires a year to complete its life cycle, numerous quantitative samples, at least on a seasonal basis, are required to calculate production.

One of the primary concerns with bioassessment in wetlands is determining appropriate metrics to measure degradation. Several chapters in this book identify numerous metrics, including information on which ones will probably respond in a systematic and linear fashion to an assortment of environmental stressors. A variety of invertebrate, macrophyte, and algal multimetrics are described in Chapters 6 to 8. (*Multimetrics* are measures that sum across several individual metrics to indicate a site's condition.). The use of similarity indices (i.e., Wright 1995) to measure degradation using the entire algal or invertebrate community, is described in Chapters 4 and 6, respectively. In Chapters 10 and 11 the use of invertebrates, macrophytes, and birds to document wetland recovery and restoration is demonstrated. How to use invertebrate functional groups to assess wetland health and integrity is described in Chapter 5, various intriguing techniques using bacteria are described in Chapter 12.

Sampling sites across a range of human intervention will allow us to determine reliable indicators of impairment. A good metric will predictably vary as a straight line through sites impacted by a range of human influence because we can separate increasing or decreasing trends from natural fluctuations (Karr and Chu 1999). This is called the *signal-to-noise ratio* (SNR). If data are available from many sites, the SNR can be determined by plotting the mean value for a metric (signal) averaged across sites divided by the individual value representing a single site (noise). If the average is equal to individual values (sites), we have a perfect signal and there is no noise. A plot of these ratios (mean of all sites divided by a value for individual sites) would be a straight line through one on the *Y* axis. Metrics with SNR values that cluster tightly around 1, have the least amount of noise, will require the fewest number of replicates, and will be best at detecting early signs of degradation. Similarly, ratio metrics (Odonata/Hemiptera–Odonata) may also be useful in detecting early signs of degradation because they vary between 0 and 1 and therefore also show a low amount of variation.

QUALITATIVE VERSUS QUANTITATIVE SAMPLES

The difference between qualitative and quantitative samples is based on the ability of the sampling device to estimate actual population abundances accurately. Bias in estimating population abundances at a site has two components: (1) the ability to collect the actual number of individuals in an area delineated by the size of the sampling device, and (2) the ability to collect enough samples to determine abundances accurately for the entire site. Quantitative samples seek to minimize bias associated with collecting the actual number or biomass of organisms in a sampling device. However, quantitative samples collect individuals from only a small proportion of the area at a site. Therefore, without numerous expensive samples, they also incur important bias in estimating the actual number or biomass of individuals for an entire site. Qualitative samplers are not designed to estimate the actual number of individuals in a sampler and therefore cannot attempt density estimates for an entire site. There are no samplers that are completely free of bias and can determine abundances perfectly either within a sampler or for an entire site. With respect to estimating the number of individuals within a sampler, various sampling devices can be arranged along a continuum from completely qualitative (important biases present) to reasonably quantitative (important biases eliminated). Ultimately, the investigator needs to justify their selection of a sampler and explain why they treated their data in a quantitative or qualitative fashion.

Figure 2.2 shows the trade-off between information content and variability associated with qualitative versus quantitative data. Although quantitative samples provide more information, they show greater variability and therefore require more samples. Typically, qualitative samples provide less information (species composition without estimates of abundance) but require fewer replicates and less processing time than quantitative samples (but see the section "Transect-Based Point Sampling"). For this reason, production data, which require numerous quantitative samples, are rarely used in bioassessment but may be necessary when the finest degree of resolution is required to detect differences between reference and disturbed conditions. That is, the production of sensitive taxa may decrease and that of tolerant taxa increase prior to changes detectable by qualitative samples such as species deletions resulting in changes in community richness or composition. (However, the section on Habitat Based Point Sampling suggests that numerous qualitative samples may be better able to detect differences in abundance than quantitative samples.) Also, production estimates for species insensitive to human disturbances may provide less information than qualitative estimates of community richness and composition. Quantitative estimates may best be applied when we understand the impact of disturbances on specific indicator species that are them sampled quantitatively. Unfortunately, the effects of many disturbances on potential indicator taxa are lacking for wetland ecosystems.

Because of associated costs, screening wetlands to determine their environmental condition usually involves rapid assessment techniques (Barbour et al. 1999), whereas more detailed quantitative data can be applied to the lower number of sites involved in some site-specific tests (see Chapters 10 and 11). Rapid assessment techniques depend on a few qualitative samples that estimate the presence or absence of taxa. Presence–absence data can define the richness and composition of an entire assemblage (algae, macrophytes, invertebrates, fish). Richness and taxonomic composition are less variable than quantitative estimates of abundance and production and are also important expressions of community complexity that provide valuable information concerning ecosystem health and integrity. Many biotic indices (e.g., community similarity; Chapter 4) and most ratio metrics (number of tolerant taxa/total number of taxa) require only presence–absence data. Depending on the severity

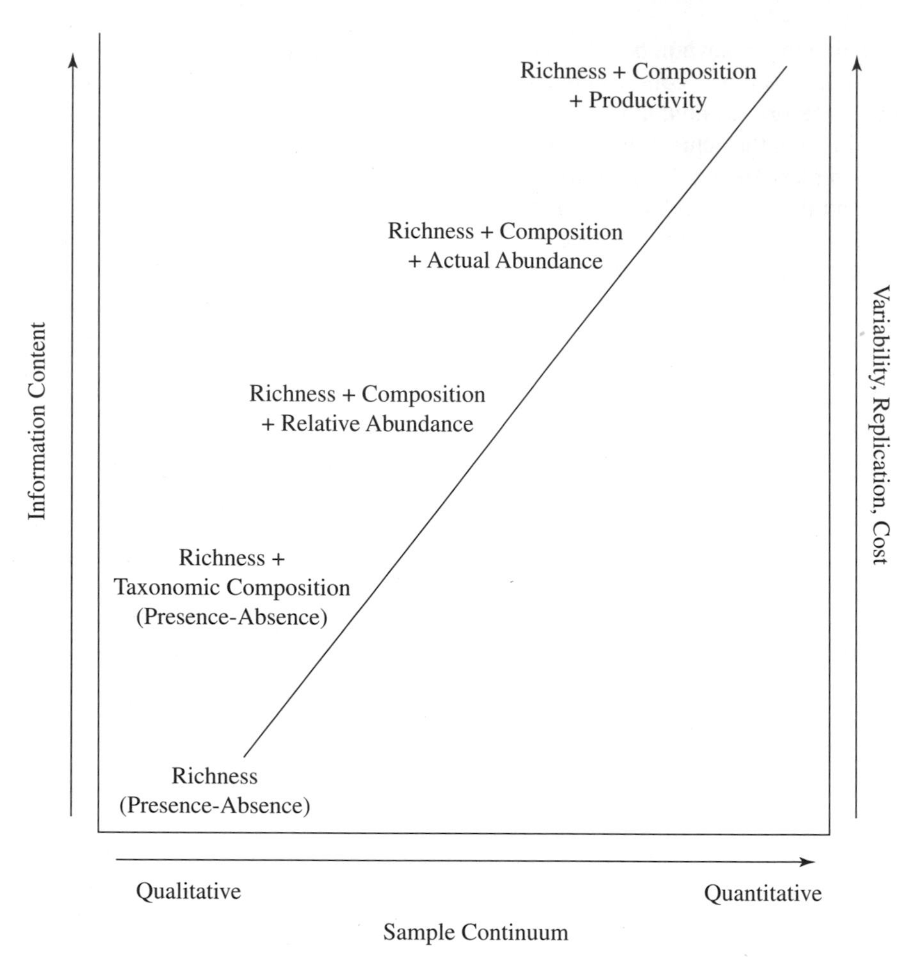

FIGURE 2.2. Conceptual relationship between type of sample (qualitative to quantitative continuum) and information content versus variability.

of impairment, objectives 1 and 2 can both be accomplished using qualitative data. However, detecting early signs of impairment may require quantitative techniques (e.g., Growns et al. 1997).

Quantitative and qualitative samples can also be used in combination. For example, a limited number of quantitative samples from a single habitat type (e.g., submergent macrophyte beds) might be used to estimate density differences among sites for dominant taxa without extreme levels of sample variation, whereas several qualitative samples from one or more habitat types can document species richness. Both types of samples require standardized procedures because both can be influenced by variability in sampling effort. For example, we expect taxonomic richness to be less in a 2-minute than in a 20-minute qualitative sample.

WHERE TO SAMPLE

Earlier we discussed procedures for stratifying the landscape for classification. In this section we discuss where to sample within a wetland once individual sites have been chosen. Within a site, most organisms show a contagious distribution because of habitat preferences. Deciding where to sample depends on a number of decisions, including what organisms are to be sampled (e.g., macrophytes versus invertebrates) and available time and money. In general, there are two ways to proceed: (1) sample one or a few habitat types, or (2) sample all habitat types. The stream literature suggests that estimates of richness and abundance at a site differ depending on the type of habitat sampled (Kerans et al. 1992, Parsons and Norris 1996). That is, we should sample the same habitat type(s) with equal effort (reference versus test) because the detection of impairment may be confounded by the type and variety of habitats sampled. Therefore, we might choose the single habitat type that contains the greatest abundance and diversity of taxa in order to reduce the cost of sampling several habitat types. However, stressors may have a different effect on different types of habitat. If we sample macrophytes when the sediment habitat has been disturbed by human intervention (heavy metals, pesticides), we may miss the effects of degradation. A potential solution is to sample the same two or three habitat types at each site (reference versus test). A composite sample from each of the habitats chosen (two or three qualitative samples per site) would reduce costs while maintaining potentially important habitat-specific data. However, each habitat type chosen would need to occur in all wetlands (reference and test).

Randomization is an important safeguard against sampling biases. For example, the data from samples (e.g., richness) can vary simply because different people vary with respect to sampling experience. Randomization can eliminate the potential for inexperienced technicians to select inadvertently for certain habitat types. Transects or grids are often used to choose specific locations randomly within a site to position samplers. In the laboratory, regularly spaced transects or grids can be superimposed over pictures, drawings, or maps of individual sites. Transects and specific positions along transects (or grids) can be randomly chosen for sampling. Otherwise, we can randomly choose sampling locations in the field. The longest length of a site can be divided into regular intervals which are chosen randomly for locating transect positions. Each transect can also be divided into regular intervals (e.g., 1-m sections) which can be chosen randomly as sampling locations. Stratifying by habitat types can be accomplished by sampling locations that correspond only to the habitat type chosen. Stratifying by habitat types also has the advantage of reducing among-site variation. More information on randomization as well as other aspects of sampling design are provided in Chapter 3.

Transect-Based Point Sampling

When habitat-specific changes (reference versus test) in the relative abundance of taxa are required (e.g., detecting early signs of degradation), the location of samples and cover proportions of each habitat type being sampled will influence the quality of the data. Depending on the size of the habitat, samples collected on the edge will be different from samples collected from the middle. This is called the *edge effect*. For instance, samples collected only a few meters from the edge of a cattail stand surrounded by a dense mat of floating vegetation (e.g., duckweed) will differ from samples collected toward the center of the cattails. The edge effect may override the actual community differences between duckweed and cattail. Samples collected at test sites from the interior of the stand compared to reference samples collected at the edge of the stand will provide erroneous habitat-specific results. Additionally, if the cattails were widely dispersed in small patches, all samples would be dominated by edge effect and should not be compared to samples at test sites composed of a large solid cattail stand. Decisions on the location of habitat edges must be made, and in very patchy wetlands such decisions can be difficult even for experienced investigators.

A qualitative sampling procedure, *point sampling,* can resolve this problem. Point sampling has been used successfully to examine microhabitat utilization of microcrustaceans in streams (Shiozawa 1986, 1991) and is easily extended to sampling both lakes and wetlands. This procedure attempts to decompose the grain of the environment to its smallest possible size (see Palmer and White, 1994). If properly applied (randomly or systematically), point samples can generate an unbiased image of the community, whether it is made up of single or multiple habitat types. We recommend applying this technique systematically along transects. Samples can be taken every 2 or 3 m along a single transect, or preferably, along multiple transects, thus sampling different habitats in proportion to their relative abundances. The need to measure habitat patchiness or coverage to account for edge effects is eliminated. For the data to estimate population abundances by habitat type accurately, a high number of separate samples should be taken. However, these samples can be pooled by habitat types in the field to avoid excessive processing time in the laboratory.

Two additional factors should be considered. First, each sample should be relatively small. This sampling procedure is modeled after point transect sampling in plant communities (see Chapter 14), and the ideal sample is actually a single point. The crew should not attempt to sample each point on the transect exhaustively. A few qualitative samples is sufficient. One of us (Shiozawa), uses a small, shallow net measuring 10 cm wide and a total of 200 samples are taken at a single site to document invertebrate habitat-specific richness and relative abundance in streams. The probability of detecting rare organisms, potentially useful in determining the early signs of impairment, will increase as a function of the number of sample points.

Second, the field crew should record physical data at each sample point. This can be accomplished easily by having a checklist where vegetation type, water depth, and so on, are recorded as each sample is taken. This information can then be used to regenerate the habitat composition along the transects. If 20% of the area is covered by cattails, the most probable frequency of cattail habitat hits will be 20%. The more samples that are taken, the more accurate the habitat estimate will become. Further, by examining the lengths of habitat runs along each transect, one can determine whether, for instance, the cattail samples occurred in large solid cattail stands (several consecutive points composed of cattails) or in scattered smaller stands (cattail hits interspersed sporadically with other habitat types). This will allow an accurate estimate of the edge effect at that particular site.

WHEN TO SAMPLE

Most wetland populations fluctuate on an annual basis in response to changing seasonal conditions (temperature, rainfall, daylength). To reduce seasonal variability, it is important to collect samples at different sites during the same time of the year or during the same season. Understanding the phenology of a region is necessary to estimate the length of each season as it relates to changes in species composition or abundance. This will determine how many sites can be sampled before the season begins to shift, which will determine how many total sites can be sampled in a year. This may be particularly important at higher elevations and latitudes where ice-free seasons last for a short proportion of the year.

Seasonal fluctuations in water level can also influence the location of samples within wetlands because of the relationship between inundation and colonization. Recently inundated parts of a wetland may contain lower densities and fewer species than parts that are inundated continuously. Colonization and growth of aquatic organisms does not begin until dried portions refill with water. In the temperate region, this may be particularly important during the spring, as this is the time that many wetlands are re-filled. If possible, samples should be located in areas known to be inundated throughout the year. When sampling wetlands that dry on an annual basis, variability among wetlands caused by differences in inundation–colonization dynamics can be reduced by positioning samples in the deepest parts that have been inundated the longest.

TAXONOMIC RESOLUTION

What is the best taxonomic resolution for detecting degradation? This depends on the severity of impairment and variation in sensitivity among the taxa within a group, which depends on their morphological, physiological, behavioral, and ecological similarity. For instance, if a family is comprised of four species and each species responds identically to the same stressor, species identification provides no more information than family designations. However, if two of the four species decline in abundance and two remain unchanged or increase, the family level response to degradation would record no change and fail to detect the impact. Since most species by definition share different gene pools and are at least somewhat different phenotypically, we might expect that the finer the taxonomic resolution, the greater the available information. This may, however, depend on the severity of the stress. All species will respond if degradation is sufficiently severe, but only sensitive taxa may respond to moderate degradation.

Although recent studies suggest that species- and genus-level data provide no additional information to detect impairment than higher levels of taxonomic resolution (e.g., Bowman and Bailey 1997, Hewlett 2000), there are rarely concurrent estimates of the level of degradation. Certainly, community change through time resulting from persistent stress may be manifest at higher levels of taxonomy (resistant species will also eventually decline), but this does little to detect early signs of degradation. This implies a positive relationship between the information needed to detect early signs of degradation and finer levels of taxonomic resolution. Figure 2.3 shows the conceptual relationship between information content/variation and taxonomic resolution. If higher levels of taxonomy are represented by more than one species, variation should increase with increasing taxonomic resolution because species within a family are more patchily distributed than would be detected with family-level metrics that average over the variation in lower taxonomic distributions (e.g., species distributions). Although a reduction in variation (order or family identifications) can sharpen our ability to detect statistical differences, it also produces a loss of information (e.g., the response of species to degradation). If we knew which species were most sensitive to degradation, we could avoid costly identifications of entire assemblages and focus on a few indicator taxa within each group.

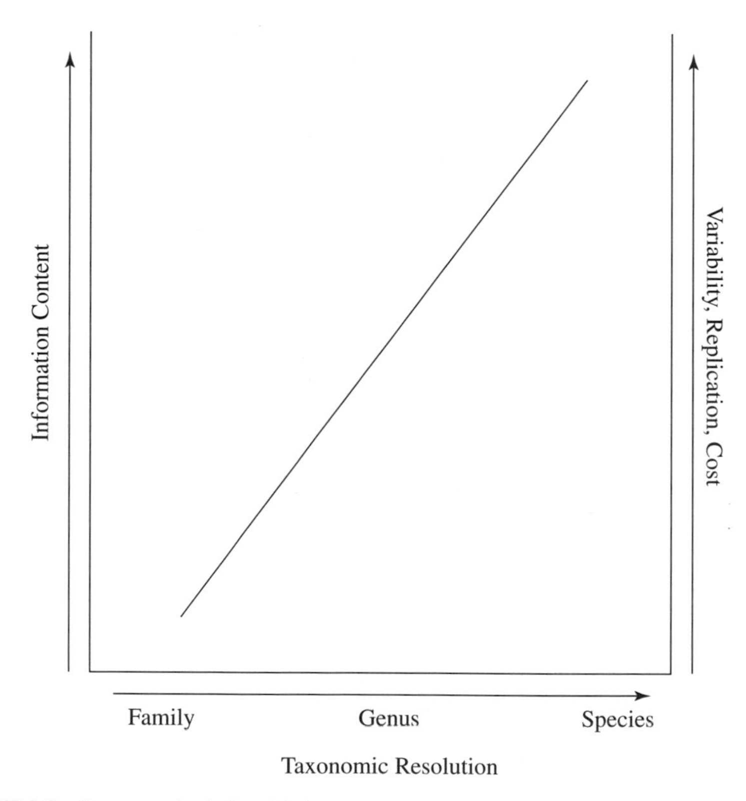

FIGURE 2.3. Conceptual relationship between taxonomic resolution and information content versus variability.

RARE TAXA

Because some taxa are more sensitive to impairment than others, and because reference sites differ/vary with respect to species composition, the greater the number of taxa used for bioassessment comparisons (reference versus test), the greater our ability to detect differences. That is, there should be a positive relationship between the information content needed to detect differences (reference versus test) and the number of taxa sampled. The chances that a specific species will be included in a sample depends on its abundance. Rare species are spatially more variable than abundant species and require greater sampling effort to detect. Figure 2.4 shows the trade-off between information content and variability as a function of the number of taxa sampled, which depends largely on their abundances. The decision to design a program to include rarer taxa will depend on associated costs and the need to detect early signs of degradation. If impairment is severe to moderate, locally abundant and regionally common taxa should be sufficient to detect differences among reference sites and test sites.

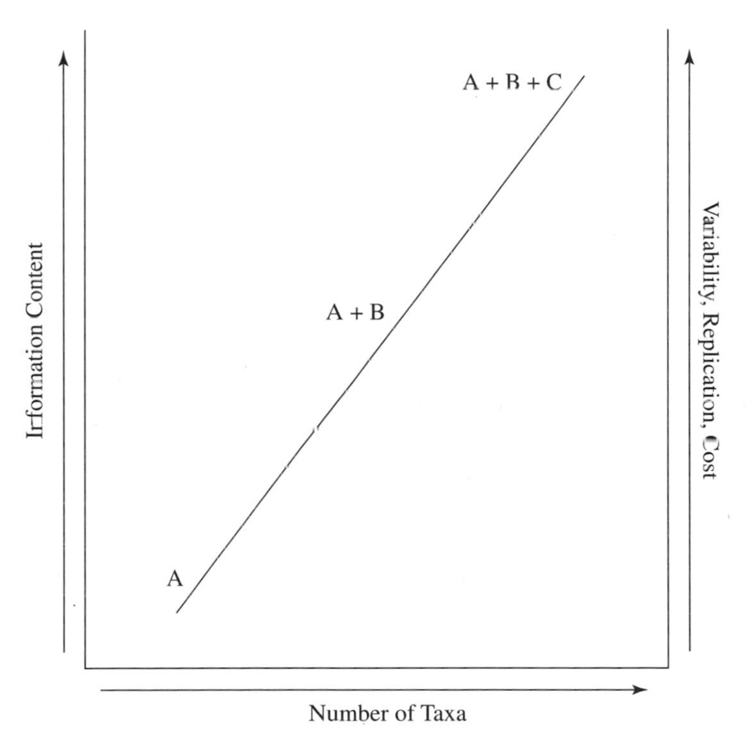

FIGURE 2.4. Conceptual relationship between number of taxa and information content versus variability. Taxa can be placed in one of three categories (e.g., Gibson et al. 1999): (A) locally abundant and regionally common (core species or dominant species), (B) intermediate abundance and occurrence (subordinate species), and (C) locally sparse and regionally rare (satellite species or transient species).

DETECTING EARLY SIGNS OF DEGRADATION

When degradation is severe, most biological metrics and bioassessment procedures will indicate alterations. Unfortunately, at this point the damage has already been done and restoration efforts, if possible, will probably be expensive. It would be much more cost-effective if we could detect the early signs of degradation. In Chapters 6 and 12 it is shown that bacteria and algae may be particularly useful in detecting the early signs of degradation. Also, Lemley and King (2000) have shown that the coverage of filamentous bacteria on insect body surfaces provide a valuable diagnostic tool for rapid detection of eutrophication in wetlands. This technique could also be useful in detecting early signs of degradation. Since small organisms have a high surface area/volume ratio and a rapid growth rate, both bacteria and algae are very responsive to changes in their environment. Perhaps quantitative estimates of abundance of sensitive bacteria and algae will be the best means of detecting early signs of degradation. However, these same size-related attributes cause bacteria and algae to show a great deal of spatial and temporal variability, which may require more replicates than it is feasible to sample. It is important to remember, however, that bioassessment is not always concerned with accurate estimates of population abundance, which require a large number of quantitative samples. Usually, we are interested only in obtaining enough information to distinguish reference conditions from test sites. Depending on a variety of other decisions, qualitative data may be sufficient to detect early signs of degradation (e.g., transect-based point sampling). Future research might examine the relative merits of using multiple groups of higher-level taxa (bacteria, invertebrates, etc.), quantitative versus qualitative samples, finer taxonomic resolution, and the importance of rare species in detecting early signs of impairment.

Is there an optimal strategy for collecting the data that will maximize our ability to detect early signs of degradation while minimizing costs associated with high variability? Figures in this chapter show the conceptual relationship between information content and variation of qualitative versus quantitative samples (Fig. 2.2), number of species (Fig. 2.3), and taxonomic resolution (Fig. 2.4). It would be useful if we knew more than the general direction of these relationships. For instance, the slope of the line would tell us if either information or variation increased faster with increasing sample type (qualitative, quantitative), number of species, or taxonomic resolution. It would also be valuable to know if there were inflection points where the slope changed, indicating a rapid increase or decrease in information content or variation. If inflection points existed, they might help us decide which sample type or taxonomic resolution would optimize information versus variation.

In addition to biological metrics, measurements of changes in habitat or physical features of the environment should be taken to corroborate bioassessment results and evaluate the causal factors associated with degradation (Resh et al. 1995). Barbour et al. (1995) suggested measuring channel morphology, bank stability, erosion potential, riparian vegetation, availability of microhabitat refugia, to help identify the factors causing degradation in streams. To date, there is no corollary or list of habitat features that would indicate causes of degradation in wetlands. However, such variables should be (1) ecologically important to the biota, (2) responsive to anthropic disturbance, and

(3) amenable to quantitative measurement. Helgen and Gernes (Chapter 8) found that water column and sediment chemistry, particularly the concentration of chloride and nutrients (nitrogen and phosphorus) were useful in identifying wetlands affected by stormwater input and agriculture, respectively. Conversely, Apfelbeck (Chapter 7) found that water column and sediment chemistry were too variable to assist in identifying impacted conditions. Future research should attempt to identify appropriate variables for indicating habitat degradation in wetlands to complement biological assessments and indicate the causal factors associated with human intervention.

ACKNOWLEDGMENTS

We thank Mark Vinson for reviewing an earlier version of this chapter. His comments helped us to refine our thinking on several points. However, the opinions offered herein are the sole responsibility of the authors.

REFERENCES

Allen, T. F. H. and T. W. Hoekstra. 1992. Toward a unified ecology. Columbia University Press, New York.

Anderson, D. J., and B. Vondracek. 1999. Insects as indicators of land use in three ecoregions in the Prairie pothole region. Wetlands 19:648–664.

Bailey, R. G., P. E. Avers, and T. King. 1994. Ecoregions and subecoregions of the United States. Forest Service Publication 73-94. U.S. Department of Agriculture, Washington, DC.

Barbour, M. T., J. B. Stribling, and J. R. Karr. 1995. Multimetric approach for establishing biocriteria and measuring biological condition. Pages 63–77 *in* W. S. Davis and T. P. Simon (ed.), Biological assessment and criteria: tools for water resource planning and decision making. Lewis Publishers, Boca Raton, FL.

Barbour, M. T., J. Gerritsen, B. D. Snyder, and J. B. Stribling. 1999. Rapid bioassessment protocols for use in streams and wadeable rivers: periphyton, benthic macroinvertebrates, and fish, 2nd ed. Report EPA 841-B-99-002. U.S. Environmental Protection Agency, Office of Water, Washington, DC.

Batzer, D. P., and S. A. Wissinger. 1996. Ecology of insect communities in nontidal wetlands. Annual Review of Entomology 41:75–100.

Beyers, D. W. 1998. Causal inference in environmental impact studies. Journal of the North American Benthological Society 17:367–373.

Bowman, M. F., and R. C. Bailey. 1997. Does taxonomic resolution affect the multivariate description of the structure of freshwater benthic macroinvertebrate communities? Canadian Journal of Aquatic Sciences 54:1802–1807.

Brinson, M. 1993. A hydrogeomorphic classification for wetlands. Wetlands Research Program Report TR-WRPDE-4. U.S. Army Corps of Engineers, Waterways Experiment Station, Vickburg, MS.

Corkum, L. D. 1990. Intrabiome distributional patterns of lotic macroinvertebrate assemblages. Canadian Journal of Fisheries and Aquatic Sciences 47:2147–2157.

Corkum, L. D. 1991. Spatial patterns of macroinvertebrate distribution along rivers in eastern deciduous forest and grassland biomes. Journal of the North American Benthological Society 10:358–371.

Covich, A. P., M. A. Palmer, and T. A. Crowl. 1999. The role of benthic invertebrate species in freshwater ecosystems. Bioscience 49:119–127.

Cowardin, L. M., V. Carter, F. C. Golet, and E. T. LaRoe. 1979. Classification of wetlands and deepwater habitats of the United States. Report FWS/OBS-79-/31. U.S. Fish and Wildlife Service, Washington, DC.

Driscoll, R. S., J. W. Russell, and M. C. Meier. 1978. Recommended national land classification system for renewable resource assessments. U.S. Forest Service, Rocky Mountain Station, Fort Collins, CO (unpublished manuscript).

Edwards, P. J., and C. Abivardi. 1998. The value of biodiversity: where ecology and economy blend. Biological Conservation 83:239–246.

Ehlich, P. R., and B. Walker. 1998. Rivets and redundancy. Bioscience 48:387.

Frissell, C. A., W. J. Liss, C. E. Warren, and M. D. Hurley. 1986. A hierarchical framework for stream habitat classification: viewing streams in a watershed context. Environmental Management 10:199–214.

Gibson, D. J., J. S. Ely, and S. L. Collins. 1999. The core-satellite species hypothesis provides a theoretical basis for Grime's classification of dominant, subordinate, and transient species. Journal of Ecology 87:1064–1067.

Golet, F. C., and J. S. Larson. 1974. Classification of freshwater wetlands in the glaciated Northeast. Resource Publication 116. U.S. Fish and Wildlife Service, Washington, DC.

Goodwin, R. H., and W. A. Niering. 1975. Inland wetland of the United States. Natural History Theme Studies, No. 2. National Park Service, Washington, DC.

Gosselink, J. G., and R. E. Turner. 1978. The role of hydrology in freshwater wetland ecosystems. Pages 63–78 *in* R. E. Good, D. F. Whigham, and R. L. Simpson (ed.), Freshwater wetlands: ecological processes and management potential. Academic Press, New York.

Green, R. H. 1979. Sampling design and statistical methods for environmental biologists. Wiley, New York.

Growns, J. E., B. C. Chessman, J. E. Jackson, and D. G. Ross. 1997. Rapid assessment of Australian rivers using macroinvertebrates: cost and efficiency of 6 methods of sample processing. Journal of the North American Benthological Society 16:682–693.

Hannaford, M. J., and V. H. Resh. 1995. Variability in macroinvertebrate rapid-bioassessment surveys and habitat assessments in a northern California stream. Journal of the North American Benthological Society 14:430–439.

Harding, J. S., M. J. Winterbourn, and W. F. McDiffett. 1997. Stream faunas and ecoregions in South Island, New Zealand: do they correspond? Archiv Fuer Hydrobiologia 140:289–307.

Hawkins, C. P. 1986. Pseudo-understanding of pseudo-replication: a cautionary note. Bulletin of the Ecological Society of America 67:184–185.

Hawkins, C. P., R. H. Norris, J. Gerrisen, R. M. Hughes, S. K. Jackson, R. K. Johnson, and R. J. Steveson. 2000. Evaluation of the use of landscape classifications for the prediction of freshwater biota: synthesis and recommendations. Journal of the North American Benthological Society 19:541–556.

Hawkins, C. P., and M. R. Vinson. 2000. Weak correspondence between landscape classifications and stream invertebrate assemblages: implications for bioassessment. Journal of the North American Benthological Society 19:501–517.

Heffner, R. A., M. J. Butler IV, and C. K. Reilly. 1996. Pseudoreplication revisited. Ecology 77:2558–2562.

Hewlett, R. 2000. Implications of taxonomic resolution and sample habitat for stream classification at a broad geographical scale. Journal of the North American Benthological Society 19:353–361.

Hildrew, A. G., and P. S. Giller. 1994. Patchiness, species interactions and disturbance in the stream benthos. Pages 21–62 *in* P. S. Giller, A. G. Hildrew, and D. G. Raffaelli (ed.), Aquatic ecology: scale, pattern, and process. Symposia of the British Ecological Society. Blackwell, Oxford.

Hills, J. M., K. J. Murphy, I. D. Pulford, and T. H. Flowers. 1994. A method for classifying European riverine wetland ecosystems using functional vegetation groups. Functional Ecology 8:242–252.

Hruby, T. 1999. Assessments of wetland functions: what they are and what they are not. Environmental Management 23:75–85.

Hughes, R. M. 1995. Defining acceptable biological status by comparing with reference conditions. Pages 31–47, *in*, W. S. Davis and T. P. Simon (ed.), Biological assessment and criteria. Lewis Publishers, Boca Raton, FL.

Hughes, R. M., T. R. Whittier, C. M. Rohm, and D. P. Larsen. 1990. A regional framework for establishing recovery criteria. Environmental Management 14:673–683.

Hughes, J. M., P. B. Mather, A. L. Sheldon, and F. W. Allendorf. 1999. Genetic structure of the stonefly, *Yoraperla brevis,* populations: the extent of gene flow among adjacent montane streams. Freshwater Biology 41:63–72.

Hurlbert, S. H. 1984. Pseudoreplication and design of ecological field experiments. Ecological Monographs 54:187–211.

Karr, J. R., and E. W. Chu. 1999. Restoring life in running waters: better biological monitoring. Island Press, Washington, DC.

Keddy, P. 1993. Do ecological communities exist? A reply to Bastow Wilson. Journal of Vegetation Science 4: 135–136.

Kerans, B. L., J. R. Karr, and S. A. Ahlstedt. 1992. Aquatic invertebrate assemblages: spatial and temporal differences among sampling protocols. Journal of the North American Benthological Society 11:377–390.

Krebs, C. J. 1989. Ecological methodology. Harper & Row, New York.

Lawton, J. 1994. What do species do in ecosystems? Oikos 71:367–374.

Lemley, A. D., and R. S. King. 2000. An insect–bacteria bioindicator for assessing detrimental nutrient enrichment in wetlands. Wetlands 20:91–100.

Maxted, J. R., M. T. Barbour, J. Gerritsen, V. Poretti, N. Primrose, A. Silvia, D. Penrose, and R. Renfrow. 2000. Bioassessment framework for mid-Atlantic coastal plain streams using benthic macroinvertebrates. Journal of the North American Benthological Society 19:128–144.

McCann, K. S. 2000. The diversity–stability debate. Nature 405:228–233.

Millar, J. B. 1976. Wetland classification in western Canada: a guide to marshes and shallow open water wetlands in the grasslands and parklands of the prairie provinces. Canadian Wildlife Service Report Series, No. 37.

Mitsch, W. J., and J. G. Gosselink. 2000, Wetlands, 3rd ed. Wiley, New York.

Mount Hood watershed analysis team. 1994. The Fish Creek watershed analysis. U.S.D.A. Forest Service, Mt. Hood National Forest, Gresham, OR.

Naeem, S. 1998. Species redundancy and ecosystem reliability. Conservation Biology 12:38–45.

Norris, R. H. 1995. Biological monitoring: the dilemma of data analysis. Journal of the North American Benthological Society 14:440–450.

Norris, R. H., and A. Georges. 1993. Analysis and interpretation of benthic macroinvertebrate surveys. Pages 234–286. *In* D. M. Rosenberg and V. H. Resh (eds.), Freshwater biomonitoring and benthic macroinvertebrates. Chapman & Hall, New York.

Norris, R. H., E. P. McElravy, and V. H. Resh. 1996. The sampling problem, Pages 184–208 *in* G. Petts and P. Calow (ed.), River Biota. Blackwell, Oxford.

Omernik, J. M. 1987. Ecoregions of the conterminous United States. Annals of the Association of American Geographers 77:118–125.

Omernik, J. M. 1995. Ecoregions: a spatial framework for environmental management. Pages 49–62 *in* W. S. Davis and T. P. Simon (ed.), Biological assessment and criteria. Lewis Publishers, Boca Raton, FL.

Palmer, M. W., and P. S. White. 1994. Scale dependence and the species–area relationship. American Naturalist 144:717–740.

Parsons, M., and R. H. Norris. 1996. The effect of habitat-specific sampling on biological assessment of water quality using a predictive model. Freshwater Biology 36:419–434.

Resh, V. H., and J. K. Jackson. 1993. Rapid assessment approaches to biomonitoring using benthic macroinvertebrates. Pages 195–233 *in* D. M. Rosenberg and V. H. Resh (ed.), Freshwater biomonitoring and benthic macroinvertebrates. Chapman & Hall, New York.

Resh, V. H., R. A. Norris, and M. T. Barbour. 1995. Design and implementation of rapid assessment approaches for water resource monitoring using benthic macroinvertebrates. Australian Journal of Ecology 20:108–121.

Reynoldson, T. B., and D. M. Rosenberg. 1996. Sampling strategies and practical considerations in building reference data bases for the prediction of invertebrate community structure: design and analysis of benthic bioassessments. Technical Information Workshop of the North American Benthological Society, Kalispell, MT.

Reynoldson, T. B., R. H. Norris, V. H. Resh, K. E. Day, and D. M. Rosenberg. 1997. The reference condition: a comparison of multimetric and multivariate approaches to assess water-quality impairment using benthic macroinvertebrates. Journal of the North American Benthological Society 16:833–852.

Ricklefs, R. E., and D. Schluter. 1993. Species diversity in ecological communities. University of Chicago Press, Chicago.

Schulze, E. D., and H. A. Mooney. 1993. Biodiversity and ecosystem function. Springer-Verlag, Berlin.

Shaffer, P. W., M. E. Kentula, and S. E. Gwin. 1999. Characterization of wetland hydrology using hydrogeomorphic classification. Wetlands 19:490–504.

Shaw, S. P., and C. G. Fredine. 1956. Wetlands of the United States. Circular 39. U.S. Fish and Wildlife Service, Washington, DC.

Shiozawa, D. K. 1986. The seasonal community structure and drift of microcrustaceans in Valley Creek, Minnesota. Canadian Journal of Zoology 64:1655–1664.

Shiozawa, D. K. 1991. Microcrustaceans from nine Minnesota streams. Journal of the North American Benthological Society 10:286–299.

Stewart, R. E., and H. A. Kantrud. 1971. Classification of natural ponds and lakes in the glaciated prairie region. Resource Publication 92. U.S. Fish and Wildlife Service, Washington, DC.

Strahler, A. N. 1957. Quantitative analysis of watershed geomorphology. Transactions of the American Geophysical Union 38:913–920.

Tansley, A. G. 1911. Types of British vegetation. Cambridge University Press, Cambridge.

Tilman, D., J. Knops, D. Wedin, P. Reich, R. Ritchie, and E. Siemann. 1997. The influence of functional diversity and composition on ecosystem processes. Science 277:1300–1305.

White, P. S., and J. L. Walker. 1997. Approximating nature's variation: selecting and using reference information in restoration ecology. Restoration Ecology 5:338–349.

Whittaker, R. H. 1962. Classification of natural communities. Botanical Review 28:1–239.

Williams, D. D. 1987. The ecology of temporary waters. Timber Press, Portland, OR.

Wilson, J. B. 1991. Does vegetation science exist? Journal of Vegetation Science 2:289–290.

Wilson, J. B., and R. J. Whittaker. 1995. Assembly rules demonstrated in a saltmarsh community. Journal of Ecology 83:801–807.

Wilson, J. B., I. Ullmann, and P. Bannister. 1996. Do species assemblages ever recur? Journal of Ecology 84:471–474.

Windell, J. T., B. E. Willard, D. J. Cooper, S. Q. Foster, C. F. Knud-Hansen, L. P. Rink, and G. N. Kiladis. 1986. An ecological characterization of Rocky Mountain montane and subalpine wetlands. Biology Report 86(11). U.S. Fish and Wildlife Service, Washington, DC.

Wright, J. F. 1995. Development and use of a system for predicting the macroinvertebrate fauna in flowing waters. Australian Journal of Ecology 20:181–197.

Wright, J. F., D. W. Sutcliffe, and M. T. Furse. 2000. Assessing the biological quality of freshwaters: RIVSPAK and other techniques. Freshwater Biological Association, Ambleside, Cumbris, England.

Yoccoz, C. O. 1991. Use, overuse, and misuse of significance tests in evolutionary biology and ecology. Bulletin of the Ecological Society of America 71:106–111.

Yoder, C. O., and E. T. Rankin. 1995. Biological criteria development and implementation in Ohio. Pages 109–144 *in* W. S. Davis and T. P. Simon (ed.), Biological assessment and criteria. Lewis Publishers, Boca Raton, FL.

3 Statistical Issues for Sampling Wetlands

STEPHEN L. RATHBUN and JEROEN GERRITSEN

Planning and implementation of wetland sample surveys requires consideration of several statistical design issues. Issues regarding the planning of a survey include the specification of what questions are to be addressed, the population of interest, and reporting units. To obtain a representative sample and unbiased estimates of environmental parameters with known precision, probability-based sampling designs should always be used to monitor wetlands or any other natural resource. Probability-based designs appropriate for monitoring wetlands include stratified random, multistage, and randomized systematic designs. Sampling design in turn influences the effectiveness of design and model-based statistical inferences. Measurement error, which affects power and precision, can be reduced in field studies by refining field sampling methods.

Ecologists have been concerned with sampling the environment at least since the Uppsala school of phytosociologists considered the use of randomly placed quadrats to quantify plant communities (e.g., Whittaker 1962). Despite this, few ecologists appear to be aware of the statistical properties of environmental sampling designs. This is particularly evident given the prevalence of judgment sampling designs for monitoring aquatic resources (Olsen et al. 1999). The failure to randomize site selection under such designs can lead to samples that are not representative of the populations for which inferences are desired, and hence biased estimates of environmental parameters. Academic statisticians are at least partially culpable, given that introductory statistics courses tend to emphasize model-based inferential procedures (e.g., multiple regression, ANOVA) and give little or no attention to sampling variation and design-based inferential procedures.

In this chapter we review some of these often neglected statistical issues regarding the planning and implementation of wetland sampling programs. Issues regarding the planning of a sampling program are considered, including specification of the questions to be addressed, the population of interest, and the reporting unit. Sampling

Bioassessment and Management of North American Freshwater Wetlands, edited by Russell B. Rader, Darold P. Batzer, and Scott A. Wissinger.
0-471-35234-9

designs appropriate for monitoring wetlands are presented followed by a discussion of the differences between design- and model-based statistical inference. The impact and control of measurement error are considered, followed by a discussion of power and precision.

SPECIFYING THE QUESTIONS

The first task is to determine and state in a specific fashion the principal questions that the sampling program will answer. For example, suppose that the objective is to establish reference conditions for plant or animal communities from wetlands in a region (e.g., ephemeral pools in a national forest). This may include the identification and characterization of classes of reference wetlands. Questions might include:

- Should the minimally disturbed wetlands in the region be divided into classes that differ in plant or animal diversity, abundance, and dynamics?
- What are the physical, chemical, and relevant biotic characteristics of each of the wetland classes?

After criteria have been developed, the questions change to encompass assessments of individual wetlands, groups of wetlands, or wetlands of an entire region or state. Specific questions might include:

- Is a specific wetland similar to reference wetlands of its class (unimpaired), or is it different from reference wetlands (altered or impaired)?
- What is the status of wetlands in the region? How many wetlands are similar to reference conditions? How many wetlands are impaired?
- Has a specific wetland, or wetlands in a region, improved or deteriorated over a certain period? Are more wetlands similar to reference conditions now than some time ago?

Finally, researchers and analysts often wish to determine the relationships among variables; that is, to develop predictive, empirical (statistical) models that can be used to design management responses to perceived problems. Examples of specific questions might include:

- Can a community attribute of a wetland be predicted by areal nutrient loading rates (e.g., Vollenweider models)?
- Can a community attribute of a wetland be predicted by watershed land use?

SPECIFYING THE POPULATION AND SAMPLE UNITS

A sample is expressed statistically as a subset of a population of objects, or *sample units*. In some cases the population is finite, countable, and easy to specify (e.g., all wetlands in a region, where each wetland is a single member of the population). In

TABLE 3.1. Examples of Sample Units and Populations

Sample Unit	Sample Population	Infinite or Finite Population?
Point in a wetland (may be characterized by single or multiple sample device deployments)	All points in the wetland	Infinite
Constant area (e.g., square meter, hectare)	All square meters of wetland surface area in a state or region	Infinite
Wetland or definable part of a wetland as a single unit (likely to be the most common sample unit)	All wetlands in a state or region; or all definable units of a single wetland	Finite

other cases, the population is more difficult to specify and may be infinite (e.g., wetland waters of the region, where any location in any wetland defines a potential member of the population) (Thompson 1992). Sampling units may be natural units (entire wetlands), or they may be arbitrary (plot, quadrat, sampling gear size; Piclou 1977). Finite populations may be sampled with corresponding natural sample units, but often the sample unit (say, a wetland) is too large to measure in its entirety, and it must be characterized with second-stage samples. For example, with wetland invertebrates, second-stage sample units might include Surber samples, benthic cores, or sweeps of a net. Other examples of sample units are given in Table 3.1.

The objective of sampling is to best characterize individual sample units (e.g., individual wetlands) in order to estimate some aspects (mean, variance, percentiles) of a population of multiple sample units. The objective is not usually to say something about a single sample unit. If a single wetland is the subject of the assessment, multiple sample units should be defined within the wetland.

SPECIFYING THE REPORTING UNIT

It is necessary to specify the units for which results will be reported. This may be the population (e.g., all wetlands) but often is a subpopulation (e.g., wetlands within a given area or wetlands of special interest). To help develop a sampling plan, it is useful to create hypothetical statements of results in the way that they will be reported; for example:

- *Status of a place:* "This wetland is degraded."
- *Status of a region:* "20% of wetlands in the state have reduced species richness (below reference expectations)."

- *Trends at place:* "Species richness in this wetland or average species richness in wetlands of the state has decreased by 20% since 1980."
- *Relationships among variables:* "Wetlands receiving runoff from large parking lots have 50% greater probability of species loss than wetlands not receiving such runoff."

DESIGNS

Once the study objectives and population have been defined and it has been determined what variables are to be measured, the next step in any field study of wetlands is the selection of sample units or locations at which observations are to be taken; that is, we must select a sampling design. The particular sampling design should be constructed so as to meet the study objectives. These objectives may call for the selection of specific sites of interest: for example, sites near point sources of environmental contamination. If, however, the objectives call for inferences about the population from which the sample is drawn, a probability-based sampling design should always be used. Probability-based designs involve some method of random selection of sample units or locations but are not restricted to simple random sampling. In contrast, nonprobability sampling designs rely on the judgment of the investigator and are not likely to yield a representative sample.

A wide variety of probability-based sampling designs are available. The *simple random sampling design* is the most basic method for selecting sample units from a population. For finite populations of discrete sample units (e.g., whole wetlands), this requires a list (called a *list frame*) of all N sample units in the population, from which a sample of n units is randomly selected in such a way that each unit is equally likely to be included in the sample, and the selection of one unit does not influence the selection of the remaining units. For sampling locations in a study region, a simple random sample is obtained by random and independent selection of X and Y coordinates from the study region (Fig. 3.1). Aside from selection of the target population from which the sample is drawn, the selection of sample units does not involve any scientific judgment.

It is usually unwise to ignore the scientific judgment of the investigator, even with probability-based designs. Under a *stratified random sampling design* (Fig. 3.2), the population is partitioned into strata, often corresponding to different habitats of interest. For example, the marshes of southern Florida might be partitioned into wet prairie, sawgrass, cattail, *Muhlenbergia,* and cypress habitats. Sample units are then selected from each stratum according to some probability-based sampling design, often a simple random sampling design. Strata should be selected in such a way that differences between strata are as large as possible, while units within strata are as uniform as possible. By controlling for differences among strata, the stratified random sampling design reduces the sampling variance and hence improves the precision of population parameter estimates. We can increase the sampling effort in ecologically important strata, ensure that an adequate sample is obtained from rare habitats, and reduce costs by reducing sampling effort in expensive strata.

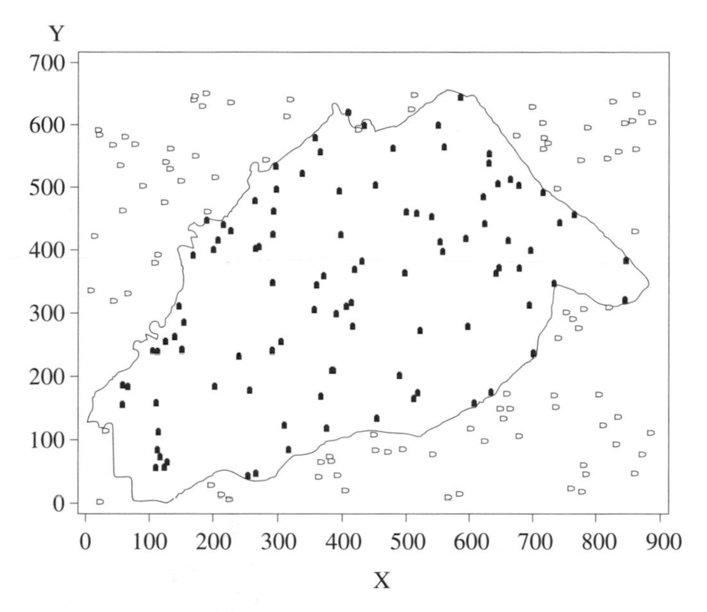

FIGURE 3.1. Simple random sample of 100 sites in Ebenezer Aquifer (filled circles), southeastern Georgia. This design was constructed by independent random selection of X coordinates (between 0 and 900) and Y coordinates (between 0 and 700). Sites falling outside the study region (open circles) were excluded from the sample.

A *multiple-stage design* does not require a list frame all of the sample units in the population and hence is more practical than simpler designs for many large-scale wetlands applications. Under a two-stage sampling design, the population is first partitioned into primary sample units, then a simple random sample of primary units is selected, and finally, a simple random sample is obtained from each of the primary units selected. Primary units should be small enough so that all sample units within each of them can easily be enumerated. The flexibility of multiple-stage designs is illustrated by the following examples:

- To investigate the effects of water removal from aquifers in central Florida on the condition of cypress trees, individual cypress domes can be treated as primary sample units. Domes can be enumerated from areal photographs, facilitating the selection of a simple random sample. Then a simple random sample of adult cypress tress can be obtained from each of the selected domes. Here, the multistage sampling design provides a random procedure without having to enumerate every tree in every dome.
- To estimate larval midge densities in a marsh, a sample of n quadrats may be randomly located in the marsh, and then m core sampling points are located within each quadrat. Here the quadrats are treated as the primary sample units. Larval densities in each quadrat may then be estimated and averaged to estimate

FIGURE 3.2. Stratified random sampling design in the wetlands surrounding a Carolina Bay. Circles are in grasslands, squares are in briars and shrubs, triangles are in vines and small trees, stars are in hardwoods and pines, and crosses are in pines.

densities over the entire marsh. Alternatively, n parallel line transects may be randomly located within the wetland, and then m core samples may be randomly selected along the length of each transect. Here, the transects are treated as the primary sample units.

Observe that quadrat and transect sampling designs familiar to ecologists are special cases of two-stage sampling designs. Two-stage sampling designs can be extended into multiple-stage designs by further partitioning each of the sampled primary units into secondary units, and so on. In the first example above, a three-stage sample design would be obtained if quadrats may be randomly located in each of the selected cypress domes, and then trees can be randomly sampled from each quadrat. Here, the quadrats are considered to be secondary sample units.

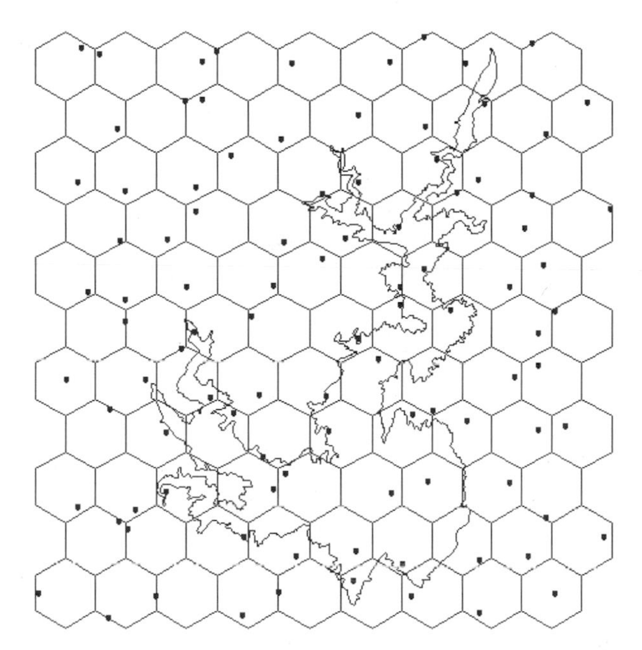

FIGURE 3.3. Randomized-tessellation stratified design for Lake Jocassee.

A *randomized-tessellation stratified design* (Stevens 1997) may be used to ensure a more regular spacing of sample sites than can be obtained under simple random sampling. Under such a design, a grid of contiguous polygons is randomly placed over the study region (Fig. 3.3), and then a single site is randomly located within each of the polygons. Only sites falling in the region of interest are included in the sample. The sampling variance under this design is smaller than that under a simple random sampling design, especially if the data show strong spatial correlation.

Thompson (1990, 1992) describes an *adaptive cluster sampling design* appropriate for a population of a rare species. Under this design, the region is partitioned into a contiguous grid of square quadrats, and a simple random sample of quadrats is selected. If the rare species is observed in a sampled quadrat, this design calls for the sampling of neighboring quadrats. Thus, the adaptive sampling design is most effective for populations of strongly clustered species (like most plants and animals). This design is biased in favor of quadrats containing the species. However, since the probability of sampling each quadrat can be quantified, a bias correction can be applied.

Study objectives may call for sampling over time as well as space. Under budgetary constraints, consideration must be given to how sampling effort is to be allocated: Should we sample a few sites at frequent intervals or a large number of sites infrequently? The answer depends on the study objectives. If they call for an investigation of seasonal trends, monthly samples may be required. Conversely, annual samples may suffice if long-term temporal trends are of interest.

The optimal allocation of sampling effort over time and space depends on the relative magnitudes of spatial and temporal correlation in the data. This spatiotemporal correlation comes from the recognition that pairs of observations that are close together in time and space tend to be more similar to one another than observations that are far apart in either time or space. For example, spatial correlations might arise from limited dispersal of organisms, resulting in clustered patterns of distribution, and/or from the correlation of their abundances with one or more unsampled environmental variables, which themselves show spatial correlations. Temporal correlations might result from dependence of population densities during a given time interval on the population densities in past time intervals. In general, the best power for detecting differences is achieved when sites are sampled in such a way as to minimize spatial and temporal correlations among the data. If temporal correlation is strong, data collected at frequent intervals will contain a large amount of redundant information. In this case, a large number of sites should be sampled at infrequent time intervals. In a similar manner, observations at neighboring locations at a given point in time will contain redundant information under strong spatial correlation, in which case a few sites should be sampled at frequent time intervals.

INFERENCE

Two types of statistical inference can be distinguished, model-based and design-based. Under model-based inference, it is assumed that the data are realized from some random model. In multiple regression, for example, the variable of interest is assumed to be a linear function of some explanatory variables plus an experimental error. Further assumptions may include the homogeneity of variance (i.e., variation in the data does not depend on location, time, or any explanatory variables that might be included in a model), and that the data are uncorrelated and normally distributed. Instead of making inferences about the population from which the data were obtained, inferences are made on model parameters under model-based inferential procedures. Such inferences may include estimates of model parameters, together with their corresponding standard errors, as well as predictions of future observations and data at unsampled sites. Model-based hypothesis testing procedures test whether or not the data are compatible with a null model.

Under design-based inference the values of the variable of interest in the population are assumed to be fixed and nonrandom. Here the source of variation comes from the random selection of sample units; so design-based inference is available only for probability-based sampling designs. Since the sampling design is specified by the investigator, and hence is known, no model assumptions are required. Design-based inferences are made on the actual population from which the sample was drawn, not on the parameters of some assumed model. Such inferences may include estimates of population parameters, together with their corresponding standard errors. Design-based hypothesis-testing procedures test how likely a sample could have been drawn from a population with the null parameter value. Since inferences are limited to the extant population from which the sample was drawn, design-based inference cannot be used to predict future observations or data at unsampled sites.

To further elucidate the differences between design-based and model-based inference, consider the following hypothetical example involving the salt marshes of the Georgia coast. Aerial photographs can be used to map the region covered by the salt marshes, from which a probability sample of sites can be selected. At each sample site, above ground biomass of plants might be measured by clipping all plants in a 1-m^2 plot, nutrients and contaminants may be assayed from a 10-cm-deep core sample, and benthic organisms may be enumerated from the same core sample. To investigate temporal trends, sample sites may be visited once per year. Here, design-based inferential procedures might be used to estimate the total biomass of plants and benthic organisms and the total mass of nutrients and contaminants in the top 10 cm of soil in the salt marshes of Georgia during each sampling interval. If the Index of Biotic Integrity (Karr 1991) can be computed for each sample, design-based inference may also be used to estimate the percentage of the surface area of salt marshes with poor integrity, or the percentage of the area with improved integrity over the years in which data are available. If, however, we wish to predict the future biotic integrity of the marshes, we would require time-series models for the data. Model-based inference would also be required to investigate relationships among soil nutrients and contaminants, plant biomass, and the community of benthic organisms. Such investigations may include multiple regression model, path analysis, and (multivariate) analysis of variance and covariance.

Design-based inferential procedures are often simpler than model-based procedures when data are spatially and/or temporally correlated. To illustrate this, consider sites drawn from a simple random sampling design. For model-based inference, a simple model may assume that an observation at a given location is equal to a mean plus a random error. Here, the sample mean is an unbiased estimator for the model mean. Since observations that are close together may be more similar to one another than observations that are far apart, we might expect the data to be spatially correlated. In this case, the standard error of the sample mean is

$$\frac{\sigma}{n}\sqrt{\sum_{i=1}^{n}\sum_{j=1}^{n}\rho ij}$$

under the model, where ρ_{ij} is the correlation between the data at sites i and j.

For design-based inference, the sample mean is also unbiased for the population mean. Since the sample sites were selected independently, however, the observations are independent from a design-based perspective. Thus, the standard error is simply

$$\sigma / \sqrt{n}$$

Here, the standard deviation can be estimated by the sample standard deviation s. More complex sampling designs require more complicated methods of analysis than those just described for the simple random sampling design (Gilbert 1987, Thompson 1992).

To illustrate the differences between model- and design-based statistical inferences, consider data collected from 416 sites in the marshes of southern Florida from 1995 to 1996 (Stober et al. 1998). Concentrations of mercury, sulfate, and phosphorus were assayed from soil samples collected at each site. Sites were selected using a randomized-tessellation stratified design, and so a different design-based estimator for the standard error is required. Here, standard errors were calculated using the Yates-Grundy estimator described in Stevens (1997). For model-based inference, the exponential correlation function

$$\rho ij(1-c_0)e^{-3d_{ij}/\alpha}$$

was used to model spatial dependence between each pair of sites $i \neq j$. This is a function of the distance d_{ij} between the pair of sites. The parameter α is the range of spatial correlation; sites farther than α apart are negligibly correlated. The parameter c_0 is the relative nugget effect; multiplication of c_0 by the population variance σ^2 yields the nugget effect, which measures small-scale spatial variation in the data. The results show that design-based estimates of standard errors are considerably smaller than their model-based counterparts (Table 3.2). This difference in magnitude is particularly notable when the range of spatial correlation is large, as was the case for mercury. This result does not mean that design-based inference is superior to model-based inference but is simply a consequence that these measure two different sources of variation. For design-based inference, the standard error measures variation due to random selection of sample sites, while for model-based inference, the standard error is a function of residual variation in the model.

In practice it is often desirable to carry out both design- and model-based inferential procedures. Design-based procedures can be employed to make inferences about the actual population from which the sample was drawn. Such procedures would be appropriate, for example, if we are attempting to estimate the total mass of a contaminant in a study region, or the proportion of the study region showing improving

TABLE 3.2. Design- and Model-Based Inference for Soil Constituents in Southern Florida[a]

		Standard Error		Model Parameter Estimates		
Variable	Mean	Design-Based	Model-Based	σ	Nugget	Range (km)
Mercury (μg/kg)	124.89	1.16	18.00	62.27	0.297	42.7
Sulfate (mg/kg)	382.46	19.05	36.25	494.47	0.000	5.8
Phosphorous (mg/kg)	395.14	7.16	27.13	232.90	0.426	14.5

[a]Concentrations are measured in units of mass for each constituent per kilogram of soil.

conditions over time. Model-based procedures can be employed to explore ecological processes and make predictions. Such inferences include the fitting of multiple regression, time series, and spatial models. Although these require model assumptions, their validity can be tested through appropriate diagnostic procedures (e.g., residual plots).

CONTROLLING MEASUREMENT ERROR

Variability has many possible sources, and the intent of many sampling designs is to minimize the variability due to uncontrolled or random effects, and conversely, to characterize the amount of variability among strata (e.g., habitat types), or caused by experimental effects. Measurement error is a component of variability and is the result of biases or variability attributed to sampling methods, gear, or instrumentation, natural spatial or temporal variation within a sampling unit that is not accounted for in the sampling design, and/or errors in proper adherence to field and laboratory protocols. Examples of natural components of measurement error include effects of habitat type (e.g., floating macrophytes, emergent sedge marsh, cattails) when the sampling unit is a whole wetland (e.g., prairie potholes); or recruitment or emergence events during the sampling season.

The basic rule of efficient sampling and measurement is to minimize measurement error, to maximize the components of variability that have influence on the study objective, and to control for other sources of variability that are not of interest (minimizing their effects on the observations). It is possible to estimate some of these components with multistage sampling and stratification, for example, by sampling several habitat types in each wetland. However, the increased effort raises sampling and laboratory costs. To reduce measurement error and cost, one should sample each wetland in the same way, place, and time to minimize variability due to location, water depth, and season if they are not of interest in the particular study. Multiple samples within a location may be combined to form a composite sample, reducing the within-location variation. In a study of benthic macroinvertebrates in nine Florida lakes, for example, Gerritsen et al. (1999) compared the coefficient of variation obtained by treating 12 petite PONAR grabs separately versus two replicates consisting of six composited PONARs for each lake (Table 3.3). Coefficients of variation were reduced under composite samples for taxa richness, percent dominance, sensitive taxa (Ephemeroptera, Trichoptera, Odonata), and abundance. A further reduction in the within-lake component of variation for taxa richness can be obtained by subsampling a fixed number of organisms from each composite sample (Barbour and Gerritsen 1996). Gear deployment, composite sampling, and subsampling should be considered when selecting sampling gear and methods (see the various sampling chapters in this volume).

Every study will need some level of repeated measurement at sampling sites to estimate measurement error. A typical level of effort in sampling and assessment programs is replication of the entire sampling protocol at approximately 10% of sites. Analysis of variance may then be used to estimate measurement error. All multiple

TABLE 3.3. Comparison of 12 PONARs Treated Separately Versus Two Replicates of Six Composited PONARs[a]

	Mean of 12 PONARs (12 samples each lake)			Mean of Two Composited PONARs (two samples of six grabs in each lake)		
	Population Mean (nine lakes)	s.d. (individual lake)	CV (average lake) (%)	Population Mean (nine lakes)	s.d. (individual lake)	CV (average lake) (%)
Taxa richness	8.85	3.62	40.9	25.7	4.36	16.9
Percent dominance	58.8	14.8	25.2	50.4	8.9	17.7
Sensitive taxa	0.39	0.628	161	1.6	1.27	79.4
Total individuals (ln)	4.13	0.717	17.4	6.12	0.145	2.4

[a]s.d., standard deviation; CV, coefficient of variation.

observations of a variable are used (from all wetlands with multiple observations), and wetlands are treated as the primary source variable. The mean squared error of the ANOVA is the estimated variance of repeated observations within wetlands. Note that a hypothesis test (*F* test) is not of interest in this application. Measurement error is minimized with methodological standardization: selection of low-variability sampling methods, proper training of personnel, and quality assurance procedures.

The presence of measurement error in explanatory variables can result in biased estimates of regression coefficients. For example, consider predicting the concentration of mercury in the tissues of mosquito fish (denoted by y) from the concentration of methyl mercury in water (denoted by x) for 96 sites in the marshes of southern Florida in September 1995 (Stober et al. 1998). A simple linear regression yields the predictive equation

$$\hat{y} = 124.6 + 113.8x$$

From duplicate water samples, the variance due to measurement error is estimated to be $\sigma_u^2 = 0.0107$, which is relatively small compared to the between-site variance of $\sigma_X^2 = 0.3193$. The percent bias of the slope is equal to

$$\frac{\sigma_X^2}{\sigma_X^2 - \sigma_u^2} \times 100\% = 3.19\%$$

(Fuller 1987), yielding the bias corrected predictive equation

$$\hat{y} = 122.1 + 117.7x$$

In this case, the effect of measurement error is small. If, however, the variance due to measurement error were larger, the bias due to measurement error could also be substantial.

POWER AND PRECISION

Developing a sampling design requires consideration of trade-offs among the measures used, the effect size that is considered meaningful, desired confidence, desired power, and resources (money and donated labor) available for the study. Statistical power is the ability of a given hypothesis test to detect an effect that actually exists and must be considered when designing a sampling program (e.g., Peterman 1990, Fairweather 1991). The power of a test is defined as the probability of rejecting the null hypothesis correctly when it is false (i.e., the probability of correctly finding a difference or an effect when one exists). For a fixed confidence level (95%), power can be increased by increasing the sample size or by altering the sample design. To evaluate power and determine sampling effort, one must decide how much change in the value of a measured variable is ecologically meaningful. For example, in studies of benthic macroinvertebrates, one might set a reduction in total taxa richness of 30%, or a reduction in relative abundance of odonates by 50%, as being ecologically meaningful changes in the ecosystem. Power analysis requires some knowledge of the variability of the measures, either from pilot research or studies done elsewhere. Numerous textbooks and journal articles provide specific guidance for power analysis (e.g., Dixon and Massey 1969, Snedecor and Cochran 1980, Green 1979, Burton and Pitt 1996).

"*Precision* is the closeness of repeated measurements to the same quantity" (Sokal and Rohlf 1969). In statistics, precision is generally expressed in terms of standard errors, variances of parameter estimates, and confidence intervals. Precision goes hand in hand with power; precise estimates yield powerful hypothesis tests. As in power analysis, determination of the sample size required to achieve a given level of precision requires knowledge of the variability of the measures. Such methods are discussed in Thompson (1992).

REFERENCES

Barbour, M. T., and J. Gerritsen. 1996. Subsampling of benthic samples: a defense of the fixed-count method. Journal of the North American Benthological Society 15:386–391.

Burton, A., and R. Pitt. 1996. Manual for evaluating stormwater runoff effects in receiving waters. Lewis Publishers, Boca Raton, FL.

Dixon, W. J., and F. J. Massey. 1969. Introduction to statistical analysis, 3rd ed. McGraw-Hill, New York.

Fairweather, P. G. 1991. Statistical power and design requirements for environmental monitoring. Australian Journal of Marine and Freshwater Research 42, 555–567.

Fuller, W. A. 1987. Measurement error models. Wiley, New York.

Gerritsen, J., B. Jessup, E. Leppo, and J. White. 1999. Development of a biological index for Florida lakes. Prepared by Tetra Tech, Inc. for the Florida Department of Environmental Protection, Tallahassee, FL.

Gilbert, R. O. 1987. Statistical methods for environmental pollution monitoring. Van Nostrand Reinhold, New York.

Green, R. H. 1979. Design and statistical methods for environmental biologists. Wiley-Interscience, New York.

Karr, J. R. 1991. Biotic integrity: a long-neglected aspect of water resource management. Ecological Applications 1:66–84.

Olsen, A. R., J. Sedransk, D. Edwards, C. A., Gotway, W. Liggett, S. Rathbun, K. H. Reckhow, and L. Young. 1999. Statistical issues for monitoring ecological and natural resources in the United States. Environmental Monitoring and Assessment 54:1–45.

Peterman, R. M. 1990. The importance of reporting statistical power: the forest decline and acid deposition example. Ecology 71:2024–2027.

Pielou, E. C. 1977. Mathematical ecology. Wiley, New York.

Snedecor, G. W., and W. G. Cochran. 1980. Statistical methods, 7th ed. Iowa State University Press, Ames, IA.

Sokal, R. R., and F. J. Rohlf. 1969. Biometry. W. H. Freeman, San Francisco.

Stevens, D. L. 1997. Variable density grid-based sampling designs for continuous spatial populations. Environmetrics 8:167–195.

Stober, J., D. Scheidt, R. Jones, K. Thornton, L. Gandy, D. Stevens, J. Trexler, and S. Rathbun. 1998. South Florida ecosystem assessment. Monitoring for adaptive management: implications for ecosystem restoration. Report EPA-904-R-98-02. U.S. Environmental Protection Agency, Washington, DC.

Thompson, S. K. 1990. Adaptive cluster sampling. Journal of the American Statistical Association 85:1050–1059.

Thompson, S. K. 1992. Sampling. Wiley, New York.

Whittaker, R. H. 1962. Classification of natural communities. Botanical Review 28:1–239.

4 Use of Predictive Models for Assessing the Biological Integrity of Wetlands and Other Aquatic Habitats

CHARLES P. HAWKINS and DAREN M. CARLISLE

Assessing the biological integrity of wetlands and other aquatic habitats requires that we compare observed conditions with those conditions expected to occur in the absence of human-caused stress. In this chapter we develop the idea of using the ratio of observed (O) taxonomic composition to the expected (E) composition as a sensitive measure of biological integrity. Obtaining an accurate measure of O/E and hence a sensitive assessment of the degree to which a wetland is biologically impaired requires that we be able to specify E as precisely as possible. We describe how multivariate statistical procedures used in conjunction with data from a network of minimally impaired reference sites allows prediction of E and hence the ability to make site-specific assessments. We demonstrate this technique by applying it to macroinvertebrate data collected from a series of wetland habitats associated with small lakes in the Uinta Mountains of Utah. These lakes were historically fishless, and the presence of introduced trout is the primary human-caused stressor.

Bioassessment is the process of determining if human activity has affected the biological integrity (Frey 1977) of a place. Biological integrity is measured by comparing how the observed value of one or more indicators (e.g., Noss 1990) differs from the value expected in ecosystems minimally affected by human activity. An intuitive and easily understood measure of biological integrity is the ratio of the observed (O) to the expected (E) condition, hereafter O/E. O/E has some desirable properties as an assessment measure. First, the units of O/E can be based on a number of ecologically relevant measures of biotic condition. Second, values near 1 imply no alteration, whereas significant deviations from 1 imply a change has occurred in biological conditions. Although it is possible that human activity might cause O/E to exceed 1 under

---

Bioassessment and Management of North American Freshwater Wetlands, edited by Russell B. Rader, Darold P. Batzer, and Scott A. Wissinger.
0-471-35234-9

some circumstances, it is the range of values from 1 (unaltered) to 0 (completely altered) that typically provides the basis for a quantitative assessment of the detrimental effect human activities have on ecosystems. Third, because *O*/*E* is a quotient, it also has the property of potentially allowing direct comparisons of assessments made on systems that differ naturally in *E*. For example, although two sites may differ markedly in *E*, *O*/*E* values of 0.5 mean both sites have lost 50% of their original condition.

Despite the intuitive appeal of this measure, putting concepts into practice can often be fraught with a variety of technical difficulties, even when the principles are simple. For example, even in situations where *E* was known previously, we need to account for any naturally occurring temporal variation in biotic condition that might have occurred since *E* was determined and *O* was measured. This issue can usually be overcome by adhering to sound sampling designs. A more critical challenge in applying this concept is that we usually lack the most fundamental piece of information necessary to conduct an assessment, a description of *E*. In contrast to the relatively simple task of estimating *O*, which we do by sampling a site, we have to predict *E* from other information.

Although the units of *O* and *E* could be those associated with any of several indicators of biological or ecological condition [e.g., measures of taxonomic composition (the specific species present), community structure (species richness and diversity), or functional processes (ecosystem production, mineral cycling, etc.)], in practice we need a measure that is both easy to make, ecologically meaningful, and relevant to both the public and decision makers. We believe a measure that describes the degree to which a site supports the specific species expected of that site has several advantages over other potential measures. For example, if communities contain functionally redundant taxa (Naeem 1998), stress may cause changes in community composition by eliminating sensitive taxa before conspicuous changes in community structure (overall richness and diversity) are apparent (Ford 1989, Gray 1989, Hawkins et al. 2000). Furthermore, if functional processes are more strongly related to species composition than raw species richness (Perry and Vanderklein 1996, Chapin et al. 1998, Tilman 1999), it follows that compositional measures may be a more useful overall measure of biological integrity than structural measures. In general, we believe a measure of *O*/*E* based on taxonomic compositional similarity (i.e., differences in the specific species that are observed and expected) has the potential of offering an early warning of other impending alterations that may occur unless corrective management practices are applied. We therefore need a way of expressing the degree to which observed species composition differs from expected species composition.

ESTIMATING *O*/*E*

The Concept of O

Although the concept of measuring the species present at a site is intuitively straightforward, for practical reasons it is usually impossible to conduct a complete census of the biota occurring in a place. Instead, we collect samples and use data from our samples to characterize the biota. However, sampling always gives an incomplete description of the species actually present at a site. The more samples we take, the more complete our description of the biota (Colwell and Coddington 1995). If we observe a species in a sample, we know it is present at the site; however, failing to observe a species does not necessarily mean it is absent. We may not have taken enough samples to detect it. In general, the probability of collecting a species will be a direct function of its numerical abundance at a site and how intensely we sample. We should therefore think about *O* in terms of the probabilities of observing (capturing) different species given the sampling methods used, with the explicit recognition that probabilities of capture (PCs) vary among taxa and with sampling intensity. We defer discussion of how we might actually use these ideas until we discuss ways of comparing *O* with *E*, the assemblage of species occurring and expected at a site.

The Concept of *E*

Observations in many types of ecosystems and habitats show that samples from individual sites always contain a subset of those taxa occurring in a larger regional species pool, thus individual sites can vary markedly from each other in biotic composition. Because of these differences in biotic composition, the PCs of many of the taxa that could potentially occur at a given site will be zero. These differences in biotic composition occur because the environment is heterogeneous at several spatial scales and different species have different resource requirements and tolerances. These processes interact to limit the specific taxa that can establish and persist at a given place, and hence taxa PCs will naturally vary across sites with differences in site environments.

Predicting Probabilities of Capture

We need a way to predict the PCs of all species in the regional species pool for any given site. Although ecological theory is too incomplete to predict PCs for different taxa from first principles, we can use empirical models that relate environmental conditions to species occurrence to make these predictions. With these models we can create a description of the assemblage expected to occur at a minimally disturbed site. This predicted assemblage will serve as the control for drawing inferences regarding whether the fauna actually occurring at a site is impaired.

To be useful for assessment purposes, PC predictions must be based on predictor variables that are generally unaffected by human activity. For example, although NO_3 concentrations may strongly influence the probabilities of finding certain taxa at a site, it would not be a good predictor of how PCs would vary naturally among sites. The reasons are that human activity has altered the nutrient concentrations of many freshwater ecosystems, which has in turn affected the densities, and hence PCs, of many taxa. In contrast, a variable such as elevation may work well as a predictor because the abundances of many taxa often vary markedly along elevation gradients. In most cases human activity will not change a site's elevation, and elevation typically is invariant over time scales relevant to bioassessment (10^0 to 10^2 years).

Need for Reference Sites. The development of empirical models for predicting PCs requires a database from which we can describe the conditions expected to occur at a potentially stressed site. For predictive modeling purposes, this database will consist of taxonomic and environmental information from unaltered or minimally altered places in the region of interest. These places serve as reference sites (Brinson and Rheinhardt 1996, Bailey et al. 1998, Reynoldson and Wright 2000) and are assumed to be representative of the environmental and biological potential of sites in the area. An additional assumption is that spatial variation in the overall biotic composition among otherwise similar sites is similar to the range of variability that an individual site might exhibit over time scales relevant for assessment purposes.

Ideally, a reference site database will consist of sites representing the range of naturally occurring conditions in the region. In constructing this database, we must be careful to specify the universe of concern for which we desire predictions. For example, if we wish to develop a model that predicts faunal composition of wetlands throughout Minnesota, we must ensure that the reference site database contains information from the range of wetland conditions that naturally occur in Minnesota. Furthermore, these sites should be distributed throughout the state in such a way to ensure all areas within the state are represented.

Selecting reference sites is not a trivial task (Reynoldson and Wright 2000). In many regions, few if any unaltered places exist. In these cases we may have to use least-impaired sites as references, recognizing that we are not truly capturing the historical potential of sites, but rather, the conditions that are practically attainable. Where minimally altered places do exist, two general approaches exist for distinguishing them from other, altered sites. One relies on the best professional judgment (BPJ) of local experts. These experts may not possess quantitative descriptions of the history of a site or the biota occurring there, but they frequently have a tacit understanding of the quality of a place based on years of observation and experience. The strength of the BPJ approach is that the knowledge base can be extensive although not easily stated in quantitative terms. Its weakness is that the evaluations are based on subjective criteria and every expert sees the world somewhat differently. For example, one expert may rate a site as excellent, whereas another may rate it as good. These differences occur because we make evaluations relative to our past experiences and observations, and everyone has a different set of experiences. The second approach uses objective criteria to identify reference sites. For example, in an analysis of the effects of logging on stream invertebrates, Hawkins et al. (2000) required that sites had less than 5% of their watersheds logged to be a reference site. The strength of this approach is that it avoids subjective differences in evaluations. Its weakness is that quantitative determination of the amount of human alteration potentially affecting a place is often time consuming, if possible at all. Furthermore, even several quantitative measures may not adequately represent the actual stress occurring at a site. For example, the number of roads, arca of agriculture, and proximity of urban development may tell us little about the condition of a site if the main stressor is the presence of exotic species.

Empirical Models. If the pool of reference sites is representative of the diversity of site conditions within the area of interest, it should be possible to relate PCs to the environmental conditions present at different reference sites and thus produce a model that can estimate E for any new site of interest. Several statistical procedures can be used to predict PCs, of which we discuss three: binary logistic regression (BLR), binary discriminant analysis (BDA), and discriminant analysis conducted in conjunction with the classification of sites into faunal classes.

Logistic regression produces an equation that predicts the probability of a new observation occurring in one of two classes. In our example, the observation would be a new site and the classes would be the presence (coded 1) and absence (coded 0) of a taxon. The form of the logistic regression equation, assuming that PC is a unimodal function of environmental features, is:

$$\log [y/(1-y)] = \beta_0 + \beta_1 X_1 + \beta_2 X_1^2 + \beta_3 X_2 + \beta_4 X_2^2 + \ldots + \beta_m X_g + \beta_n X_g^2$$

in which y is the probability of the taxon occurring at a site given certain environmental conditions specified by the predictor variables X_1 to X_g. The advantage of this approach is that it is conceptually straightforward; its disadvantage is that it is mechanically cumbersome to build all of the individual models needed to calculate E. Separate models are needed for each taxon in the regional species pool and the number of taxa can be large (e.g., 10^2 to 10^3) in even relatively small regions. Although the form of the analysis and the equations produced are different, binary discriminant analysis can be used in exactly the same way, with PCs being estimated as the posterior probabilities of group occurrence, which are calculated as a function of environmental features.

Although the advent of powerful microcomputers has obviated some of the computational problems associated with the development and application of models based on either BLR or BDA, we can take advantage of the tendency for groups of taxa to co-occur and then use BDA to substantially reduce the computational complexity associated with estimating E. For example, Moss et al. (1987) developed a procedure to estimate probabilities of capture of individual taxa, and hence E, by (1) classifying reference sites into biologically similar groups, (2) developing discriminant functions that estimated the probability of a new site belonging to each of the biotic groups, and (3) estimating the frequency of occurrence of each taxon within the biotic groups. This procedure was developed explicitly as a means of estimating E and thus providing a basis for assessing the biological integrity of running waters. The authors called their procedure the River Invertebrate Prediction and Classification System (RIVPACS). Wright (1995) provides an overview of the development and application of this procedure as a means of assessing the biological condition of stream ecosytems in Great Britain. The technique has subsequently been implemented in Australia as a standard procedure for assessing the biological condition of that nation's running waters (Simpson and Norris 2000). It has also been evaluated as a means of detecting effects of logging practices on invertebrate assemblages in mountainous streams of California (Hawkins et al., 2000).

Comparing *O* and *E*

Comparing *O* and *E*, when they represent probabilistic statements about the occurrence of taxa requires that samples be collected in a standard and consistent way. Assuming we can develop and implement standard sampling protocols, comparing probabilities of capture is possible, although such comparisons are less straightforward than simply comparing species lists, especially when an assemblage-level ratio such as *O*/*E* is desired. Although a measure of departure from expected conditions is possible by comparing the degree to which empirically measured PCs agree with expected PCs with a goodness-of-fit test, this approach lacks the intuitive appeal of a simple, standardized, assemblage-level ratio. Furthermore, many replicate samples are required to empirically estimate PCs at a site, and we will seldom have the resources to take such samples at most sites.

To create a measure of *O*/*E* that retains information about species composition, we will define *O*/*E* as S_O/S_E, where S_O is the number of those species collected at a site that were predicted to occur there and S_E is the number of species expected given the sampling procedure used and the environmental conditions at that site. S_E is calculated as Σ PCs for all species in the regional species pool. A simple way of understanding this calculation is as follows. Assume that we take 100 standard samples at a site. If a species is abundant and has a PC of 1, we would expect to collect it in 100 subsequent samples. If a species is much less abundant and has a PC of 0.1, it will, on average, be collected in only 10 of 100 new samples. If 10 species have PCs of 0.1, we would expect to collect 1 of them, on average, in all 100 samples but not the same species in each sample. Thus given the simple situation in which one species has a PC = 1 and 10 other species have PCs = 0.1, we should collect two species, on average, in any given sample.

It is also important to understand that S_O/S_E is not a comparison of raw species richness; instead, it constrains *S* to include only those taxa predicted to occur at a site. The following example from Table 4.1 will illustrate this idea. In this hypothetical example, there are 12 species (A to L) in the regional species pool that could potentially occupy a site. At the hypothetical site examined in Table 1, only six species are predicted to occur with PC > 0 (see data under column E), and their PCs vary from 0.1 to 1.0, presumably as a consequence of species-specific differences in environmental requirements and tolerances. For example, from these data we would conclude that this site has optimal conditions for species A, intermediate conditions for species B to D, marginal conditions for species E and F, and unsuitable conditions for species G to L. These PC estimates are derived from our empirical model, which specifies how the PCs of different taxa should vary given the environmental conditions present at this site. As explained above, the sum of these PCs give the number of species we would expect to collect (in this case, 3.1) given the standard sample procedures used.

O_1 and O_2 represent two different hypothetical scenarios that could arise if the site experienced different environmental stresses. Note that although we would not typically estimate observed PCs, we have included hypothetical examples for these two scenarios to aid explanations. For scenario O_1, note that replicate sampling would detect a total of six species, but these species are completely different from those pre-

TABLE 4.1. Hypothetical Example Showing How *O/E* Can Differ Depending on Whether *O* Is Constrained to Include Only Those Taxa Predicted to Occur with Probability >0

	Taxon	E	O_1	O_2
	A	1.0	0	0.8*
	B	0.8	0	0.9*
	C	0.7	0	0.3
	D	0.4	0	0.2
	E	0.1	0	0.2
	F	0.1	0	0
	G	0	1.0*	0.2*
	H	0	0.8*	0.6*
	I	0	0.7*	0
	J	0	0.4	0
	K	0	0.1	0
	L	0	0.1	0
$S = \Sigma\, PC_i$		3.1		
Taxa collected in sample			3	4
O/E unconstrained			0.97	1.29
O/E constrained			0	0.65

[a]Values under *E* represent probabilities of capture predicted from an empirical model. Values under O_1 and O_2 represent two different hypothetical scenarios in which the fauna of a site has been altered by environmental stress. The probabilities of capture given under O_1 and O_2 are not needed for calculating *O/E* but are included to illustrate how stress might affect the abundance and hence probability of capturing different species (see the text for details). The taxa actually recorded from a single sample for scenarios O_1 and O_2 are identified with asterisks. The fauna in scenario O_1 has the same number of taxa with the same difference among taxa in probabilities of capture as *E*, but O_1 shares no taxa in common with *E*. The fauna in scenario O_2 contains most of the taxa as *E* plus some additional taxa. When *O* is not constrained to include only those taxa in *E*, *O/E* ratios are similar to or greater than 1 for O_1 and O_2, but when *O* is constrained to taxa A to F, O_1 and O_2 are different from *E* with O_1 deviating completely from *E* and O_2 being moderately different than *E*.

dicted to occur (compare columns E and O_1). Given the PCs measured here, we would expect, on average to collect the same number of species as in *E* (i.e., 3.1). Furthermore, if we calculated *O/E* based on the raw number of species actually collected in a standard sample (illustrated by species marked with an *), we would obtain an *O/E* value of 0.97, which would imply that this site was very similar to that expected and thus not impaired. However, if we constrain *O* to include only those taxa that were predicted to occur with PC > 0, then *O/E* = 0, which is clearly a more realistic representation of the profound differences that exist between *E* and O_1 and that are obvious from inspecting the table.

In scenario O_2 we present a more typical example. Here, some species that were predicted to occur were present as were some species that were not. Furthermore, if we took replicate samples, we would find that the PCs of those species predicted to occur are somewhat different from expected. If we sum PCs for all taxa present, we would conclude that, on average, we would collect 3.0 species from an individual standard sample. Calculating *O*/*E* from this estimate of *S* would give an *O*/*E* value of 1.0, again implying no biological impairment. If we constrain *O* to include only those species collected that were predicted to occur, our estimate of *O*/*E* is 0.65. Under this scenario, although the site has some of the taxa that were predicted to occur, their actual probabilities of capture (as estimated hypothetically from replicate samples) are generally lower than those predicted, as might be expected if the fauna were stressed. Hence the number of these taxa that we would expect to capture would be lower than expected. We would therefore conclude that this site is also biologically impaired, although not as severely as under scenario O_1.

Although calculating *O*/*E* in this way also obscures some taxon-specific information, it is a more specific and relevant measure than a comparison of raw taxa richness. Furthermore, *O*/*E* expressed in this way provides a convenient summary of the agreement between observed and expected biota, and if *O*/*E* is found to deviate substantially from 1, it is a simple matter to determine the specific taxa for which observed and expected PCs substantially differ.

APPLICATION OF THE RIVPACS APPROACH TO THE BIOLOGICAL ASSESSMENT OF AQUATIC HABITATS

In the rest of this chapter we use data from fringing wetland habitats associated with historically fishless mountain lakes of Utah to describe how the RIVPACS approach can be used to assess the biological integrity of aquatic invertebrate assemblages. In doing so, we describe the individual steps involved in building RIVPACS-type models and show that the technique can be used to assess the effects of a stressor (in this case, trout introductions) on these systems.

Study Area

The Uinta Mountains of Utah contain over 1000 small lakes and ponds, which are classed by Cowardin et al. (1979) as either lacustrine (limnetic and littoral) or palustrine wetlands. These lakes and associated wetland habitats vary considerably in size, dominant environmental features, and the biota they support and thus contribute greatly to the biological diversity of this area. However, because of this high natural heterogeneity, assessing the biological integrity of individual sites is particularly challenging and requires methods that can account for potentially large differences in expected conditions.

The fringing wetlands associated with these lakes support various assemblages of plants, invertebrates, and vertebrates, although nearly all of these systems historically lacked fish because of lack of hydrological connections with lower-elevation lakes and streams that harbor fish. During the last 100 years, however, trout have been introduced into nearly all of these lakes and > 70% of lakes > 4 ha contain either naturally reproducing populations of trout or populations that are maintained by periodic stocking. Trout introductions into these and other historically fishless lakes represent the most obvious human-caused stress to these systems and have greatly affected populations of amphibians and invertebrates (Bradford 1989, Liss and Larson 1991, Carlisle and Hawkins 1998). The presence of trout represent a chronic stress on these systems and thus provide a good test of this method. Furthermore, no attempt has been made to date to measure these effects in terms of numerical biological criteria that could be used to judge the degree to which these introductions have caused biological impairment.

Data Set

We used data from Carlisle and Hawkins (1998) to illustrate how a RIVPACS-type model can be used in wetland assessment. The study area was a 5000-km^2 region encompassing the Uinta Mountains of northeastern Utah. Sites were selected based on the presence and type of trout that were stocked: three fish treatment types = no fish, cutthroat trout present, and brook trout present. No fish sites served as reference sites. For each fish treatment, approximately equal number of sites were selected that differed in dominant bottom substrate: sand, cobble, and vegetation. Site selection was also stratified to ensure that all treatment combinations occurred in each of the major westward- and northward-flowing drainages of the Uinta Mountains within a 5000-km^2 area.

Data consisted of collections of benthic macroinvertebrates from 47 lakes (sites) along with selected environmental data (Table 4.2). Ten 0.05-m^2 samples (500-µm mesh) were collected with a "T" Hess-type sampler at approximately equally spaced intervals around the periphery of each lake in water 1 to 1.5 m deep. These 10 samples from each site were then combined prior to laboratory sorting, specimen identification, and analyses. Data from the 16 fishless lakes were used to build a predictive model for the 84 taxa (mostly identified to genus) that were collected at these sites. To evaluate the error of the model, we compared the fauna actually collected at each reference site with that predicted by the model. We then compared the fauna from lakes containing trout with model predictions to examine the degree to which trout introductions have altered the benthic invertebrate faunas of each test lake.

General Procedures

Because RIVPACS-type models are designed to predict the biota expected to occur under different naturally varying conditions, this approach holds promise as a means of assessing the biological integrity of wetlands and lakes as well as streams and rivers. Development and application of a RIVPACS model requires eight primary steps:

1. Classification of reference sites into biologically similar groups
2. Development of a discriminant model with data collected from sites without fish (reference sites) to estimate the probabilities of a new site belonging to each of the site groups defined in step 1
3. Calculation of the probabilities of all taxa in the regional taxa pool occurring within each reference site group based on presence–absence data
4. Calculation of the probabilities that each taxon will occur at a new site based on steps 2 and 3

TABLE 4.2. Environmental Features of the Reference Wetlands

	Latitude (°N)	Longitude (°W)	Elevation (m)	Area (ha)	Maximum Depth (m)	Mean Depth (m)	Date[a]	Outflow (m^3/s)
Mean	40.82	–110.53	3198	3.6	9.4	5.0	230	0.024
Minimum	40.67	–110.98	2765	1	2	1.5	190	0.000
Maximum	40.97	–110.20	3490	7.9	27	15	288	0.184

Substrate Characteristics[b]

	>512 mm	128–512 mm	32–128 mm	8–32 mm	2–8 mm	< 2 mm	Detritus	Macrophytes
Mean	6.3	7.9	7.4	3.9	13.4	21.9	21.5	17.3
Minimum	0	0	0	0	0	0	0	0
Maximum	36	24	25	15	77	80	60	70

[a]Sampling date in days past January 1.

[b]All substrate measurements are in percent cover in the littoral zone.

5. Summation of the estimated probabilities of capture of all taxa to estimate the number of taxa expected (E) at a new site
6. Calculation of O/E
7. Estimation of model error
8. Assessing the degree of impairment of a new site given the error in the model

Site Classification (Step 1). We classified sites into biologically similar groups based on the presence and absence of taxa. Classification required two substeps. First, we calculated pairwise similarities among all reference sites with the Czekanowski coefficient of dissimilarity, where $C = 1 - 2a/(2a + b + c)$ and a = the number of taxa that occurred at both sites, b = the number of taxa occurring only at one site, and c = the number of taxa occurring only at the other site. Values of this measure vary from 0 to 1, with 0 implying the two sites had identical taxa and 1 implying no shared taxa. We then performed a cluster analysis with these similarities as data. We used the flexible unweighted pair group using arithmetic averages (UPGMA) method with β set to -0.1 because it had previously been shown to reveal real patterns better than some other methods (Belbin and McDonald 1993). Other clustering techniques could potentially be used in this step as described by Moss et al. (1999). After creating the cluster diagram, we visually identified four groups of sites that appeared to be somewhat distinct from one another (Fig. 4.1). These groups contained five, three,

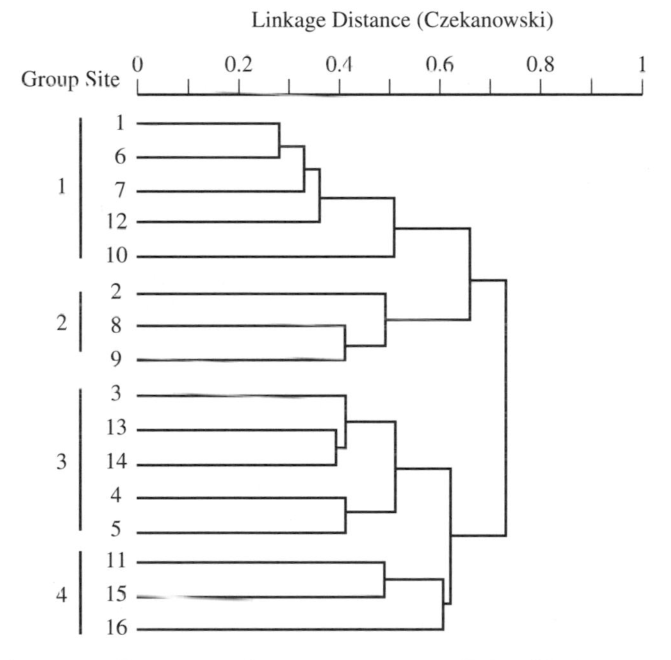

FIGURE 4.1. Cluster diagram showing how groups were distinguished among the 16 reference sites. Because we used a disimilarity measure (Czekanowski coefficient), sites that link at low values are more similar to one another.

five, and three sites, respectively. The subjective nature of group identification associated with this last step is generally not a problem, because probabilities of capture are weighted by the probability of new sites belonging to each group (see below) and small errors in classification would not therefore result in large errors in predicting E. However, both the total number of sites and the number of sites per group are much lower than would be normally used. Ideally the number of sites per group would be ≥ 10, but use of a small data set here allows us to easily illustrate the potential of this approach.

Discriminant Model (Step 2). We used forward stepwise discriminant analysis to select the subset of environmental features out of those sampled that best discriminated among sites. That analysis showed that four variables were useful in separating groups: latitude, elevation, stream flow out of the lake, and the log of the percent of the bottom substrate present in the 2 to 8 mm size range (Table 4.3). Both reference sites and those with trout varied considerably in these four factors but, with the exceptions discussed below, were similar in the range of these factors (Fig. 4.2).

For these data, latitude and elevation were strongly correlated, but latitude was slightly better than elevation in distinguishing groups and hence was used in the model instead of elevation. The discriminant model developed from these three predictor variables was highly significant (Wilk's lamda = 0.053, $F = 6.410$, 9 and 24 degrees of freedom, $p < 0.0001$) with latitude being most important in discriminating groups (F value = 14.89) followed by substrate ($F = 3.04$) and outflow (2.56). Most of the variation among groups was related to discriminant factor 1, which described a climatic gradient (probably temperature) related to latitude and elevation. The importance of the other two factors was probably related to the fact that substrate is a well-known characteristic influencing the distribution of aquatic benthic invertebrates (Minshall 1984) and that discharge out of a lake is related to the retention of water and nutrients.

An internal (jackknifed) test of the model showed that it reclassified 81% of the sites correctly, implying that the model was generally accurate in discriminating groups. This model was then used to predict the probabilities of group membership of each of the sites containing trout. Although these calculations require matrix alge-

TABLE 4.3. Mean Values of the Four Predictor Variables for Each of the Groups of Sites Identified from the Cluster Analysis

Group	Latitude (°N)	Elevation (m)	Outflow (m^3/s)	Log % 2–8 mm Substrate
1	40.79	3382	0.070	1.15
2	40.84	3258	0.000	1.29
3	40.90	3024	0.000	0.30
4	40.70	3121	0.009	0.84

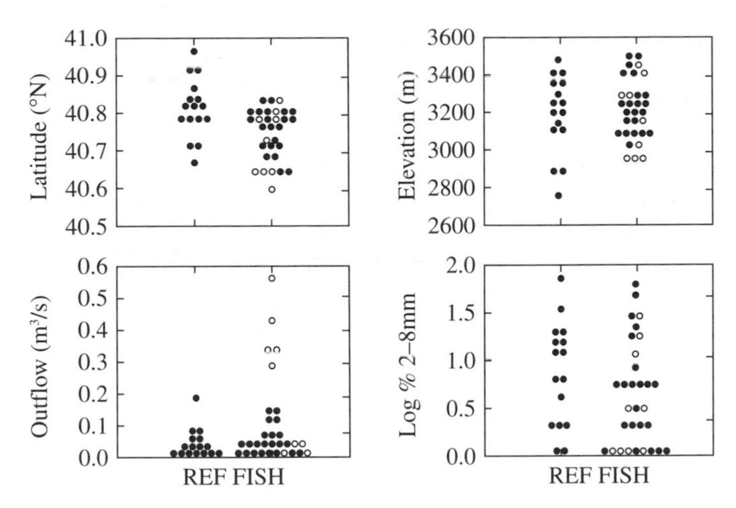

FIGURE 4.2. Range of values of latitude, elevation, outflow from lakes, and log(% sediment in the range 2 to 8 mm) for the reference (fishless) and stressed (fish) lakes. Filled symbols for the fish lakes are those sites falling within the range of reference values, and open symbols are those fish lakes falling outside the range of reference values and were therefore not assessed (see text for details).

bra, they can easily be conducted within most modern spreadsheet programs. The reader should consult a basic text that describes discriminant analysis (e.g., Manly 1994) for details regarding how to make these calculations.

Probabilities of Capture (Steps 3 and 4). Calculation of the probabilities of capture of each taxon at a new site requires three steps that we illustrate in Table 4.4. We must first estimate the probabilities of observing each taxon within each of the groups (F_t). This is a simple calculation with PC being the number of sites within a group in which a taxon was observed divided by the total number of sites within a group.

TABLE 4.4. Illustration for *Lumbriculus variegatus at Site* T011 of How Probabilities of Occurrence Are Calculated for Each Taxon for a New Site

Group	Probability of Site T011 Belonging to Group, P_g	Frequency of Occurrence of *Lumbriculus variegatus* Within Group F_t	Contribution of Group to Probability of Occurrence at the Site $P_g \times F_t$
1	0.33	0.80	0.26
2	0	0	0
3	0	0.60	0
4	0.67	0.33	0.22

Probability of *Lumbriculus variegatus* occurring at site T011 = $\Sigma(P_g \times F_t)$

These estimates are then weighted by the probabilities that a new site will belong to each of the groups (P_g) to calculate the overall probability that a taxon will occur at a new site (Table 4.4).

Estimation of* E *(Step 5). The sum of all probabilities of capture at a site estimates the number of taxa we expect to observe at a site (i.e., $E = \Sigma PC_i$, where i = each of the different taxa collected from the reference sites). In theory, E should be calculated from all nonzero PCs. However, as described in step 6, for assessment purposes, it may be best to base E on only those taxa with PCs greater than some nonzero threshold. In Table 4.5 we illustrate how E is calculated for two sites, one that was not affected by trout (T008) and another that was strongly affected (T011).

TABLE 4.5. Calculation of *E*, the Number of Expected Taxa, and *O/E* from Taxon Probabilities of Occurrence (PC) and the Taxa Actually Observed for Two Sites[a]

Site T008		Site T011	
Taxon	Probability of Occurrence	Taxon	Probability of Occurrence
*Sphaerium nitidum**	1.00	*Tanytarsus**	1.00
Nematoda*	0.98	Nematoda	0.93
*Procladius**	0.98	*Procladius*	0.93
*Paratanytarsus**	0.90	*Psectrocladius**	0.93
Gammarus lacustris	0.87	*Sphaerium nidium**	0.77
*Tanytarsus**	0.71	*Bezzia*	0.67
*Nais communis**	0.70	*Microtendipes*	0.67
*Ostracoda**	0.67	*Dicrotendipes*	0.58
Dicrotendipes	0.64		
Lebertia (Mite)	0.60		
*Mcrotendipes**	0.59		
*Nephelopsis obscuris**	0.58		
$E = \Sigma PC_i$	9.22		6.50
O	9		3
O/E	0.98		0.46

[a]The taxa at site T008 were not different from expected, whereas the taxa at site T011 were substantially different from expected. Taxa flagged with an asterisk are those that were collected. All others were not collected. For these calculations, a PC threshold of 0.5 was used. Taxa are ranked in order of their predicted probabilities of occurrence at each site.

Calculation of* O/E *(Step 6). *O/E* is calculated by first deciding on what threshold probability will be used in the calculations. RIVPACS calculates *O/E* from all PCs > 0, based on the assumptions that more information provides a more representative assessment and that rare taxa (i.e., those with low PCs) are typically those most sensitive to human-caused stress. AUSRIVAS calculates *O/E* only from those taxa with PCs ≥ 0.5 based on the assumptions that (1) rare taxa are often "accidentals" that contribute no information to the assessment, or (2) that even if these taxa come from viable populations, they cannot be modeled accurately and thus contribute primarily to model error. Two empirical analyses conducted to date (Simpson and Norris 2000, Hawkins et al. 2000) support the latter view, and as such we recommend use of the PC = 0.5 threshold until additional analyses suggest otherwise.

Once a threshold PC value is selected, *E* is expressed as the sum of those PCs ≥ the threshold value (Table 4.5). *O* is then calculated by summing the number of taxa that were actually collected at the site (Table 4.5). In general, at an unimpaired site, we would expect to collect taxa in relation to their PCs. For example, as described earlier, if 10 taxa have PCs of 0.10, we would expect to collect one of those taxa. In an impaired site, the actual probability of capturing at least some taxa will be lower than that predicted, and hence the overall number of taxa actually collected should be less than *E*, as illustrated in Table 4.3.

Estimation of Model Error (Step 7). Model error can be estimated in two ways. The most rigorous estimate of model error is made by treating data from a series of reference sites that were not used in model construction as test data and then applying the calculations in steps 3 to 6 to these data. Alternatively, when the number of reference sites available is limited, as in this example, the same reference data can be used. In both cases, the resulting distribution of *O/E* values typically exhibits a near-normal frequency distribution with values centered on 1. The spread of these *O/E* values represents model error. The more accurately that steps 1 to 6 predict true PCs, the narrower the distribution of reference *O/E* values. For the 16 reference sites used in this example, the range of *O/E* values was 0.74 to 1.15, with a mean *O/E* of 0.98 and a standard deviation of 0.14 (Fig. 4.3).

Site Assessment (Step 8). If model predictions were perfect, all reference site *O/E* values would have estimates of 1, and test sites showing any deviation from 1 would be considered impaired to some degree. However, because of model error, an *O/E* value below 1 may fall within the range of values observed for reference sites and hence would be considered equivalent to reference (i.e., unimpaired). Inferences of impairment must therefore be made in the context of model error, and *O/E* thresholds must be determined below or above which a site would be considered as different from the reference. As mentioned earlier, this threshold would ideally be set to balance the likelihood of making type I and II errors. A banding scheme could then be used to describe degrees of impairment. These bands are created by dividing the interval from *O/E* = 0 to *O/E* = threshold value into a series of equal-length bands, the number of which is most rigorously assigned by determining the error associated with individual estimates of *E* (see Clarke et al. 1996). In general, however, it appears that

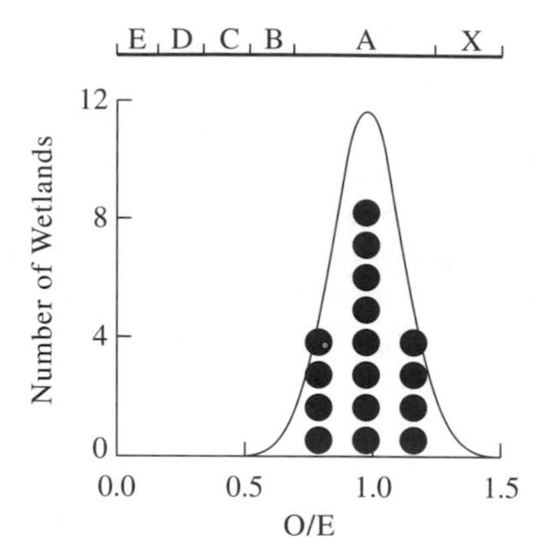

FIGURE 4.3. Frequency distribution of *O/E* values derived from the reference site data. A normal distribution curve is superimposed on the data to facilitate interpretation. Each dot represents an individual reference wetland. For this model, assessed sites with values falling below 0.7 would be considered impaired, whereas those with values greater than 1.26 would be flagged as being either slightly impaired or as having potentially high conservation value. For application purposes, this distribution, and hence the threshold criteria for inferring impairment, would be based on a larger number of reference sites. The top scale illustrates how impairment bands could be established as described in the text.

four bands can reliably be distinguished for models that produce standard deviations of reference site *O/E* values of ~0.2. For the Uinta data, we illustrate this idea by using ±2 standard deviations (0.28 *O/E* units) to set threshold values at *O/E* of 0.70 and 1.26. We then divided the interval from 0 to 0.70 into four bands, labeled B, C, D, and E. The interval within 2 standard deviations is considered equivalent to reference and is labeled band A.

Sites with *O/E* values greater than 2 standard deviations above the mean are placed in a separate band, X, which denotes sites with more taxa than expected. Large *O/E* values may occur if the effect of certain habitat features were not captured by the model. These habitat features may be unique to the site and hence create sites of potentially high conservation value (Wright et al. 1996). Simpson and Norris (2000) offer an alternative interpretation for high *O/E* values, when they are observed in systems with naturally low nutrient levels. The initial stages of eutrophication (i.e., mild nutrient enrichment) may cause an increase in population densities without a loss of taxa. Because the actual PCs would be greater than the modeled PCs, *O/E* values would tend to be >1. The AUSRIVAS program considers these high values to be a measure of the initial stages of impairment.

Effects of Trout Introductions on Wetland Invertebrates

We applied the model to 31 small lakes to assess the effect introductions of two species of trout have had on benthic invertebrates. Of these lakes, nine were statistically outside the experience of the model (i.e., values of at least one predictor variable at these lakes were substantially outside the range of values measured at the reference sites). We therefore could not assume that the reference lakes adequately represented the faunas that should occur in these lakes in the absence of fish, hence we made no assessment on these systems. For most of these lakes, estimates of flow out of the lake were much higher than the range of flows observed at reference lakes (see Fig. 4.2). For the remaining 22 lakes, we calculated *O*/*E* values, determined if values depended on trout species and overall type of bottom substrate, and determined which taxa were most strongly affected by trout introductions.

Individual Site Assessments. *O*/*E* values for the assessed systems varied from 0.38 to 1.12 (Table 4.6), implying that trout introductions had variable effects on the benthic invertebrate fauna, depending on the site. Thirteen sites fell within the A band, as defined above, and would be considered equivalent to reference conditions. Seven sites fell within the B band and would be considered significantly impaired. Two sites fell within the C band and would be considered substantially impaired, with < 50% of the fauna that was expected.

TABLE 4.6. Observed (*O*), Expected (*E*), and *O*/*E* Values for the 22 Test Sites That Were Within the Experience of the Model

Site[a]	*O*	*E*	*O*/*E*
T003	9	7.98	1.13
T008	9	9.21	0.98
T021	9	9.61	0.94
T023	9	9.62	0.94
T016	8	8.67	0.92
T013	9	10.53	0.85
T007	9	10.59	0.85
T024	7	8.30	0.84
T028	7	8.67	0.81
T004	7	8.67	0.81
T019	7	8.67	0.81
T025	7	8.67	0.81
T002	8	10.39	0.77
T014	6	8.67	0.69
T026	6	8.67	0.69
T022	6	8.98	0.67
T020	6	9.87	0.61
T029	5	8.67	0.58
T030	6	10.60	0.57
T031	6	10.60	0.57
T006	5	10.53	0.47
T011	3	6.50	0.46

[a]Ranked by *O*/*E* value.

Regional Effects of Trout on Wetland Invertebrates. The effect of trout introductions on the regional population of wetlands was assessed by comparing mean O/E values from sites with and without trout. The results of the regional analysis showed that on average, trout introductions have had a measurable and substantial overall effect on the invertebrate assemblages of these wetlands. ANOVA showed that the mean O/E value from wetlands with trout was about 25% lower than that of reference sites (mean reference $O/E = 0.98$, mean test $O/E = 0.76$, $r^2 = 0.38$, $p < 0.0001$). Trout tended to affect certain invertebrate taxa to a much greater extent than others (Table 4.7). Five of the six taxa missing from ≥ 80% of the wetlands at which they were predicted to occur were relatively large, conspicuous, and mobile animals; the one exception was the chironomid midge *Diplocladius*. In contrast to these negative effects, one taxon was found far more often than predicted in the presence of trout. Although predicted to be absent at PC < 0.50, the oligochaete *Lumbriculus variegatus* was found at 93% of those sites at which its predicted probability of capture was < 0.5.

Sensitivity of O/E Relative to Other Types of Assessments

Comparing how O/E values and raw taxa richness varied among sites with and without trout illustrates how a predictive modeling approach based on site-specific reference data can increase the sensitivity of assessments over that possible when natural differences among sites in environmental conditions and taxonomic composition are not taken into account. Raw taxa richness was not statistically different between sites with and without trout (mean for sites without trout = 18.0, mean for sites with trout = 15.3, $p = 0.070$), and richness was unrelated to O/E for both just those lakes with fish (Fig. 4.4, $r^2 = 0.06$, $p = 0.27$, $n = 22$) and among all lakes ($r^2 = 0.00$, $p = 0.81$, $n = 38$). The effectiveness of constraining richness (O) to include only those taxa predicted to occur at a specific site (E) is apparent from how much more separated the two normalized O/E distributions were for sites with and without trout compared with the distributions estimated for raw taxa richness (Fig. 4.4).

The potential confounding effect of natural variability among sites is even more apparent when fish and fishless sites are used as treatments and ANOVA is used to test for effects of trout on taxon densities. ANOVA designs are common in both basic and applied ecological research and are a potentially powerful means of detecting effects of environmental stressors. However, when data were analyzed with a completely randomized design that averages across the four reference groups, only two of the seven taxa identified in Table 4.7 as sensitive to trout had statistically significant differences in mean densities between fish and fishless lakes, even though the magnitude of differences between the two treatments was as high as 40× (Table 4.8). Lack of significance in the other cases was almost certainly associated with the high within-treatment variances in densities, as revealed by the number of zero counts within both reference and treatment lakes (Table 4.8).

TABLE 4.7. Taxa Either Consistently Missing from Wetlands with Trout or Present When Not Predicted

Taxon[a]	Number of Wetlands in Which the Taxon Was Predicted (≥0.5) to Occur	Number of Wetlands the Taxon Was Missing When Predicted to Occur	Percent of Sites in Which the Taxon Was Missing for Which It Was Predicted to Occur
Diplocladius (Chi)	6	6	100
Hesperophylax (Tri)	6	6	100
Callibaetis (Eph)	8	7	88
Limnephilus (Tri)	8	7	88
Gammarus (Amp)	6	5	83
Nephelopsis obscuris (Hir)	5	4	80
	Number of Wetlands in Which the Taxon Was Not Predicted (≥0.5) to Occur	**Number of Wetlands the Taxon Was Present When Predicted not to Occur**	**Percent of Sites in Which the Taxon Was Present for Which It Was Not Predicted to Occur**
Lumbriculus variegatus (Oli)	15	14	93

[a]Chi, Chironomidae; Tri, Trichoptera; Eph, Ephemeroptera; Amp, Amphipod; Hir, Hirudinea; Oli, Oligochaeta.

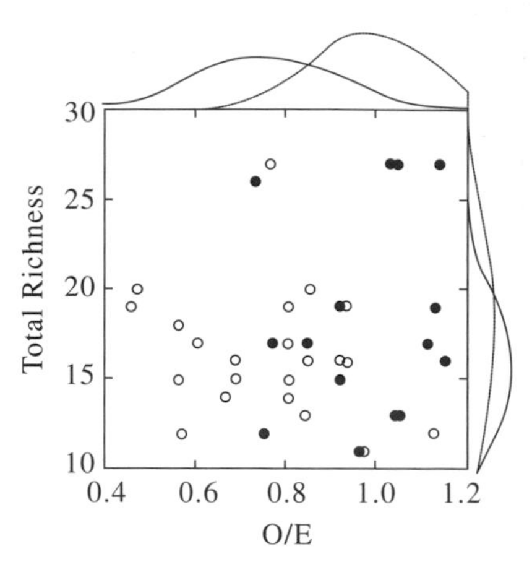

FIGURE 4.4. Relationship between raw taxa richness and *O/E* values. Filled circles are reference sites and open circles are sites with trout. Normal distribution curves were estimated for both variables for each group of sites. Note that within-group variance is smaller for *O/E* than for taxa richness, and as a consequence, *O/E* is much more effective at distinguishing test sites from reference sites than is raw taxa richness.

Given the manner in which frequencies of occurrence varied among the four reference groups identified in Figure 4.1 (Table 4.8) and the success of the discriminant model in predicting group membership, these patterns of presence and absence of taxa appear to be related primarily to differences in temperature, flushing rate, and bottom substrates among lakes. The classic approach to this problem in ANOVA would be to block the lakes to account for naturally occurring spatial heterogeneity (e.g., Dutilleul 1993). However, when we blocked lakes by dominant substrate type (sand, cobble, vegetation), as was done in the analyses performed by Carlisle and Hawkins (1998), differences in abundance between reference and fish sites were significant for only one additional taxon (the caddis *Hesperophylax*), implying that a priori blocking was not highly effective. Moreover, post hoc examination of frequencies of occurrence of those taxa sensitive to trout (Table 4.8) showed that the most effective blocking strategy would have differed for nearly every individual taxon, implying that single-factor blocking may often be ineffective for assessment purposes. Given the results of the discriminant model, we would have needed to block sites in multivariate environmental space (i.e., latitude, outflow, and substrate), but it is unlikely that we could ever apply such a blocking scheme during site selection and prior to sampling. Classic ANOVA approaches, even with blocking, may therefore be of limited use in the assessment of the biological integrity of physically heterogeneous ecosystems, in which we need assessments of how entire assemblages and their component taxa respond to stress.

TABLE 4.8. Results of ANOVA Tests on Those Taxa Identified in Table 4.7 as Sensitive to the Presence of Fish[a]

	Taxon						
Treatment	*Callibaetis*	*Diplocladius*	*Gammarus*	*Hesperophylax*	*Limnephilus*	*Nephelopsis*	*Lumbriculus*
Reference	1.53	1.00	9.91	0.60	0.89	1.98	0.21
Fish	1.17	0.22	0.18	0.03	0.22	0.37	35.43
p	0.808	0.216	0.004	0.059	0.147	0.106	0.007
	Number of Sites with Zero Individuals						
Reference	13	13	10	13	12	11	8
Fish	18	21	21	21	18	20	2
	Frequency of Occurrence						
Group 1	0	0.60	0.20	0.6	0	0	0.80
Group 2	0	0	1.00	0	0	0.67	0
Group 3	0.2	0	0.40	0	0.4	0.60	0.60
Group 4	0.67	0	0	0	0.67	0	0.33

[a]Tests were conducted on $\log(X + 1)$ density data and means are back-transformed from log data. Fish sites used in the analyses included only those sites that were considered similar to reference sites by the chi-square distance test conducted during the discriminant functions analysis, so tests were conducted on 16 reference and 22 fish sites. p values are not corrected for multiple comparisons. The number of sites in which no individuals of each taxon were collected in both reference and fish lakes are listed below the ANOVA results as are the frequencies of occurrence within each of the groups identified in Fig. 4.1.

SUMMARY

Assessments based on predictive models have been shown to work well in stream and large lake ecosystems (Reynoldson et al. 1995, Wright 1995, Hawkins et al. 2000, Simpson and Norris 2000), and our example shows that they have great potential as a means of conducting wetland assessments at both site and regional levels. The strength of the predictive modeling approach stems from its ability to match the biological potential of an assessed site with an appropriate reference condition (Bailey et al. 1998). As such, predictive models may provide a more sensitive appraisal of biological integrity than that afforded by methods that either compare biotic conditions at a test site with those observed across a range of reference sites (aggregate reference condition, Barbour et al. 1995) or use ANOVA-like designs without controlling for the naturally occurring factors that most strongly affect biota (Carlisle and Hawkins 1998). Also, if species replacements occur in response to light to moderate stress (e.g., Hawkins et al. 2000), methods that use collective measures of assemblage structure (e.g., raw richness of different taxonomic groups) will not be as sensitive as methods that account for differences in assemblage composition.

The predictive modeling approach as described here allows both an overall assessment of the biological condition of a site as well as the ability to identify those taxa most strongly affected by a stressor. Because the method allows us to identify the responses of individual taxa, we can potentially derive hypotheses regarding either the general causes of or the specific mechanisms causing biological impairment. Considering the ease with which predictive models can now be built and implemented, the utility of this approach for the assessment of wetland invertebrates as well as other wetland biota should be explored in depth. The critical status of the remaining wetlands in this and other countries demands that we use accurate and scientifically defensible methods of assessments. *O/E* as an assessment measure is especially attractive because it is easily interpretable by all researchers, managers, the public, and policymakers.

At this time, the main constraint in applying predictive models for biological assessment of aquatic ecosystems in the United States is that few models have been built for either different areas or habitat types, and no standard or easily available software packages exist with which to easily construct models. We have developed models that run within a spreadsheet environment for streams in California, Oregon, and Washington and small lakes in both California and Utah. However, as was illustrated in this chapter, developing and implementing the models require some statistical and programming expertise. Ideally, models should be built that apply to large areas, as has been done in Great Britain and Australia. For example, one model was developed for stream ecosystems that applies to the whole of Great Britain. Models can also be housed in a central location, which would promote standardization and consistency in application. In such cases, field biologists could obtain assessments of local sites by uploading required data to a server or submitting the data to agency personnel. The number of models applicable to freshwater ecosystems in the United States should increase as the potential value of this approach is more widely recognized by water resource management agencies. Until that time, it is possible for individuals to create their own models with standard statistical and spreadsheet software.

ACKNOWLEDGMENTS

We thank Mark Vinson of the U.S. Department of the Interior, Bureau of Land Management. National Aquatic Monitoring Center for several initial discussions that led to the eventual writing of this chapter, and members of the Aquatic Ecology Lunch Bunch at Utah State University for constructive criticisms that greatly improved the clarity of this manuscript. Editorial suggestions by Russ Rader helped us clarify some remaining particularly fuzzy ideas.

REFERENCES

Bailey, R. C., M. G. Kennedy, M. Z. Dervish, and R. M. Taylor. 1998. Biological assessment of freshwater ecosystems using a reference condition approach: comparing predicted and actual benthic invertebrate communities in Yukon streams. Freshwater Biology 39:765–774.

Barbour, M. T., J. B. Stribling, and J. R. Karr. 1995. Multimetric approach for establishing biocriteria and measuring biological condition. Pages 63–77 *in* W. S. Davis and T. P. Simon (eds), Biological assessment and criteria. Lewis Publishers, Boca Raton, FL.

Belbin, L., and C. McDonald. 1993. Comparing three classification strategies for use in ecology. Journal of Vegetation Science 4:341–348.

Bradford, D. F. 1989. Allotopic distribution of native frogs and introduced fishes in high Sierra Nevada lakes of California: Implication of the negative effect of fish introductions. Copeia 1989:775–778.

Brinson, M. M., and R. Rheinhardt. 1996. The role of reference wetlands in functional assessment and mitigation. Ecological Applications 6:69–76.

Carlisle, D. M., and C. P. Hawkins. 1998. Relationships between invertebrate assemblage structure, 2 trout species, and habitat structure in Utah mountain lakes. Journal of the North American Benthological Society 17:286–300.

Chapin, F. S., III, O. E. Sala, I. C. Burke, J. P. Grime, D. U. Hooper, W. K. Laurenroth, A. Lombard, H. A. Mooney, A. R. Mosier, S. Naeem, S. W. Pacala, J. Roy, W. L. Steffen, and D. Tilman. 1998. Ecosystem consequences of changing biodiversity. BioScience 48:45–52.

Clarke, R. T., M. T. Furse, J. F. Wright, and D. Moss. 1996. Derivation of a biological quality index for river sites: comparison of the observed with the expected fauna. Journal of Applied Statistics 23:311–332.

Colwell, R. K., and J. A. Coddington. 1995. Estimating terrestrial biodiversity through extrapolation. Pages 101–118 *in* D. L. Hawksworth (ed), Biodiversity: measurement and estimation. Chapman & Hall, London.

Cowardin, L. M., V. Carter, F. C. Golet, and E. T. LaRoe. 1979. Classification of wetlands and deepwater habitats of the United States. Report FWS/OBS-79/31. U.S. Fish and Wildlife Service, Office of Biological Services, Washington DC.

Dutilleul, P. 1993. Spatial heterogeneity and the design of ecological field experiments. Ecology 74:1646–1658.

Ford, J. 1989. The effects of chemical stress on aquatic species composition and community structure. Pages 99–144 *in* S. A. Levine, M. A. Harwell, J. R. Kelly, and K. D. Kimball (eds), Ecotoxicology: problems and approaches. Springer-Verlag, New York.

Frey, D. G. 1977. Biological integrity of water: an historical approach. Pages 127–147 *in* R. K. Ballentine and L. J. Guarria, symposium coordinators, The integrity of water. U.S. Environmental Protection Agency, Office of Water and Hazardous Materials, Washington, DC.

Gray, J. S. 1989. Effects of environmental stress on species rich assemblages. Biological Journal of the Linnean Society 37:19–32.

Hawkins, C. P., R. H. Norris, J. N. Hogue, and J. W. Feminella. 2000. Development and evaluation of predictive models for measuring the biological integrity of streams. Ecological Applications. 10:1456–1477.

Liss, W. J., and G. L. Larson. 1991. Ecological effects of stocked trout on North Cascades naturally fishless lakes. Park Science 11:22–23.

Manly, B. F. J. 1994. Multivariate statistical methods, 2nd Ed. Chapman & Hall, London.

Minshall, G. W. 1984. Aquatic insect–substratum relationships. Pages 358–400 *in* V. H. Resh and D. M. Rosenberg (eds), The ecology of aquatic insects. Praeger, New York.

Moss, D., M. T. Furse, J. F. Wright, and P. D. Armitage. 1987. The prediction of the macroinvertebrate fauna of unpolluted running-water sites in Great Britain using environmental data. Freshwater Biology 17:41–52.

Moss, D., J. F. Wright, M. T. Furse, and R. T. Clarke. 1999. A comparison of alternative techniques for prediction of the fauna of running-water sites in Great Britain. Freshwater Biology 41:167–181.

Naeem, S. 1998. Species redundancy and ecosystem reliability. Conservation Biology 12:39–45.

Noss, R. F. 1990. Indicators for monitoring biodiversity: a hierarchical approach. Conservation Biology 4:355–364.

Perry, J., and E. Vanderklein. 1996. Water quality: management of a natural resource. Blackwell, Oxford.

Reynoldson, T. B., and J. F. Wright. 2000. The reference condition: problems and solutions. Pages 303–313 *in* J. F. Wright, D. W. Sutcliffe, and M. T. Furse (eds), Assessing the biological quality of freshwaters: RIVPACS and other techniques. Freshwater Biological Association, Ambleside, Cumbria, England.

Reynoldson, T. B., R. C. Bailey, and R. H. Norris. 1995. Biological guidelines for freshwater sediment based on BEnthic Assessment of SedimenT(the BEAST) using multivariate approach for predicting biological state. Australian Journal of Ecology 20:198–219.

Simpson, J. C., and R. H. Norris. 2000. Biological assessment of river quality: development of AUSRIVAS models and outputs. Pages 125–142 *in* J. F. Wright, D. W. Sutcliffe, and M. T. Furse (eds), Assessing the biological quality of freshwaters: RIVPACS and other techniques. Freshwater Biological Association, Ambleside, Cumbria, England.

Tilman, D. 1999. The ecological consequences of changes in biodiversity: a search for general principles. Ecology 80:1455–1474.

Wright, J. F. 1995. Development and use of a system for predicting the macroinvertebrate fauna in flowing waters. Australian Journal of Ecology 20:181–197.

Wright, J. F., J. H. Blackburn, R. J. M. Gunn, M. T. Furse, P. D. Armitage, J. M. Winder, and K. L. Symes. 1996. Macroinvertebrate frequency data for RIVPACS III sites in Great Britain and their use in conservation evaluation. Aquatic Conservation: Marine and Freshwater Ecosystems 6:141–167.

5 Application of Invertebrate Functional Groups to Wetland Ecosystem Function and Biomonitoring

KENNETH W. CUMMINS and RICHARD W. MERRITT

Invertebrates have been used extensively for biomonitoring in aquatic ecosystems, but the approach has been based primarily on taxonomic differences associated with a particular set of environmental conditions. The functional analysis of invertebrate communities presented here focuses the effort required to make taxonomic separations at the level of resolution that enables accurate evaluation of the roles played by the invertebrates in community organization. Invertebrates of large river riparian marsh (defined here as littoral aquatic macrophytes plus those of the coupled broadleaf marsh) have been assigned to functional feeding groups, habit groups, and voltinism (generation time) categories. The designations that have been widely used in studies of lotic systems should apply equally well in riparian marsh river floodplain systems and other wetlands. The distinction between the functional organization of lotic and lentic invertebrate communities is largely based on the different relative proportions of the same basic nutritional resource and microhabitat types. The Kissimmee and Caloosahatchee Rivers in southern Florida have been used as case studies to describe how functional analyses can be used to predict ecosystem attributes. Because the ratios that are calculated using invertebrate data, categorized by function and life cycle, are dimensionless, they are relatively independent of sample size. This allowed semiquantitative timed samples to be used in the analyses. In these two riparian marsh ecosystems, dissolved oxygen conditions are of major significance and both riparian marshes were shown to be heterotrophic. The method of analysis using invertebrate community organization applied in the case studies provided insight as to possible outcomes of the efforts to restore remnant channel habitats in both rivers.

Bioassessment and Management of North American Freshwater Wetlands, edited by Russell B. Rader, Darold P. Batzer, and Scott A. Wissinger.
0-471-35234-9

For over three decades, invertebrates have been used extensively to evaluate the ecological condition of aquatic habitats. Many aquatic invertebrate taxa have received special attention because they serve as key components in the food webs, leading to higher trophic levels involving water birds and sport or commercial fish. Despite all the advantages associated with using invertebrates for the analysis of ecological condition of aquatic ecosystems—such as their abundance, diversity, large average individual size, short life cycles, and habitat and food specificity—the incomplete status of the taxonomy of many groups often precludes the use of detailed taxonomic diversity in such analyses. Significant strides have been made in developing the requisite keys to make separations to the generic level possible (e.g., Pennak 1989, Merritt and Cummins 1996a). However, even these are often difficult and, as has been pointed out by others, those who can use keys most efficiently probably don't need them.

The use of invertebrate diversity, or taxonomic richness, to compare ecological condition between sites or seasons has been based on the usually unstated, and almost never met, assumption that identification of invertebrates will be to species. In practice, diversity measures are based most frequently on a mixture of levels of taxonomic refinement. For example, assigning the same generic value of 1 in diversity calculations to the limnephilid caddisfly *Anabolia*, with its potentially five species but with only one usually represented in any given littoral lentic-type habitat, and the orthoclad midge *Cricotopus*, with its 20 or more species, often with many species in the same littoral habitat, clearly produces a distortion in the analysis. Such distortions highlighted the need for functional analyses of aquatic invertebrate communities (e.g., Cummins 1974) for the purpose of assessing their ecological condition or, as it is often termed, integrity.

Wetland ecosystems are certainly amenable to functional analysis of invertebrate communities, and such analyses can provide important insights into fundamental attributes of these systems (e.g., Merritt et al. 1996, 1999; Cummins et al. 1999). The coverage here focuses on the wetlands associated with large rivers or floodplain wetlands; these are defined as riparian marshes (the littoral fringe and coupled broadleaf marsh; Merritt et al. 1999).

ASSIGNING INVERTEBRATES TO FUNCTIONAL GROUPS

To concentrate the analysis of aquatic invertebrate communities on the ecological roles played by the organisms in the ecosystems and habitats they occupy, a functional approach was adopted over 20 years ago and has been modified a number of times since then (e.g., Cummins 1973, 1974, 1988, 1993; Cummins and Klug 1979; Merritt et al. 1984; Cummins and Wilzbach 1985; Merritt and Cummins 1996a,b). The general concept is to cluster invertebrate taxa based on functional relationships that transcend taxonomic categories. Ratios of the numerical abundance, or better yet, the biomass, of various functional groups can be used as indicators, or surrogates, of ecosystem attributes. The analysis concentrates on expending the effort at making taxonomic separations where it will yield the most information about function (e.g., ordinal level for odonates but generic level for mayflies and caddisflies). The discussion below concentrates on four functional categorizations that can be used to establish ratios that indicate ecological condition (integrity) of a given riparian marsh ecosystem. These are: functional feeding groups (FFGs), habit functional groups (habit), drift propensity (drift), and life-cycle patterns (voltinism).

Functional Feeding Groups

The FFG categories are summarized in Table 5.1. The procedure entails separation of invertebrate taxa based on the morphological and behavioral adaptations that they use to harvest particular food resource categories such as vascular plant litter or attached algae (periphyton). At least middle to late instars (or growth stages) tend to be functionally consistent at the species level, and most often at the generic level. In many cases, there is functional feeding group integrity at the family (e.g., most Trichoptera), subfamily (some Chironomidae), subordinal (Plecoptera), or even ordinal (e.g., all Odonata) levels. Merritt and Cummins (1996a) provide a summary in which essentially all genera of North American aquatic insects have been at least tentatively assigned to functional groups. It is essential to note that the analysis of gut contents is not a reliable predictor of functional group assignment, because the separations are based on the morphological–behavioral adaptations of food acquisition, not ingestion (Cummins 1984). This fixed relationship between the morphobehavioral capability and the food resource to be harvested does not change much, but the food actually harvested often does (e.g., Anderson and Cummins 1979).

Within functional feeding groups, additional designations can be made as to whether there is either obligate or facultative relationships with particular food resources (e.g., Merritt et al. 1996, 1999). These assignments are optimally made based on food resource–specific growth data (e.g., Cummins and Klug 1979). Facultative forms are able to utilize a wider range of food resource categories; for example, the grass shrimp *Palaemonetes* can function as a CPOM (coarse particulate organic matter) shredder or an FPOM (fine particulate organic matter) gathering collector. However, the facultative forms may grow slightly better on one of the food resource categories. The obligate forms have much less flexibility in food acquisition, but they can maximize their efficiency in converting a food resource to their own tissue. The expectation is that in stable environments, in which the relative abundance of food resource categories remains fairly constant, obligate forms would have a competitive advantage over facultative forms. In less stable, changing environments, with shifting availability of resource categories, the more flexible linkage between the invertebrate taxon and its food resources would often give the facultative forms the competitive advantage, despite their less efficient conversion of food to growth on any one resource. At present, the general database is not often sufficient to allow distinctions to be made between obligate and facultative forms. But when the data are available, the obligate forms provide more reliable surrogates for ecosystem conditions.

TABLE 5.1. Invertebrate Functional Feeding Groups Used in Riparian Marsh Ecosystem Analysis

Functional Feeding Group	Nutritional Resources	Particle Size or Other Characteristics	Common Invertebrate Examples[a]
Detritivore Shredders	Coarse particulate organic matter detritus	>1 mm diameter in benthos	Amphipod, isopod, and decapod crustaceans (e.g., *Palaemonetes*), limnephilid caddisflies
Herbivore Shredders	Live macrophyte tissue	Live vascular plants and macroalgae (generally >1 cm diameter)	Pyralid and noctuid lepidopterans, chrysomelid and curculionid beetles
Scrapers	Periphyton	Attached algae and associated material trapped in interstices (generally <1 mm diameter)	Gastropods, *Hyalella* amphipod, psychomyiid caddisflies, corixid hemipterans
Plant piercers	Macrophyte cell fluids	Cell fluid contents	Hydroptilid caddisflies, haliplid beetles
Filtering collectors	Fine particulate organic matter detritus	<1 mm diameter in suspension	Bivalves, psychomyiid and polycentropodid caddisflies, culicids, Tanytarsini midges
Gathering Collectors	Fine particulate organic matter detritus	<1 mm diameter in benthos	Oligochaetes, baetid and caenid mayflies, leptocerid caddisflies, Chironomini midges
Predators	Live prey	Engulf or pierce prey	Odonates, *Oecetis* leptocerid caddisflies, dytiscid beetles, larval hydrophilid beetles, tanypod midges

Sources: Modified from Merritt and Cummins (1996a) and other references of Cummins and of Merritt.

[a]Most or many, but not necessarily all, species in taxon belong to the functional group indicated; some representatives may be facultative (i.e., can also function as another group, but less efficiently).

Habit Groups

The habit functional group separations (Table 5.2) are based on locomotion (e.g., swimmers), attachment (e.g., clingers), and concealment (e.g., burrowers). Designation of habit groups is particularly useful in predicting the availability of invertebrates as prey for wading birds and fish. In both groups the distinction can be made between

TABLE 5.2. Habit (Mode of Existence) Functional Group Categorizations

Functional Habit Group	Characteristics	Common Invertebrate Examples
Clingers	Behavioral and/or morphological adaptations for attachment to resist dislodgement in currents or waves	Philopotamidae, Tanytarsini
Climbers	Opposable legs or suction surface for moving vertically on plant stems or woody debris	Aeschnidae, Coenagrionidae
Sprawlers	Adaptations for staying on top of fine sediments and protecting respiratory surfaces or on the surface of floating vascular plant leaves with no special attachment adaptations	Physidae,harpacta-coid Copepoda, Libellulidae, Caenidae, Tricorythidae
Burrowers	Penetrate sediment interstices or construct discrete tubes in sediment or within vascular plant tissue	Oligochaeta, Bivalvia, Gomphidae, Ephemeridae, Pyralidae, Noctuidae
Swimmers	Fishlike swimming, usually as short bursts from resting positions on submerged vascular plants or woody debris	Amphipoda, Baetidae
Skaters/divers	Skate and/or rest on the surface film; diving, regularly or occasionally, beneath surface to feed or escape predators	Gerridae, Corixidae, Notonectidae, adult Dytiscidae and Hydrophilidae
Planktonics	Suspended in the water column with some ability to move vertically but little competency for horizontal movement	Cladocera, cyclopoid Copepoda, Leptoceridae, Chironomidac pupae

Source: Modified from Merritt and Cummins (1996).

tactile- and visual-feeding species. For example, invertebrate swimmers are vulnerable to the same wading bird and fish visual predators that feed on small water column–inhabiting fishes, whereas burrowers and clingers would be more vulnerable to tactile-feeding birds and fish. Invertebrate climbers would be more vulnerable to water column–inhabiting visual fish predators and to a lesser extent, to site-feeding wading birds. Invertebrate sprawlers would be most vulnerable to tactile bottom feeders such as catfishes and benthic site-feeding fishes such as darters and gobies.

Drift Propensity

Invertebrates are separated into taxa known to exhibit behavioral drift (i.e., directed behavior in which animals can be found up in the water column on a predictable schedule, namely, dawn and dusk) and taxa that appear in the water column only as a consequence of dislodgment due to increased flow or wave action (Wilzbach et al. 1988). The link to fish is that this availability of invertebrate prey in the water column on a predictable diurnal schedule constitutes a reliable (i.e., predictable) food source. Thus, the ratio of behavioral to accidental drifters can be used as a predictor of food supplies for water column–feeding fish (Table 5.3). Although behavioral drifters can appear in the water column at times of high flows or wave action, this would not follow a predictable diurnal pattern. However, it is interesting to speculate on artificial diurnal patterns in accidental drift that can be created below power dams that maintain daily release schedules that are predictable (i.e., at peak power load times in the morning and at night).

Voltinism (Life Cycles)

The length of time from egg to adult, expressed as a number of generations per year, or voltinism, was divided into two categories: (1) those taxa with an annual life cycle (univoltine) and those having more than one generation per year (two per year or bivoltine, and more than two per year or multivoltine); and (2) those requiring more than one year to complete a generation (semivoltine). A ratio of the short-life-cycle taxa (usually less than a year) to the long-life-cycle taxa (more than a year) provides an expression of the pioneer nature of the invertebrate community (Table 5.3). A ratio above 1.0 indicates that the invertebrate community is dominated by rapid turnover (short life cycle), pioneer species that would be predicted to be rapid and early colonizers of new habitats as they become available, for example following rehabilitation of riparian marshes along river channels.

USING RATIOS TO PREDICT ECOSYSTEM ATTRIBUTES

Functional feeding group, habit group, drift propensity, and voltinism ratios that can serve as surrogates for various ecosystem attributes are summarized in Table 5.3. Because the direct measurement of the ecosystem parameters is difficult, time consuming, often requires specialized equipment, and is usually not integrated over long time periods, employing aquatic macroinvertebrates as surrogates for estimating the attributes can be very useful. As indicated above, the functional categorization of invertebrates normally holds over at least the last two-thirds of the growth period of their life cycle. This means that the estimate of the ecosystem attribute is integrated over that time interval, which is normally three to six months or more.

The best example in which macroinvertebrate functional groups have been used as surrogates for ecosystem function involves the ratio that estimates daily (24-hour) gross primary production as a proportion of total community respiration (P/R, Table

5.3). It is widely held that this measure of community metabolism provides one of the very best integrative measures of aquatic ecosystem condition. (e.g., Vannote et al. 1980). This measures the balance between autotrophy and heterotrophy that determines the pathway of the major energy transfers in a given system, such as the Kissimmee and Caloosahatchee River riparian marshes considered as case studies below. Essentially pure littoral plant bed stands of either spatterdock (*Nuphar luteum*) or smartweed (*Polygonum densiflorum*) in the Kissimmee River pool B remnant channel or spatterdock and pennywort (*Hydrocotyle umbellata*) in the Caloosahatchee River oxbows were among the habitats sampled for invertebrate surrogates representing P/R and other riparian marsh ecosystem attributes. In the Kissimmee River pool B remnant channels, direct measurements of P/R were made by monitoring 24-hour changes in dissolved oxygen (DO) using YSI 600 series sondes either positioned in profile on PVC rods in the spatterdock or smartweed plant beds and attached to data loggers, or in closed, recirculating 20-L chambers with submerged portions of plants and their attached periphyton inside (Dodds and Brock 1998, Cummins et al. 1999). Recirculation within the chambers was maintained with axial impellers. Supersaturation and complete oxygen depletion were avoided by opening and flushing the chambers with ambient water when needed. When ambient river water was too high in oxygen, nitrogen gas was bubbled into a head box to create lower-oxygen water which was then used to purge the chambers. The recirculating chambers could also be fitted on to stainless steel bottoms with a flanged opening to the sediment surface. When used, this allowed the plants and roots to remain intact and for the exchange of subsurface water with the chamber water (Cummins et al. 1999). Water depth in the beds ranged from 0.75 to 1.25 m. The expected (predicted) ratios for other ecosystem attributes (Table 5.3) are based on values calculated from the literature for stream-river and littoral fringe habitats and those measured directly by the authors in other studies.

METHODS OF SAMPLING RIPARIAN MARSH WETLANDS

Timed (45-second) samples were taken with a D-frame dip net with an 800-μm mesh size. Collections in plant beds included the surficial sediments around the bases of the plants and organisms on the plant stems and submerged leaves. Sampling consisted of a series of sweeps from the sediments along the plant stems to the water surface. The net was moved back and forth and to and fro vigorously during these vertical sweeps to ensure the dislodgment of attached/associated organisms. For the snag and sediment samples, the net was vigorously bumped along the snag or in the surface sediments in a series of sweeps lasting 45 seconds. Once a sample was collected, the net contents were washed into a 14-L (5-gal) bucket, field-sorted to remove large debris, and washed several times through a 250-mm mesh sieve to remove silt and very fine detritus. The washed samples were then placed in Whirl-pak bags and preserved with 95% ethanol so that dilution with sample water yielded about 70% concentration.

In the lab, invertebrate samples from the plant bed and other habitats were sorted by taxa (generic level in most cases) and categorized according to functional feeding

TABLE 5.3. Functional Group, Drift Propensity, and Voltinism Ratios That Can Serve as Surrogates for Ecosystem Parameters

Ecosystem Parameters Measured Directly	Methods of Direct Measurement	Functional Group Ratios	Expected Ratios[a]
Gross primary production *as a proportion of* community respiration (P/R)	P/R measurements per unit area, biomass, or chlorophyll over 24-hour periods usually based on dissolved oxygen data	Shredders (live vascular plants) + scrapers + piercers *as a proportion of* CPOM shredders + total collectors	Autotrophic system >0.75 <0.75 Heterotrophic system
Coarse particulate organic matter *as a proportion of* fine particulate organic matter (CPOM/FPOM)	CPOM/FPOM measurements per unit area of benthic storage in/on the sediments	Total CPOM shredders *as a proportion of* total collectors	Riparian system by season *Dry/cold (fall–winter)* Supports normal shredder populations >0.50 <0.50 Shredder populations insignificant *Wet/warm (spring–summer)* Supports normal shredder populations >0.25 <0.25 Shredder populations insignificant
Suspended (in Transport) FPOM *as a proportion of* Benthic (Deposited) FPOM (SFPOM/BFPOM)	SFPOM measured per unit volume and BFPOM measured per unit area of bottom	Filtering Collectors *as a proportion of* gathering collectors	System Enriched in suspended FPOM >0.50 System dominated by Benthic FPOM <0.50
Habitat (substrate) stability FFG ratio	Stable surfaces (e.g., large woody debris, sediments not moved at high flow, rooted vascular hydrophytes)	Scrapers + filtering collectors *as a proportion of* shredders + gathering collectors	Stable substrates for attachment not limiting >0.50 <0.50 Stable substrates for attachment limited

Habit ratio	Measured as area (wood and sediments) or stem density (vascular plants)	Clingers + climbers *as a proportion of* burrowers[b] + sprawlers + swimmers	Stable substrates for attachment not limiting >0.60 <0.60 Stable substrates for attachment limited
Top-down (predator) control of prey populations	Direct measure of predator–prey food links	Predators *as a proportion of* total of all other functional feeding groups	Over abundance of predator populations and very abundant rapid turnover prey populations >0.15 <0.15–>0.05 Normal predator populations
Short life cycle (rapid turnover) vs.long life cycle	Determine phenology of adult populations by light and emergence traps and sweep netting	Taxa with one or more generations per year *as a proportion of* taxa requiring More than one year for a generation	Pioneer, early-colonizer populations dominant >1.0 <1.0 Established, slow-colonizer populations dominant
Predictable food supply for water column-feeding fish (drift food)	Measure diurnal patterns of invertebrate drift together with fish gut analysis	Behavioral drifters *as a proportion of* accidental drifters	Reliable food supply for water column–feeding fish >0.50 <0.50 Below-optimal food supply for water column–feeding fish
Good food supply for wading birds and benthic-feeding fish (benthic food)	Measure benthic invertebrate standing crop together with wading bird and benthic fish gut analyses	Sprawlers *as a proportion of* clingers + climbers + burrowers + swimmers	Good food availability for wading birds and benthic-feeding fish >6.0 <0.60 Poor food availability for wading birds and benthic-feeding fish

[a] Based on biomass.

[b] The presence of large, long-lived pelecypod mollusks may indicate stable sediments and their biomass should be tallied in the numerator.

group (FFG), habit, or life-cycle groups as given in Merritt et al. (1996, 1999) and Merritt and Cummins (1996a,b; Tables 5.1 and 5.2). Samples were sorted under a dissecting microscope and all invertebrates were identified, to the generic level in most cases. Sorting of animals from detritus, plant parts, and sediments was facilitated by staining with rose bengal (Mason and Yevich 1967). Samples that contained very large amounts of detritus, plant material, and/or sediment were subsampled using the method of Waters (1969).

Biomass of the invertebrates sorted to FFGs and measured to length were calculated using Invertcalc, a software program (K. W. Cummins and M. A. Wilzbach, unpublished) that uses 33 length–mass regressions developed for specific invertebrate taxonomic groupings (see also Smock 1980). As developed in Merritt et al. (1996, 1999), various ratios of FFGs and other functional groups can be used as surrogates for various aquatic ecosystem attributes. For example, the autotrophic index [i.e., P/R (autotrophy/heterotrophy)] can be predicted by the ratio of live plant-consuming invertebrates as a proportion of the detrital-consuming invertebrates (Merritt et al. 1996, 1999). Table 5.1 summarizes data from the lower remnant channel of pool B of the Kissimmee River for the P/R surrogate. P/R ratios have been measured directly as ≥ 1 for *Nuphar* beds and as <1 for *Polygonum* beds in the riparian marsh of that reach of river.

CASE STUDIES: TWO RIVERS IN SOUTHERN FLORIDA

The littoral wetland portion of riparian marsh was investigated in two southern Florida rivers: the Kissimmee River, which heads in the Kissimmee Lakes south of Orlando in south-central Florida and flows into Lake Okeechobee, and the Caloosahatchee River, which heads in Lake Okeechobee by virtue of a canal built by the U.S. Army Corps of Engineers connecting the historical wetland headwaters to the lake, and flows into the Lower Charlotte Harbor/Pine Island Sound estuary system (Figs. 5.1*A*, 5.2*A*). The flows in both rivers are regulated by dams and locks. The major water management mechanism for all of southern Florida is regulation of releases into and out of Lake Okeechobee. The primary water supply for Lake Okeechobee comes from the Kissimmee River at the northern end of the lake (Fig. 5.1*A*). In addition to supplying water to the Everglades and the lower east coast, the lake supplies the east and west coast estuaries. Coupled with the hydrologic restoration efforts that are specifically directed at the Everglades, there are projects to rehabilitate the middle third of the Kissimmee River (Fig. 5.1*B*) and oxbows in the lower third of the Caloosahatchee River (Fig. 5.2*A*). Major efforts have been mounted to establish baseline data (i.e., information on prerestoration conditions for the Kissimmee River) so as to allow continued evaluation of the success of the rehabilitation efforts (e.g., Toth 1993, 1996; Toth et al. 1995; papers in Dahm 1995). If restoration is defined as returning a system to presettlement conditions, the projects are more properly referred to as rehabilitation (i.e., the return of the system back to some intermediate, presumably improved, condition between the present and the historical state). The continued evaluation of the ecological parameters that are identified as critical is viewed as providing the data necessary to make "midcourse corrections" in a restoration project as

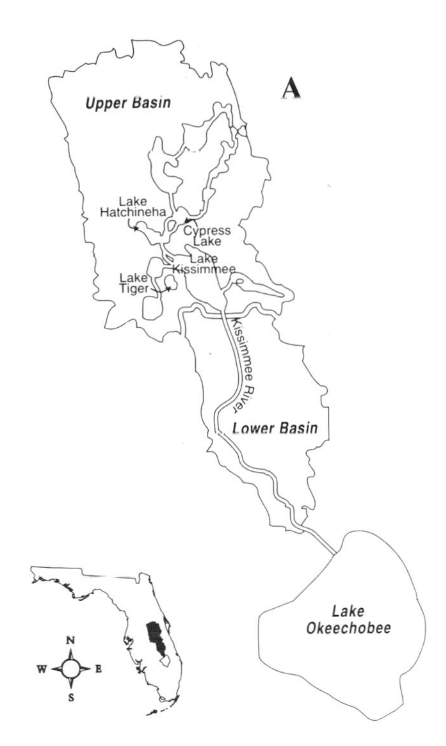

FIGURE 5.1. (*A*) Map of the Kissimmee River basin from the upper basin headwaters in the Kissimmee Lakes through the lower basin and into Lake Okeechobee. Location of the basin in south-central Florida is indicated in the inset. (B) Channelization of the Kissimmee River into the C-38 canal, converted 166 km of meandering river into a series of five impoundments (pools A to E) behind structures S-65A to S-65E. The star indicates the sampling area of remnant channel. (Modified from South Florida Ecosystem Restoration Plan, Ecosystem Restoration Department, South Florida Water Management District, West Palm Beach, FL.)

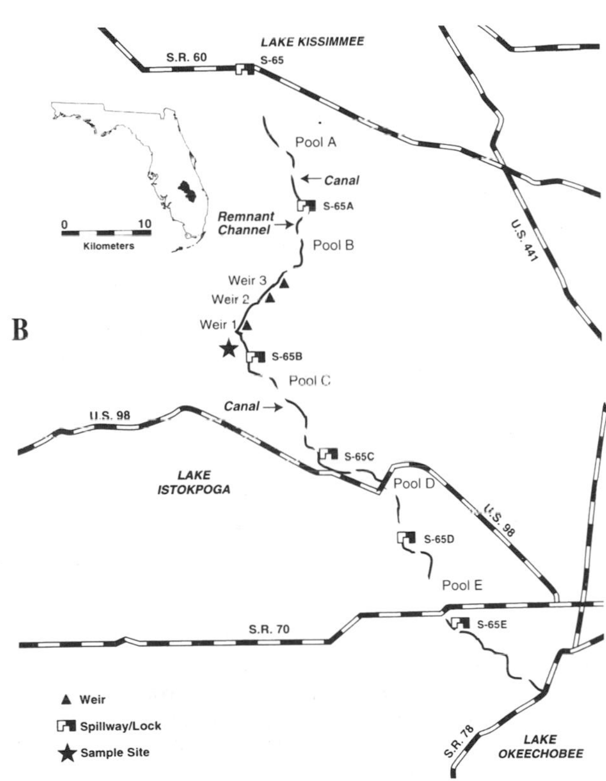

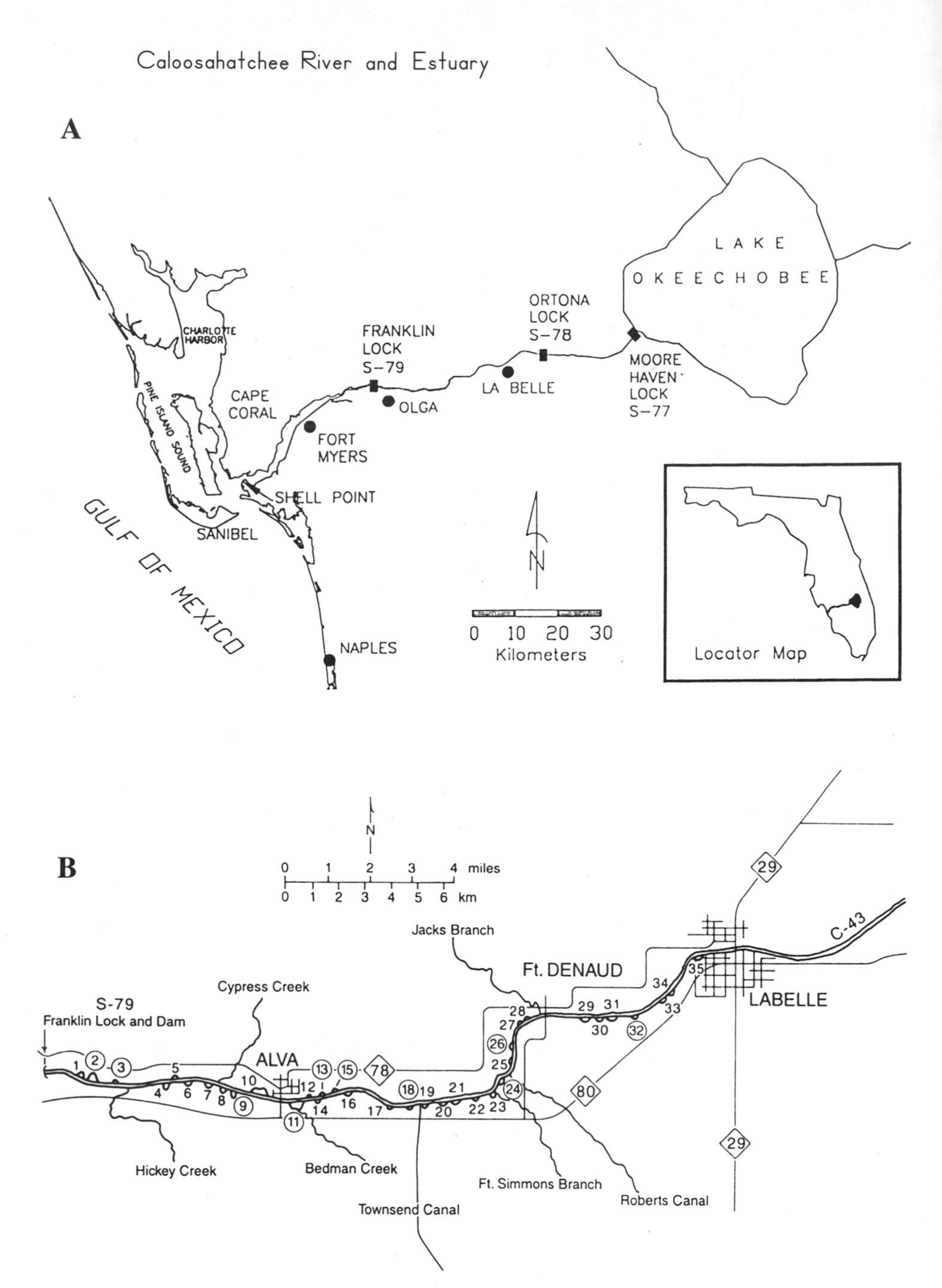

FIGURE 5.2. (*A*) Map of the Caloosahatchee River, which heads in Lake Okeechobee by virtue of a canal connecting the wetland headwaters to the lake and flows into the Lower Charlotte Harbor/Pine Island Sound Estuary system. Location of the river system in southwestern Florida is indicated in the inset. (*B*) Map of Caloosahatchee River oxbows (35 total) along the channelized river from LaBelle down to the Franklin Lock. Circled numbers indicate the 10 oxbows sampled.

dictated by an adaptive management strategy. The Kissimmee River restoration will involve backfilling of the channelized river using the existing spoil levee material. This will result in reconnection of the main cannel with the associated historical floodplain and its remnant wetlands and braided channels. In addition, the regulated hydrologic regime will be adjusted to better reflect normal historical annual flow patterns: namely, more water released in the wet season and less in the dry season, rather than the reversed pattern that now exists.

Using the case studies below, ecosystem attributes have been predicted for the riparian marsh habitats of the Kissimmee and Caloosahatchee Rivers using the functional group/voltinism relationships described above. Because river-side wetlands, or riparian marsh habitats, have been severely altered for almost all major low-gradient rivers, such case study evaluations assume importance in the never-ending restoration/rehabilitation issues facing river managers in the future.

Kissimmee River

Prior to channelization, the Kissimmee River flowed about 161 km (100 miles) through the 1200-km^2 (1600-square mile) Lower Basin (Fig. 5.1*A*) from its headwaters in the Kissimmee Lakes to Lake Okeechobee. The river was broad and meandering, flowing through a 1.5- to 3-km-wide floodplain. U.S. Geological Survey records indicate that historically, over 90% of the floodplain was inundated over 50% of the time (see papers in Dahm 1995). Channelization of the river resulted in the loss of approximately 14,000 hectares (35,000 acres) of riparian marsh (Koebel 1995). Undoubtedly, many plant and animal species were lost or reduced to relict populations (Toth et al. 1995).

The Kissimmee River in its present configuration is a channelized canal (C-38) divided into a series of five pools (impoundments) separated by lock and dam structures that were constructed by the Corps from 1962 through 1971 (Fig. 5.1*B*). In the mid-1980s, a series of three weirs was constructed to augment the flow through three remnant channels along Pool B (Fig. 5.1*B*). These partially restored remnant channels, with their increased circulation under high-flow conditions, represent the best example of some of the characteristics that would be expected in the rehabilitated channel system (papers in Dahm 1995, Toth 1996).

The invertebrate studies reported here were conducted within a 1000-m reach of the lower (farthest down river) of these remnant channels (Fig. 5.1*B*). The riparian marsh along this section of river is dominated by the littoral plants *Nuphar luteum* (spatterdock) and *Polygonum densiflorum* (smartweed) and the broadleaf marsh plants *Pontederia cordata* (pickerelweed), *Sagittaria lancifolia* (arrowhead), and *Panicum hemitomum* (maidencane) (Fig. 5.3). Floating plants variously found interspersed in open spaces in the riparian marsh included the native *Pistia stratiotes* (water lettuce) and the exotics *Eichornia crassipes* (water hyacinth) and *Salvinia minima* (common salvinia fern).

The majority of the studies on community metabolism (photosynthesis/respiration) were conducted in the *Nuphar* and *Polygonum* plant beds (J. Brock and K. W. Cummins, unpublished data). These two plant types presented very different habitats for invertebrate populations. For example, the stem densities and percent foliar coverage at the surface differed significantly between the two. The mean stem density of *Nuphar* was

FIGURE 5.3. Riparian marsh plant communities and other habitats sampled for macroinvertebrates from the reach of the lower remnant meander channel of Pool B of the Kissimmee River. Species pictured are as follows: (*a*) *Nuphar luteum*, (*b*) *Polygonum densiflorum*, (*c*) *Pontederia cordata*, (*d*) *Sagittaria lancifolia*, (*e*) snag (dead live-oak, *Quercus virginiana*), (*f*) sediments. (From Merritt et al. 1996.)

measured as 15 and as 46 for *Polygonum*. The effect of cover on light penetration into the bed (e.g., to reach the periphyton on the plant stems) was correspondingly different between the two. Measurements made in dense and sparse beds of *Nuphar* and *Polygonum* revealed that the attenuation of light [photosynthetically active radiation (PAR)] in *Polygonum* beds was 10 to 20% greater than in *Nuphar* beds. The *Nuphar* beds exhibited light extinction profiles similar to those of the open channel (Cummins et al. 1999). The periphytic dissolved oxygen (DO) production was significantly greater in *Nuphar* than in *Polygonum* beds, differing by between 1 and 2 mg/L (Fig. 5.9). Because DO in the river often reaches critical levels for game fish such as centrarchids, this difference in DO availability was probably also quite significant for invertebrate survival.

The results of analyses of macroinvertebrate functional ratios used as surrogates for ecosystem attributes in the Kissimmee River during two seasons and over six habitat types are summarized in Figures 5.4. The results indicate that the lower remnant channel of Pool B of the Kissimmee River was heterotrophic except in *Nuphar* beds in the autumn. In particular, the invertebrate P/R surrogate ratio indicated that the sediments were highly heterotrophic (Fig. 5.4*A*). The system exhibited normal summer shredder ratios in all habitats, especially the plant beds. Only *Pontederia* beds and sediments showed low autumn shredder ratios (Fig. 5.4*B*). Thus, in general, the lower remnant channel would be characterized as having a normal shredder component for a floodplain river. With the exception of the *Nuphar* beds, all other habitats sampled registered as having adequate suspended FPOM loading to support good filtering collector populations (Fig. 5.4*C*). In general, the spring samples indicted more FPOM in transport than in the autumn (Fig. 5.4C). Habitat stability is estimated using invertebrate ratios in two ways: using functional feeding and habit groups. In the majority of cases (8 of 12 and 9 of 12), the habitats were judged to be deficient in stable substrates for attaching or embedding (Fig. 5.4*D* and *E*). Attaching filtering collectors were uncommon in the habitats, including the snags (Fig. 5.3*e*), and the high ratios, on a biomass basis, in the sediments (Fig. 5.3*f*) were due to populations of bivalves. Using the FFG ratios to estimate the top-down control (i.e., predator effects) in the six habitats indicated that predator dominance, on a biomass basis, was generally higher than the predicted 15% (8 of 12 cases, Fig. 5.4*F*). The large biomass of predators was attributable primarily to odonates and *Belastoma* and some large numbers of tanypod midges. This agrees well with the life-cycle (voltinism) ratios, which indicated that most habitats in both seasons were dominated by short, rapid turnover taxa (8 of 12 cases) that could support a large predator biomass (Fig. 5.4*G*). In particular, the *Nuphar* and *Polygonum* beds were heavily dominated by such taxa. These data on life-cycle length also lead to the prediction that colonization of new habitats (e.g., following restoration of riparian marsh) would be rapid. Drift food for fish was estimated to be favorable in the majority of habitats in both seasons (9 of 12 cases, Fig. 5.4H). This ratio is based on a comparison of invertebrates that drift behaviorally, primarily at dusk and dawn, to those that appear in the water column largely accidentally. The ratios calculated to represent benthic food availability to site-feeding fishes and wading birds, which is based on invertebrate habits that would make them vulnerable to such predation, was characterized as adequate in only three habitats, all in the autumn (and judged inadequate in 9 of all 12 cases, Fig. 5.4I).

Thus, in overview, the partially restored remnant channel along Pool B of the Kissimmee River was evaluated with regard to ecosystem attributes in the autumn and spring, as viewed through the use of macroinvertebrate functional group ratios (Fig. 5.4). The system was judged to (1) be heterotrophic and, as such, supported normal levels of CPOM shredders and a sufficient level of water column FPOM to support filtering animals; (2) probably be limiting to invertebrates seeking stable attachment sites except for burrowing bivalves; (3) combine large predator biomass with the expected abundance of short-life-cycle rapid-turnover prey taxa; (4) have the potential for rapid colonization by macroinvertebrates of newly restored habitat; and (5) be favorable for drift (water column)–feeding fishes but poor for benthic site–feeding fish and wading birds.

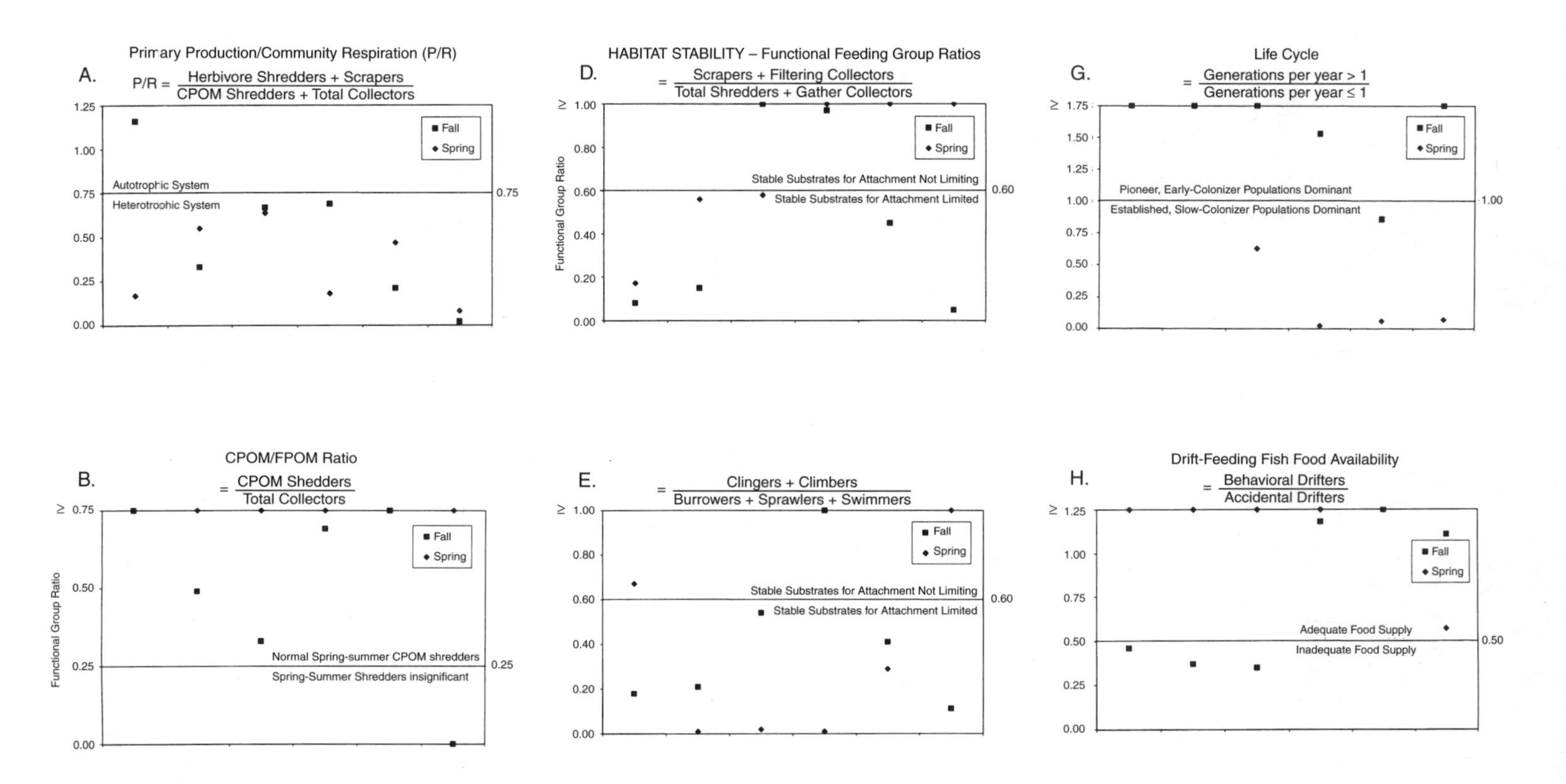

A.
Primary Production/Community Respiration (P/R)
P/R = Herbivore Shredders + Scrapers / CPOM Shredders + Total Collectors
Fall
Spring
Autotrophic System
Heterotrophic System
0.75
1.25
1.00
0.75
0.50
0.25
0.00
D.
HABITAT STABILITY – Functional Feeding Group Ratios
= Scrapers + Filtering Collectors / Total Shredders + Gather Collectors
Functional Group Ratio
Fall
Spring
Stable Substrates for Attachment Not Limiting
Stable Substrates for Attachment Limited
0.60
≥ 1.00
0.80
0.60
0.40
0.20
0.00
G.
Life Cycle
= Generations per year > 1 / Generations per year ≤ 1
Fall
Spring
Pioneer, Early-Colonizer Populations Dominant
Established, Slow-Colonizer Populations Dominant
1.00
≥ 1.75
1.50
1.25
1.00
0.75
0.50
0.25
0.00
B.
CPOM/FPOM Ratio
= CPOM Shedders / Total Collectors
Functional Group Ratio
Fall
Spring
Normal Spring-summer CPOM shredders
Spring-Summer Shredders insignificant
0.25
≥ 0.75
0.50
0.25
0.00
E.
= Clingers + Climbers / Burrowers + Sprawlers + Swimmers
Fall
Spring
Stable Substrates for Attachment Not Limiting
Stable Substrates for Attachment Limited
0.60
≥ 1.00
0.80
0.60
0.40
0.20
0.00
H.
Drift-Feeding Fish Food Availability
= Behavioral Drifters / Accidental Drifters
Fall
Spring
Adequate Food Supply
Inadequate Food Supply
0.50
≥ 1.25
1.00
0.75
0.50
0.25
0.00

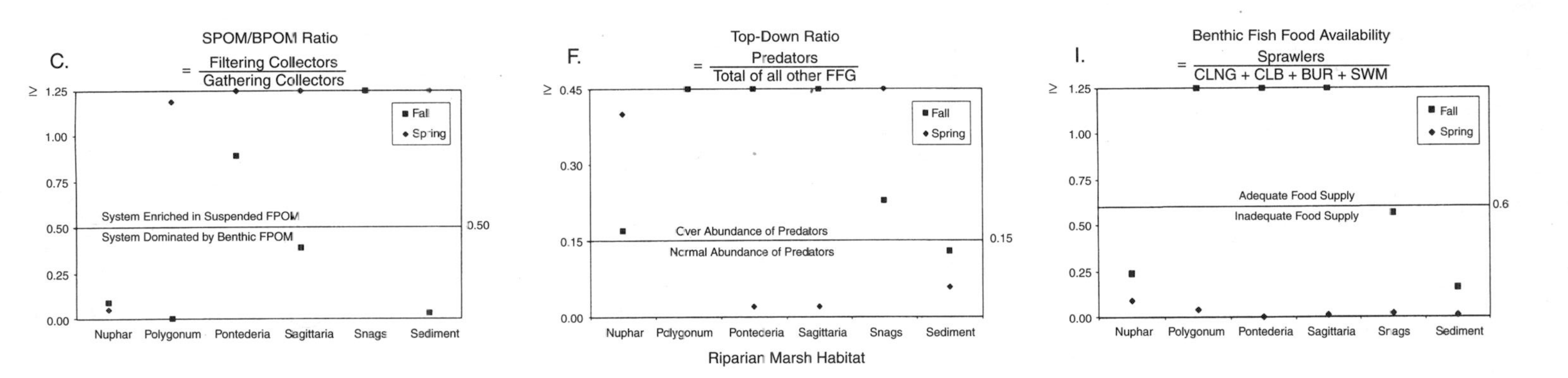

FIGURE 5.4. Graphical analysis of macroinvertebrate biomass ratios by habitat in fall and spring for Pool B of the Kissimmee River based on four functional categorizations: functional feeding group (FFG), habit functional groups (habit), drift propensity (drift), and life-cycle patterns (voltinism).

Caloosahatchee River

The headwaters of the Caloosahatchee River arise at present from Lake Okeechobee, due to a channelized connection built by the U.S. Army Corps of Engineers (USACOE) in the mid- to late 1960s (Fig. 5.2*A*). The river drains a watershed of approximately 2600 km^2 (1000 square miles) dominated by agriculture in the upper portion and urban development in the lower portion. Historically, the river headed in a series of lakes and wetlands between the present Ortona Lock and Dam and Lake Okeechobee and flowed over a natural water fall downstream from Lake Flirt (the waterfall was dynamited early in the nineteenth century as part of the channelization project by Hamilton Disston). The USACOE project channelized the river from Lake Okeechobee to the estuary system, approximately 100 km, (62 miles), and established three lock and dam systems, one at Lake Okeechobee (Morehaven), one 24 km (15 miles) below the lake (Ortona), and one 66 km (44 miles) from the lake at Olga (Franklin) (Fig. 5.2*A*). The Franklin lock and dam (S-79) serves as a tidal barrier, excluding most upstream movement of saline water. The city of LaBelle, which lies 13 km (8 miles) downriver from the Ortona lock and dam (S-78), marks the upstream boundary of the area of study reported here. The present-day river is a major waterway in southwestern Florida as part of the USACOE Intra-Coastal Waterway used for the transport of petroleum, heavy equipment and manufacturing supplies, irrigation, potable water, and recreation.

There are 35 remnant side channels, or oxbows (Fig. 5.2B), along the 29 km (18 miles) of channelized river from LaBelle down to the Franklin lock. The ecological condition of these oxbows varies from reasonably good in the few with significant flowthrough, to very poor in those where flow is restricted or blocked and significant organically rich sediments have accumulated. From the 35 oxbows, 10 were selected for intensive study. The 10 oxbows were chosen to represent the full range of ecological condition seen in all 35 (Fig. 5.2*B*). The selection was based on preliminary surveys conducted by John Capece (University of Florida, Institute of Food and Agricultural Science) and Rae Ann Wessel (Caloosahatchee River Citizens' Association) (Capece and Scholle [Wessel] 1996). The oxbows were dominated by two plant bed types, spatterdock (*Nuphar luteum*) and pennywort (*Hydrocotyle umbellata*) (Fig. 5.5).

The long-term management objective for the Caloosahatchee River remnant oxbows is the rehabilitation of ecological integrity of as many of the 35 as possible. The target for this rehabilitation would be conceptual reference conditions based on the ecological literature, historical conditions (Milleson 1979, Miller et al. 1982, Camp, Dresser and McKee, Inc. 1995), and the existing condition in the oxbow(s) judged to be presently in the best ecological condition (Capece and Scholle [Wessel] 1996 and the study reported here). The goal of this strategy would be to improve the recreational quality of the oxbows and to enhance their capability as water quality filters and for off-channel storage of water during wet periods.

The results of analyses of macroinvertebrate functional ratios used as surrogates for ecosystem attributes in the Caloosahatchee River oxbows in two plant bed types in the summer are summarized in Figure 5.6. The results indicated that all the *Nuphar* beds were heterotrophic. Only three of the oxbows (especially number 11) had *Hydrocotyle* beds that were rated as autotrophic, and in all but one (number 32) *Hydrocotyle* beds

FIGURE 5.5. Sampling locations in the Caloosahatchee River oxbows: (*a*) sampling boat in oxbow 24; typical (*b*) *Nuphar* (spatterdock) and (*c*) *Hydrocotyle* (pennywort) beds.

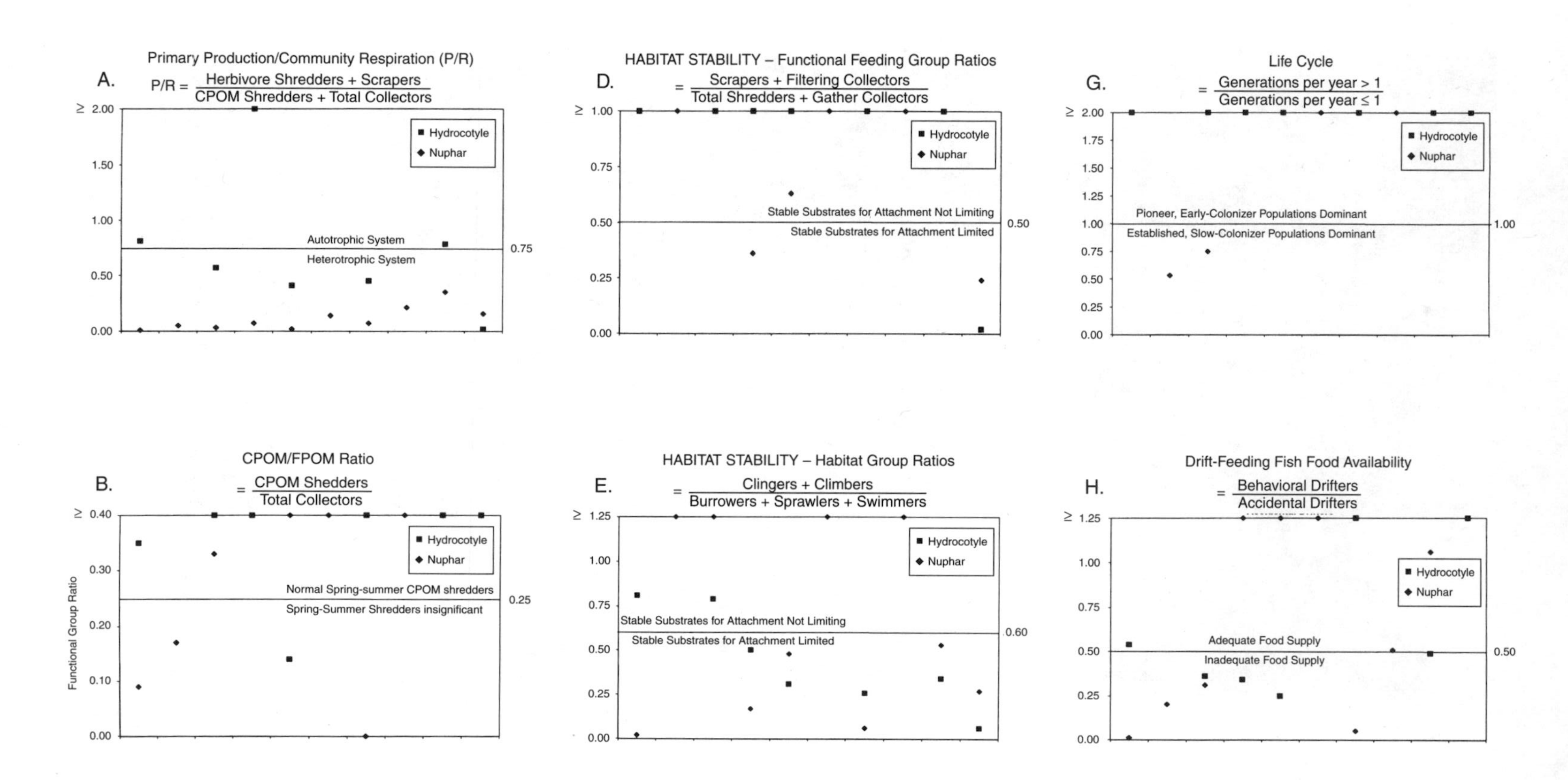

A.
Primary Production/Community Respiration (P/R)
P/R = Herbivore Shredders + Scrapers / CPOM Shredders + Total Collectors
≥ 2.00
1.50
1.00
0.50
0.00
Hydrocotyle
Nuphar
Autotrophic System
Heterotrophic System
0.75
D.
HABITAT STABILITY – Functional Feeding Group Ratios
= Scrapers + Filtering Collectors / Total Shredders + Gather Collectors
≥ 1.00
0.75
0.50
0.25
0.00
Hydrocotyle
Nuphar
Stable Substrates for Attachment Not Limiting
Stable Substrates for Attachment Limited
0.50
G.
Life Cycle
= Generations per year > 1 / Generations per year ≤ 1
≥ 2.00
1.75
1.50
1.25
1.00
0.75
0.50
0.25
0.00
Hydrocotyle
Nuphar
Pioneer, Early-Colonizer Populations Dominant
Established, Slow-Colonizer Populations Dominant
1.00
B.
CPOM/FPOM Ratio
= CPOM Shedders / Total Collectors
Functional Group Ratio
≥ 0.40
0.30
0.20
0.10
0.00
Hydrocotyle
Nuphar
Normal Spring-summer CPOM shredders
Spring-Summer Shredders insignificant
0.25
E.
HABITAT STABILITY – Habitat Group Ratios
= Clingers + Climbers / Burrowers + Sprawlers + Swimmers
≥ 1.25
1.00
0.75
0.50
0.25
0.00
Hydrocotyle
Nuphar
Stable Substrates for Attachment Not Limiting
Stable Substrates for Attachment Limited
0.60
H.
Drift-Feeding Fish Food Availability
= Behavioral Drifters / Accidental Drifters
≥ 1.25
1.00
0.75
0.50
0.25
0.00
Hydrocotyle
Nuphar
Adequate Food Supply
Inadequate Food Supply
0.50

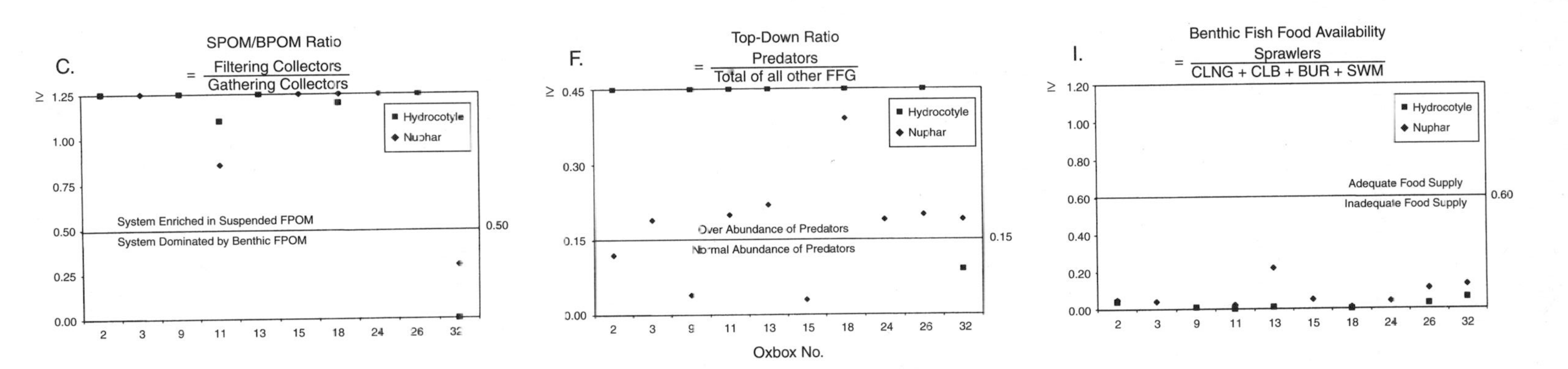

FIGURE 5.6. Graphical analysis of macroinvertebrate biomass ratios by oxbow for two plant bed types, *Nuphar* and *Hydrocotyle*, in the Caloosahatchee River based on four functional categorizations: functional feeding group (FFG), habit functional groups (habit), drift propensity (drift), and life-cycle patterns (voltinism).

were less heterotrophic than *Nuphar* beds (Fig. 5.6*A*). The autotrophic ratios in oxbows 2, 11, and 26 in the *Hydrocotyle* beds were due primarily to the biomass of the facultative scraper amphipod *Hyallela.* Although *Hyallela* can also function as a facultative FPOM gathering collector, the correlation with high DO levels measured in those three oxbows suggest that the system was autotrophic with periphyton cover sufficient to support algal scraping (Fig. 5.7*A*). In general, both spatterdock and pennywort beds were rated as having sufficient CPOM to support summer shredders (13 out of 17 cases, Fig. 5.6*B*). This evaluation was due largely to the presence of the large facultative shredder grass shrimp *Palaemonetes* in most samples and again reflects the general heterotrophic nature of the system, in which DO saturation levels above 75% at the sediment–water interface were measured in only two oxbows (numbers 2 and 13, Fig. 5.7*B*). With the exception of number 32, both plant bed types in the remaining oxbows were predicted to have sufficient suspended FPOM to support filtering collectors, in this case largely bivalve mollusks which would be filtering primarily near the sediment–water interface (Fig. 5.6*C*). The results of the habitat stability analysis do not agree between the FFG and habit ratios because both calculations are dominated by bivalve mollusks (*Corbicula* and *Sphaerium*), which are burrowing obligate filtering collectors, which places them in the numerator (filtering collectors) in the FFG calculation but in the denominator (burrowers) in the habit calculations (Table 5.3). If the bivalves are treated as a special case, designating stable substrates, and combined with clingers and climbers (Table 5.3), the majority of habitats would be classified as having abundant stable substrates (Fig. 5.6*D,E*). Predator abundance ratio, by biomass, was above the expected level (<0.15) for both bed types in most oxbows (13 of 17 cases, Fig. 5.6*F*). This correlates with the high abundance of short-life-cycle rapid-turnover taxa that dominated both bed types in most oxbows (14 of 17 cases, Fig. 5.6*G*) which would be capable of supporting the high predator biomass. The majority of beds in the 10 oxbows were predicted to be inadequate in supplying a reliable food base for water column–feeding fish (10 of 17 cases, Fig. 5.6*H*). Although half of the *Nuphar* beds (oxbows 11, 13, 15, 26, and 32) were evaluated as supporting a reliable food supply for water column–feeding fish species, water quality in oxbow 32 may not have been sufficient to support the fish (Fig. 5.7*B*). All beds sampled in all 10 oxbows were predicted to provide poor food supplies for visual-feeding benthic fish and wading birds (Fig. 5.6*I*). This corresponds with observations made during invertebrate sampling that few wading birds and benthic invertivorous fish were present.

An overall evaluation of the oxbows was made by ranking them from 1 to 10 for each of 12 categories: DO at the bottom, 30-second sampling of density and diversity, plus the nine functional categories given in Table 5.3. The two plant bed types (*Nuphar* and *Hydrocotyle*) were ranked separately because the latter occurred in only 7 of the 10 oxbows (the first two columns Table 5.4). Combined rankings were calculated only for the seven oxbows in which both plant bed types occurred. In addition, the rankings for DO levels at the sediment–water interface were combined with the *Nuphar* bed rankings, both of which were measured in all 10 oxbows. The results of these rankings are given in Table 5.4. Either oxbow 2 or oxbow 26 received the highest ranking in all five categories. Oxbows 13 and 11 received the lowest plant bed rankings, individually or combined (Table 5.4). For the total data set, oxbow 2 received the highest number of

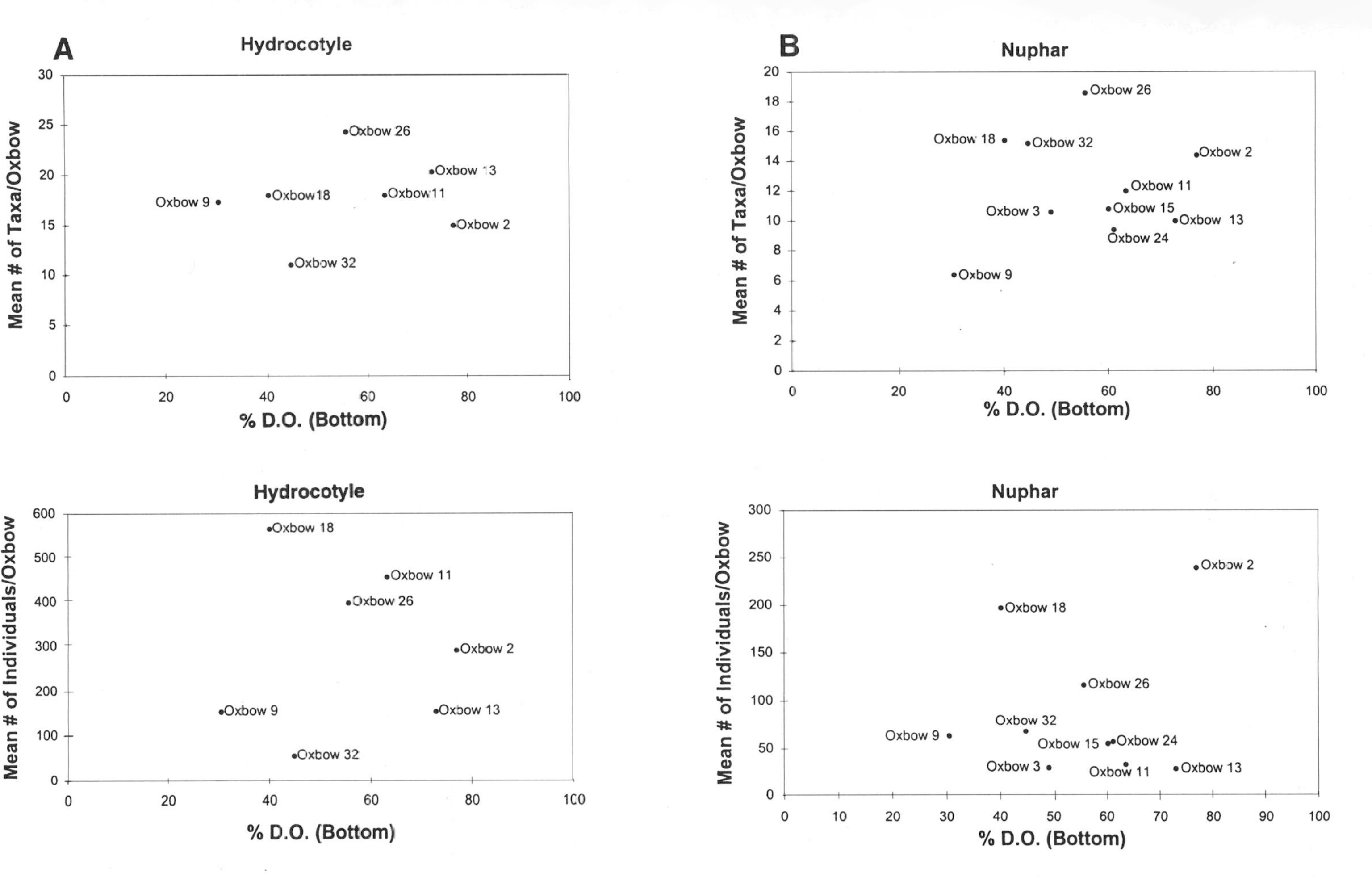

FIGURE 5.7. Relationship between dissolved oxygen (DO) at the sediment–water interface with diversity and density of invertebrates in (*A*) *Hydrocotyle* beds in seven oxbows and (*B*) *Nuphar* beds in 10 oxbows in the Caloosahatchee River.

TABLE 5.4. Summary Rankings for 10 Study Oxbows[a]

Rank[b]	*Nuphar* Beds	*Hydrocotyle* Beds[c]	Combined Estimate[d]	Dissolved Oxygen At Bottom	*Nuphar* Beds + Dissolved Oxygen At Bottom
1	26	2	26	2	2
2	24	9	2	13	24
3	32	26	32	11	26
4	2	18	9	24	15
5	15	32	18	15	32
6	18	11	11	26	11/13
7	9	13	13	3	—
8	3			32[d]	3/18
9	11			18[d]	–
10	13			9[d]	–

[a]The table entries are the identification numbers for the oxbows (see Fig. 5.6). Values are based on oxygen levels at the bottom at the time of sampling, density, diversity, and nine functional ratios serving as surrogates for ecosystem attributes in two plant bed types (*Nuphar* and *Hydrocotyle*).

[b]1, Best condition; 10, worst condition.

[c]Excludes oxbows 3, 15, and 24 which had no *Hydrocotyle beds*

[d]Below 50% saturation

first-place rankings (6/24) and oxbows 11 and 32 the highest number of last-place rankings (2/24). The two bed types were different, as is clear in Figure 5.6. For example, oxbow 26 was ranked in the top three for both *Nuphar* and *Hydrocotyle* beds. However, oxbows 24 and 32 were in the top three as ranked for *Nuphar* beds, while oxbows 2 and 9 were in the top three for *Hydrocotyle* beds. The evaluation presented here would lead to the conclusion that efforts to rehabilitate oxbows 2, 26, and probably 24 would be most likely to succeed because they are presently in reasonable condition, whereas rehabilitation of oxbows 11 and 13 is probably most needed because of their present poor condition. Of course, because they are presently in poor condition with regard to ecological integrity as measured in this study, their rehabilitation would be most difficult, require the most significant alterations, and might have the lowest probability of desired success.

CONCLUSIONS

As more riparian marsh habitat is lost or severely modified, the need increases for techniques for rapid assessment of their ecological condition. This is true for evaluation of both deterioration or loss as well as for evaluation of rehabilitation progress resulting from restoration efforts. Our case studies presented here, together with investigations reported in the literature, indicate that the aquatic macroinvertebrates can be used in such evaluations. The emphasis here has been on functional characteristics of the macroinvertebrate populations, which can serve as surrogates for riparian marsh ecosystem attributes. In the case studies presented, the surrogate ratios have been used to predict the outcome of rehabilitation of Kissimmee River floodplain, or riparian marsh, depending on which vascular hydrophyte community dominates following the restoration work and to identify which Caloosahatchee River oxbows would be most likely to respond to rehabilitation efforts. The methods appear to provide valuable insight into ecosystem function and, with further validation, could be applied to a wide range of river margin habitat types. The basic tenet is that the way the invertebrate biomass is organized in riparian marsh communities does reflect fundamental features of these riverine wetland environments. The method is based on the tight linkages that the invertebrates exhibit toward food resource types, features of the habitat, and environmental cues (e.g., diurnal light pattern). Future research needs include increasing the number of test sites where the method is employed, field enclosure experiments to establish obligate versus facultative relationships for key invertebrate taxa, and the simultaneous measurement of macroinvertebrate functional ratios and the ecosystem parameters for which the ratio are to serve as surrogates (e.g., 24-hour oxygen profiles to estimate P/R directly).

ACKNOWLEDGMENTS

We would like to acknowledge the following people for field work on various aspects of the Caloosahatchee River research: M. Berg, J. Novak, M. Higgins, K. Wessell, and J. Lessard. Thanks go to B. Graham for assistance with graphics.

REFERENCES

Anderson, N. H., and K. W. Cummins. 1979. The influence of diet on the life histories of aquatic insects. Journal of the Fisheries Research Board of Canada 36:335–342.

Camp, Dresser and McKee, Inc. 1995. Caloosahatchee River basin assessment water quality data analysis report (phase II). South Florida Water Management District, West Palm Beach, FL.

Capece, J. C., and R. A. Scholle [Wessell]. 1996. Locations of the Caloosahatchee oxbows. University of Florida Institute of Food and Agricultural Science, Southwest Florida Research and Education Center, Immokalee, FL and the Caloosahatchee River Citizens Association, Ft. Myers, FL (*http://www.imok.ufl.edu*).

Cummins, K. W. 1973. Trophic relations of aquatic insects. Annual Review of Entomology 18: 183–206.

Cummins, K. W. 1974. Structure and function of stream ecosystems. Bioscience 24:631–641.

Cummins, K. W. 1984. Invertebrate food resource relationships. Bulletin of the North American Benthological Society 1:44-45.

Cummins, K. W. 1988. The study of stream ecosystems: a functional view. Pages 247-262 *in* L. R. Pomeroy and J. J. Alberts. (eds.), Concepts of ecosystem ecology: a comparative view. Springer-Verlag, New York.

Cummins, K. W., 1993. Invertebrates. Pages 234-250 *in* P. Calow and G. E. Petts. (eds.), Rivers handbook. Blackwell, Oxford.

Cummins, K. W., and M. J. Klug. 1979. Feeding ecology of stream invertebrates. Annual Review of Ecology and Systematics 10:147–172.

Cummins, K. W., and M. A. Wilzbach. 1985. Field procedures for the analysis of functional feeding groups of stream invertebrates. University of Pittsburgh, Pymatuning Laboratory of Ecology, Linesville, PA.

Cummins, K. W., M. A. Wilzbach, R. W. Merritt, and J. Brock. 1999. Evaluation of the restoration of the Kissimmee River in South Florida. Pages 98–118 *in* D. S. White and S. W. Hamilton (eds.), The natural history of the lower Tennessee and Cumberland River valleys, 8th Symposium. Austin Peay University Press, Clarksville, TN.

Dahm, C. N. (ed.). 1995. Restoration Ecology 3:145–238. (Entire issue devoted to the Kissimmee River restoration).

Dodds, W. K., and J. Brock. 1998. A portable flow chamber for in situ determination of benthic metabolism. Freshwater Biology 39:49–59.

Koebel, J. W., Jr. 1995. An historical perspective on the Kissimmee River restoration project. Restoration Ecology 3:149–159.

Mason, W. T., and P. P. Yevich. 1967. The use of phloxine B and rose bengal stains to facilitate sorting benthic samples. Transactions of the American Microscopical Society 89:221–223.

Merritt, R. W., K. W. Cummins, and T. M. Burton. 1984. The role of aquatic insects in the processing and cycling of nutrients. Pages 144–163 *in* V. H. Resh and D. M. Rosenberg (eds.), The ecology of aquatic insects. Praeger, New York.

Merritt, R. W., and K. W. Cummins (eds.). 1996a. An introduction to the aquatic insects of North America, 3rd. Kendall/Hunt, Dubuque, IA.

Merritt, R. W., and K. W. Cummins. 1996b. Trophic relations of macroinvertebrates. Pages 453–474 *in* F. R. Hauer and G. A. Lamberti, (eds.), Methods in stream ecology. Academic Press, San Diego, CA.

Merritt, R. W., J. R. Wallace, M. J. Higgins, M. K. Alexander, M. B. Berg, W. T. Morgan, K. W. Cummins, and B. VandenEeden. 1996. Procedures for the functional analysis of invertebrate communities of the Kissimmee River–floodplain ecosystem. Florida Scientist 59:216–274.

Merritt, R. W., M. J. Higgins, K. W. Cummins, and B. VandenEeden. 1999. The Kissimmee River–Riparian marsh ecosystem, Florida: seasonal differences in invertebrate functional feeding group relationships. Pages 55–79 *in* D. P. Batzer, R. B. Rader, and S. A. Wissinger (eds.), Invertebrates in freshwater wetlands of North America: ecology and management. Wiley, New York.

Miller, T. H., A. C. Federico, and J. F. Milleson. 1982. A survey of water quality characteristics and chlorophyll *a* concentrations in the Caloosahatchee River system, Florida. Technical Publication 82-4. South Florida Water Management District, West Palm Beach, FL.

Milleson, J. F. 1979. Caloosahatchee River oxbows: environmental inventory. Report 8758. South Florida Water Management District, West Palm Beach, FL.

Pennak, R. W. 1989. Fresh-water invertebrates of the United States: Protozoa to Mollusca. Wiley, New York.

Smock, L. A. 1980. Relationships between body size and biomass of aquatic insects. Freshwater Biology 10:375–383.

Toth, L. A. 1993. The ecological basis of the Kissimmee River restoration plan. Florida Scientist 56:25–51.

Toth, L. A. 1996. Restoring the hydrogeomorphology of the channelized Kissimmee River. Pages 369–383 *in* A. Brookes and F. D. Shields, Jr. (eds.), River channel restoration: guiding principles for sustainable projects. Wiley, New York.

Toth, L. A., D. A. Arrington, M. A. Brady, and D. A. Muszick. 1995. Conceptual evaluation of factors potentially affecting restoration of habitat structure within the channelized Kissimmee River ecosystem. Restoration Ecology 3:160–180.

Vannote, R. L., G. W. Minshall, K. W. Cummins, J. R. Sedell, and C. E. Cushing. 1980. The river continuum concept. Canadian Journal of Fisheries and Aquatic Sciences 37:130–137.

Waters, T. F. 1969. Subsampler for dividing large samples of stream invertebrate drift. Limnology and Oceanography 14:813–815.

Wilzbach, M. A., K. W. Cummins, and R. A. Knapp. 1988. Toward a functional classification of stream invertebrate drift. Verhandlungen der Internationalen Vereinigung fuer Limnologie 23:1244–1254.

6 Using Algae to Assess Wetlands with Multivariate Statistics, Multimetric Indices, and an Ecological Risk Assessment Framework

R. JAN STEVENSON

Algae, important drivers of many wetland functions, have been shown to be valuable in wetland assessment. Algae are well suited for use in wetland assessment because of the high diversity of algae, known sensitivity of individual species to different environmental factors, and well-tested protocols that have been used in wetlands, but particularly in other aquatic habitats. In this chapter, multivariate statistics are recommended for classifying wetlands and determining the strength of relationships between changes in species composition and environmental factors. An ecological risk assessment framework is described that clearly distinguishes metrics for determining whether a problem exists in wetland function (e.g., supporting biotic integrity) and metrics that diagnose stressors that threaten or impair wetland function. These metrics and resulting multimetric indices can also be used to describe wetland quality for purposes of regulating wetland use. The risk assessment framework also emphasizes developing stressor–response relations, with indicators to better establish criteria as targets for wetland protection or wetland restoration. Algal protocols are provided for developing an algal monitoring and assessment program that utilizes an ecological risk assessment framework.

Algae are important elements in wetlands and occur in many types of wetlands, both freshwater and estuarine (see reviews by Goldsborough and Robinson 1996, Sullivan 1999; Chapter 13, this volume). They can be important sources of primary production

Bioassessment and Management of North American Freshwater Wetlands, edited by Russell B. Rader, Darold P. Batzer, and Scott A. Wissinger.
0-471-35234-9

and dissolved oxygen, transform and retain nutrients, stabilize substrata, and serve as habitat for other organisms (e.g., Sullivan and Moncreiff 1988, Sundbäck and Granéli 1988, Browder et al. 1994, MacIntyre et al. 1996, Miller et al. 1996, Wetzel 1996, McCormick et al. 1997, 1998). Algae are an important base of wetland food webs (Sullivan and Moncreiff 1990, Murkin et al. 1992, Browder et al. 1994, Campeau et al. 1994, Mihuc and Toetz 1994).

Algae can be extremely helpful in assessing ecological conditions in aquatic ecosystems (McCormick and Cairns 1994), particularly wetlands (Adamus 1996, McCormick and Stevenson 1998). Algae are sensitive to many physical and chemical factors that vary naturally among wetlands (Stevenson et al. 1999) and also to the stressors that human activities produce in wetlands (Pan and Stevenson 1996). A high diversity of species are found in most wetlands, so that a great amount of information about ecological conditions can be inferred based on presence and relative abundance of algal species and their autecological characteristics (species-specific sensitivity to different environmental conditions; Stevenson et al. 1999). Algal taxonomy is relatively well established for most common algal genera and diatom species. The ecological preferences of many algal taxa are also known. Lists of references for algal taxonomy and autecology can be found in Stevenson and Bahls (1999). Algae are particularly valuable indicators in wetlands compared to other groups of organisms because they are present in most (if not all) wetlands, diatom remains may persist even when the wetland is dry, and they have high species richness. Therefore, using algae in bioassessment enables use of sensitive and informative metrics from the same organisms in all types of wetlands, even at times of the year when assessment is important for regulation but no water remains.

In this chapter I provide guidelines for the development and implementation of an algal biomonitoring program for wetlands. First, I present a general framework for bioassessment. Then I discuss how to gather the information to develop an algal biomonitoring program, such as classifying wetland types, characterizing relations between important ecological attributes and environmental stressors, and developing indicators of environmental stressors. Finally, I introduce some fundamental elements of implementing an algal biomonitoring program. Unfortunately, it is beyond the scope of this chapter to review the ecology of algae in wetlands, even though a fundamental understanding is important for developing a monitoring program. General reviews of algal ecology in wetlands can be found in Vymazal (1994) and Goldsborough and Robinson (1996). Throughout this presentation, the complementary use of multivariate statistics, metrics, and multimetric indices is emphasized.

RISK ASSESSMENT FRAMEWORK

Generally, the goal of almost all bioassessment programs is to provide characterizations of environmental conditions with which management decisions can be made. The risk assessment–risk management framework of the U.S. Environmental Protection Agency (1996a) provides a valuable first step in how to standardize bioassessment programs and predicting the kinds of information that will be important for solving environmental problems. Although recent guidelines for ecological risk assessment emphasize two fundamental pieces of information (USEPA 1992), field-oriented bioassessments require three pieces of information (USEPA 1996a, Stevenson 1998) which requires some revision of definitions. First we need to know if there is a problem, such as impairment of designated use (e.g., fishable, swimmable, potable, support of warm-water life), signs of eutrophication or unusually high impact from human activities (hydrologic alteration or siltation), or impairment of biotic or ecological integrity (Cairns 1977, 1995; Karr and Dudley 1981; Angermeier and Karr 1994). In risk assessment, this could be referred to as *hazard assessment* (Fig. 6.1). We also need to assess exposure of the ecosystem to stressors (e.g., concentrations of pollutants and duration of high concentrations) and the relationships between stressors and targeted ecological attributes of designated use or biotic integrity.

In this chapter a *stressor* is defined as any physical, chemical, or biological factor that results from human activities and that affects targeted ecological conditions (Fig. 6.1). Management options (e.g., best management practices for pollution prevention and waste treatment) regulate human activities, which should in turn regulate stressors. The goal of management is to limit specific stressors to protect or restore a designated use or biotic integrity. Since one or two stressors may be the primary causes of impairment, human activities associated with those stressors may be the only activities that need regulation.

Characterization of stressor conditions is referred to as *exposure assessment* in the risk assessment framework (Fig. 6.1). Algal indicators of stressor levels may be valuable complements to direct measurement of physical, chemical, and biological stressors

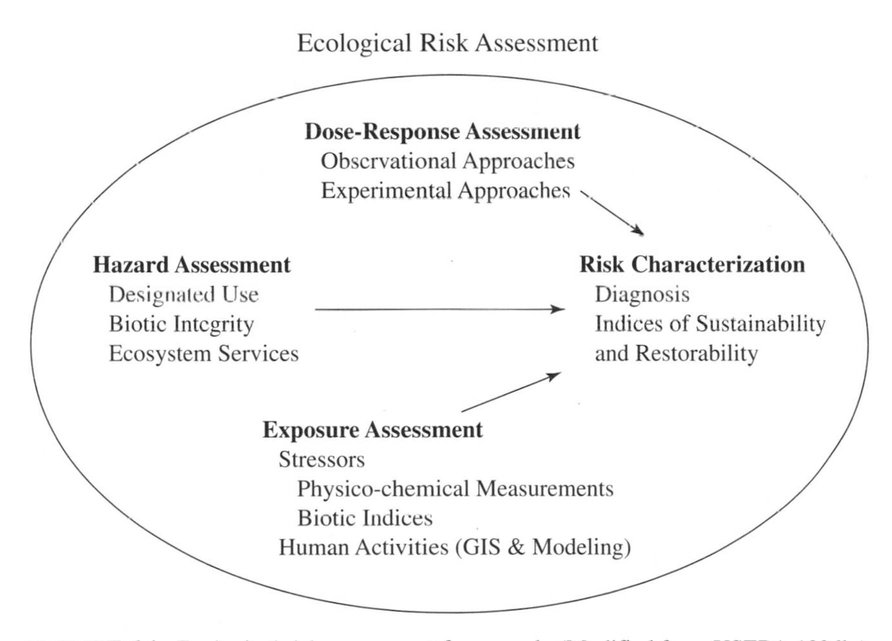

FIGURE 6.1. Ecological risk assessment framework. (Modified from USEPA 1996b.)

in a habitat. Levels of many stressors vary greatly from time to time. In addition, the frequency and duration of high stressor levels may be important in exposure assessment. Therefore, inferences of stressor levels based on biological attributes, which are regulated by historical as well as present environmental conditions, may be more precise and reliable indicators of ecological conditions than one-time assessments of physical and chemical characteristics. Thus, algal indicators of ecological conditions can complement physical and chemical measurements and confirm the intensity of different stressors in a wetland.

The relationships between designated use or elements of biotic integrity and stressor levels can be determined from the literature or can be generated as part of the development of a bioassessment program if adequate information is not already known. These relationships, referred to as *stressor–response relationships* in the risk assessment framework (USEPA 1998), may be developed using observed changes in algal assemblages among wetlands with different environmental conditions (Fig. 6.2), but experimental manipulation of stressors may be important for confirming cause–effect relationships (Fig. 6.1).

An environmental assessment can be completed with the hazard assessment, exposure assessment, and stressor–response relationships (Fig. 6.1). If designated use is impaired, then the stressor or stressors that are most likely responsible can be inferred from the exposure assessments and the stressor–response relationships. Management strategies would then call for regulating the human activities that cause those stressors. If designated use is not impaired, it may be very important for management to protect further degradation. In this case, identifying the stressors that most threaten designated use and regulating causes of those stressors becomes a priority.

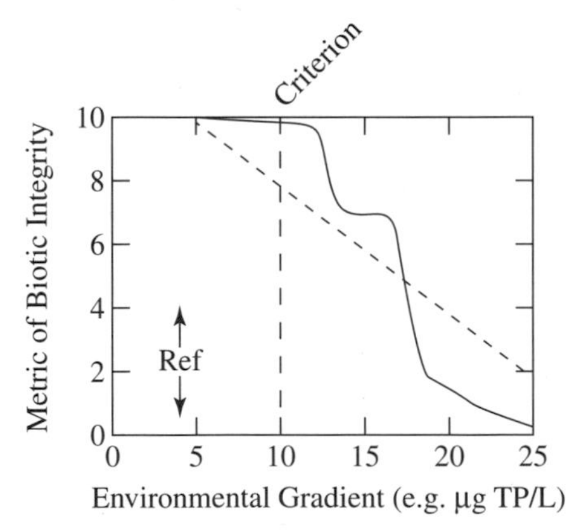

FIGURE 6.2. Potential stressor–response relationship showing 100% biotic integrity at a low level of environmental stress in reference condition and decreasing biotic integrity with increasing environmental stress for two measures of biotic integrity, one responding linearly and one nonlinearly. A criterion for stressor level was set below the threshold for change in the measure of biotic integrity that changed nonlinearly along the stressor gradient.

DEVELOPING RISK ASSESSMENT METHODS

The steps for developing a biomonitoring program and developing environmental criteria are similar in many habitats. First, goals for management must be identified so that they can be used to establish designated use and to guide the selection of attributes for study. Wetlands within ecoregions should be classified, based on their potential for supporting different designated uses and the potential need for assessing different attributes of the wetland ecosystem. Then, based on designated uses for different wetland classes and probable effects of human disturbance, a set of biological, physical, and chemical attributes should be chosen for study. Data should be gathered (either from databases, literature, or new studies) and analyzed:

- To test wetland classifications
- To develop metrics to assess ecological conditions
- To develop stressor–response relationships between valued ecological attributes and environmental stressors
- To develop biological indicators of ecosystem stressors
- To develop biological and physicochemical criteria for protecting and restoring each class of wetlands with their respective designated uses

Goals

Many goals have been described for wetland management and assessment, ranging from increasing wetland area to restoring and maintaining the physical, chemical, and biotic integrity that is called for in the Clean Water Act. An algal bioassessment program could be designed to assess the important attributes of wetlands that algae provide directly (such as primary production, nutrient uptake and transformation, habitat for other microorganisms, wetland aesthetics, and biotic integrity of algal assemblages). An algal bioassessment could also indirectly indicate the status of other wetland attributes that are difficult to measure, such as eutrophication and potential for low DO, general water quality characteristics, or biotic integrity of other organisms. Defining goals for bioassessment has been an iterative process between discovering the values that the public and managers have for wetlands and evaluating which attributes of algal assemblages could be used to characterize those values (Stevenson 1998).

Wetland Classification

Classification or categorization of wetlands by type should improve the sensitivity and precision of wetland assessments. In addition, different goals and designated uses may be established for different classes of wetlands. We would expect to find different algal assemblages in wetlands with different hydrology, geology, geomorphology, and climate (Apfelbeck, this volume). We would also expect to find different responses of algae to human disturbance in different wetland classes. Therefore, metrics should be developed and tested while controlling for wetland class.

Many wetland classification frameworks have been developed (e.g., Cowardin et al. 1979, Brinson 1993, Finlayson and van der Valk 1995). However, often many classes of wetlands will have similar diatom assemblages. Therefore, reclassification of wetlands by the types of diatom assemblages in them and the similarity in their response to stressors can simplify the development and use of algal bioassessment methods (see Table 2.1).

Choosing Variables

Many attributes of algal assemblages can be measured to evaluate whether or not goals of wetland management have been met (see Stevenson 1996 for a recent review). Structural attributes of algal assemblages, such as biomass and species composition, are usually recommended (McCormick and Cairns 1994, Kelly et al. 1998) because they are less variable and require less field time to measure than functional attributes such as production and nutrient dynamics. Structural attributes can be used to infer algal function in wetlands because algal metabolism is highly related to algal biomass and growth form (e.g., Hill and Boston 1991, Steinman et al. 1992). Structural attributes also require less field time to sample than do in situ measurement of algal metabolism. However, functional attributes, such as primary production, respiration, and nutrient uptake may be important in situations where these are important management goals or in developing structural indicators of these functional algal attributes.

Structural Attributes. Structural attributes of algal assemblages can be classified in three major categories (Stevenson 1996): biomass, taxonomic composition, and chemical composition. Algal biomass can be estimated with such measures as chlorophyll *a* (chl *a*), cell numbers and biovolume, and ash-free dry mass (AFDM) in an area or volume of habitat for benthic or planktonic algae, respectively. Each of these measures of algal biomass has pros and cons. Chl *a* per unit cytoplasm (biomass) varies with light and nutrient concentrations, but is usually a good estimate of algal biomass. Estimates of biomass with cell numbers and biovolume can be biased because small cells have less mass than large cells, but large cells also have proportionally large vacuoles than small cells (Sicko-Goad et al. 1977). Biovolume can be corrected for vacuole area by subtracting vacuole volume from the total cell volume. Because most algal cytoplasm is arranged near the cell wall, algal surface area should be a good indicator of cytoplasm volume in a thin layer near the cell wall. Even though problems exist with these measures of algal biomass, cell numbers and biovolume should be calculated whenever taxonomic composition is assessed by identifying and counting cells. After initial investment in developing a database of taxa and biovolumes, calculating biovolumes simply becomes another step in the data analysis. Ash-free dry mass, although easy to measure, usually overestimates algal biomass because bacteria, fungi, and fine particles of detritus can be collected in algal samples. Chl *a*/AFDM ratios provide indicators of the relative importance of algae versus bacteria and detritus in samples. Because of difficulty in quantitatively sampling sediments, Zheng and Stevenson (unpublished data) found that dividing algal biomass assessments (e.g., chl *a*) by dry mass (an estimate of sediment quantity) provided better relation to environmental factors in salt marshes than did area-based estimates of algal biomass.

Visual semiquantitative assessments of algal biomass have proven to be useful in streams (e.g., Holmes and Whitton 1981, Stevenson and Bahls 1999), and similar methods may provide valuable estimates of algal biomass in the wetlands. For example, percent cover within a habitat and thickness of benthic algae or metaphyton could be ranked and used to quantify relative algal biomass. Secchi disk depth or other measures of water transparency (Wetzel and Likens 1991) could be used to estimate phytoplankton biomass.

Taxonomic attributes of algal assemblages also provide sensitive and precise assessments of environmental conditions in wetlands (McCormick and O'Dell 1996, Pan and Stevenson 1996, Stevenson et al. 1999, Pan et al. 2000). Useful taxonomic attributes can be obtained simply by identification and size classification of algae at class and order levels to determine the biovolume and relative biovolume of different growth forms and classes of algae (e.g., McCormick and Odell 1996, Pan et al. 2000). All major taxonomic groups of algae provide valuable characterizations of biotic integrity and water quality of wetlands (VanderBorgh 1999), but assessments of algal assemblages are often simplified to identifying and counting diatoms. Diatom taxonomy and species autecologies have been widely distributed in the literature. Valuable assessments of biotic integrity and water quality can be obtained with identification of genera (VanderBorgh 1999), but much greater sensitivity and greater certainty of a complete assessment can be obtained when species are identified and counted. Simpler taxonomic analyses may be warranted during early stages of a program when technical expertise in a lab is being developed or at later stages when reliable indicators have been established. As a standard, I recommend a two-stage counting process to observe changes in the entire algal assemblage and get detailed assessment of highly sensitive species responses with the diatoms, which have the most easily and certainly identifiable species. First, identify and count genera of all algae in wet mounts on microscope slides or in Palmer cells. Then identify and count diatom species for a thorough analysis of taxonomic composition of algal assemblages. To reduce cost of assay, focus analysis on diatom assemblages. Many projects in other aquatic habitats have shown that species-level identification of diatom assemblages is sufficient for most goals of bioassessment (Kelly et al. 1998, Stevenson and Bahls 1999, Stoermer and Smol 1999, Pan et al. 2000).

Many metrics of the biotic integrity and water quality of wetlands have been calculated with the relative abundances of algal taxa in samples (see reviews in McCormick and Cairns 1994, Stevenson and Pan 1999, Stevenson and Smol in press). Differences in similarity of species composition in reference and test wetlands can indicate deviations in biotic integrity of wetlands (McCormick and O'Dell 1996, Pan et al. 2000). Indices of water quality (e.g., pH, conductivity, total phosphorus and total nitrogen concentration) can be calculated based on relative abundances of algal species and species optima and tolerance for these conditions (Pan and Stevenson 1996, Stevenson et al. 1999). Completing assessments of biotic integrity and water quality enable a determination of impairment of wetlands and a diagnosis of probable causes of impairment. The methods for developing these metrics will be described later in the section on data analysis.

The diversity of algal assemblages, calculated after identification and counting of algal cells, may be valuable metrics of biotic integrity, water quality, and the human disturbance of wetlands. Both richness and evenness of algal assemblages or integrated diversity indices (e.g., Shannon diversity index, Shannon 1948) may indicate changes in wetlands. However, algal diversity can increase and decrease in response to different environmental stressors and along different parts of some gradients of human disturbance, so interpretation of changes in diversity can be ambiguous (Stevenson 1984, Stevenson and Lowe 1986).

Chemical composition of algal assemblages may help determine the nutrient status of wetlands or the level of contamination by toxic substances. The N and P concentrations in algal samples have been used to infer level of trophic status and nutrient limitation in wetlands (Vymazal et al. 1994, Vymazal and Richardson 1995, McCormick and O'Dell 1996, Pan et al. 2000). Deviations in molar N/P ratios from 16 indicate N or P limitation. Assuming that AFDM is a proper estimate of microbial biomass in a sample, N/AFDM and P/AFDM may be good indicators of N or P limitation and saturation in a habitat. Heavy metal concentrations in samples may also provide evidence for toxicity. Chemical contaminant analyses of algal and invertebrate samples may provide insight into stressor levels and food web dynamics. These attributes of algal assemblages could be valuable metrics for diagnosing causes of impairment.

Morpological attributes may also be valuable metrics from algal assessments. Percent of diatoms with abnormal valves has been related to heavy metal concentrations (McFarland et al. 1997). Percent live diatoms, those with cytoplasm in them, may be negatively related to some types of human disturbance (Hill et al. 2000). Chloroplast size may be related to nutrient limitation. The percentage of motile diatoms in samples has been investigated as an indicator of silt loading in streams (Bahls 1993, Hill et al. 2000).

Functional Attributes. Functional attributes are processes occurring in wetland ecosystems. Primary production, respiration, nutrient uptake and cycling, nitrogen fixation, algal decomposition, and alkaline phosphatase activity are examples of microbial (algal, bacterial, and fungal) attributes that could be used in wetlands to assess effects of stressors and human activities on wetland function. In wetlands particularly, distinction between algal, bacterial, and fungal metabolism is difficult, so most functional assessments broadly characterize microbial activities. Even though assay of functional attributes may be more time consuming and variable (both temporally and spatially) than rapid sampling and laboratory assessment of algal assemblage structure, microbial function in wetlands is a valuable ecosystem service that should be better understood and evaluated (McCormick et al. 1998). The possibility that microbial functions could be inferred based on quantitative assessment of algal biomass, (Bannister 1974, Hill and Boston 1991), should be investigated to better understand algal ecology in wetlands and to help broaden the utility of algal bioassessments.

Selecting Habitats to Sample. There can be many distinct algal assemblages in different habitats in wetlands, and each may respond somewhat differently to environmental stress. Algal assemblages suspended in the water, on plants, and floating entangled with macrophytes (metaphyton) can respond to factors related to human disturbance in wetlands (McCormick and O'Dell 1996, Pan and Stevenson 1996, VanderBorgh 1999). Diatoms accumulating in surface sediments may provide a temporally integrated characterization of biotic integrity and water quality (Charles 1985, Stevenson et al. 1999). Diatoms buried deeper in sediments have commonly been used to infer historical conditions and recent changes in lakes and wetlands by using paleoecological methods (Fritz et al. 1991; Smol 1992; Cooper 1995, 1999; Dixit et al. 1999; Slate and Stevenson 2000).

At this time, the pros and cons of using one assemblage versus another have not been rigorously evaluated but can be postulated. Water column samples are easy to collect if water occurs in the habitat. When collecting water, exercise caution to prevent contamination of plankton samples by benthos that can easily become resuspended or dislodged during sampling. Epiphytic algae may be most sensitive to nutrient gradients, because nutrients taken up by plants from sediments are released and readily absorbed by epiphytic biofilms (Moeller et al. 1988). However, epiphytic algae may vary seasonally, with plant type, and may not occur in drier wetlands. Edaphic (soil) algae or algae on sediments occur in all wetlands that we have observed, and may therefore be the best standard sample; however, they may not be as sensitive to environmental gradients because algae settle into the sediments from many other habitats. Composite multihabitat samples may provide a more thorough assessment of all possible impairments of the biotic integrity of algae in a wetland, but differential responses to stressors in different habitats may again reduce the sensitivity and precision of metrics of human disturbance and specific stressors. In a survey of algal assemblages, environmental conditions, and human disturbance of 20 Maine wetlands, we found a similar number of metrics for algae suspended in the water, on plants, and in surface sediments (R. J. Stevenson, P. R. Sweets, B. Wang, and J. L. DiFranco unpublished data).

A complete assessment of biotic integrity of algae in a wetland may call for sampling more than one habitat. Optimizing sensitivity to environmental stress may call for sampling plankton or epiphytes. However, since algae on sediments are found in all wetlands, algae on sediments would be the best assemblage to compare among many different types of wetlands. In a survey of algal assemblages, environmental conditions, and human disturbance of 20 Maine wetlands, we found similar number of metrics for diatoms in the water column, on plants, and in surface sediments, but fewer metrics for multihabitat samples (R. J. Stevenson, P. R. Sweets, Wang, and J. L. DiFranco unpublished data). Future studies should continue to evaluate the trade-offs of targeted versus multihabitat sampling and sampling of plankton or epiphytes or algae in sediments.

Gathering and Analyzing Data

Data acquisition and analysis should be targeted toward analyzing algal attributes that are clearly related to goals of the program. Data can be gathered from the literature or as a result of ongoing studies. The methods for sampling and laboratory analysis of algal samples have been described in Chapter 13 and elsewhere (Aloi 1990, Wetzel and Likens 1991). Therefore, in this chapter I concentrate on the sampling designs and data analysis that could be used to classify wetlands, identify metrics, establish stressor–response relationships between metrics and environmental factors, and develop biological and physical-chemical criteria for wetland protection. As more wetland bioassessments are done and results in different ecoregions and classes of wetlands are compared, we may find that many metrics and bioassessment methods can be used in many kinds of wetlands. Any time that metrics or stressor–response relations are assumed from literature findings, their application should be reassessed in regional projects.

Classifying Wetlands and Defining Reference Conditions

Wetlands are classified for several reasons. One is to establish a proper reference to compare to other sites that will be assessed in the future. In addition, different types of wetlands may have different management goals and responses to environmental stresses. Two basic approaches can be used to classify wetlands. One is used when *reference wetlands* can be identified. Reference wetlands are those with the least amount of human impact. An extended discussion of the importance of using reference habitats in bioassessment (Jackson and Davis 1994) and solutions to problems in identifying reference conditions (Hughes 1995) is beyond the scope of this chapter. Because of the complexity of routes of water supply to wetlands and relating land use in a region to a prediction of level of human impact, other approaches, such as paleolimnology, can be valuable. If sediments have accumulated with relatively little disturbance in wetlands, then reference condition can be determined by using a paleoecological approach and inferring historical conditions in the wetlands before regional changes in land use occurred (Cooper 1995, Slate 1998, Slate and Stevenson 2000).

The second approach for defining reference wetlands is used when it is difficult to identify the level of human disturbance. The least impacted wetlands are defined as having conditions that would be expected if human disturbance was low. For example, wetlands with lower levels of human disturbance would be predicted to have lower nutrient concentrations, low turbidity, and more natural hydrology than wetlands impacted by fertilizer, sediments, and hydrologic draining.

In practice, both approaches should be used to select reference wetlands. After a preliminary selection of reference wetlands based on little human activity near the wetland and in its watershed, environmental stressors that are commonly associated with human activity should be assessed. If conditions common to human disturbance are identified in a wetland selected to be a reference site, careful consideration should be given to reclassifying its status.

When reference wetlands can be identified, algae in a random subset of wetlands should be selected and sampled. Reference wetlands with similar algal assemblages are then grouped by cluster analyses (e.g., UPGMA; Faith et al. 1987, Belbin and McDonald 1993) or Twinspan (Hill 1979). Discriminate analysis can be performed to determine the statistical significance of differences among groups of wetlands that were identified with Twinspan or cluster analysis to have the most similar species composition (e.g., Pan et al. 2000). A RIVPACS-like approach (see Wright 1995, Chessman et al. 1999; Chapter 4, this volume) could be used to develop a predictive model of different wetland classes based on physical, chemical, and biological features that are not affected by humans. The latter model would be used to classify wetlands during monitoring so that test wetlands could be compared to appropriate reference wetlands (see Reynoldson et al. 1997 for discussion).

When reference wetlands cannot be identified, wetlands with low levels of stressors that are typical of human disturbance can be assumed to be least affected. Thus, wetlands with low nutrients and inorganic turbidity or freshwater wetlands with low conductivity could be chosen as least impacted wetlands. Then analysis of covariance or regression analysis that include nonhuman environmental gradients could be used to identify groups of wetlands with similar algal responses to environmental gradients related to human stressors (such as nutrients, inorganic turbidity, and conductivity). Classification in this case would be based on groups of wetlands with similar algal responses to environmental stressors.

Ordinationlike multivariate statistical techniques are valuable for illustrating differences among habitats, for preliminary analyses of how thorough assessments were, and for identifying major environmental gradients in the regions studied (Lowe and Pan 1996, McCormick et al. 1996, Pan and Stevenson 1996, Pan et al. 2000). Simple ordinations using just species data or environmental data enable identification and easy illustration of the major sources of variability in the data. Comparing the amount of variation in species data that can be explained by ordination axes and the amount of that variation explained by environmental factors provides an indication of how well assessments included measurement of important environmental conditions. The amount of variability in species assemblages in a study can increase with the number and diversity of wetland types studied. If only small amounts of the variation in species composition can be related to the environmental factors measured in the study, some environmental factors that are important determinants of species composition may not have been measured or included in the analysis. If sufficient variability in species composition is related to environmental factors, the environmental factors that are highly correlated with changes in species composition can be identified (Stevenson et al. 1999).

Results of classification, as well as other goals of data analysis, may vary with different forms of data used in the analysis. We would expect that species composition of assemblages described based on presence and absence of species would be less variable temporally than relative abundances of taxa, which may vary weekly. Thus, longer-term ecological influences, such as wetland type, may be observed more precisely with presence–absence data than with more "noisy" relative abundance data. On the other hand, more short-term environmental stresses may be detected more readily with the more highly sensitive relative abundance data. An effort should be made in classification, as well as other data analyses, to choose the form of data that would be predicted to respond most precisely and sensitively to the environmental factors of interest, based on the spatial and temporal variability of the environmental factors.

SELECTING METRICS FROM ANALYSES OF ATTRIBUTES

Metrics are developed for two reasons in risk assessment: to detect ecological impairment and to diagnose the environmental causes of impairment (Stevenson 1998). In addition, relations between metrics and physical and chemical conditions in a habitat are used to establish criteria for protection and restoration of wetlands. To review, metrics are attributes of biological assemblages that respond to human disturbance and multimetric indices are quantitative summaries of many metrics in a single number (Karr and Chu 1999). In this section, two methods for developing metrics are discussed, one with categorical classification of impairment in sites and another when impairment is quantified with a continuous variable of human disturbance. Even though different analytical methods can be used to develop metrics, the sampling design is basically the same. Reference and impaired sites should be sampled. At least 10 sites should be sampled, but larger sample size enables more precise assessment of metrics and stressor–response relationships.

All attributes of algal assemblages are potential metrics (Stevenson and Smol in press), but the metrics that will probably work during different seasons and in many different kinds of wetlands will be those that could be predicted with ecological theory and knowledge of the natural history of algae. Algal attributes can be evaluated using box plots or regression analysis. When using box plots, if the interquartile ranges of an attribute in reference and test habitats do not overlap, that attribute is considered a good metric (Fig. 6.3) (Barbour et al. 1996, 1999). Another less rigorous test is if interquartile ranges of an attribute in either the reference or test site do not overlap the mean of that attribute in the other condition, an attribute is considered to be an acceptable metric (Fig. 6.3). Alternatively, metrics are attributes that are significantly related by linear regression to differences in level of human disturbance near wetlands (Fig. 6.2) or to an environmental characteristic that is caused by human disturbance (e.g., total phosphorous concentration, total suspended solids, toxic contamination).

Metrics that respond sensitively and precisely to gradients of human disturbance are the most valuable metrics for hazard assessment. Metrics should also be biologically relevant, socially relevant, cost-effective, integrative, or possess other characters deemed valuable in the assessment process (Cairns et al. 1993). In addition, the best metrics for hazard assessment respond linearly throughout the range of human disturbance so that changes in hazard assessment (designated use status, biotic integrity, ecosystem services, etc.) can be detected at low, medium, and high levels of human disturbance. Nonlinear but monotonic changes in metrics along gradients of human disturbance are more valuable for development of environmental criteria than

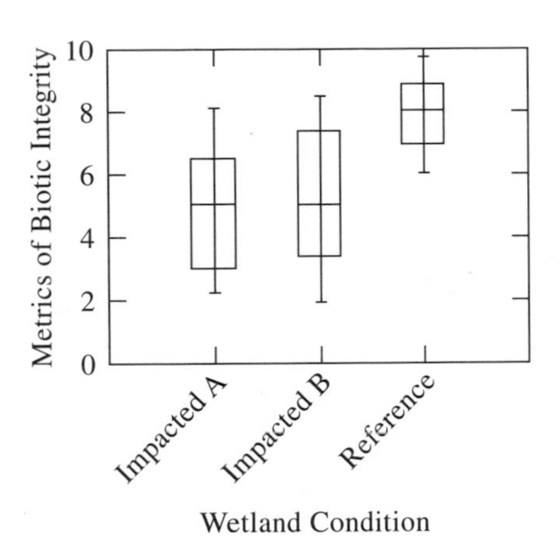

FIGURE 6.3. Example of variation in a metric of biotic integrity in reference wetlands as compared to groups of impacted wetlands. Box plots are used to show the mean, interquartile ranges, and range of the metric in the three classes of wetlands. Interquartile ranges of reference wetlands and impacted A wetlands do not overlap. Interquartile ranges of reference wetlands and impacted B wetlands overlap, but the mean of one wetland type is not included in the interquartile range of the other.

for hazard assessment (see Fig. 6.2). For example, in a project to develop metrics for wetland assessment in Maine (R. J. Stevenson, P. R. Sweets, B. Wang, and J. L. DiFranco unpublished data), the number of diatom taxa in the genus *Eunotia* was negatively affected by human disturbance in the most impacted wetlands only, whereas a weighted-average indicator of wetland trophic status increased sensitively throughout the gradient of human disturbance (Fig. 6.4). The decreased biodiversity of *Eunotia* may be a cause of concern and target of protection, whereas the change in trophic condition could be used as an early warning of habitat degradation before impairment was observed for algal diversity of native Maine wetlands.

Slight alterations in the calculation of metrics can make substantial differences in their performance (sensitivity and precision). Biomass can be log-transformed, and relative abundances can be transformed with an arcsine–square root function to increase normality and homogeneity of variances. In addition, estimates of algal biomass (chl *a* or mg C estimated from biovolume) can be divided by all organic biomass (AFDM) to detect effects of stressors on algae versus just physical disturbances that would affect all sampled biomass (e.g., algae, fungi, bacteria, and detritus). When testing attributes based on species composition of algal assemblages, some differences in sensitivity of stressor effects would be predicted if densities, relative abundances, log-transformed relative abundances, or presence–absence of taxa were used in analyses. Differences would also be predicted if metrics were based on numbers of species of a sensitive genus versus relative abundances of a genus. As an example, many *Pinnularia* species may be sensitive to anthropogenic disturbance, but if one species is adapted to disturbance [e.g., *Pinnularia borealis* may dominate in the moist soils (Round 1981) of a hydrologically impaired wetland],

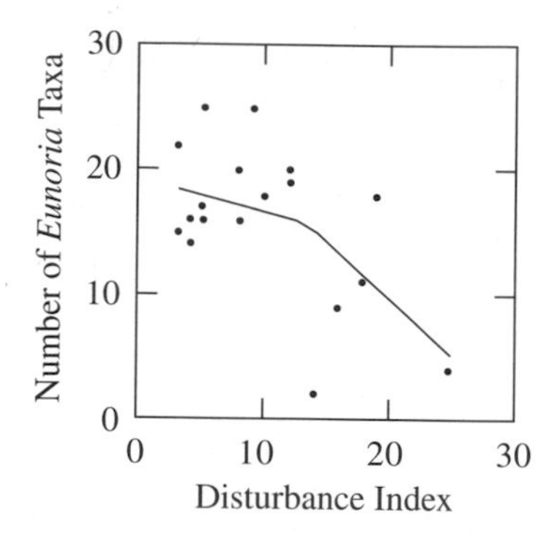

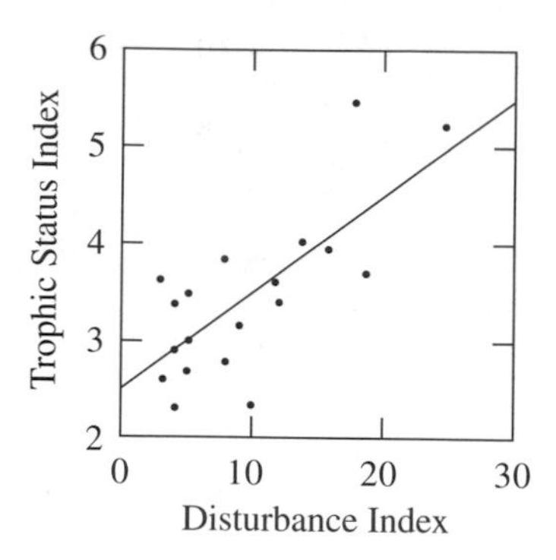

FIGURE 6.4. Change in two metrics for planktonic algae in Maine wetlands, number of *Eunotia* taxa and a weighted-average trophic status index, along a gradient of human disturbance. The trophic status index was calculated as the sum (for all taxa) of the products of relative abundances of taxa and a rank of the trophic status in which they are most often observed. The disturbance index was a multimetric index of separate metrics-based field assessments of vegetation removal, hydrologic alteration, percent impervious surface, and non-point-source pollution. (From R. J. Stevenson, P. R. Sweets, B. Wang, and J. L. DiFranco unpublished data.)

relative abundance of the *Pinnularia* genus may be high in a sample. Again, understanding the basic ecology of algae in wetlands will be important for developing reliable and transferable metrics for wetland assessment.

Few attributes of algal assemblages have been assessed in more than one habitat to determine if expectations are transferable. However, known responses of some metrics show that metrics respond differently in different classes of wetlands. For example, algal biomass commonly increases in response to nutrients and can have negative effects on other ecosystem attributes, such as eelgrass beds (Short and Wyllie-Echeverria 1996, Wear et al. 1999), but benthic algal biomass decreases with increasing P enrichment and human disturbance in the Everglades (McCormick and O'Dell 1996, McCormick et al. 1996, Pan et al. 2000).

With greater knowledge of algal ecology in wetlands, better predictions of habitat-specific and wetland class-specific responses should be possible. The problem of different algal responses to different stressors in different habitats within wetlands and among wetland classes can then be compensated for by relating metrics to expected values in reference conditions or target conditions in wetlands managed for specific purposes (waste treatment). Any deviation between expected and observed values then becomes an indication of impairment. The sign of the deviation, positive or negative, may not matter with respect to deviations in biotic integrity if reference (expected) condition has been defined well. However, the direction of deviation will surely matter with respect to many other designated uses (fishability, nutrient retention, etc.). Substantially more research is needed to determine which algal attributes are good metrics in wetlands and to determine which attributes are good metrics in more than one region.

Diagnostic Metrics: Biotic Indices of Stressors and Stressor Level

In shallow-water habitats such as streams and wetlands, physical and chemical factors vary greatly. Biotic indices have been argued to be more precise indicators of the physical and chemical conditions in such habitats than one-time sampling and assay of water quality. The organisms in a habitat have survived historic conditions. As a result, species composition reflects the complex responses of organisms to the past variability in resource availability and environmental constraints on acquiring those resources. This basic principle has been applied to development of algal indices of physical and chemical conditions in lakes, streams, and wetlands (e.g., Palmer 1962, Charles 1985, Pan and Stevenson 1996, Pan et al. 1996). If we know the environmental conditions in which species or taxa have their own maximum relative abundances, we should be able to infer environmental conditions (EC) in a test site with the relative abundances (p_i) of species that are in that habitat and their autecological characteristics (Θ_i) (ter Braak and van Dam 1989). If we know the autecologies of all S taxa in the habitat, we should be able to infer levels of a stressor as (Zelinka and Marvan 1961)

$$\text{EC} = \Sigma_{i=1,S}\, p_i \Theta_i$$

This use of weighted-average inference (WAI) models to infer water quality has been broadly applied with categorical classifications of environmental gradients, such as the saprobien, halobien, and pollution tolerance classifications of taxa (Palmer 1962, Lowe 1974, Whitton et al. 1991, Grimshaw et al. 1993, Raschke 1993, van Dam et al. 1994, Whitton and Rott 1996). WAI models have also been used with accurate quantitative assessments of species optima and tolerance to infer specific environmental conditions precisely. The latter approach has lead to the development of algal indices that assess pH, nutrient concentrations, salinity, heavy metal contamination, and biological characteristics of ecosystems (e.g., fish presence-absence) and that address major issues of global change: acidification and eutrophication of surface waters and climate change (Fritz 1990, Fritz et al. 1991, Agbeti 1992, Cumming et al. 1992, Dixit et al. 1992a,b, Hall and Smol 1992, Kingston et al. 1992, Sweets 1992, Cumming and Smol 1993, Reavie et al. 1995).

Computer programs are available to develop and test WAI models using the optima and tolerances of algae. Three steps are often involved. First, the environmental factors that explain most variability in species composition among samples are determined by some form of correspondence analysis. These factors are the best candidates for developing WAI models. Then WACALIB (Line et al. 1994) can be used to determine optima and tolerances of taxa and to construct inference equations based on weighted-average calibration and regression with species optima and tolerance for environmental conditions and species relative abundances. Finally, Calibrate (Juggins and ter Braak 1992) is used to test the precision of WAI models with a jackknifing procedure (i.e., repeatedly developing WAI models with one or more sites excluded in the development process and then assessing the precision of models by inferring conditions at the excluded sites). After these models are developed, tested, and used in monitoring programs, continued testing and redevelopment can become an ongoing process in a monitoring program as new data are collected.

WAI model approaches with precise quantitative autecological characterizations (species optima and tolerances) could be used in wetlands to infer levels of specific stressors. WAI models for pH, conductivity, and total phosphorous (TP) have been developed using diatoms in the phytoplankton and surface sediments with relatively high precision [root-mean-square error (RMSE), based on bootstrapped or jackknifed WAI models]: ±0.7 pH (Stevenson et al. 1999, VanderBorgh 1999), ±0.3 log(μS/cm conductivity) (Stevenson et al. 1999, VanderBorgh 1999), and ±0.1 to 0.4 log(μg TP/L) (Pan and Stevenson 1996, Slate 1998, Stevenson et al. 1999, VanderBorgh 1999).

To address the value of WAI models to assess stressor levels in wetlands, I compared variation in TP concentration in a region of the Everglades with a WAI model of TP based on the relative abundances and autecologies of diatoms. I used the standard deviation of water column TP as the measure of precision for one-time observation of water column TP concentration. I used the RMSE of the WAI model as the measure of precision for the inferred TP concentration based on one-time sampling and assay of diatom assemblages and inference of TP with a WAI model. The standard deviation of TP concentration measured repeatedly at a single site in the Everglades (data from the South Florida Water Management District, courtesy of Paul McCormick) was 0.20 log(μg TP/L). The RMSE associated with the jackknifed prediction of TP concentration with a WAI model using diatom relative abundances and autecologies was 0.06 log(μg TP/L) (Slate 1998). Clearly, the precision in estimating TP concentration in a habitat can be greater if WAI models are constructed with good estimates of diatom autecologies than with one-time sampling and assay of water chemistry.

Some evidence indicates that improvement in the precision of WAI models used to predict stressor levels can be related to more precise assessments of water chemistry characteristics in the habitats. There are two sources of error in the development of WAI models of stressors with algal autecological information: the precision associated with estimating water quality and the precision of our characterizations of algal autecologies and how that might vary among habitats. Slate (1998) found that RMSE decreased with increasing number of water samples taken over time prior to sampling algae, which should have provided a more accurate assessment of the environmental conditions in which the algal assemblage developed than one-time sampling of water quality at the time of sampling. In addition, Slate (1998) compared RMSE of WAI models for TP based on changes in algal assemblages along a P gradient in the Everglades, where many variables affect P concentration versus models based on algal changes in mesocosms, where P dosing is constant and thus, concentrations should be more constant. She found lower, more precise RMSE for WAI models developed from the controlled environments of mesocosms. Thus, better estimates of water quality increase precision (lower RMSE) of WAI models.

Multimetric Indices

Multimetric indices of biotic integrity are commonly used in stream bioassessment (Karr 1981; Kerans and Karr 1994; Barbour et al. 1996, 1999). Multimetric indices combine more than one metric (typically, 6 to 10) by summation into an index (Karr and Chu 1999). They are useful for summarizing and communicating complex ecological data. Many argue that multimetric indices bury ecological data and are less sensitive than individual metrics. Although both these problems may exist in reports where results of only multimetric indices are provided, individual metric values are always available in more complete reports and data sets where these values are preserved. Results of multimetric indices have been useful in summarizing and communicating results in a way that a nontechnical audience can understand, have been incorporated into state water quality standards (Yoder and Rankin 1998), and are being used by an increasing number of states (USEPA 1996b).

Multimetric indices are developed by selecting between 6 and 10 metrics that would respond to a diversity of possible environmental stresses in a region (e.g., hydrologic alteration, eutrophication, siltation). Values of the metrics must be normalized to a standard scale. Although metrics are commonly normalized to an ordinal scale from 1 to 5 (Kentucky Division of Water 1993, Karr and Chu 1999), others recommend a more objective continuous scale ranging from 0 to 10 (Hill et al. 2000). High values of metrics should always reflect high-quality habitat conditions. Traditionally, multimetric indices have been calculated by summation of all metrics, but that is not necessary. A highly sensitive multimetric index could be based on the value of the lowest metric rather than the average metric value, but that has not been put into practice as far as this author knows. Summation of 10 metric values with each ranging from 0 to 10 yields a multimetric index with a range from 0 to 100.

Diagnostic metrics can be used in development of a multimetric index of biotic integrity or to complement physical and chemical measures of stressors. Percent eutrophic species, low-pH species, or high-conductivity species may provide indicators of changes in biotic integrity as well as diagnose causes of impairment. These kinds of diagnostic metrics are more likely and appropriately used in a multimetric index of biotic integrity than metrics that precisely estimate stressor levels. Weighted-average inference models using the optima and tolerances of algal species that precisely estimate stressor levels should be used to complement physical and chemical assessments of stressors. A problem may arise, however, if diagnostic metrics are used in hazard and exposure assessment and linked in stressor–response relationships. In other words, the hazard assessment may indicate a problem because there are too many eutrophic species in a habitat and exposure assessment based on algal inferences of nutrients would automatically indicate high-nutrient conditions. This circular argument can be reduced if diagnostic metrics are used primarily in exposure assessment and risk characterization and not in hazard assessment.

Stressor–Response Relationships from Metric Development

The relationship between ecological attributes and the physical, chemical, or biological factors that affect those attributes can be referred to as a stressor–response relationship (USEPA 1998). Stressor–response relationships may be found in the literature but should be tested regionally with existing or new data. If sampling is required, emphasis should be placed on stratified sampling of each class of wetlands and choosing wetlands within each class that represent a range of conditions. Wetlands along the entire range of environmental conditions should be sampled. Many stressor–response relations will probably be established during metric development. However, extra effort should be targeted to establish a broad range of potentially important relationships because they are essential for diagnosing causes or threats to wetland impairment and for establishing criteria for protection and remediation.

A distinction should be made between establishing stressor–response relationships based on results from observational and experimental studies. Observational studies are those typically employed in field surveys in which observations of algal attributes and environmental conditions are made in many wetlands or along a gradient in one large wetland. Significant relationships between algal attributes and environmental factors observed in field surveys of many habitats indicate a "correlative" relationship, but manipulation of single factors in well-controlled experiments is important for establishing cause-and-effect relations. For example, McCormick and O'Dell (1996) showed that changes in algal communities were highly correlated to phosphorus concentrations in field surveys of the Everglades. These changes were also correlated to other factors that change along this Everglades phosphorus gradient (e.g., hydroperiod or vegetation change). They confirmed the probable cause–effect relationship between observed changes in algal communities and phosphorus concentration in an experiment in which manipulating phosphorus caused many of the same changes in algal communities as observed in the field survey.

Multivariate/Ordination Analyses

Multivariate statistics have been used widely in ecology to summarize complex ecological data (Pielou 1984, Jongman et al. 1995). They are particularly useful in preliminary stages of data analysis to quantify relationships between ecological attributes and specific environmental stressors. Correspondence analysis and detrended correspondence analysis are ordination techniques that can be used to relate shifts in species composition among wetlands to shifts in environmental characteristics among wetlands. The uses of these techniques are described in detail in other publications and only a short review is practical here.

Basically, ordination techniques can be used to relate changes in species composition and environmental characteristics and ordination axes. Locations of wetlands along ordination axes can be used to determine similarity in species composition of algal assemblages in wetlands. Locations of wetlands along ordination axes can also be related to the strength of loading of species scores and environmental factor scores to determine which species and environmental factors were most important in distinguishing a test site from reference wetlands, for example. Distances of test sites from reference sites in ordination space could be used as a metric of biotic integrity. In addition, the correlation between environmental variables and direction (axis) of change in species composition between test and reference sites could be used to diagnose causes of change in biotic integrity.

The main problem with using multivariate statistics in hazard assessment and diagnosing probable stressors is that results are difficult to communicate to a nontechnical audience. In addition, adding data to an analysis can alter relations among species and environmental factors that have been used in previous analyses. The latter problem can be solved by establishing relations among species and environmental factors in an a priori analysis and subsequently evaluating biotic integrity and diagnosing threats or causes of impairment by plotting samples in species space using ordination scores from the a priori analysis. Despite some problems with application, multivariate statistics continue to be an indispensable tool for exploring relationships between samples, species, and environmental factors.

DEVELOPING CRITERIA

Criteria can be established for physical, chemical, or biological conditions in habitats and as either stressors, human activities, or ecological responses to stressors (Jackson and Davis 1994). The fundamental objectives of criteria are to protect existing conditions in habitats that meet goals of management or to provide targets for restoration of impaired wetlands. Thus, goals of management must be established and measures of hazard (metrics of designated uses, biotic integrity, or ecosystem services) become the first priority for establishing criteria. How much algae is too much, or how many species may be lost before biotic integrity is considered impaired? Criteria that limit stressors and human activity causing those stressors are established to prevent or remediate hazards and are best set when stressor–response relationships are well established and goals or protection and levels of acceptable hazard are established. In other words, first we must know what we consider acceptable levels of metrics, and then we establish criteria for the environmental stressors that endanger or impair wetlands.

Defining important levels of environmental or biological condition can be facilitated if thresholds occur along environmental gradients at which sudden changes are observed in some ecological metrics (Fig. 6.2). Along linear stressor–response curves, arguments could be made that criteria should be set a little higher or lower, depending on perspectives of stakeholders. If great changes in a metric occur at thresholds along environmental gradients, these punctuated changes and their environmental thresholds may provide a basis for setting criteria. Changes in algal assemblages should be reviewed for those changes that occur nonlinearly along environmental gradients. Theoretically, lower-level biological attributes, such as species relative abundances, presence, or even community similarity, would respond more sensitively and potentially nonlinearly along environmental gradients than community attributes. Functional redundancy in species of communities may enable replacement of function and resulting manifestations in community structural attributes (biomass and diversity) along these same environmental gradients that restrict distributions of specific species (Schindler 1990). Again, the form of data and scale of spatial and temporal variability may affect whether species relative abundances, community similarity based on relative abundance or presence–absence, or a community parameter responds most sensitively or even nonlinearly along environmental gradients.

ECOLOGICAL RISK CHARACTERIZATION

The ecological risk assessment process results in a risk characterization that indicates the probable stressor or stressors that are threatening or causing impairment (Fig. 6.1). It provides information for making decisions for protecting or restoring ecosystems. The first step in ecological risk assessment is hazard assessment and determining whether or not there is a problem. If hazard assessment shows that metric values are below acceptable levels, a diagnosis of the responsible stressor or stressors is called for (Fig. 6.5*A*–*C*). Diagnosis of the responsible stressor(s) requires relating stressor levels to their respective stressor–response relationships to see if the stressor is sufficient to generate the observed change in designated use (response variable). The most likely stressor responsible for impairment is the stressor that generates the highest response. In the ideal case illustrated in Figure 6.5*A* and *B*, all stressors except

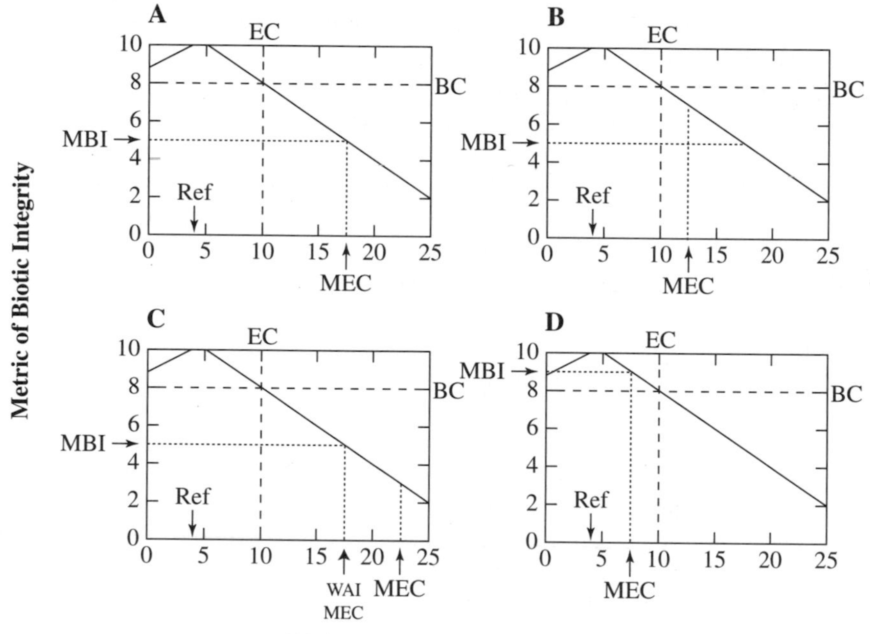

FIGURE 6.5. Examples of risk characterization where measured biotic integrity (MBI) and a measured environmental condition (MEC) in a wetland are compared to a stressor–response relationship to identify a likely stressor that impairs (*A* to *C*) or threatens impairment (*D*) of the wetland. EC marks an environmental criterion that would be established to protect the designated minimum acceptable level of the biotic criterion (BC). *A* and *D* Measured environmental condition predicts measured biotic integrity according to dose–response relationship; (*B*) measured environmental condition predicts higher biotic integrity than observed; (*C*) measured environmental condition predicts lower biotic integrity than observed. WAI MEC in (*C*) marks an example of a weighted-average inference model (biotic indicator) of environmental stress that predicts the measured biotic integrity according to the stressor–response relationship.

one are too low to cause the observed impairment in biotic integrity. Theoretically, only one or a combination of two or more stressors with synergistic or antagonistic effects should cause impairment (Fig. 6.5*A*). If all stressor levels are more than sufficient (Fig. 6.5*C*) or less than sufficient to cause the response (Fig. 6.5*B*), then error in stressor–response relationships may be a problem. Alternatively, imprecision in hazard assessment or exposure assessment may have been a problem. If physical and chemical conditions are variable in the habitat, biological inference of abiotic conditions may provide a more reliable assessment of the recent history of those conditions than can one-time sampling (Fig. 6.5*C*).

If no hazard (impairment) is found and criteria for designated use are met (Fig. 6.5*D*), knowing stressor levels and stressor–response relationships for unimpaired habitats can identify the most important stressors that threaten impairment, which could be focuses of regulation to protect endangered habitats. Thus, the stressor that is causing the greatest effects on designated use becomes the most important target of remediation.

Metrics of sustainability and restorability of habitats could be developed based on differences in measured levels of environmental stressors and environmental criteria for those stressors (Stevenson 1998). If stressors levels are below criteria and ecosystems conditions should be preserved, the sustainability or resistance to impairment increases with the difference between stressor level and its respective criterion. Restorability of designated use in an impaired wetland would be inversely related to the difference between criterion and measured levels of stressors in the habitat. These metrics may provide valuable summary statistics for comparative risk assessment and choosing management options to protect or restore wetlands.

SUMMARY

Algal assemblages are rich with information about environmental conditions in wetlands, and their biotic integrity responds sensitively to human disturbances. During the long history of using algae to assess environmental conditions, especially in streams and lakes, many methods and indicators has been developed and applied. Traditionally, most algal methods have been developed to infer changes in ecological conditions based on changes in relative abundances of algae among habitats and autecologies of those algal species. Substantial regulatory emphasis is now being placed on effects of human activities on designated use and biotic integrity of aquatic ecosystems. In this chapter I have tried to integrate the classical approaches of bioassessment with algae and an ecological risk assessment framework that borrows many approaches from the concepts that have been used to assess aquatic habitats with macroinvertebrate and fish assemblages. Organizing these many indicators and concepts in a risk assessment framework provides a means of standardizing the types of information gathered during bioassessment and the ways in which that information is used.

The ecological risk assessment framework also helps us better understand the challenges ahead for improving bioassessments of wetlands with algae and other organisms. Although we know that WAI indicators of environmental conditions can be highly precise and accurate, we know little about using them in more than one region and wetland class. In the same way, we know that many algal attributes respond to environmental conditions in wetlands, but we have little knowledge about how similarly those metrics respond to gradients of human disturbance and specific stressors in different classes of wetlands. When algal attributes are shown to be transferable metrics from one wetland class to another, the sensitivity in metric response to specific environmental stressors may still differ in different wetland classes. This becomes a problem in risk characterization when trying to predict which stressor is causing impairment or threatening impairment of designated use. Developing a better understanding of algal ecology in wetlands is fundamental to all these challenges so that bioassessment methods can be more certainly transferred from region to region and among different classes of wetlands.

REFERENCES

Adamus, P. R. 1996. Bioindicators for assessing ecological integrity of prairie wetlands. Report EPA/600/R-96/082. U.S. Environmental Protection Agency, National Health and Environmental Effects Laboratory, Corvallis, OR.

Agbeti, M. D. 1992. Relationship between diatom assemblages and trophic variables: a comparison of old and new approaches. Canadian Journal of Fisheries and Aquatic Sciences 49:1171–1175.

Aloi, J. E. 1990. A critical review of recent freshwater periphyton methods. Canadian Journal of Fisheries and Aquatic Sciences 47:656–670.

Angermeier, P. L., and J. R. Karr. 1994. Protecting integrity versus biological diversity as policy directives. BioScience 44:690–697.

Apfelbeck, R. No date. Development of preliminary bioassessment protocols for Montana wetlands. State of Montana, Department of Environmental Quality, Helena, MT (draft report).

Bahls, L. L. 1993. Periphyton bioassessment methods for Montana streams. Department of Health and Environmental Sciences, State of Montana, Water Quality Bureau, Helena, MT.

Bannister, T. T. 1974. Production equations in terms of chlorophyll concentration, quantum yield, and upper limit to production. Limnology and Oceanography 19:1–12.

Barbour, M. T., J. Gerritsen, G. E. Griffith, R. Frydenborg, E. McCarron, J. S. White, and M. L. Bastian. 1996. A framework for biological criteria for Florida streams using benthic macroinvertebrates. Journal of the North American Benthological Society 15:185–211.

Barbour, M. T., J. Gerritsen, B. D. Snyder, and J. B. Stribling. 1999. Rapid bioassessment protocols for use in streams and wadeable rivers: periphyton, benthic macroinvertebrates and fish, 2nd Ed. Report EPA 841-B-99-002. U.S. Environmental Protection Agency, Office of Water, Washington, DC.

Belbin, L., and C. McDonald. 1993. Comparing three classification strategies for use in ecology. Journal of Vegetation Science 4:341–348.

Brinson, M. M. 1993. A hydrogeomorphic classification of wetlands. Technical Report WRP-DE-4. Wetlands Research Program, U.S. Army Corps of Engineers, Waterways Experimental Station, Vicksburg, MS.

Browder, J. A., P. J. Gleason, and D. R. Swift. 1994. Periphyton in the Everglades: spatial variation, environmental correlates, and ecological implications. Pages 379–418 *in* S. M. Davis and J. C. Ogden (ed.), Everglades: the ecosystem and its restoration. St. Lucie Press, Delray Beach, FL.

Cairns, J., Jr. 1977. Quantification of biological integrity. Pages 171–187 *in* R. K. Kallentine and L. J. Guarraia (eds.), The integrity of water. U.S. Environmental Protection Agency, Office of Water and Hazardous Substances, Washington, DC.

Cairns, J., Jr. 1995. Ecological integrity of aquatic ecosystems. Regulated Rivers: Research and Management 11:313–323.

Cairns, J., Jr., P. V. McCormick, and B. R. Niederlehner. 1993. A proposed framework for developing indicators of ecosystem health. Hydrobiologia 263:1–44.

Campeau, S., H. R. Murkin, and R. D. Titman. 1994. Relative importance of algae and emergent plant litter to freshwater marsh invertebrates. Canadian Journal of Fisheries and Aquatic Sciences 51:681–692.

Charles, D. F. 1985. Relationships between surface sediment diatom assemblages and lakewater characteristics in Adirondack lakes. Ecology 66:994–1011.

Chessman, B., I. Growns, J. Currey, and N. Plunkett-Cole. 1999. Predicting diatom communities at the genus level for the rapid biological assessment of rivers. Freshwater Biology 41:317–331.

Cooper, S. R. 1995. Chesapeake Bay watershed historical land use: impact on water quality and diatom communities. Ecological Applications 5:703–723.

Cooper, S. R. 1999. Estuarine paleoenvironmental reconstructions using diatoms. Pages 334–351 *in* E. F. Stoermer and J. P. Smol (ed.), The diatoms: applications for the environmental and earth sciences. Cambridge University Press, Cambridge.

Cowardin, L. M., V. Carter, F. C. Golet, and E. T. LaRoe. 1979. Classification of wetlands and deepwater habitats of the United States. Report FWS/OBS-79/31. U.S. Fish and Wildlife Service, Corvallis, OR.

Cumming, B. F., and J. P. Smol. 1993. Development of diatom-based salinity models for paleoclimatic research from lakes in British Columbia (Canada). Hydrobiologia 269/270: 575–586.

Cumming, B. F., J. P. Smol, J. C. Kingston, D. F. Charles, H. J. B. Birks, K. E. Camburn, S. S. Dixit, A. J. Uutala, and A.R. Selle. 1992. How much acidification has occurred in Adirondack region lakes (New York, USA) since pre-industrial times? Canadian Journal of Fisheries and Aquatic Sciences 49:128–141.

Dixit, A. S., S. S. Dixit, and J. P. Smol. 1992a. Long-term trends in lake water pH and metal concentrations inferred from diatoms and chrysophytes in the lakes near Sudbury, Ontario. Canadian Journal of Fisheries and Aquatic Sciences 49(Suppl. 1):17–24.

Dixit, S. S., J. P. Smol, J. C. Kingston, and D. F. Charles. 1992b. Diatoms: powerful indicators of environmental change. Environmental Science and Technology 26:23–33.

Dixit, S. S., J. P. Smol, D. F. Charles, R. M. Hughes, S. G. Paulsen, and G. B. Collins. 1999. Assessing water quality changes in the lakes of the northeastern United States using sediment diatoms. Canadian Journal of Fisheries and Aquatic Sciences 56:131–152.

Faith, D. P., P. R. Minchin, and L. Belbin. 1987. Compositional dissimilarity as a robust measure of ecological distance. Vegetatio 69:57–68.

Finlayson, C. M., and A. G. van der Valk (eds.) 1995. Classification and inventory of the world's wetlands. Kluwer Academic Publishers, Dordrecht, The Netherlands.

Fritz, S. C. 1990. A twentieth-century salinity and water-level fluctuations in Devils Lake, North Dakota: test of a diatom-based transfer function. Limnology and Oceanography 35:1771–1781.

Fritz, S. C., S. Juggins, R. W. Battarbee, and D. R. Engstrom. 1991. Reconstruction of past changes in salinity and climate using a diatom-based transfer function. Nature 352:706–708.

Goldsborough, L. G., and G. G. C. Robinson. 1996. Pattern in wetlands. Pages 77–117 *in* R. J. Stevenson, M. L. Bothwell, and R. L. Lowe (eds.), Algal ecology: freshwater benthic ecosystems. Academic Press, San Diego, CA.

Grimshaw, H. J., M. Rosen, D. R. Swift, K. Rodbern, and J. M. Noel. 1993. Marsh phosphorus concentration, phosphorus content and species composition of Everglades periphyton communities. Archiv fuer Hydrobiologie 128:257–276.

Hall, R. I., and J. P. Smol. 1992. A weighted-averaging regression and calibration model for inferring total phosphorus from diatoms in British Columbia (Canada) lakes. Freshwater Biology 27:417–437.

Hill, M. O. 1979. TWINSPAN: a FORTRAN program for detrended correspondence analysis and reciprocal averaging. Cornell University, Ithaca, NY.

Hill, W. R., and H. L. Boston. 1991. Community development alters photosynthesis–irradiance relations in stream periphyton. Limnology and Oceanography 36:1375–1389.

Hill, B. H., A. T. Herlihy, P. R. Kaufmann, R. J. Stevenson, and F. H. McCormick. 2000. The use of periphyton assemblage data as an index of biotic integrity. Journal of the North American Benthological Society. 19:50–67.

Holmes, N. T. H., and B. A. Whitton. 1981. Phytobenthos of the river Tees and its tributaries. Freshwater Biology 11:139–163.

Hughes, R. M. 1995. Defining acceptable biological status by comparing with reference conditions. Pages 31–47 *in* W. S. Davis and T. P. Simon (eds.), Biological assessment and criteria: tools for water resource planning and decision making. Lewis Publishers, Boca Raton, FL.

Jackson, S., and W. Davis. 1994. Meeting the goal of biological integrity in water-resource programs in the U.S. Environmental Protection Agency. Journal of the North American Benthological Society 13:592–597.

Jongman, R. H. G., C. J. F. ter Braak, and O. F. R. van Tongeren. 1995. Data analysis and community and landscape ecology. Cambridge University Press, Cambridge.

Juggins, S., and C. J. F. ter Braak. 1992. CALIBRATE: a program for species-environment calibration by [weighted averaging] partial least squares regression. University College, Environmental Change Research Center, London.

Karr, J. R. 1981. Assessment of biotic integrity using fish communities. Fisheries 6:21–27.

Karr, J. R., and E. W. Chu. 1999. Restoring life in running waters. Island Press, Washington, DC.

Karr, J. R., and D. R. Dudley. 1981. Ecological perspective on water quality goals. Environmental Management 5:55–68.

Kelly, M. G., A. Cazaubon, E. Coring, A. Dell'Uomo, L. Ector, B. Goldsmith, H. Guasch, J. Hürlimann, A. Jarlman, B. Kawecka, J. Kwandrans, R. Laugaste, E. A. Lindstrom, M. Leitao, P. Marvan, J. Padisák, E. Pipp, J. Pyrgiel, E. Rott, S. Sabater, H. van Dam, and

J. Viznet. 1998. Recommendations for the routine sampling of diatoms for water quality assessments in Europe. Journal of Applied Phycology 10:215–224.

Kentucky Division of Water. 1993. Methods for assessing biological integrity of surface waters. Kentucky Natural Resources and Environmental Protection Cabinet. Frankfort, KY.

Kerans, B. L., and J. R. Karr. 1994. A benthic index of biotic integrity (B-IBI) for rivers of the Tennessee Valley. Ecological Applications 4:768–785.

Kingston, J. C., H. J. B. Birks, A. J. Uutala, B. F. Cumming, and J. P. Smol. 1992. Assessing trends in fishery resources and lake water aluminum from paleolimnological analyses of siliceous algae. Canadian Journal of Fisheries and Aquatic Sciences 49:127–138.

Line, J. M., C. J. F. ter Braak, and H. J. B. Birks. 1994. WACALIB version 3.3: a computer program to reconstruct environmental variables from fossil assemblages by weighted averaging and to derive sample-specific errors of prediction. Journal of Paleolimnology 10: 147–152.

Lowe, R. L. 1974. Environmental requirements and pollution tolerance of freshwater diatoms. Report EPA-670/4-74-005. U.S. Environmental Protection Agency, Cincinnati, OH.

Lowe, R. L., and Y. Pan. 1996. Benthic algal communities and biological monitors. Pages 705–739 *in* R. J. Stevenson, M. Bothwell, and R. L. Lowe. (eds.), Algal ecology: freshwater benthic ecosystems. Academic Press, San Diego, CA.

MacIntyre, H. L., R. J. Geider, and D. C. Miller. 1996. Microphytobenthos: the ecological role of the "secret garden" of unvegetated, shallow water marine habitats. I. Distribution, abundance, and primary production. Estuaries 19:186–201.

McCormick, P. V., and J. Cairns, Jr. 1994. Algae as indicators of environmental change. Journal of Applied Phycology 6:509–526.

McCormick, P. V., and M. B. O'Dell. 1996. Quantifying periphyton responses of phosphorus in the Florida Everglades: a synoptic-experimental approach. Journal of the North American Benthological Society 15:450–468.

McCormick, P. V., P. S. Rawlik, K. Lurding, E. P. Smith, and F. H. Sklar. 1996. Periphyton–water quality relationships along a nutrient gradient in the northern Florida Everglades. Journal of the North American Benthological Society 15:433–449.

McCormick, P. V., and R. J. Stevenson. 1998. Periphyton as a tool for ecological assessment and management in the Florida Everglades. Journal of Phycology 34:726–733.

McCormick, P. V., M. J. Chimney, and D. R. Swift. 1997. Diel oxygen profiles and water column community metabolism in the Florida Everglades. Archiv fuer Hydrobiologie 140: 117–129.

McCormick, P. V., R. B. E. Shuford III, J. B. Backus, and W. C. Kennedy. 1998. Spatial and seasonal patterns of periphyton mass and productivity in the northern Everglades, Florida, U.S.A. Hydrobiologia 362:185–208.

McFarland, B. H., B. H. Hill, and W. T. Willingham. 1997. Abnormal *Fragilaria* spp. (Bacillariophyceae) in streams impacted by mine drainage. Journal of Freshwater Ecology 12:141–149.

Mihuc, T., and D. Toetz. 1994. Determination of diets of alpine aquatic insects using stable isotopes and gut analysis. American Midland Naturalist 131:146–155.

Miller, D. C., R. J. Geider, and H. L. MacIntyre. 1996. Microphytobenthos: The ecological role of the "secret garden" of unvegetated, shallow-water marine habitats. II. Role in sediment stability and shallow-water food webs. Estuaries 19:202–212.

Moeller, R. E., J. M. Burkholder, and R. G. Wetzel. 1988. Significance of sedimentary phosphorus to a rooted submersed macrophyte (*Najas flexilis*) and its algal epiphytes. Aquatic Botany 32:261–281.

Murkin, E. J., H. R. Murkin, and R. D. Titman. 1992. Nektonic invertebrate abundance and distribution at the emergent vegetation–open water interface in the Delta Marsh, Manitoba, Canada. Wetlands 12:45–52.

Palmer, C. M. 1962. Algae in water supplies. U.S. Department of Health, Education and Welfare. Washington, DC.

Pan, Y., and R. J. Stevenson. 1996. Gradient analysis of diatom assemblages in western Kentucky wetlands. Journal of Phycology 32:222–232.

Pan, Y., R. J. Stevenson, B. H. Hill, A. T. Herlihy, and G. B. Collins. 1996. Using diatoms as indicators of ecological conditions in lotic systems: a regional assessment. Journal of the North American Benthological Society 15:481–495.

Pan, Y., R. J. Stevenson, P. Vaithiyanathan, J. Slate, and C. J. Richardson. 2000. Using experimental and observational approaches to determine causes of algal changes in the Everglades. Freshwater Biology 43:1–15.

Pielou, E. C. 1984. The interpretation of ecological data: a primer on classification and ordination. Wiley, New York.

Raschke, R. L. 1993. Diatom (Bacillariophyta) community response to phosphorus in the Everglades National Park, USA. Phycologia 32:48–58.

Reavie, E. D., R. I. Hall, and J. P. Smol. 1995. An expanded weighted-averaging model for inferring past total phosphorus concentrations from diatom assemblages in eutrophic British Columbia (Canada) lakes. Journal of Paleolimnology 14:49–67.

Reynoldson, T. B., R. H. Noris, V. H. Resh, K. E. Day, and D. M. Rosenberg. 1997. The reference condition: a comparison of multimetric and multivariate approaches to assess water quality impairment using benthic macroinvertebrates. Journal of the North American Benthological Society 16:833–852.

Round, F. E. 1981. The ecology of algae. Cambridge University Press, Cambridge.

Schindler, D. W. 1990. Experimental perturbations of whole lakes as tests of hypotheses concerning ecosystem structure and function. Oikos 57:25–41.

Shannon, C. F. 1948. A mathematical theory of communication. Bell System Technical Journal 27:37–42.

Short, F. T., and S. Wyllie-Echeverria. 1996. Natural and human-induced disturbance of seagrasses. Environmental Conservation 23:17–27.

Sicko-Goad, L., E. F. Stoermer, and B. G. Ladewski. 1977. A morphometric method for correcting phytoplankton cell volume estimates. Protoplasma 93:147–163.

Slate, J. 1998. Inference of present and historical environmental conditions in the Everglades with diatoms and other siliceous microfossils. Ph.D. dissertation, University of Louisville, Louisville, KY.

Slate, J. E., and R. J. Stevenson. 2000. Recent and abrupt environmental change in the Florida Everglades indicated from siliceous microfossils. Wetlands 20:346–356.

Smol, J. P. 1992. Paleolimnology: an important tool for effective ecosystem management. Journal of Aquatic Ecosystem Health 1:49–58.

Steinman, A. D., P. J. Mulholland, and W. R. Hill. 1992. Functional responses associated with growth form in stream algae. Journal of the North American Benthological Society 11:229–243.

Stevenson, R. J. 1984. Epilithic and epipelic diatoms in the Sandusky River, with emphasis on species diversity and water quality. Hydrobiologia 114:161–175.

Stevenson, R. J. 1996. An introduction to algal ecology in freshwater benthic habitats. Pages 3–30 *in* R. J. Stevenson, M. Bothwell, and R. L. Lowe. (eds.), Algal ecology: freshwater benthic ecosystems. Academic Press, San Diego, CA.

Stevenson, R. J. 1998. Diatom indicators of stream and wetland stressors in a risk management framework. Environmental Monitoring and Assessment 51:107–118.

Stevenson, R. J., and L. L. Bahls. 1999. Periphyton protocols. Pages 6-1 through 6-22 *in* M. T. Barbour, J. Gerritsen, and B. D. Snyder (eds.), Rapid bioassessment protocols for use in wadeable streams and rivers: periphyton, benthic macroinvertebrates, and fish, 2nd ed. Report EPA 841-B-99-002. U.S. Environmental Protection Agency, Office of Water, Washington, DC.

Stevenson, R. J., and R. L. Lowe. 1986. Sampling and interpretation of algal patterns for water quality assessment. Pages 118–149 *in* B. G. Isom (ed.), Rationale for sampling and interpretation of ecological data in the assessment of freshwater ecosystems. Report ASTM STP 894. American Society for Testing and Materials, Philadelphia.

Stevenson, R. J., and Y. Pan. 1999. Assessing ecological conditions in rivers and streams with diatoms. Pages 11–40 *in* E. F. Stoermer and J. P. Smol (eds.), The diatoms: applications to the environmental and earth sciences. Cambridge University Press, Cambridge.

Stevenson, R. J., and J. P. Smol. In press. Use of algae in environmental assessments. *in* J. D. Wehr and R. G. Sheath (eds.), Freshwater algae in North America: classification and ecology. Academic Press, San Diego, CA.

Stevenson, R. J., P. R. Sweets, Y. Pan, and R. E. Schultz. 1999. Algal community patterns in wetlands and their use as indicators of ecological conditions. Pages 517–527 *in* A. J. McComb and J. A. Davis (eds.), Proceedings of INTECOL's 5th International Wetland Conference. Gleneagles Press, Adelaide, Australia.

Stoermer, E. F. and J. P. Smol. 1999. The diatom: applications for the environmental and earth sciences. Cambridge University Press. Cambridge.

Sullivan, M. J. 1999. Applied diatom studies in estuaries and shallow coastal environments. Pages 334–351 *in* E. F. Stoermer and J. P. Smol (eds.), The diatoms: applications for the environmental and earth sciences. Cambridge University Press, Cambridge.

Sullivan, M. J., and C. A. Moncreiff. 1988. Primary production of edaphic algal communities in a Mississippi salt marsh. Journal of Phycology 24:49–58.

Sullivan, M. J., and C. A. Moncreiff. 1990. Edaphic algae are an important component of salt marsh food-webs: evidence from multiple stable isotope analyses. Marine Ecology Progress Series 62:149–159.

Sundbäck, K., and W. Granéli. 1988. Influence of microphytobenthos on the nutrient flux between sediment and water: A laboratory study. Journal of Experimental Marine Biology and Ecology 18:79–88.

Sweets, P. R. 1992. Diatom paleolimnological evidence for lake acidification in the Trail Ridge region of Florida. Water, Air and Soil Pollution 65:43–57.

ter Braak, C. J. F., and H. van Dam. 1989. Inferring pH from diatoms: a comparison of old and new calibration methods. Hydrobiologia 178:209–223.

USEPA. 1992. Framework for ecological risk assessment. Report EPA 630/R-92/001. U.S. Environmental Protection Agency, Washington, DC.

USEPA. 1996a. Strategic plan for the Office of Research and Development. Report EPA/600/R-96/059. U.S. Environmental Protection Agency, Washington, DC.

USEPA. 1996b. Summary of state biological assessment programs for streams and rivers. Report EPA 230/R96/007. U.S. Environmental Protection Agency, Washington DC.

USEPA. 1998. Guidelines for ecological risk assessment. Report EPA 630/R-95/002F. U.S. Environmental Protection Agency, Washington DC.

van Dam, H., A. Mertenes, and J. Sinkeldam. 1994. A coded checklist and ecological indicator values of freshwater diatoms from the Netherlands. Netherlands Journal of Aquatic Ecology 28:117–133.

VanderBorgh, M. A. 1999. The use of phytoplankton assemblages to assess environmental conditions in wetlands. M.S. thesis, University of Louisville, Louisville, KY.

Vymazal, J. 1994. Algae and element cycling in wetlands. Lewis Publishers, Boca Raton, FL.

Vymazal, J., and C. J. Richardson. 1995. Species composition, biomass, and nutrient content of the periphyton in the Florida Everglades. Journal of Phycology 31:343–354.

Vymazal, J., C. B. Craft, and C. J. Richardson. 1994. Periphyton response to nitrogen and phosphorus in Florida Everglades. Algological Studies 3:75–97.

Wear, D. J., M. J. Sullivan, A. D. Moore and D. F. Millie. 1999. Effects of water-column enrichment on the production dynamics of three seagrass species and their epiphytic algae. Marine Ecology Progress Series. 199:201–213

Wetzel, R. G. 1996. Benthic algae and nutrient cycling in lentic freshwater ecosystems. Pages 641–667 *in* R. J. Stevenson, M. L. Bothwell, and R. L. Lowe (eds.), Algal ecology: freshwater benthic ecosystems. Academic Press, San Diego, CA.

Wetzel, R. G., and G. E. Likens. 1991. Limnological analyses. 2nd ed. Springer-Verlag, New York.

Whitton, B. A., and E. Rott (eds.), 1996. Use of algae for monitoring rivers II. E. Rott, Publisher, Institut für Botanik, Universität Innsbruck, Innsbruck, Austria.

Whitton, B. A., E. Rott, and G. Friedrich (eds.). 1991. Use of algae for monitoring rivers. E. Rott, Publisher, Institut für Botanik, Universität Innsbruck, Innsbruck, Austria.

Wright, J. F. 1995. Development and use of a system for predicting the macroinvertebrate fauna of flowing waters. Australian Journal of Ecology 20:181–197.

Yoder, C. O., and E. T. Rankin. 1998. The role of biological indicators in a state water quality management process. Environmental Monitoring and Assessment 51:61–88.

Zelinka, M., and P. Marvan. 1961. Zur Präzisierung der biologischen Klassifikation des Reinheit fliessender Gewässer. Archiv fuer Hydrobiologie 57:389–407.

7 Development of Biocriteria for Wetlands in Montana

RANDALL S. APFELBECK

A goal of the Clean Water Act is to restore and maintain the chemical, physical, and biological integrity of the nation's waters. Attainment of this goal includes development and implementation of wetland water quality standards. In an effort to create wetland-specific water quality standards, the Montana Department of Environmental Quality is developing biocriteria that are sensitive and responsive to changes in wetland water quality. We sampled diatom and macroinvertebrate communities and associated environmental variables from 80 Montana wetlands. We designed the study to sample approximately 75% reference sites and 25% impaired sites having known anthropogenic impacts. Diatoms were collected as a composite grab sample, identified to the lowest taxonomic level feasible, and analyzed using multivariate procedures. Macroinvertebrates were collected using a 1-mm-mesh D-net, identified to a standardized taxonomic, level and assessed using multimetric techniques. Prior to sampling we classified wetlands using ecoregions and hydrogeomorphology and later delineated other classes using water-column chemistry. Diatoms and macroinvertebrates were useful for evaluating the biological integrity of perennial wetlands with stable surface water habitats that were not excessively alkaline or saline. We concluded that multivariate analysis was a useful tool for developing a wetland classification system and that hydrogeomorphology and ecoregions were practical approaches to classifying wetlands for the development of biocriteria. Both the multimetric and multivariate techniques were valuable for developing wetland biocriteria. In most cases, the multimetric and multivariate identified the same wetlands as being impaired.

Montana initiated the development of wetland biocriteria in 1992 with funding from the Environmental Protection Agency's State Wetlands Protection Program, as defined in Section 104(b)(3) of the Clean Water Act. At that time, no regulatory agencies had information concerning the water quality of Montana's wetlands. Further,

---

Bioassessment and Management of North American Freshwater Wetlands, edited by Russell B. Rader, Darold P. Batzer, and Scott A. Wissinger.
0-471-35234-9

Montana's water quality standards were developed to protect the beneficial uses (e.g., aquatic life) of lakes, rivers, and streams. Wetlands were not considered state waters when Montana's water quality standards were developed. Therefore, many of Montana's water quality standards are not applicable for most wetlands. For this reason we are attempting to develop bioassessment protocols and water quality standards that will protect the aquatic life that live in wetlands.

The Montana Department of Environmental Quality (DEQ) initiated the collection of wetland water quality data for several reasons:

- To determine the status and trends in wetland water quality
- To acquire an understanding of how climate, hydrologic controls, and geomorphic settings influence wetland biological communities so that they can be classified for the development of successful biocriteria
- To develop biological measurements that could be used in developing biocriteria to define the extent and degree of anthropogenic impacts to wetland water quality

BACKGROUND OF WETLAND BIOLOGICAL CRITERIA

The main objective of the Clean Water Act (CWA) is to restore and maintain the chemical, physical, and biological integrity of the nation's waters, including wetlands (Adamus 1996). Historically, methods of assessing water quality and developing standards for aquatic habitats have typically focused on chemical and physical parameters. However, the assessment of biological systems is essential for determining ecosystem health and should provide a basis for environmental management. The role of chemical and physical attributes of wetlands is also important in the identification of factors causing impairment and the selection of appropriate remedial actions (Reynoldson and Metcalfe-Smith 1992). The EPA recognizes that assessing biological criteria contributes data needed to evaluate the condition of aquatic habitats. The agency is encouraging states to monitor the general health of the biota in aquatic habitats as an assessment tool in gathering the necessary information to meet the goals of the CWA. The EPA has also mandated that states develop standards for wetlands (Adamus and Brandt 1990). Minimum requirements for these standards include the following:

- Include wetlands in the definition of state waters.
- Establish beneficial uses for wetlands.
- Adopt existing narrative and numeric criteria for wetlands.
- Apply antidegradation policies to wetlands.
- Adopt narrative biocriteria for wetlands. (*Note:* DEQ is attempting to surpass this requirement by developing numeric biological criteria for wetlands.)

Biocriteria can be either narratives or numbers that describe the aquatic communities of a healthy ecosystem and provide a means to evaluate and protect aquatic life

(USEPA 1990). The term *reference condition* refers to the least disturbed sites that can be found. Researchers cannot lump all reference sites into one group, they must be classified by their physical, chemical, and biological characteristics (Reynoldson et al. 1997).

Biocriteria are developed to protect biological integrity, which is the ability of an aquatic ecosystem to support and maintain a balanced, integrated, adaptive community of organisms having a species composition, diversity, and functional organization comparable to that of the natural habitats of the region (Karr and Dudley 1981). Assessing biological integrity requires comparing the biological communities of reference sites to the biological community of the wetland being evaluated. Theoretically, the two should be similar if the community is undisturbed.

Classification is an important component in developing biocriteria. Classification is used to reduce the natural variation in biological communities by identifying similar wetlands located in different areas. States often classify aquatic resources by regions that are ecologically similar (Omernik and Bailey 1997). Temporal variability is the variance occurring over time in the same or similar wetlands and is usually controlled by sampling during the same time period.

STUDY DESIGN

Montana's approach to developing biocriteria currently includes several studies aimed at developing tools to help detect human influence on wetland water quality. Our study was incorporated into DEQ's ambient water quality monitoring program, where the data collected were also used to determine the status and trends of the state's water quality and for making aquatic life use-support determinations. This study involved sampling 80 wetlands throughout Montana from April through September during 1993 and 1994 (Fig. 7.1). The study included the collection of macroinvertebrate (e.g., aquatic insects) and diatom (algae) samples from wetlands in the following ecoregions (Omernik 1986): rocky mountains (23 sites), intermountain valleys and prairie foothills (20 sites), glaciated plains (19 sites), and unglaciated plains (18 sites). All wetlands within the same ecoregion were sampled during similar time periods. wetlands of the plains ecoregion from early April through mid-June, wetlands of the intermountain valleys and prairie foothills ecoregion from mid-June until early August, and wetlands of the rocky mountain ecoregion from early July through September.

The development of biocriteria requires that sampling and analysis focus within similar classes, with sites ranging from minimal to severe human disturbance (Karr and Chu 1999). In the beginning, however, it was not clear to us which wetland environments had similar biological communities. Often, wetlands that appeared to be similar, such as prairie potholes, showed a wide variety of very different biological communities. For that reason, our study emphasized characterizing the biological communities of Montana's least-impaired wetlands in order to develop a reference classification system. We made an effort to sample the full spectrum of wetland types in Montana in order to develop an understanding of the natural factors that affect the

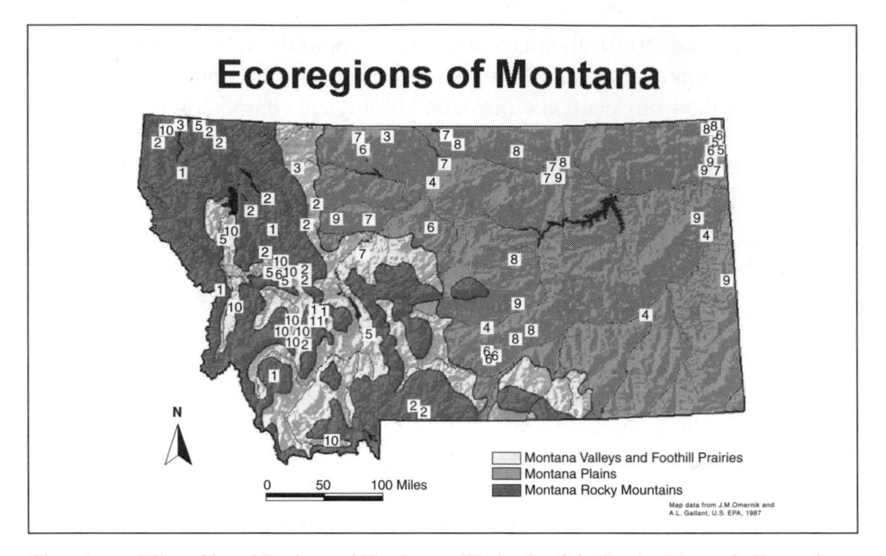

Class 1 Dilute Closed Basins and Headwater Wetlands of the Rocky Mountain Ecoregion
Class 2 Riparian Wetlands of the Rocky Mountain and Intermountain Valley Ecoregion
Class 3 Groundwater Recharge Closed Basin Wetlands
Class 4 Riparian Wetland of the Plains Ecoregion
Class 5 Alkaline Closed Basin Wetlands
Class 6 Saline Wetlands
Class 7 Surface Water Supported Closed Basin Wetlands
Class 8 Ephemeral Wetlands
Class 9 Open Lake Wetlands of the Plains Ecoregion
Class 10 Open Lake Wetlands of the Rocky Mountain and Intermountain Valley Ecoregions

FIGURE 7.1. Sites used by Montana Department of Environmental Quality for characterizing wetlands.

composition of their biological assemblages. We defined wetlands as having surface water for at least one season and habitat that supports diatom and macroinvertebrate communities. Several of the sites were large shallow reservoirs or deep glaciated potholes that did not support fish. On topographical maps these sites were often identified as lakes. We sampled approximately 75% reference sites and 25% impaired sites. Requiring the majority of the wetlands to be reference sites gave us the ability to determine the reference condition of a wide variety of wetland types. Sampling of wetlands that were impaired provided the opportunity to test the ability of our biological measurements to detect water quality impairment. Those biological measurements that were consistently and noticeably different when comparisons were made between impaired wetlands and reference wetlands had the ability to detect anthropogenic impacts on water quality. Numerous wetland types were sampled with differing anthropogenic impacts, such as irrigation or logging. This allowed us to determine if our biological measurements could detect an assortment of anthropogenic impacts on water quality for a variety of wetland types. This was important because we wanted to develop a consistent approach to assessing water quality for a wide variety of wetlands. We felt that metrics specific to each wetland class and anthropogenic impact could be developed at a later time. Such metrics would probably be more sensitive to slight alterations or early signs of degradation.

We considered a wetland to be impaired if extensive anthropogenic activities occurred within its watershed. These activities included dryland agriculture, irrigation, feedlots, grazing, silviculture, road construction, hydrologic manipulation, urban runoff, wastewater, mining, and oil and natural gas production. Direct causes of impairment included elevated nutrients, high salinity, excessive organic carbon, excessive sediment, metals, and elevated water temperature, exotic species, destruction of habitat, and unnaturally fluctuating water levels. Other variables were also included in site selection, such as the availability of historical data, special interests by other entities, cooperation by landowners, and accessibility. Wetlands evaluated for the study included those located within U.S. Fish and Wildlife Service National Wildlife Refuges and Waterfowl Production Areas (41%), U.S. Forest Service Research Natural Areas, and special-interest areas (24%), the Nature Conservancy (TNC) (4%), Montana Department of Fish, Wildlife and Parks waterfowl production areas and Montana Department of Transportation mitigation sites (13%), and industrial and individual private lands (18%).

SAMPLING METHODS

We designed the sampling methods to allow 1 to 2 hours for the collection of chemical, macroinvertebrate, and algae samples for each wetland. Water and sediment samples for chemical analysis and macroinvertebrates and diatoms were collected from each site and returned to the laboratory for processing. Samples were collected from locations representing the various habitats within each wetland. These locations were easily accessible with hip boots. I also recorded field chemical measurements, observations, and photographs at this location. To reduce sample variability, I collected all the samples myself and made all the observations for all the wetlands evaluated.

Water column samples were analyzed for common ions, nutrients, total organic carbon, and total recoverable metals. pH, conductivity, salinity, dissolved oxygen, turbidity, and temperature were measured in the field using a Horiba U-10 water quality checker. Composite surface sediment grab samples (six per site) were collected for metal analyses. I documented the date and time samples were collected, wetland identification and location, ownership, ecoregion, and potential sources, causes, and degree of impairment, and approximate wetland area and maximum depth. I also recorded field observations such as substrate texture, percent open water, water color, vegetation in the upper watershed, a broad description of riparian and aquatic vegetation, anthropogenic activities within the watershed, hydrogeomorphic features, and the relative abundance of wildlife.

Following methods of Bahls (1993), I collected diatoms as composite grab samples from the natural substrate, such as vegetation and sediment. I collected the diatom samples in a 250-mL plastic container and preserved them with Lugol's solution. The Academy of Natural Sciences of Philadelphia (ANSP) subsampled and identified the diatoms to the lowest feasible taxonomic level (Charles et al. 1996). Using a 1-mm-mesh D-net in a sweeping motion, I collected macroinvertebrates from all habitat types at each sampling location for approximately 1 minute. Macroinvertebrates and associated materials were combined into a simple composite and placed into a 1-L plastic container with ethanol.

Because it took up to 18 hours to sort individual macroinvertebrate samples as a 300-organism subsample, we modified the subsampling protocol to a 200-organism subsample. We identified a lower threshold of 125 organisms based on a scatter plot of number of individuals versus total taxa. At 17 of the 80 sites, an insufficient number of organisms were present in the sample to permit subsampling, so the entire sample was sorted and identified. Seven wetland sites had fewer than 125 organisms. These sites were saline, highly alkaline, ephemeral, or impaired.

The taxonomic level of identification is based on Montana stream protocols (Bukantus 1994). Organisms found in the sample were identified to the genus level whenever possible unless a different level of identification was specified. Species-level identifications were made for Amphipoda. Identifications were made to the family or order level only when appropriate (e.g., for the Empididae) or when the individual was either too damaged or such an early instar that certain identification was not possible (Stribling et al. 1995). Several taxa were eliminated from consideration for metric development, so only estimates were made of their abundance. These taxa were either too difficult to sort and identify, are known to have widely fluctuating populations, or were considered to be uninformative in reflecting water quality (Stribling et al. 1995). These taxa represent nonbenthic adults of taxa with benthic larva (Dytiscidae and Hydrophilidae), semiaquatic surface dwellers (Gerridae, Collembola), nonbenthic crustacea (Ostrocoda, Anostraca, Copepoda, Cladocera), and other nonbenthic Hemiptera (Notonectidae, Corixidae) (Stribling et al. 1995). For example, many true bugs and adult beetles that are found in wetland habitats may have recently colonized, and for that reason may not reflect the environment from which they were found (Kantrud et al. 1989, Peckarsky et al. 1990). Freshwater crustaceans such as a fairy shrimp are difficult to use for biocriteria because they only inhabit temporary pools or saline environments and often complete their life cycles in as few as 16 days, which enables them to lay their eggs before predaceous aquatic insects colonize the habitat or reach sufficient size to decimate their populations (Peckarsky et al. 1990). It should be noted that these organisms were not considered to be unimportant to wetland ecology and that they may be useful for future development of wetland biocriteria. For instance, many of the invertebrates that we omitted were found to be useful for classifying wetlands in western Australia (Growns et al. 1992) and the Grande Teton and Yellowstone National Parks (Duffy 1999). Also, Helgen (1999) has determined that the relative abundance of Corixidae to all other aquatic Hemiptera and Coleoptera is a useful metric to consider for depressional wetlands in Minnesota. Also, ostracods have been useful for reconstructing palaeoenvironmental records (De Decker and Forester 1988).

WETLAND CLASSIFICATION

A priori Classification

We developed a wetland classification system based on physical and chemical data (Shapley 1995). Using topographic maps, field observations, and information gathered from land management agencies, we documented geomorphic characteristics such as the area of the wetland, area of contributing drainage basin, elevation of the wetland surface, maximum elevation within the drainage basin, and net relief of the contributing basin. We also recorded the existence of surface water inlets and outlets, artificial impoundments and significant water imports, and collected geologic setting information such as the primary surficial and bedrock geologic units. Interpolated mean annual climatic data, including annual precipitation, estimated mean annual evaporation, and net precipitation, were extracted from the Montana Agricultural Potentials System database. A hydrogeomorphic database was created that included the geomorphic, geologic, and climatic data. A bibliography was assembled that had all known sources of geologic, hydrologic, and water quality information. Maps were generated for each wetland using a geographic information system (GIS). Map features included were hydrologic delineations, land management areas, counties, and cities, major transportation corridors, wetland watershed boundaries, and sampling locations.

Post-sampling Classification

We found that wetland biological communities varied widely among the reference sites. Hydroperiod, hydrologic functions, the source of the water, geomorphology, and climate were the major causes of the variability. We used this information to further stratify the natural variability of biological communities. According to Meeks and Runyon (1990), hydrologic content affects primarily the chemical and physical aspects of wetlands, which, in turn, affect the biological component. Kantrud et al. (1989) also found that the source of the water plays an important role in determining wetland water quality and biology. For example, atmospheric water tends to be low in dissolved salts, surface runoff tends to be intermediate, and groundwater, depending on the characteristics of the substrate, tends to be high. Swanson et al. (1988) determined that wetland hydrologic functions control the chemical characteristics of wetlands, and as a result, plant and invertebrate communities. For example, closed basin wetlands (those without surface water outlets) that receive precipitation predominantly tend to be dilute. Groundwater flow-through systems that are closed basins tend to be higher in dissolved salts. Closed basin wetlands, which lose water to the atmosphere by evaporation generally have the highest concentration of dissolved salts. According to Winter (1977), the most important variables to consider in the hydrologic classification of lentic water bodies, such as lakes and wetlands, to predict water-column chemistry are concentration of dissolved substances in groundwater, precipitation–evaporation balance, streamflow inlet and outlet, the ratio of drainage basin area to water level area, depth, local relief, regional slope, and regional position. Ecoregions are also often used by states to group or classify water bodies from regions with similar variables such as climate, landforms, hydrology, and so on (Omernik and Bailey 1997).

In developing the Montana wetland classification system, we first used ecoregions to explain and sort out the natural variability of macroinvertebrate and diatom communities. Then we incorporated hydrogeomorphology to help explain the variation within each of the assigned ecoregions. Then we used water chemistry, such as pH, alkalinity, and conductivity (salinity), to assist in the final delineation of wetland classes. For example, Duffy (1999) found conductivity and pH useful to differentiate three wetland classes in the Grande Teton and Yellowstone National Parks. However, we often found chemistry difficult to use for classification because we could not always determine if the salinity was the result of natural conditions or human activities, and because photosynthesis/respiration often cause pH to fluctuate dramatically over a 24-hour period. Finally, we also used vegetation, and the presence or absence of amphipoda, which cannot tolerate hydroperiods of less than 1 year (Euliss et al. 1999), to confirm that wetlands were ephemeral; and the presence or absence of snails, which cannot tolerate excessive salinity (Euliss et al. 1999), to confirm that wetlands were saline.

Barbour (1992) has determine that metrics derived from stream macroinvertebrates are often able to discriminate between montane and valley/plains groupings of ecoregions. Bahls et al. (1992) determined that reference streams within each of Montana's three major geographical regions were similar to one another. These regions included the mountain, valley and foothill, and plains regions. Both Stribling et al. (1995) and Charles et al. (1996) determined that wetlands located in the rocky mountain, intermountain valley and prairie foothills, and plains ecoregions had significantly different biological communities. However, they detected no significant differences among ecoregion subclasses.

Hydrogeomorphology has been used extensively to classify water bodies (Winter 1977, Cowardin et al. 1979, Rosgen 1996). Brinson (1993) has developed and is continuing to refine a hydrogeomorphic approach (HGM) to classify wetlands. Our approach to wetland classification follows concepts similar to those of HGM, but our wetland classes were also evaluated to determine if reference sites within each class had similar macroinvertebrate and diatom communities.

We currently have delineated 10 wetland classes using ecoregions, hydrogeomorphology, and water-column chemistry (Figure 7.1). This classification system is likely to change as more data are collected and additional wetland reference sites are evaluated. We classified wetlands using the following ecoregions: rocky mountain, intermountain valleys and prairie foothills, and plains ecoregions. Using hydrogeomorphic features, we further stratified the wetlands as headwater wetlands, riparian wetlands, open lake wetlands, and closed basin wetlands. Our wetland classes are synonymous with HGM classes or a hybrid of HGM plus other criteria (Table 7.1). Headwater wetlands were located at regionally high elevations such as the headwaters of a stream, receive precipitation primarily from snowmelt and have low pH and conductivity. Riparian wetlands were associated with a stream, large spring, or are calcareous fens that had relatively large watersheds or received groundwater. Water

TABLE 7.1. Comparison of Biocriteria and HGM Wetland Classes

Biocriteria Wetland Class	HGM Wetland Classes	Description
Headwater wetlands	Extensive peatland	High elevation wet meadow or bog
Riparian wetlands	Riverine	Floodplain
	Slope	Break in slope where groundwater is discharged (e.g., spring)
	Extensive peatland	Rich fen
Open lake wetland	Fringe	Shoreline of lacustrine
Closed basin wetland	Depressional	Pothole
	Fringe	Large shallow reservoir or playa lake

received by riparian wetlands has longer contact with the surrounding soil and geology, so these wetlands tend to have higher pH and conductivity values than those of headwater wetlands. Open lake wetlands were ponded sites having a surface water inlet and outlet and relatively stable water levels. Closed basin wetlands were lacking surface water outlets or had outlets that were poorly developed.

Closed basin wetlands ranged from depressional (pothole) wetlands to shallow closed basin reservoirs. Therefore, we refined the wetland classification system to further stratify closed basin wetlands using several chemical and hydrogeomorphic variables such as salinity, alkalinity, their landscape position, hydroperiod, and watershed/wetland area ratios. The following are examples of the rationale for classifying closed basin wetlands using chemical and hydrogeomorphic variables. Closed basin wetlands at regionally high positions in the landscape tend to receive predominantly surface water are dilute, and tend to recharge to the groundwater (Euliss et al. 1999). Closed basin wetlands that predominantly receive groundwater generally are at regionally low positions in the landscape and are saline due to evaporation (Euliss et al. 1999). Closed basin wetlands that have groundwater flowing through them are often perennial, deep, and tend to be alkaline. Closed basin wetlands that have large watershed/wetland area ratios tend to receive predominantly surface water and often have widely fluctuating water levels during the course of a year.

Several wetland types were difficult to classify. Wetlands such as potholes located in the arid west, including Montana, have high spatial and temporal variability. For these wetlands, the hydrology, water chemistry, and biology can change dramatically throughout a season or from year-to-year as a result of climate change (Euliss et al. 1999). For example, the biological community often changes due to a shorter hydroperiod or an increase in salinity caused by drought (Hershey et al. 1999). The classification system will need to address year to year variability for those wetlands where a change in climate dramatically changes the biological communities being measured. It is also likely that macroinvertebrate and algal communities of forested wetlands will change with seral stage. Another type of wetland that is difficult to classify is large shallow reservoirs. These wetlands are hydrologically manipulated with controlled outlets such as dams or dikes. Large shallow reservoirs can be managed as closed basins (no surface water outlets) or as open lakes (with surface water outlets). These wetlands were difficult to classify because the hydrologic features that were used for classification were subject to human manipulation.

ANALYTICAL METHODS

We used the multimetric approach to evaluate wetland macroinvertebrate and diatom communities. The multimetric approach incorporates many attributes into the assessment process and has the ability to integrate biological information to provide an overall indication of biological condition or ecological health (Barbour et al. 1995). Multivariate analysis complements the development of multimetric indices, as it can be used by researchers to detect ecologically meaningful patterns for the development of metrics or for classification. Therefore, it is advantageous to use both approaches in developing biocriteria (Gerritsen 1995, Norris 1995, Cao et al. 1996, Reynoldson et al. 1997, Karr and Chu 1999).

Multivariate analysis is a statistical approach used by biologists to determine relationships among biota such as diatoms or macroinvertebrates, and environmental variables such as water-column chemistry. Multivariate analyses require the sampling of many reference sites for two things: biota, and a wide range of environmental variables such as water-column chemistry. These sites are then classified or grouped based on biota uniformity, which are described by the environmental variables (Norris 1995). Multivariate analysis is a useful exploratory approach that can help uncover patterns when only a little is known about the natural history of a place or biological community and can play a significant role in the classification of aquatic resources and for developing biocriteria (Reynoldson and Metcalfe-Smith 1992, Gerritsen 1995). For example, Growns et al. (1992) used multivariate pattern analysis of invertebrate communities and chemical variables to classify wetlands in western Australia, and Kerans and Karr (1994) used principal components analysis to determine the association of individual invertebrate attributes and water quality for rivers of the Tennessee Valley. Hawkins et al. (1997) also used multiple multivariate regression analysis to both quantify differences in insect assemblages among California streams and to determine the degree to which assemblage structure was related to temperature. Multivariate methods are attractive because they require no prior assumptions either in creating reference site groups or in comparing test sites with reference groups. Unfortunately, potential users may be discouraged by the complexity of multivariate techniques (Reynoldson et al. 1997).

Algae

Detrended canonical correspondence analysis (DCCA) and two-way indicator analysis (Twinspan) were used to investigate relationships between wetland diatom assemblages and environmental variables (mostly water-column chemistry). We graphically displayed clusters of diatoms with similar composition using DCCA (Charles et al. 1996). Longer vectors show a stronger correlation among diatom assemblages and environmental variables (Fig. 7.2*a*). Using DCCA, we determined that phosphorus, conductivity, and pH were strongly correlated with differences in diatom assemblages (Fig. 7.2*a*). There also appeared to be a correlation between diatom assemblages and calcium and lead. However, the vectors are shorter (Fig. 7.2*a*). Diatoms were strongly correlated with phosphorus and conductivity in wetlands of western Kentucky (Stevenson and Yangdong 1996), with total phosphorus in the Florida Everglades (McCormick et al. 1996) and with pH in streams of the Mid-Atlantic Highlands region of the United States (Pan et al. 1996).

Both DCCA and Twinspan were again used to test and refine our wetland classification procedure and for detecting water quality impairment. The sites were a priori classified based on ecoregions and hydrogeomorphic data. For most of the a priori wetland classes, diatom assemblages clustered fairly well using DCCA (Fig. 7.2*b*); most of the polygons enclose symbols of the same type, showing that they are biologically similar even when impaired sites were included in the analysis (Charles et al. 1996). Ploygons based on water chemistry data enclosed all of the reference sites within each postsampling wetland class (Fig. 7.3). I evaluated outlier sites, such as WET59 and WET63, to determine if they were impaired or misclassified (Fig. 7.3*a*). I did this by examining my field notes and the chemical data. For example, I was able to determine that WET59 was receiving acid mine drainage and had elevated metals, and WET 63 was saline, had no vegatative buffer, and its immediate watershed was farmed intensively. I further evaluated wetland classes with highly variable diatom assemblages to determine if wetlands were misclassified or if additional class delineations were necessary. I did this by examining my field notes, and chemical, geomorphic, hydrologic, and climatic data. Outlier reference sites that had different macroinvertebrate communities and hydrogeomorphic features or chemistry that were not the result of anthropogenic impairment were considered to be misclassified. If considerable overlap occurred between classes (e.g., Fig. 7.3*b*), I evaluated them further to determine if they could be combined. Wetland a priori classes were combined if the diatom and macroinvertebrate communities, and the water chemistry of their reference sites, were similar.

The reference sites of three wetland classes—riparian, open lakes, and groundwater recharge closed basin—had significant overlap (Fig. 7.3*b*). Although these wetland classes had different hydrology, their diatom and macroinvertebrate assemblages were similar. For this reason, I concluded that these classes could be combined, although I recommend that more wetlands should be sampled and evaluated to confirm this decision. The alkali closed basin wetland class had three reference sites (i.e., WET47, WET53, and WET69) that appeared to be misclassified (Fig. 7.3*c*). These sites were hydrologically similar but biologically and chemically different from the rest of the reference sites in their class; consequently, this class may need further refinement.

Sites that were accurately classified but fell outside the enveloped reference sites were considered impaired (Fig. 7.3). The greater the distance between the outlier site and the centroid of the enveloped reference sites, the greater the impairment. Field notes and the location of the outlier sites often indicated the cause of impairment. For

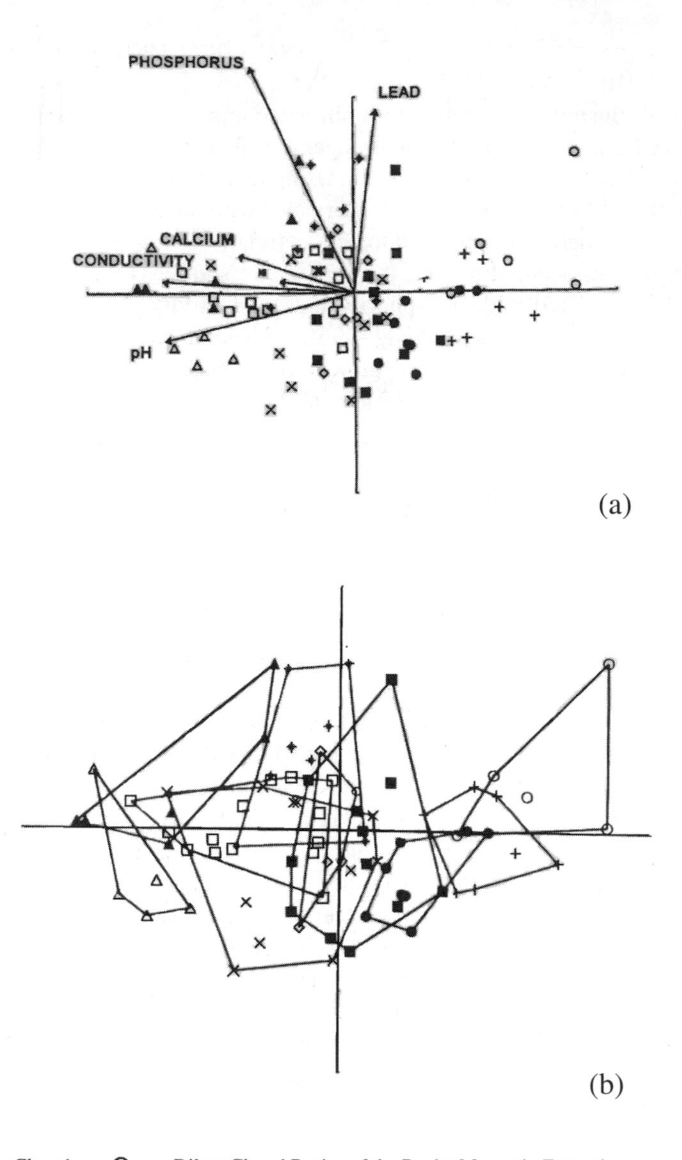

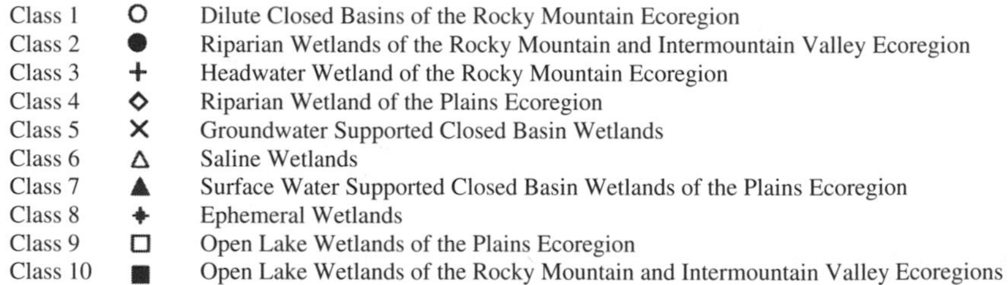

FIGURE 7.2. (*a*) Ordination of Montana wetland diatom assemblages and environmental variables. Conductivity, pH, and phosphorus had the longest vectors, indicating their strong correlation with the diatom assemblage. (*b*) Envelopes enclosed at least 95% of all sites in each a priori wetland class.

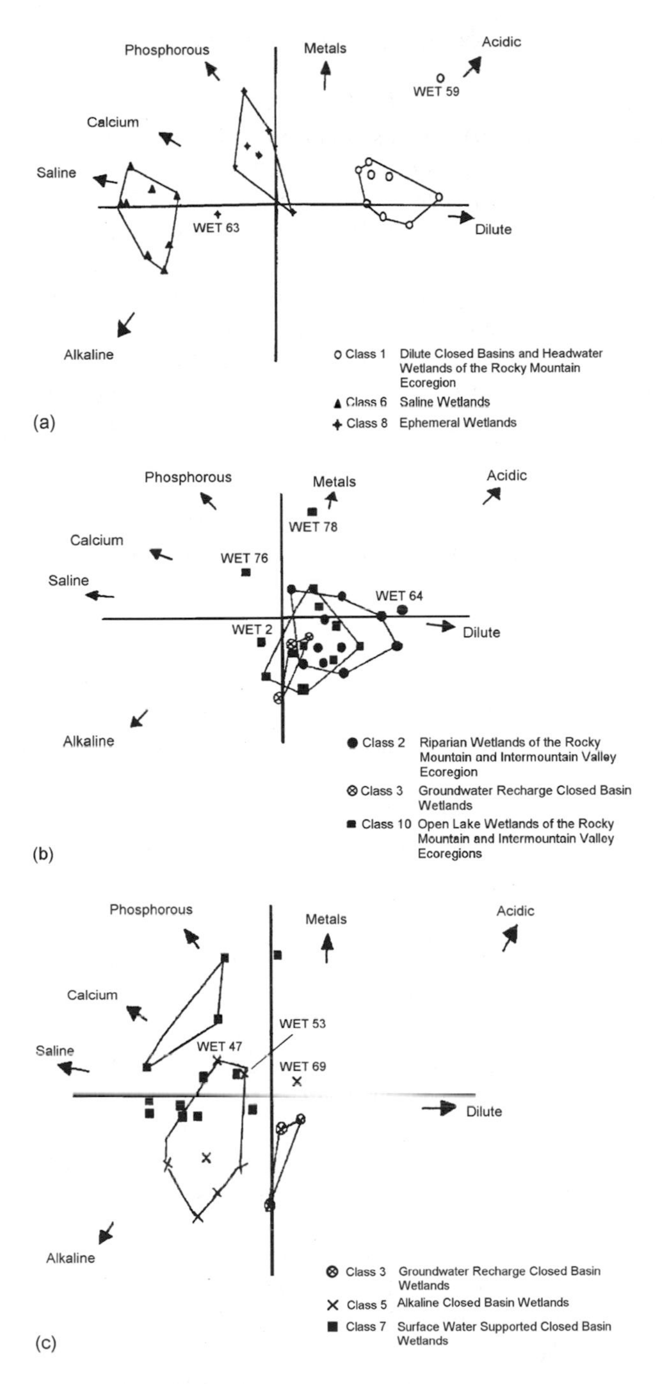

FIGURE 7.3. Diatom assemblage clusters and environmental variables that accounted for most of the assemblage variability. Reference sites of each wetland class are enveloped. Outlier sites are impaired or misclassified. Impaired sites were also detected using macroinvertebrate multimetric indices (Fig. 7.4). *Note:* Outlier sites located below class 7 reference sites (*c*) are hydrologically manipulated reservoirs with controlled surface water outlets.

example, outlier sites located in the far upper right quadrant may have received acid mine drainage (e.g., WET59), while salinity might have affected outlier sites (e.g., WET63) located toward the left side of the enveloped reference sites (Fig. 7.3*a*).

Twinspan has been used successfully in the United Kingdom for classifying macroinvertebrate communities in the River Trent system (Cao et al. 1996), and for wetlands in western Australia (Growns et al. 1992). Our Twinspan analysis produced seven distinct diatom groupings, similar to the DCCA clusters. The Twinspan analysis, however, did not distinguish ephemeral from perennial wetlands. Probably, the presence of water throughout the year is less important to diatoms than to macroinvertebrates since diatoms have shorter life spans and are ecological generalists with respect to hydroperiod (Adamus 1996). Twinspan also appeared to group wetlands that had high sediment. The sediment probably came from two sources: highly variable water levels that expose the sediment through wave action, or intensive agriculture that resulted in sediment from runoff. This Twinspan group was characterized by an abundance of motile diatom taxa, particularly *Nitzschia* and *Navicula* (Charles et al. 1996).

Macroinvertebrates

A *metric* is defined as an enumerated or calculated term representing some aspect of a community, such as structure, function, or some other measurable feature that changes in a predictable way in response to anthropogenic stress (Stribling et al. 1995). For example, a possible metric could be the number of taxa living in the wetland or the relative abundance of a taxon that tolerates pollution. *Multimetrics* are additive biological indices, meaning that they are the sum of several metrics (Gerritsen 1995). The combination of metrics provides a more consistent response to a broad range of human impacts and is used to determine the degree of impairment. Individual metrics can also be examined to develop a better understanding of the nature of the impairment (Barbour et al. 1995). Rapid bioassessment protocols that incorporate multimetrics are commonly used in the United States to assess the water quality of streams and rivers (Reynoldson and Metcalfe-Smith 1992, USEPA 1999).

One of the advantages of multimetric indices is that they are relatively simple to calculate from collected data (Gerritson 1995). According to Norris (1995), one of the weaknesses of using multimetric indices is that it requires a thorough understanding of each individual metric's ecological relationship to anthropogenic stress in order to evaluate water quality conditions. For example, a high diversity of macroinvertebrates may indicate the reference condition for most wetland types, but for some types, such as forested wetlands, a higher diversity of macroinvertebrates may indicate impairment. Further, metrics are often redundant in a combination index and errors can be compounded, lessening the ability of the multimetric approach to detect impairment accurately (Reynoldson et al. 1997). Because of this we evaluated both the proposed metric's and associated environmental data in an attempt to develop an understanding of ecological relationships, to test each proposed metric's ability to predict various anthropogenic stressors, and to determine redundancy. Several metrics were eliminated because they were redundant or because they did not have the ability to detect impairment (Stribling et al. 1995). Other metrics were combined in an effort to develop ecologically meaningful metrics that would predict impairment with greater consistency (Table 7.2). For example, the average was used for the "percent 1, 2, and 5 dominant taxa" metrics because these metrics were highly correlated ($r > 0.75$), and the average of the "percent dominance" metrics was more consistent in responding to stress. Proposed metrics that were expressed in percent did not detect stress consistently and were difficult to interpret. For example, the "percent chironomidae" and "percent crustacea and mollusca taxa" metrics would often increase with moderate stress but decrease with severe stress. For this reason, an effort was made to lessen the response of the metrics that were derived from percentages and to combine them into metrics that counted the number of taxa within the taxonomic group of interest. For example, the "percent Crustacea and Mollusca taxa" metric was combined with the "number of Crustacea and Mollusca taxa" metric. The resulting metrics were far less likely to give false signals and were designed to decrease with stress, which they usually did, although their response to stress was marginal for some wetland types.

Seven macroinvertebrate measurements appeared to be useful for developing multimetric indices for detecting wetland water quality impairment (Table 7.2). I chose metrics that appeared to detect impairment of several types of anthropogenic stressors for the majority of the wetland classes. However, these metrics did not work for saline or highly alkaline wetlands for two major reasons: (1) the community composition of saline wetlands were very different from the other wetland types, and (2) alkaline closed basin wetlands need further stratification because the biological communities of the reference sites were highly variable, due, in part, to fluctuating salinities and water levels. Similarly, it was also difficult to detect impairment for shallow depressional wetlands because they also tended to have highly variable macroinvertebrate communities caused by fluctuating water levels and salinities. Macroinvertebrates within these wetlands are exposed to harsh environmental conditions, ranging from frigid winter temperatures to hot summers, seasonal drying, and steep salinity gradients. Consequently, it is difficult to develop biocriteria for these wetlands because the macroinvertebrates that live there tend to be ecological generalists that possess the necessary adaptations to tolerate environmental extremes (Euliss et al. 1999). Our macroinvertebrate indices were most useful for detecting impairment of perennial wetlands with relatively stable water levels and with low alkalinity and salinity.

I did not score the proposed metrics using 1–3–5 where severely impaired is 1, slightly impaired is 3, and least impaired is 5, as in Gerritson (1995) and Karr et al. (1986). Instead, I designed the metric calculations and scores to reflect slight changes in the macroinvertebrate community and weighted them according to their ability to respond to anthropogenic stressors (Table 7.2). The weighting of the metrics was a two-step approach that included scaling and scoring which was determined by trial and error until the desired effect was obtained (Table 7.3). We examined how consistently and to what degree each metric responded in wetlands that had known stress. We also

TABLE 7.2. Proposed Metrics, Proposed Metric Calculations, and Score Calculations Used for Developing Wetland Macroinvertebrate Indices

Proposed Metrics	Theorized Direction of Change in Presence of Stressor	Proposed Metric Calculation	Score Calculation
Number of taxa	Decrease	Count taxa	(Number of taxa) × 0.75
Percent dominance	Increase	Average of percent 1, 2, and 5 most dominant taxa	(100 – percent dominance) × 0.36
Percent 1 dominant taxon	Increase		
Percent 2 dominant taxa	Increase		
Percent 5 dominant taxa	Increase		
POET	Decrease	Count Plecoptera, Odonata, Ephemeroptera, and Tricoptera taxa	(POET) × 3
Number of individuals	Decrease	Count individuals in total sample (maximum count of 300)	(Number of individuals) ÷ 33
Chironomidae	Decrease	(Number of Chironomidae taxa × 100) – (% Chironomidae + 50) × (% Orthocladiinae of total Chironomidae/100 + 0.5)	(Chironomidae) ÷ 83
Number of Chironomidae taxa	Decrease		
Percent Chironomidae taxa	Increase		
Percent Orthocladiinae/Chironomidae	Decrease		
Crustacea/mollusca	Decrease	(Number of Crustacea and Mollusca taxa × 100) – (% Crustacea/Mollusca taxa + 50)	(Crustacea/Mollusca) ÷ 33
Number of Crustacea and Mollusca taxa	Decrease		
Percent Crustacea and Mollusca taxa	Increase		
Leech/sponge/clam	Decrease	Count Hirudinea, Porifera, and Sphaeriidea taxa	(Leech/sponge/clam) × 3

TABLE 7.3. Example of Metric Calculations, Weighting, Scaling, and Scoring for Pine Butte Swamp, and the Average and Maximum Scores for the Entire Wetland Data Set[a]

Metric	Example of Proposed Metric Calculation, Scaling, and Scoring (WET60, Pine Butte Swamp)					Average and Maximum Metric Scores for the Entire Data Set (80 Wetlands)	
		Metric Weighting					
Metric	Initial Score	Metric Calculation[b]	Scaling Factor	Scoring Factor	Score	Average Score	Maximum Score
Number of taxa	NA	26	NA	× 0.75	20	15.4	28
Percent dominance		(100 – 42)	÷ 8.33	× 3	21	16.0	25
Percent 1 dominant taxon	25			Note:			
Percent 2 dominant taxa	38			÷ 8.33 × 3 = × 0.36			
Percent 5 dominant taxa	62						
POET	NA	7	NA	× 3	21	9.9	24
Number of individuals	NA	300	÷100	× 3	9	6.9	9
				Note:			
				÷ 100 × 3 = ÷ 33[c]			
Chironomidae		13 × (100 – 33 + 50) ×	÷ 250	× 3	16	9.7	22
Number of Chironomidae	13	[(36 ÷ 100) + 0.5]		Note:			
Percent Chironomidae taxa	33			÷ 250 × 3 = ÷83[c]			
Percent Orthocladiinae/ Chironomidae	36						
Crustacea/Mollusca		5 × (100 – 42 + 50)	÷100	× 3	16	8.8	22
Number of Crustacea and Mollusca taxa	5			Note:			
				÷ 100 (3 = ÷ 33[c]			
Percent Crustacea and Mollusca taxa	42						
Leech/sponge/clam	NA	2	NA	× 3	6	2.7	18

[a]NA, not applicable.

[b]Refers to metric calculations proposed in Table 7.2.

[c]Note refers to the combination of scaling and scoring in the metric score calculations in Table 7.2.

examined a metric to determine how frequently and to what degree it gave a false response which indicated that a reference site was impaired. For example, the "number of taxa" and "percent dominance" metrics consistently provided a strong response to stress and were least likely to give a false response for all wetland types. Therefore, their scores were given the greatest weight; that is, the average and maximum scores for these metrics were highest when applied to all wetlands (Table 7.3). The "number of individuals" metric was the least responsive to stress and it frequently provided a false response. For example, a number of impaired wetlands had an abundant number of organisms, which were taxa that were tolerant of stress. However, other severely impaired wetlands had only a few organisms. Also, the "Leech/sponge/clam" metric was not very useful for assessing many of the wetland classes, as the taxa within this metric did not inhabit all the wetlands sampled. Therefore, scores for these metrics were weighted the least; their average and maximum scores were the lowest when applied to all the wetlands (Table 7.3). We were careful not to provide too much weight to any one metric because we wanted a biological index that truly integrated multiple factors. The calculations in Table 7.2 were used to combine and weight the metrics and to scale their scores. Table 7.3 gives an example of how the metrics were calculated, scaled, and scored, and also provides the average and maximum scores for the entire data set. All metrics were designed to decrease with stress. For example, the inverse was taken of the "percent dominance" metric, which was hypothesized to increase with stress.

For several of the metrics, the maximum scores were theoretically infinite, but they did tend to level off as they approached their maximum potential (reference condition). For example, the maximum "number of taxa" for a riparian wetland and "ephemeral" wetland was approximately 28 and 15, respectively. Since each metric had a different maximum for each wetland class, the macroinvertebrate index scores for reference conditions also differed between wetland classes. Eventually, we intend to normalize the scoring so that the reference condition for all wetland classes will have the same score.

I developed decision thresholds of optimal, good, fair, and poor to determine wetland biological conditions (Fig. 7.4). The decision thresholds incorporated the evaluation of sites with documented impairment and the variability of reference site index scores within each wetland class. Physicochemical and historical data, field notes, and photographs assisted in the delineation of the decision thresholds. Our confidence that our decision thresholds represented real differences in condition varied with each wetland class because it depended, in part, on the number of reference and impaired sites sampled. The threshold for optimal conditions was set near the reference site with the lowest score. If a reference site score was notably lower than those of other sites, I further evaluated the macroinvertebrate, diatom, and physiochemical data to determine if the site was misclassified, or impaired, or if there were sampling or analytical errors. The threshold for poor was placed slightly above the score of wetlands that were severely impaired. When there were no severely impaired sites within a wetland class, we used the scores from severely impacted sites of other classes where reference site scores were similar. The remaining good and fair decision thresholds were evenly spaced between the poor and optimal decision thresholds.

Class 1 Dilute Closed Basins and Headwater Wetlands of the Rocky Mountain Ecoroegion

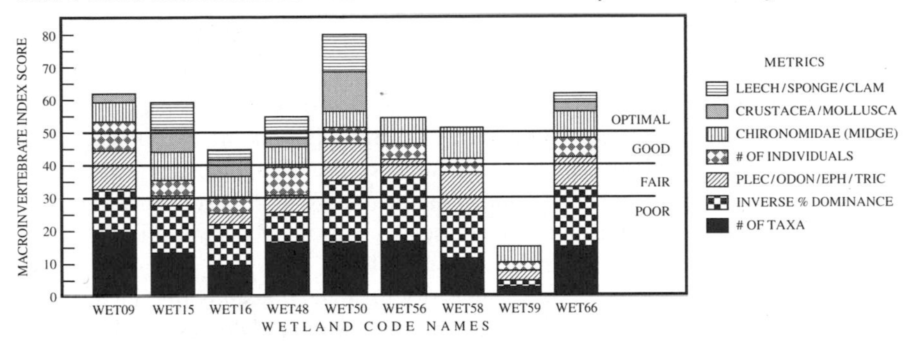

Class 8 Ephemeral Wetlands.

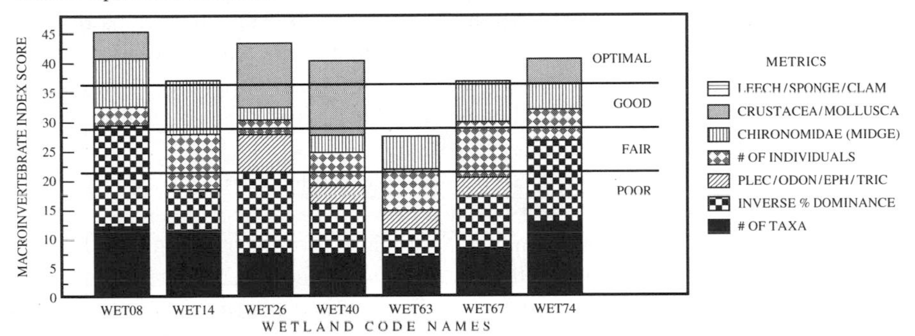

Class 10 Open Lake Wetlands of the Rocky Mountain and Intermountain Valley Ecoregions.

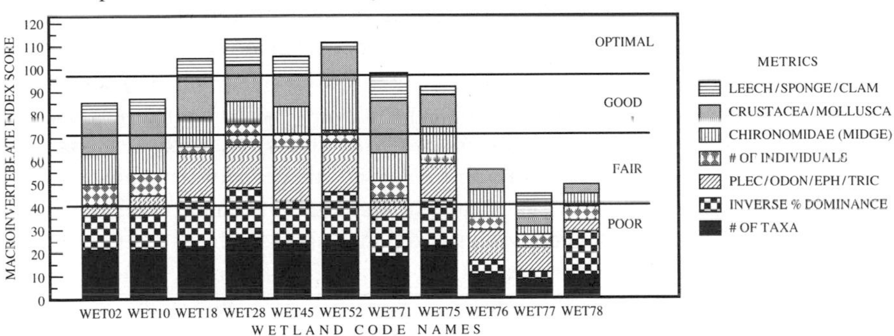

FIGURE 7.4. Macroinvertebrate multimetric index scores for wetland classes 1, 8, and 10. In class 1, site WET59 is impaired by acid mine drainage. In class 8, site WET63 is impaired by saline seeps. In class 10, sites WET76 and WET78 are impaired by elevated trace metal concentrations in the water column, and WET77 was sampled in an area that had poor habitat conditions.

Supplementary Research

Three other studies were conducted in Montana with the purpose of developing wetland biocriteria. Two of the studies involve sampling depressional wetlands in the Ovando and Mission valleys of western Montana. We selected depressional wetlands for more intensive research because of the highly variable hydrology, water-column chemistry, and biological communities. We also investigated how the diatoms in wetland sediment core samples can be used to infer past histories of water quality (Charles 1997), which involves the evaluation of paleolimnological approaches to biological monitoring (Charles et al. 1994).

Researchers from Montana State University conducted a study that included the development of vegetation biocriteria for western Montana depressional wetlands (Borth 1998). They maintain that vegetation is easier to assess than macroinvertebrates or diatoms for depressional wetlands that are seasonally dry. Their study design included the sampling of vegetation, macroinvertebrates, and diatoms from 24 depressional wetlands with similar climate, hydrology, and water chemistry. Their study also involved sampling across three levels of human impairment—minimally impacted, slightly impacted, and moderately impacted—involving two anthropogenic disturbances—dryland agriculture and grazing.

Researchers from the University of Montana conducted a study to determine the effects of natural variability on the use of macroinvertebrates as bioindicators of disturbance in intermontane depressional wetlands in northwestern Montana (Ludden 2000). Their study design included the collection of macroinvertebrate samples and physiochemical data from 15 pristine and six disturbed intermontane prairie potholes. Their study also included the analysis of macroinvertebrate samples that were previously collected by Borth (1998). The researchers collected macroinvertebrate samples across three seasons and from three wetland zones. They used multivariate Detrended Correspondence Analysis to ordinate the raw macroinvertebrate data and physicochemical variables as secondary matrices to establish vectored biplots of correlation. Candidate metrics were analyzed using univariate analysis. They determined that no environmental variables were strongly correlated with the macroinvertebrate data, and many of the metrics varied across wetland zones and across seasons. Nevertheless, 35-candidate metrics were able to discriminate between minimally and highly disturbed sites. These wetlands are also being sampled intensively to develop and test a model for assessing depressional wetlands using the hydrogeomorphology (HGM) functional assessment approach (Brinson and Rheinhardt 1996, Federal Register 1996, Rheinhardt et al. 1997, Hauer et al. 1999). After this study, researchers could share data and link biological criteria to HGM.

CONCLUSIONS

The bioassessment protocols that we are developing appear to be able to detect anthropogenic impacts on wetland water quality for several wetland types. These wetlands tended to be permanent and had relatively stable water levels. Except for three wetland sites, both the multimetric approach using macroinvertebrates and the multivariate approach using diatoms identified the same wetlands as having a similar biological condition in eight of the wetland classes. Neither approach was useful for determining the biological condition of saline or highly alkaline wetlands with fluctuating water levels. However, diatom biocriteria could be more promising for detecting impairment in saline (Blin 1993) and alkaline environments because the diatom communities are more diverse and might therefore provide better ecological resolution.

Aquatic macroinvertebrate and diatom biocriteria are not very useful for detecting impairment in ephemeral wetlands that usually lack surface water. Vegetation biocriteria are likely to be more appropriate for assessing the biological condition of these wetland types (Borth 1998). Due to longer life spans, macroinvertebrates are more useful for detecting habitat impairment because they are more likely than diatoms to integrate impairment over time (Adamus 1996). Diatoms appeared to be most responsive to high levels of nutrients, salinity, sediment, acidity, and metals in the water column that existed during the time of sampling or that occurred 2 to 3 weeks prior to sampling (Charles et al. 1996). Diatoms that are found in sediment core samples may also be useful for inferring historical changes in wetland water quality for some wetland types (Fritz 1990, Charles et al. 1994, Charles 1997).

We were unable to determine if wetlands were affected by elevated trace metal concentrations in the sediment by assessing the diatom or macroinvertebrate communities. Apparently, elevated trace metal concentrations in the sediment do not necessarily have obvious impacts on the community structure of diatoms or macroinvertebrates that live in the water column or on the surface of the bottom sediment. Yet benthic invertebrates are known to accumulate metals through dietary ingestion of sediment-associated metals (Moore et al. 1991, Schlekat et al. 2000) or exposure to metals in sediment porewater (Besser et al. 1996), and others have found that macroinvertebrate community structure exhibited a predictable response to elevated sediment metal concentrations (Canfield et al. 1994, Diggins and Stewart 1998, Gurrieri 1998). We also had difficulty determining if physical disturbances such as sedimentation or habitat alteration were affecting a wetland's biological community because the physical disturbances were only qualitatively described, not measured. A more quantitative approach is needed for the measurement of wetland physical characteristics if we are to link physical disturbances to changes in biological communities.

We believe that multimetric indices are the most appropriate tool for portraying biocriteria, as they are easier than multivariate analysis to interpret and explain to the public. Nonetheless, the multivariate approach was useful for detecting wetland water quality impairment, and therefore may also be useful for detecting ecologically meaningful patterns that can be used for developing metrics. We found DCCA to be useful, as it often indicated which measured environmental variables were causing impairment, such as elevated nutrients or trace metals. It was, however, difficult for us to evaluate the degree of impairment.

We feel that classification is one of the most important components for developing successful wetland biocriteria. Wetland hydrology, water-column chemistry, and biological communities tend to be highly variable both spatially and temporally.

It is difficult to develop a classification system that will stratify the natural variability of wetland biological communities so that anthropogenic impacts can be detected consistently. Hydrogeomorphology appears to be a useful approach to wetland classification. Still, the relationship between wetland biological communities such as diatoms and macroinvertebrates and hydrogeomorphology is not well understood.

Multivariate analysis is an important tool for wetland classification because it is an objective approach that detects patterns in biological assemblages caused by natural variability, such as wetland site groupings. Consequently, a researcher can use multivariate analysis to initiate the development of a classification system without bias by detecting patterns in biological data. These patterns can be used to test a priori preconceived conceptual approaches to wetland classification, such as the use of ecoregions, hydrogeomorphic characteristics, and natural history, to investigate the change of biological community structure along a pollution gradient (Cao et al. 1996). However, the multivariate techniques can be complicated, and results can be difficult to interpret (Charles et al. 1996, Reynoldson et al. 1997). When multivariate analyses are used inappropriately, they may fail to discriminate among important sources of variability, such as investigator-induced variation caused by sampling, subsampling, and error (Karr and Chu 1999). Thus a number of natural factors that affect biological assemblages may not be detected, or some of the patterns may have no biological meaning. For example, investigators may be tempted to eliminate rare species when performing multivariate analysis, which can damage the sensitivity of community-based methods to detect ecological changes (Cao et al. 1998, Karr and Chu 1999). Regardless of the analytical approach used, meticulous selection of least impaired reference sites is crucial, and there is no substitute for careful application of biological and ecological knowledge (Karr and Chu 1999).

ACKNOWLEDGMENTS

First and foremost, I thank Loren Bahls, Montana Department of Environmental Quality, for his assistance in designing this study and for his insights on the biocriteria process and appropriate bioassessment approaches. Thanks also to James Stribling, Michael Barbour, Joyce Lathrop-Davis, Jeffery White, and Eric Leppo of Tetra Tech, who identified the macroinvertebrates and initiated the development of the multimetric indices; Donald Charles, Frank Acker, and Norma Roberts, who identified the diatoms and performed the multivariate analysis, and Mark Shapley, who collected the hydrogeomorphic data and provided the interpretations for developing the wetland classification procedure. I would also like to thank Susan Jackson, Tom Danielson, and others from the EPA for providing funding and encouragement; my support staff, Kim Hoy and Christy Dighans and interns Kathryn Williams and Russ Gates, who helped me with the figures and formatting; and my supervisors, Gary Ingman and Bob Bukantus, who gave me the freedom to pursue this endeavor even when there were other burning issues.

REFERENCES

Adamus, P. R. 1996. Bioindicators for assessing ecological integrity of prairie wetlands. Report EPA/600/R-96/082. U.S. Environmental Protection Agency, Environmental Research Laboratory, Corvallis, OR.

Adamus, P. R., and K. Brandt. 1990. Impacts on water quality of inland wetlands of the United States: a survey of indicators, techniques, and applications of community-level biomonitoring data. Report EPA/600/3-90/073. U.S. Environmental Protection Agency, Environmental Research Laboratory, Corvallis, OR.

Bahls, L. L. 1993. Periphyton bioassessment methods for Montana streams. Montana Department of Health and Environmental Sciences, Water Quality Bureau, Helena, MT.

Bahls, L. B., R. Bukantis, and S. Tralles. 1992. Benchmark biology of Montana reference streams. Montana Department of Health and Environmental Sciences, Water Quality Bureau, Helena, MT.

Barbour, M. T., J. B. Stribling, and J. R. Karr. 1995. Multimetric approach to establishing biocriteria and measuring biological condition. Pages 63–77 *in* W. S. Davis, and T. P. Simon (eds.), Biological assessment and criteria: tools for water resource planning and decision making. Lewis Publishers, Boca Raton FL.

Barbour, M. T., J. L. Plafkin, B. P. Bradley, C. G. Graves and R. W. Wisseman. 1992. Evaluation of EPA's rapid bioassessment benthic metrics: method redundancy and variability among reference stream sites. Environmental Toxicology and Chemistry 11:437–449.

Besser, J. M., C. G. Ingersoll, and J. P. Giesly. 1996. Effects of spatial and temporal variation of acid-volatile sulfide on the bioavailability of copper and zinc in freshwater sediments. Environmental Toxicology and Chemistry 15(3):286–293.

Blin, D. W. 1993. Diatom community structure along physiochemical gradients in saline lakes. Ecology 74(4):1246–1263.

Borth, C. 1998. Effects of land use on vegetation in glaciated depressional wetlands in western Montana. M.S. thesis, Montana State University, Bozeman, MT.

Brinson, M. 1993. A hydrogeomorphic classification of wetlands. Technical Report WRP-DE-4, U.S. Army Corps of Engineers. Wetlands Research Program, Washington, DC.

Brinson, M. M., and R. Rheinhardt. 1996. The role of reference wetlands in functional assessment and mitigation. Ecological Applications 6:69–76.

Bukantis, R. 1994. Rapid bioassessment macroinvertebrate protocols: sampling and sample analysis SOPs. Montana Department of Environmental Quality; Planning, Prevention and Assistance Division, Helena, MT.

Canfield, T. J., N. E. Kemble, W. G. Brumbaugh, F. J. Dwyer, C. G. Ingersoll, and J. F. Fairchild. 1994. Use of benthic invertebrate community structure and the sediment quality triad to evaluate metal-contaminated sediment in the upper Clark Fork River, Montana. Environmental Toxicology and Chemistry 13(12):1999–2012.

Cao, Y., A. W. Bark, and W. P. Williams. 1996. Measuring the response of water communities to water pollution: comparison of multivariate approaches, biotic and diversity indices. Hydrobiolgia 341:1–19.

Cao, Y., D. D. Williams, and N. E. Williams. 1998. How important are rare species in aquatic community ecology and bioassessment? Limnology and Oceanography 43(7):1403–1409.

Charles, D. 1997. Analysis of sediment and periphyton diatom assemblages from Montana wetlands and assessment of ecological condition. Proposal submitted to Montana Department of Environmental Quality. Patrick Center of Environmental Research, Environmental Research Division, Academy of Natural Sciences, Philadelphia.

Charles, D. F., J. P. Smol, and D. R. Engstrom. 1994. Paleoreconstruction of the environmental status of aquatic ecosystems. Pages 233–293 *in* S. L. Loeb, and A. Spacie (eds.), Biological monitoring of aquatic systems. Lewis Publishers, Boca Raton, FL.

Charles, D. F., C. D. Acker, and N. A. Roberts. 1996. Diatom periphyton in Montana lakes and wetlands: ecology and potential as bioassessment indicators. Report submitted to the Montana Department of Environmental Quality. Patrick Center of Environmental Research, Environmental Research Division, Academy of Natural Sciences, Philadelphia.

Cowardin, L. M., V. Carter, F. C. Golet, and E. T. LaRoe. 1979. Classification of wetlands and deepwater habitats of the United States. Report FWS/OBS-79/31. U.S. Fish and Wildlife Biological Service Program, Washington, DC.

De Decker P., and P. M. Forester. 1988. The use of ostrocods to reconstruct continental palaeoenvironmental records. Pages 175–198 *in* P. De Deckker et al. (eds.), Ostrocoda in the earth sciences. Elsevier, New York.

Diggins, T. P., and K. M. Stewart. 1998. Chironomid deformities, benthic community composition, and trace elements in the Buffalo River (New York) area of concern. Journal of the North American Benthological Society 17(3):311–323.

Duffy, W. G. 1999. Wetlands of Grand Teton and Yellowstone National Parks. Pages 733–756 *in* D. Batzer, R. B. Rader, and S. A. Wissinger (eds.) Invertebrates in freshwater wetlands of North America. Wiley, New York.

Euliss, N. H., Jr., D. A. Wrubleski, and D. M. Mushet. 1999. Wetlands of the prairie pothole region: invertebrate species composition, ecology, and management. Pages 471–514 *in* D. Batzer, R. B. Rader, and S. A. Wissinger (eds.), Invertebrates in freshwater wetlands of North America. Wiley, New York.

Federal Register. Aug. 16, 1996. Regulatory Program of the U.S. Army Corps of Engineers. National action plan to develop the hydrogeomorphic approach for assessing wetland functions. Federal Register 61(160):42593–42603.

Fritz, S. C. 1990. Twentieth-century salinity and water-level fluctuations in Devils Lake, North Dakota: test of a diatom-based transfer function. Limnology Oceanography 35(8):1771–1781.

Gerritsen, J. 1995. Additive biological indices for resource management. BRIDGES: Integrating basic and applied benthic science. Journal of the North American Benthological Society 14(3):440–450.

Growns, J. E., J. A. Davis, F. Cheal, L. G. Schmidt, R. S. Rosich, and S. J. Bradley. 1992. Multiple pattern analysis of wetland invertebrate communities and environmental variables in Western Australia. Australian Journal of Ecology 17:275–288.

Gurrieri, J. T. 1998. Distribution of metals in water and sediment and effects on aquatic biota in the upper Stillwater River basin, Montana. Journal of Geochemical Exploration 64:83–100.

Hauer, F. R., B. J. Cook, M. C. Gilbert, E. C. Clairain, and D. R. Smith. 1999. A regional guidebook: assessing the functions of intermontane prairie pothole wetlands in the northern Rocky Mountains. University of Montana Flathead Lake Biological Station, Polson, MT.

Hawkins, C. P., J. N. Hogue, L. M. Decker, and J. W. Feminella. 1997. Channel morphology, water temperature, and assemblage structure of stream insects. Journal of the North American Benthological Society 16(4):728–749.

Helgen, J. 1999. Draft technical method for biological assessment of depressional wetlands, sampling and scoring metrics for a wetland index biological integrity (WIBI), invertebrate methods. Minnesota Pollution Control Agency, St. Paul, MN.

Hershey, A. E., L. Shannon, G. J. Niemi, A. R. Lima, and R. R. Regal. 1999. Prairie wetlands of south-central Minnesota: effects of drought on invertebrate communities. Pages 515–542 *in* D. Batzer, R. B. Rader, and S. A. Wissinger (eds.), Invertebrates in freshwater wetlands of North America. Wiley, New York.

Kantrud, H. A., G. L. Krapu, and G. A. Swanson. 1989. Prairie basin wetlands of the Dakotas: a community profile. Biological Report 85(7.28) U.S. Fish and Wildlife Service, Northern Prairie Wildlife Research Center, Jamestown, ND.

Karr, J. R., and E. W. Chu. 1999. Restoring life in running waters. Island Press, Washington, DC.

Karr, J. R., and D. R. Dudley. 1981. Ecological perspective on water quality goals. Environmental Management 5:55–68.

Karr, J. R., K. D. Fausch, P. L. Angermeier, P. R. Yant, and I. J. Schlosser. 1986. Assessing biological integrity in running water: a method and its rationale. Special Publication 5. Illinois Natural History Survey, Campaign, IL.

Kerans, B. L., and J. R. Karr. 1994. A benthic index of the biotic integrity (B-IBI) for rivers of the Tennessee Valley. Ecological Applications 4:768–785.

Ludden, V. E. 2000. The effects of natural variability on the use of macroinvertebrates as bioindicators of disturbance in intermontane depressional wetlands in Northwestern Montana, USA. M.S. Thesis. The University of Montana, Missoula. pp. 163.

McCormick, P., P. S. Rawlik, K. Lurding, E. P. Smith, and F. H. Sklar. 1996. Periphyton–water quality relationships along a nutrient gradient in the northern Florida Everglades. Journal of the North American Benthological Society 15(4):433–449.

Meeks, G., and C. L. Runyon. 1990. Wetland protection and the states. National Conference of State Legislatures. ISBN 1-55516-425-0.

Moore, J. N., S. N. Luoma, and D. Peters. 1991. Downstream effects of mine effluent on an intermontane riparian system. Canadian Journal of Fisheries and Aquatic Sciences 48:222–232.

Norris, R. H. 1995. Biological monitoring; the dilemma of data analysis. BRIDGES: integrating basic and applied benthic science. Journal of the North American Benthological Society 14(3):440–450.

Omernik, J. M. 1986. Ecoregions of the coterminous United States. Report RID 8769170 (map) U.S. Environmental Protection Agency, Corvallis, OR.

Omernik, J. M., and R. G. Bailey. 1997. Distinguishing between watersheds and ecoregions. Journal of the American Water Resources Association 33(5):935–949.

Pan, Y., R. J. Stevenson, B. H. Hill, A.T. Herlihy, and G. B. Collins. 1996. Using diatoms as indicators of ecological conditions in lotic systems: a regional assessment. Journal of the North American Benthological Society 15:481–494.

Peckarsky, B. L., P. R. Fraissinet, M. N. Penton, and D. J. Conklin. 1990. Freshwater macroinvertebrates of northeastern North America. Cornell University Press, Ithaca, NY.

Reynoldson, T. B., R. H. Norris, V. H. Resh, K. E. Day, and D. M. Rosenberg. 1997. The reference condition: a comparison of multimetric and multivariate approaches to assess water-quality impairment using benthic macroinvertebrates. Journal of the North American Benthological Society 16(4):833–852.

Reynoldson, T. B., and J. L. Metcalfe-Smith. 1992. An overview of the assessment of aquatic ecosystem health using benthic invertebrates. Journal of Aquatic Ecosystem Health 1:295–308.

Rheinhardt, R. D., M. M. Brinson, and P. M. Farley. 1997. Applying wetland reference data to functional assessment, mitigation, and restoration. Wetlands 17:195–215.

Rosgen, D. 1996. Applied river morphology. Library of Congress card no. 96-60962. Wildland Hydrology Consultants, Pagosa Springs, CO.

Shapley, M. D. 1995. Geologic, geomorphic and chemical characteristics of wetlands selected for use in biocriteria development by the Montana Department of Environmental Quality. Report submitted to the Montana Department of Environmental Quality. Montana Natural Heritage Program, Helena, MT.

Schlekat, C. E., A. W. Decho, and G. T. Chandler. 2000. Bioavailability of particle-associated silver, cadmium, and zinc to the estuarine amphipoda *Leptocheirus plumulosus* through dietary ingestion. Limnology and Oceanography 45(1):11–21.

Stevenson, R. J., and P. Yangdong. 1996. Gradient analysis of diatom assemblages in western Kentucky wetlands. Journal of Phycology 32:222–232.

Stribling, J. B., J. Lathrop-Davis, M. T. Barbour, J. White, and E. W. Leppo. 1995. Evaluation of environmental indicators for the wetlands of Montana: the multimetric approach using benthic macroinvertebrates. Report submitted to Montana Department of Environmental Quality Tetra Tech, Inc., Owings Mills, MD.

Swanson, G. A., T. C. Winter, V. A. Adomaitis, and J. W. LaBaugh. 1988. Chemical characteristics of prairie lakes in south-central North Dakota: their potential for influencing use by fish and wildlife. Fish and Wildlife Technical Report 18. U.S. Fish and Wildlife Service, Northern Prairies Research Center, Jamestown, ND.

USEPA. 1999. Rapid bioassessment protocols for use in wadeable streams and rivers: periphyton, macroinvertebrates and fish, 2nd ed. Report EPA 841-B-99-002. U.S. Environmental Protection Agency, Office of Water Regulations and Standards, Washington, DC.

USEPA. 1990. Biological criteria: national program guidance for surface waters. Report EPA 440-5-90-004. U.S. Environmental Protection Agency, Office of Water Regulations and Standards, Washington, DC.

Winter, T. C. 1977. Classification of the hydrologic settings of lakes in the north-central United States. Water Resources Research 13(4).

8 Monitoring the Condition of Wetlands: Indexes of Biological Integrity Using Invertebrates and Vegetation

JUDY C. HELGEN and MARK C. GERNES

Biological integrity or the condition of wetlands can be assessed using two discrete indexes of biological integrity (IBIs), each composed of 10 attributes of wetland vegetation or invertebrates. The IBIs, and the metrics that comprise them, show graded responses to a range of human disturbance and to specific stressors such as total phosphorus in water or metals in sediments. The metrics were developed by analyzing sets of high-quality, least impaired depressional wetlands along a gradient of influence from agriculture or urban stormwater runoff. Both the vegetation and invertebrate IBIs are significantly related to chloride. Invertebrates were sampled in the nearshore emergent vegetation zone with dipnets and activity traps. The most sensitive invertebrate metrics were intolerant taxa, Odonata, ETSD, and total taxa. Next strongest were metrics using the proportion of the leech Erpobdella *and the proportion of the three most dominant taxa. Vegetation was sampled in the emergent and submergent zones using relevé methods. The strongest vegetation metrics were the sensitive species, percent tolerant taxa, persistent litter, vascular genera, and non-vascular taxa. Both IBIs rated 17 of 24 study sites with the same condition assessment. Seven sites had wider ranges of IBI scores between the two indexes. These results suggested that vegetation and invertebrates respond to different stressors.*

The biological integrity of wetlands can be evaluated by adapting the framework of the well-established method using indexes of biological integrity (IBIs) to assess the condition of streams (Gernes and Helgen 1999). Evaluating the biological integrity of wetlands may necessitate a paradigm shift from the prevailing view that wetlands serve primarily to protect the water quality of lakes and streams downstream by functioning as filters or sinks for pollutants from the landscape. In the future, resource

---

Bioassessment and Management of North American Freshwater Wetlands, edited by Russell B. Rader, Darold P. Batzer, and Scott A. Wissinger.
0-471-35234-9

managers who acknowledge that wetlands are waters of the nation may apply a level of protection to wetlands similar to that which is currently given to streams and lakes (Yoder and Rankin 1998, Rankin and Yoder 1999).

In the stream IBI framework, there are 12 measures, or metrics, of the fish community such as percent top carnivores and the number of species of darters. Each metric receives a score of 1, 3, or 5. The scores of the 12 metrics are totaled for the overall IBI score, which then indicates a condition of excellent, moderate, or poor quality. Streams with high biological integrity have been shown to have high IBI scores; streams affected by pollution have low IBI scores (Yoder and Rankin, 1998). The scoring criteria are derived from baseline studies of the fish and invertebrates in streams over a range of conditions. We report here the development of two indexes of biological integrity for depressional wetlands, one based on attributes of the vegetation in the emergent zone and one using aquatic invertebrates. Both IBIs exhibit significant responses to stormwater and agricultural input to wetlands, but they are intended to be used more broadly to evaluate the effects of a wide range of human disturbances on the biological integrity of depressional wetlands.

To develop IBIs for wetlands, both high-quality and disturbed sites must be evaluated. Six wetlands with the least amount of human disturbance were selected to serve as biological reference sites. Nine wetlands affected by agricultural input and 11 wetlands receiving urban stormwater inputs were selected as potentially disturbed sites. The wetlands were located within the North Central Hardwood Forest Ecoregion in Minnesota (Fig. 8.1; Omernik 1995).

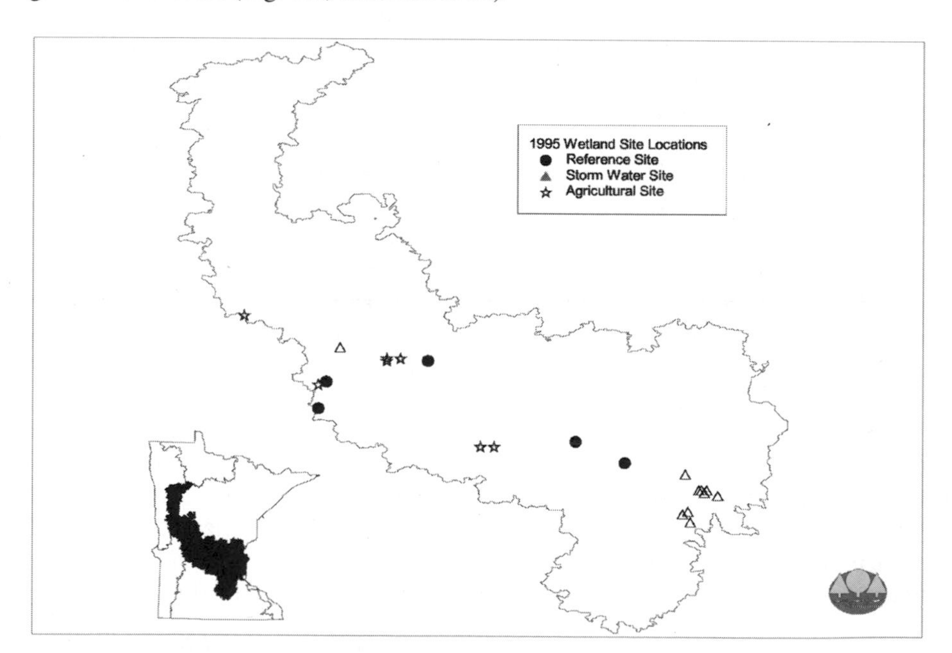

FIGURE 8.1. Locations of the 26 depressional wetlands in three categories of condition: least impaired reference wetlands and wetlands influenced by stormwater or agricultural input. Wetlands were located in the North Central Hardwood Forest Ecoregion in Minnesota.

WATER AND SEDIMENT CHEMISTRY

Sampling of surface water and surficial sediments provided basic information on wetland chemistry in the reference, stormwater-, and agriculture-affected wetlands and was done to determine if there was a range of disturbance in the study sites. Water chemistry means and standard deviations are shown in Table 8.1. Similarly, means and standard deviations for sediment nutrients and sediment metals are shown in Table 8.2. Statistical significance ($p < 0.05$) between the reference site results and results from agricultural or stormwater sites was computed using a one-tailed *t*-test assuming unequal variances between samples. Chloride concentration was a very strong indicator of stormwater input. On average, stormwater chloride concentrations were over 30 times greater than reference site chloride concentrations. Turbidity, total suspended solids, and conductance were significantly lower in the reference sites than in the agricultural or the stormwater sites. Total phosphorus concentrations were also significantly lower in the reference sites. However, nitrogen concentrations were significantly higher in both the water and sediments of the agricultural wetlands compared with the reference sites, whereas total organic carbon was significantly lower. Furthermore, concentrations of heavy metals (copper, zinc, and lead) in the reference sites were also lower than in the agricultural and stormwater sites.

Dissolved oxygen (DO) was measured at dawn to capture the minimum overnight values in August. Although the data were variable, the median DO was over 4 mg/L in the reference sites (range 0.3 to 7.5 mg/L), 2.1 mg/L in the stormwater sites (range 0.0 to 10.1 mg/L), and 0.3 mg/L in the agricultural affected wetlands (range 0 to 4.1 mg/L). These data indicated that there was a gradient of disturbance across the wetlands in this study. It was not intended that the limited number of chemical factors measured would encompass all types of disturbances that might be present, nor that they could serve to assess the condition of wetlands. Below we review the field sam-

TABLE 8.1. Mean Concentrations of Chloride, Total Nitrogen, and Total Phosphorus and Means of Turbidity, Total Suspended Solids (TSS), and Conductivity in Surface Water[a]

Site Type[b]	Chloride (mg/L)	Nitrogen (mg/L)	Phosphorus (mg/L)	Turbidity (ntu)	TSS (mg/L)	Conductance (μS)
Ref	2.7	1.08	0.058	2.0	2.5	205
	(2.83)	(0.18)	(0.028)	(0.57)	(0.78)	(197)
Ag	18.1*	2.41*	0.395*	5.58*	9.5*	621*
	(16.1)	(0.94)	(0.244)	(2.40)	(6.73)	(282)
SW	109*	1.29	0.181*	10.04*	11.4*	639*
	(109)	(0.36)	(0.051)	(8.37)	(15.7)	(503)

[a]Standard deviations are enclosed in parentheses below each mean. Statistical differences ($p < 0.05$) between reference and agricultural or storm water results are indicated by asterisks.

[b]Reference (Ref), agricultural (Ag)-, and storm water (SW)- influenced depressional wetlands in June 1995.

TABLE 8.2. Sediment Chemistry Mean Concentrations or Percent Air-Dried Sample Weights

Site Type[b]	Chloride (mg/kg)	Nitrogen (mg/kg)	Total Organic Carbon (%)	Phosphorus (mg/kg)	Carbonates (mg/kg)	Sodium (mg/kg)	Copper (mg/kg)	Zinc (mg/kg)	Lead (mg/kg)
Ref	29	3,983	6.1	18.7	0.3	157	9	32	11.5
	(17.1)	(845)	(1.9)	(5.3)	(0.3)	(56)	(3.3)	(9.7)	(8.0)
Ag	232*	12,875*	13.4	84.2*	3.8*	228*	21*	104	7.5
	(223)	(7,325)	(9.5)	(50.6)	(4.8)	(73)	(12)	(123)	(2.1)
SW	950*	7,618	10.5	41.6*	4.8	1,063	40*	139*	127*
	(1,715)	(7,335)	(8.26)	(27.4)	(13)	(1,468)	(19.0)	(98)	(130.0)

[a]Standard deviations are enclosed in parentheses below each mean. Statistical differences ($p < 0.05$) among reference, agricultural or stormwater results are indicated by asterisks.

[b]Reference (Ref), agricultural (Ag)-, and stormwater (SW)- influenced depressional wetlands in June 1995.

pling techniques and the invertebrate and vegetation IBIs developed for depressional wetlands.

INVERTEBRATE IBI FOR WETLANDS

Aquatic invertebrates were sampled because they are sensitive to environmental stresses such as metal and organic chemical contaminants, organic enrichment, salinity, siltation, and acidification (Metcalfe et al. 1984, Diggins and Stewart 1998) or because of their feeding ecologies (Hungerford 1948, Saether 1979). Gernes and Helgen (1999) include additional citations concerning the response of invertebrates to environmental stresses. Existing tolerance values based on stream invertebrates (Hilsenhoff 1987, Adamus 1996) are not generally applicable for wetland invertebrates, since many are uniquely adapted to such wetland conditions as the daily fluctuations of dissolved oxygen.

Sampling took place during the same season at all sites (June), in the nearshore emergent vegetation zone, in water less than 1 m deep. Frequently, but not always, the greatest richness of invertebrates has been observed in the emergent vegetated area of wetlands (Driver 1977, Hanson and Swanson 1989). Two samples per wetland using a standardized dip net were used to capture the greatest number and types of invertebrates. Each of the two samples consisted of three to five sweeps through the vegetated water column. The contents of the net were sieved into pans through a framed hardware cloth screen (½-in. mesh) for 10 minutes. Activity traps were also used to capture the actively swimming predators, chiefly the predaceous beetles, bugs, and leeches, which tend to be underrepresented in dip nets. Ten traps, constructed from 2-L plastic beverage bottles converted into funnel traps, were placed horizontally for two nights in the same habitat (emergent vegetation) as dip net samples. Invertebrates were counted and identified to genus in most cases. The snails were identified to species where possible. A minimum of 100 larval chironomids (midges) were identified by Leonard C. Ferrington of the University of Kansas.

Ten invertebrate metrics were selected for the wetland IBI, giving a minimum possible score of 10 and a maximum score of 50. The scoring range for each metric is shown in Table 8.3 with the number of reference, agricultural, and stormwater wetlands scoring 5, 3, or 1 for each metric. Least-squares linear regression found significant negative relationships between chloride or phosphorus concentrations in the water and several of the metrics and the overall IBI score (Table 8.4). To visualize the response of each metric and the overall IBI score to disturbance, the biological data were plotted against chemical factors and against a ranking of the wetlands as an estimate of the degree of disturbance based on best professional judgment (Fig. 8.2). Sites were ranked along a disturbance gradient: reference wetlands (rank 1), low agricultural influence (rank 2), moderate stormwater impact (rank 3), high agricultural impact (rank 4), and high stormwater input (rank 5).

Based on the regression analysis with chemical factors and on the visual plots of individual metrics along the disturbance gradient, the three most sensitive metrics were intolerant taxa, the number of dragonfly and damselfly taxa, and the ETSD met-

TABLE 8.3. Scoring Criteria for 10 Invertebrate Metrics for Depressional Wetlands

			Number of Sites[b]		
Metric Description[a]	Range	Score	Ref	Ag	SW
Percent water boatmen of all bugs plus beetles from BT samples (Corixidae ratio)	<33%	5	5	4	1
	33–67%	3	1	2	4
	>67%	1	0	3	6
Percent leech *Erpobdella* in BT and DN samples of the total DN abundance	0–11%	5	6	3	6
	>11–22%	3	0	3	3
	>22–33%	1	0	2	2
Percent top three dominant taxa of total invertebrate abundance in DN samples	35–54%	5	4	1	0
	>54–72%	3	2	5	5
	>72–90%	1	0	2	6
Number of genera of Chironomidae larvae	14–>20	5	3	0	0
	7–13	3	2	5	4
	0–6	1	1	4	6
ETSD, the number of genera of mayfly and caddisfly nymphs plus presence of fingernail clams and dragonfly nymphs	4–6	5	6	0	0
	2–3	3	0	3	6
	0–1	1	0	5	5
Number of intolerant taxa: *Leucorrhinia, Libellula, Tanytarsus, Procladius, Triaenodes, Oecetis*	5–7	5	3	0	0
	3–4	3	3	0	0
	0–2	1	0	8	11
Number of genera of leeches in BT and DN samples	4–5	5	4	2	2
	2–3	3	2	5	8
	0–1	1	0	1	1
Number of genera of dragonfly and damselfly nymphs	5–6	5	3	0	0
	3–4	3	3	4	3
	0–2	1	0	4	8
Number of taxa of snails to species where possible	5–7	5	4	1	1
	3–4	3	1	5	6
	0–2	1	1	2	4
Total number of taxa (see text)	26–38	5	3	0	0
	13–25	3	3	6	7
	0–13	1	2	4	0

[a]BT, bottle trap; DN, dip net.

[b]Sites scoring as reference (Ref), agricultural (Ag)-, and storm water (SW)- affected wetlands.

ric, or the number of types of mayfly and caddisfly taxa plus the presence of fingernail clams and dragonflies (Table 8.4). The intolerant taxa, which were selected because they tended to be absent in the disturbed wetlands, consisted of two dragonflies, two midges, and two caddisflies (Table 8.3). Since the fingernail clams are filter feeders, they are also sensitive to siltation and pollutants. Dragonflies are sensitive to disturbance because they are long-lived predators in the aquatic stage and

TABLE 8.4. Invertebrate IBI Total Score and Individual Metrics (with Significant *p* Values) Regressed on Concentrations of Chloride and Phosphorus in Surface Waters in 26 Wetlands in Minnesota

	Chloride (log mg/L)		Phosphorus (total, log mg/L)	
	p	r^2	p	r^2
IBI score	0.0009	0.384	0.0010	0.383
Intolerant taxa	0.0000	0.526	0.0000	0.531
Odonata genera	0.0001	0.490	0.0003	0.445
ETSD score	0.0008	0.394	0.0001	0.484
Taxa genera	0.010	0.253	0.024	0.202
Chironomidae genera	0.035	0.179	—	—
Three dominant taxa (%)	0.041	0.169	0.0210	0.212
Corixidae ratio	0.042	0.167	—	—

therefore have a long exposure to pollutants and will respond to changes in their prey. Mayflies and caddisflies are sensitive indicators of stream water quality and appeared also to be sensitive in wetlands, even though they were not present in large numbers.

The next four strongest metrics were the total taxa metric, the proportion of the dominant three taxa, the proportion of water boatmen to beetles and bugs, and the number of taxa of larval midges. The total taxa metric includes the number of taxa of larval midges, caddisflies, mayflies, dragonflies, damselflies, leeches, snails, and macrocrustaceans, plus the presence of the phantom midge, *Chaoborus*, and fingernail clams. Generally, higher-quality aquatic habitats tend to have more taxa, although there are exceptions. The proportion of the dominant three taxa metric is the proportion of the three most abundant taxa from the two dip net samples divided by the abundance of total invertebrate taxa in the two dip net samples. Both the proportion of the dominant three taxa and the proportion of water boatmen tended to increase in more disturbed wetlands. Increased dominance by a small number of pollution-tolerant taxa such as water boatmen or Erpobdellid leeches was often observed in more polluted aquatic habitats. The increase in proportion of water boatmen was accompanied by a decrease in the numbers of predaceous beetles and bugs. A higher proportion of these predators occurred in higher-quality wetlands, just as a greater number of carnivorous fish are often found in high-quality streams. Gernes and Helgen (1999) provide a detailed description of all invertebrate and vegetation metrics.

The IBI approach offers a wetland condition index that is more meaningful than a single chemical measurement such as phosphorus. In the plot of the overall invertebrate IBI scores against phosphorus concentrations, the biological reference wetlands, with higher IBI scores and lower concentrations of phosphorus in the water,

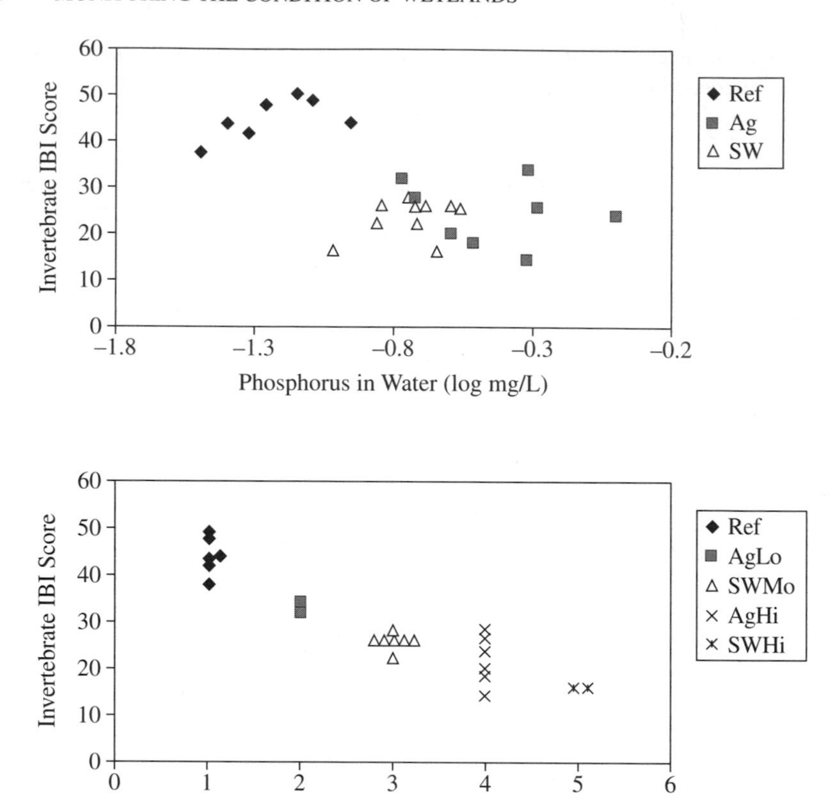

FIGURE 8.2. Invertebrate index of biological integrity (IBI) for depressional wetlands plotted against total concentration of phosphorus (upper plot) in the water and against an estimated gradient of disturbance in three classes of influence: least impaired reference sites (Ref), wetlands affected by agricultural input (Ag) and wetlands affected by stormwater discharge (SW). The invertebrate IBI is the sum total of 10 invertebrate metric scores.

were distinctly separated from the disturbed sites (Fig. 8.2). When the IBI scores varied from low to high at the similar levels of phosphorus, it was useful to ask what else was different about the site, such as local land use, the presence of other pollutants, or best management practices. For example, two agricultural sites, Tyrone and Winter, had phosphorus concentrations in the water of 0.46 mg/L. Yet Tyrone had an IBI score of 32 (moderate), whereas Winter scored 16 (poor). We later learned that although invertebrates in Tyrone were sampled near the site of a former feeding operation for poultry, this site was otherwise relatively unaffected. Winter, however, was an active farm under repossession. It was in an intensive agricultural landscape of tilled crops, with a large agricultural ditch flowing into it, and recently had cattle grazing at the water's edge. Finding explanations for differences in IBI scores when one particular type of disturbance appears similar may provide managers with useful information on what it takes to maintain or restore high-quality wetlands.

The criteria for distinguishing overall condition as excellent, moderate, or poor must be drawn from knowledge of the range of conditions within the class of wetlands within an ecogeographic region. Classifying wetlands into three groups using the invertebrate IBI scores suggested that IBI scores of 37 to 50 indicated excellent conditions; 24 to 36, moderate conditions; and 10 to 23, poor conditions. Of the 10 wetlands in the moderate IBI range, five were agricultural and five were stormwater sites. Of the eight wetlands in the poor IBI range, five were stormwater and three were agricultural wetlands.

Neither the presence of fish nor their densities appeared to influence the invertebrate IBI. The brook stickleback, fathead minnow, and central mudminnow were found in seven, five, and two of the 25 study sites, respectively. There were no significant relationships between mean fish abundances in bottletraps or the number of water boatmen (Corixidae), with the overall IBI scores, or with the number of invertebrates collected in dip nets. Among the disturbed sites with fish present, the median IBI score was 24, whereas the median IBI in disturbed sites without fish was 26. Among the six reference sites, the median IBI was 44. Only one reference site, Prairie, with an IBI of 38, had fish present. Alternatively, however, other researchers have shown a significant reduction in the biomass of invertebrates in wetlands dominated by large sunfish and bass (Fairchild et al. 1999). Wildlife managers have also observed an increase in turbity and a reduction in invertebrates available for waterfowl in high-quality wetlands stocked with fish. Also, Hanson and Riggs (1995) compared wetlands with dense populations of fathead minnows to fishless wetlands and reported a reduction in the invertebrate density, biomass, and richness.

Our study suggested that fish had little effect on invertebrate IBI results. Sampling invertebrates in the emergent vegetation and excluding zooplankton provided sufficient information for assessing the condition of depressional wetlands using the IBI approach. Further work is needed, however, to clarify the effects of minnows and larger fish, especially from stocking, and other kinds of disturbance on the invertebrate community used for wetland IBI metrics.

WETLAND VEGETATION IBI

Plants offer several advantages as indicators of wetland integrity. They effectively respond to environmental changes (Wilcox 1995, Kantrud and Newton 1996) and have been used in univariate toxicity tests because they are acutely sensitive to unbound heavy metal contaminates such as copper (Huebert et al. 1993). Plants offer a rich assemblage to provide clear and robust signals of human disturbance. Rooted plants typically must survive in the same location their entire life, often for several years, making them a good indicator of conditions at that place. Plant sampling methodologies are well defined, and many wetland biologists are familiar with aquatic vascular plant taxonomy.

The plant sampling method used in this study was a relevé approach adapted from Mueller-Dombois and Ellenberg (1974). Prior to sampling, each wetland was evaluated briefly for overall community structure. A 100-m^2 plot was then established in a typical location within the emergent plant community. Plants occurring in the plot were then inventoried, and the cover class (abundance) of each plant taxon in the plot was estimated (Almendinger 1987). All vegetation sampling occurred in July, which corresponds to the peak maturity time of the plant community. This was the period considered most appropriate for determining the community structure during a single visit. Vascular plant taxa were identified to species whenever possible.

Karr and Chu (1999) recommend validating biological metrics by looking for response signals across a gradient of human disturbance. For the plant IBI, a qualitative disturbance index based on observations by the investigator concerning the extent of human disturbance among several common factors at each site was used to evaluate and validate the vegetation metrics.

The two principal disturbance factors, stormwater and agricultural influence, were assigned a value of 0, 2, 4, or 8, where 0 represented a factor that contributed little human influence at a given site and 8 represented sites with great influences from either of the two principal factors. Three other factors—hydrologic alteration, historic influences, and buffer quality—were also evaluated, and each site was assigned a value of 0 to 4 for these factors. A value of 0 represented sites where insignificant amounts of these factors affected the wetland, and 4 represented sites where these factors were believed to have greatly compromised wetland integrity. Hydrologic alterations included human actions such as construction of low-head dams, various types of berms, improperly sized culverts, and similar actions that had observably changed the wetland hydrology. Historic influences included various practices that were observable during the project fieldwork, such as old-field communities or evidence of in-wetland excavations typically longer than 10 years earlier. Wetland buffer quality was evaluated qualitatively with respect to the relative width and type of vegetation shoreward from the wetland edge. Minimally disturbed native vegetation provided a higher-quality buffer than did any type of planted/agricultural vegetation. Continuity of the buffer was also considered.

Similar to invertebrate scores, scoring criteria for the vegetation metrics were developed by sorting metric values from high to low and then dividing the data into three groups, which often coincided with natural breaks. When natural breaks in the data were not apparent, the scoring divisions were made to include more sites in the moderate range and fewer sites scoring 5 or 1. Distribution of data for each metric plotted against the disturbance index was also helpful in developing metric scoring criteria.

The vegetation multimetric index discussed here is comprised of 10 metrics. The highest possible score for this index is 50 and the lowest is 10. One reference site, Glacial, which was in a state park, received a score of 50, whereas one agriculturally affected site, Winter, received a score of 10. Winter was an agricultural wetland with a sizable inflow from a large drainage ditch. Two of the plant metrics were counts of different plant taxa and four were based on plant life form (Table 8.5). There were two sensitive and tolerant taxa metrics and two community structure metrics. Aquatic guild species include submerged and floating aquatic plants. Persistent litter taxa that are slow to decompose included common reed (*Phragmites*) , smartweeds (*Polygonum*), bullrushes (*Scirpus*), burreeds (*Sparganium*), cattails (*Typha*), and purple loosestrife (*Lythrum salicaria*). Monocarpic taxa are those plants that reproduce once during each life cycle, they typically are considered to be annual or biennial plants. The sensitive plant species metric was based on those plants that decreased in occurrence with increasing disturbance. Sensitive species included *Asclepias incarnata, Dulichium arundinaceum, Eriophorum gracile, Scirpus validus, Iris versicolor, Iris* sp., *Scutellaria galericulata, Utricularia macrorhiza, Calamagrostis canadensis, Calamagrostis stricta, Glyceria striata, Glyceria borealis, Polygonum sagittatum, Polygonum scandens, Polygonum amphibium, Riccia fluitans, Spiraea alba*, and *Rubus occidentallis.* Tolerant taxa included 20 nonnative taxa found to occur at the study sites and 27 aggressive native plants found to increase in the sample apparently in response to human disturbance (Gernes and Helgen 1999). Examples of tolerant aggressive plants were *Impatiens capensis*, *Echinochloa muricata*, *Phalaris arundinaceae*, *Lemna minor*, and *Typha angustifolia.* A separate nonnative taxa metric was considered, but the number of nonnative taxa was not sufficient to provide a reliable measure. However, nonnative taxa were an important consideration in metric development. For metrics where a negative correlation was expected with disturbance, such as the metric for vascular genera, the strongest correlation was observed when nonnative taxa were excluded from the counts. Conversely, for metrics where a positive correlation was expected with disturbance, such as the percent tolerant taxa metric, the strongest correlation was obtained when the nonnative taxa were included in the count.

Table 8.5 presents the scoring range for the 10 vegetation metrics and the scoring distribution among the six reference, 11 stormwater, and nine agriculturally affected wetlands. All 20 impaired sites scored a 1 in at least one metric, and the six reference sites scored 3 or 5 in all metrics. Three vegetation metrics—nonvascular taxa, tolerant taxa, and aquatic guild taxa—were considered to be the strongest metrics. They each had at least four reference sites scoring 5 and at least 10 impaired sites scoring 1. "Tolerant taxa" was the only plant metric where only reference sites scored 5. Two metrics, "monocarpic" and "persistent litter," were notable since they had at least four reference sites that also scored a 5.

Plant metric values for each site were regressed against environmental chemistry data from water and sediment data summarized in Tables 8.1 and 8.2. Table 8.6 presents the regression results for the 10 vegetation metrics and the overall vegetation IBI. All of the vegetation metrics, especially "sensitive species" and "percent tolerant taxa," showed a significant relationship to the concentration of chloride in the water. Phosphorus concentration in the water was not related statistically to sedge cover, grasslike, monocarpic, dominance, or persistent litter metrics. Only the "sensitive species" metric was considered strongly related to the concentration of phosphorus in water. Significant relationships were found between nine metrics and the concentration of copper in sediments. Only the aquatic guild metric was not related

TABLE 8.5. Scoring Criteria for 10 Vegetation Metrics for Depressional Wetlands

Metric	Metric Description	Range	Score	Number of Sites[a]		
				Ref	Ag	SW
Vas genera	Number of vascular genera	>14	5	3	0	3
		9–14	3	3	4	3
		<9	1	0	5	5
Nonvas taxa	Number of nonvascular taxa	>1	5	4	1	1
		1	3	2	3	4
		0	1	0	5	6
Sedge cover	Sum of all sedge species cover classes	>3	5	2	2	2
		0.1–2.9	3	4	4	3
		0	1	0	3	6
Sensitive species	species that decrease with disturbance	>4	5	3	0	0
		1–3	3	3	4	5
		0	1	0	5	6
Tolerant taxa	Proportion of tolerant taxa	<25.0	5	4	0	0
		25.1–60	3	2	5	5
		>60	1	0	4	6
Grasslike taxa	Number of grass, sedge and rush species	>8	5	2	1	1
		2–7	3	4	5	5
		<2	1	0	3	5
Monocarpic	Proportion of monocarpic species cover and richness divided by cover	>2.6	5	4	2	1
		2–2.59	3	2	5	4
		<2	1	0	2	6
Aquatic guild	Number of aquatic guild species	>6	5	4	2	1
		3–5	3	2	3	4
		<3	1	0	6	6
Dominance	Distribution of cover in sample	<0.07	5	3	1	3
		0.08–0.2	3	3	5	3
		>0.2	1	0	3	5
Persistent litter	Sum of persistent litter taxa cover classes	<1	5	4	1	1
		1–5.9	3	2	7	4
		>6	1	0	1	6

[a]Sites scoring as reference (Ref); agricultural (Ag)-, and stormwater (SW)-affected sites.

statistically to the concentration of sediment copper. These guilds were adapted from the work of Galatowitsch and McAdams (1994). Plants in the aquatic guild were rooted submersed aquatics (e.g., water milfoils; *Myriophyllum* sp.), unrooted submersed aquatics (e.g., coontails; *Ceratophyllum* sp.), floating perennials (e.g., water lilies; *Nuphar* sp. and *Nymphaea* sp.), and floating annuals such as duckweeds (*Lemna* sp.). Unrooted submersed aquatics, floating annuals, and floating perennials often have little direct connection or exposure to metals or other contaminants in the sediments. So we were not surprised by the lack of a strong relationship between this aquatic guild and copper in the sediments.

Figure 8.3 shows a significant negative linear relationship between the vegetation IBI and copper concentrations in wetland sediments. Two agricultural sites, Winter and DaveNr, had lower biological scores than expected, probably because of copper concentration in sediments. Both of these sites were heavily impacted by agricultural

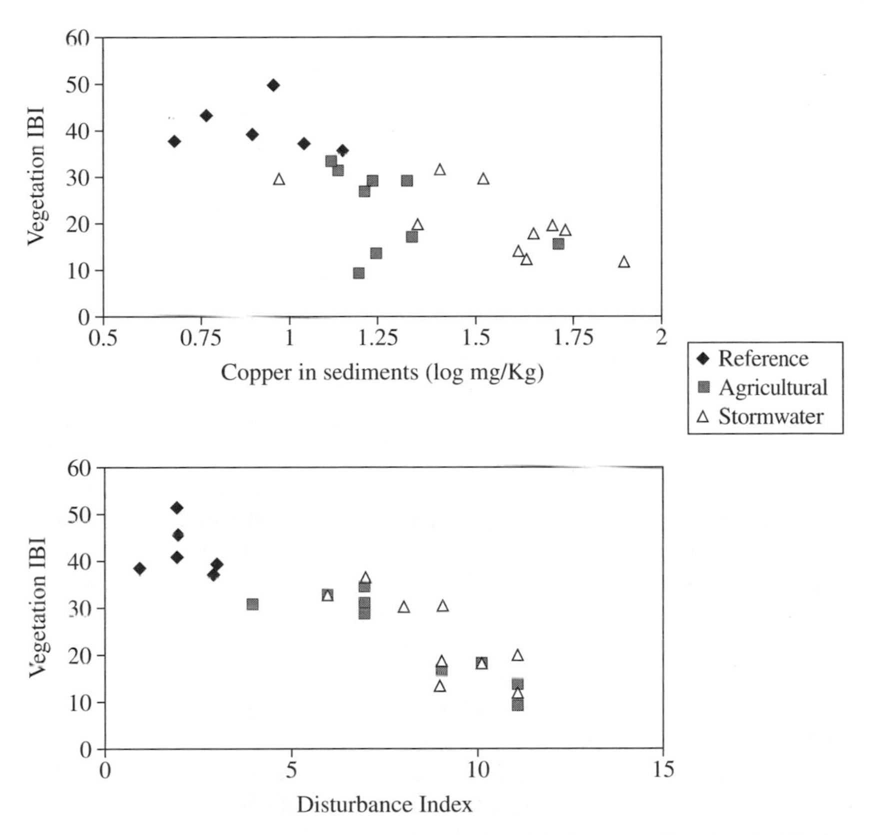

FIGURE 8.3. Vegetation index of biological integrity (IBI) for depressional wetlands plotted against the concentration of copper in wetland sediment (upper plot) and plotted against a disturbance index weighted on stormwater or agricultural influences. The 26 wetlands studied were in one of three categories of influence: reference, affected by agricultural practices, or affected by stormwater discharge. The vegetation IBI is the total score of 10 metric scores.

TABLE 8.6. Ten Vegetation Metrics Regressed Against Water and Sediment Chemistry from 26 (n = 25 for Sediment Results) Depressional Wetlands

	Water				Sediments	
	Chloride (log mg/L)		Phosphorus (log mg/L)		Copper (log mg/kg)	
Metric	r^2	p	r^2	p	r^2	p
Vascular genera	0.19	0.025	0.23	0.014	0.31	0.004
Nonvascular genera	0.32	0.003	0.24	0.011	0.42	0.000
Sedge cover (log)	0.25	0.009	0.13	0.072	0.39	0.001
Grasslike species	0.22	0.015	0.13	0.071	0.44	0.000
Monocarpic (log)	0.26	0.008	0.05	0.284	0.38	0.001
Aquatic guild	0.21	0.020	0.32	0.003	0.08	0.181
Sensitive species	0.42	0.000	0.46	0.000	0.41	0.001
Percent tolerant taxa	0.43	0.000	0.33	0.002	0.47	0.000
Dominance (log)	0.30	0.004	0.10	0.124	0.32	0.003
Persistent litter	0.36	0.001	0.10	0.119	0.48	0.001
Vegetation IBI	0.52	0.000	0.40	0.0005	0.58	0.000

inputs, suggesting that these wetlands may have been affected by factors in addition to the concentration of copper in the sediment. For example, field observations indicated that hydrologic factors influenced the plants at some sites. Figure 8.3 also illustrates the strong linear relationship between the vegetation IBI and a qualitative disturbance index which included several disturbance factors. These results demonstrate the predictive response of the plant IBI to a variety of human disturbance. Distribution of the impaired sites against the disturbance gradient suggested that the urban stormwater sites were more biologically impoverished than the agricultural sites.

Three categories of site condition—excellent, good, and poor—can be derived from the IBI scoring ranges. These categories would relate directly to the biological integrity of the wetland. Appropriate ranges of condition for the plant IBI were 36 to 50 (excellent), 20 to 34 (good), and 10 to 20 (poor). Using these scoring ranges, all six reference and one stormwater site, Springbrook, were found to be in excellent condition. Five agricultural and five stormwater sites were considered to be in good condition, and four agricultural and five stormwater sites were considered to be in poor condition. Slightly different results were found using the invertebrate index. Using the invertebrate index, all six reference sites scored excellent, 11 sites scored good, and eight sites scored poor. For both indexes it would, however, be appropriate to have larger data sets before establishing actual criteria or condition assessments for public policy.

DISCUSSION

Local government staff and state and federal resource managers have expressed a need for alternative ways to evaluate wetlands. They want baseline information on the quality of wetlands, and they want to know what effects various human activities have on wetlands and how to monitor them. They want to prioritize high-quality wetlands and understand whether a created or restored wetland adequately replaces a wetland altered as a result of a mitigation agreement. They are dissatisfied with the current types of evaluations, which assess the wetlands indirectly using physical surrogates as predictive indicators of wetland quality, or employ a mixed list of assessed functions and values which seem tangential to the quality of the wetland or its ecological condition.

Measurements of single chemicals such as chloride or phosphorus, although direct measures useful as indicators of the degree of urban or agricultural influence, are very limited in scope and insufficient for describing the condition of a wetland. The IBI gives a more comprehensive indication of the quality of wetlands by directly measuring the life within the wetland, the plants and invertebrates. Because they are exposed to the water and sediment in the wetland over a period of time, the plants and animals, properly measured, give the most direct assessment of wetland condition and quality. Using several direct measures of biology, as opposed to single indicators, gives the IBI a robust ability to respond to the many kinds of human disturbances to wetlands.

Using two biological indices, in this case invertebrates and vegetation, we expect the results to be similar at most, but not all sites. Using a two-sample t-test assuming equal variances we found no significant ($p < 0.05$, $n = 24$) difference between the site scores using the two indexes. However, invertebrate index scores were noticeably different from the vegetation index scores at seven out of 24 sites as determined by a separation of 8 or more points. Plants and invertebrates respond differently to various environmental stressors. Invertebrates were more strongly related to nutrient concentrations than were the plants, whereas plants showed a stronger relationship to sediment metal concentrations.

In making wetland condition assessments using two or more assemblages, we anticipate the need to apply a "weight of evidence" approach. Additional data, including chemical or physical measurements at study sites, can contribute to the weight of evidence indicating a site's condition. Broadly, the wetlands that receive input from stormwater and agriculture showed varying degrees of impairment based on the emergent vegetation and the aquatic invertebrate metrics. There were fewer sensitive plant taxa, such as *Iris*, *Riccia*, and bladderwort in the disturbed wetlands, and a greater abundance of the more tolerant taxa such as cattails, duckweed, and reed canary grass. Similarly, plants with persistent standing stems such as purple loosestrife, cattails, bullrush, and burreed, were more abundant in disturbed sites. There were fewer kinds of aquatic invertebrates in the impaired wetlands, especially among the dragonflies, damselflies, mayflies, caddisflies, fingernail clams, and midges. Also in the impaired wetlands, the leech *Erpobdella* and the water boatmen tended to be relatively more abundant, while there were fewer predaceous aquatic beetles and bugs.

Future needs for wetland IBIs will involve improving the taxonomic rigor for nonvascular plants and comparing differences between open and wooded landscapes on the vegetation IBI. Differentiating between open versus wooded landscapes for small depressional wetlands may help reduce the variation in metric scores. Also, a metric that measures individual plant health, analogous to the metric for anomalies in fish (Karr et al. 1986), should be considered. For invertebrates, metrics should be developed using the species composition of the aquatic beetles and bugs. Also, additional work on the impact of fish on the invertebrate IBI should be explored. For both IBIs, further validation and identification of sensitive and tolerant taxa will be important, but difficult. Existing information from the literature is limited and often contradictory, so sensitive and tolerant taxa will need to be derived empirically by comparing the taxa found in biological reference sites with those in polluted sites. Since there may be regional differences in responses to stress, it may be important to develop sensitive and tolerant taxa on a regional scale.

The results reported here have established a sound foundation for continuing development of biological criteria for evaluating wetland water quality. The invertebrate and vegetation IBIs are effective multimetric indexes for monitoring and assessment of small to moderate-sized depressional wetlands. Work is continuing on testing and developing IBIs for large depressional wetlands, and plans are in place to work on other classes of wetlands. The existing 10 metrics for vegetation and the 10 metrics for invertebrates will be used or modified and new metrics will be derived as appropriate for a particular wetland class. Nationally, several other research groups are also working to develop biological tools for monitoring wetlands.

Both the vegetation and invertebrate IBIs presented here support many useful applications. The vegetation may respond more to contamination of sediments and the invertebrates to pollutants in the water. Analysis of the vegetation community may be most useful in the drier wetlands, such as sedge meadows, fens, and certain riparian wetlands. Analysis of the invertebrates will be useful in lakes and vernal pools and certain riparian wetlands. In addition, the vegetation and invertebrate IBIs have already been adapted for use by citizen teams in one county in Minnesota. Linked to their local government and high school biology teachers, these adult citizens are assessing the biology of their wetlands and reporting back to their local boards or commissions.

The IBI assessment approach fosters a new way of looking at wetland quality. It excites citizens and interests local government staff because it measures what people want to know: Is the environment around them, wetlands included, in good health? It measures what they want to protect, the living environment they inhabit. They are aware of the biological richness of good wetlands, they see the dragonflies and the sedges, and they hear the frogs. The IBI approach gives us a tool to measure what people value, a tool that shows responses to pollution in the wetlands, a tool that people understand.

CONCLUSIONS

1. Biological reference wetlands had lower phosphorus, nitrogen, and chloride concentrations and clearer water than the disturbed wetlands. Stormwater wetlands had higher chloride in both water and sediments and more heavy metals in the sediments. Agricultural wetlands had more phosphorus and nitrogen.
2. Wetlands impaired by stormwater or agricultural input show distinct and measurable changes in the community composition of the aquatic vegetation and invertebrates. They show degrees of reduced biological integrity. The reference wetlands had the highest IBI scores.
3. Development of metrics for wetland IBIs necessitates inclusion of reference wetlands with the least amount of human disturbance as well as wetlands experiencing a wide range of human disturbance. The multimetric approach builds a robust response to a wide range of potential stresses to wetlands.
4. IBIs for wetlands will be a useful tool for direct measurement of the integrity of wetlands, for and evaluating restored and created wetlands, and for conducting inventories of high-priority wetlands. IBIs will be useful for understanding the degree of impact from various land-use and management activities.

ACKNOWLEDGMENTS

This work was done under partial support from the U.S. Environmental Protection Agency assistance grant CD995525-01. We acknowledge excellent field and laboratory work of Joel Chirhart and Jon Haferman; Leonard C. Ferrington for the work on Chironomidae; technical feedback from James R. Karr and other members of the Biological Assessment Working Group for Wetlands (BAWWG) organized by the Environmental Protection Agency; and the assistance of Sue Elston, Susan Jackson, Doreen Vetter, and Tom Danielson from the Environmental Protection Agency.

REFERENCES

Adamus, P. R. 1996. Bioindicators for assessing ecological integrity of prairie wetlands. Report EPA/600/R-96/082. U.S. Environmental Protection Agency, Office of Research and Development, Corvallis OR.

Almendinger, J. C. 1987. A handbook for collecting releve data in Minnesota. Department of Natural Resources, Minnesota Natural Heritage Program, St. Paul, MN.

Diggins, T. P., and K. M. Stewart. 1998. Chironomid deformities, benthic community composition and trace elements in the Buffalo River (New York) area of concern. Journal of the North American Benthological Society 17(3):311–323.

Driver, E. A. 1977. Chironomid communities in small prairie ponds: some characteristics and controls. Freshwater Biology 7:121–133.

Fairchild, G. W., A. M. Faulds and L. L. Saunders. 1999. Constructed marshes in southeast Pennsylvania. Page 423–446 *in* D. P. Batzer, R. R. Rader, and S. A. Wissinger (eds.), Invertebrates in freshwater wetlands of North America: ecology and management. Wiley, New York.

Galatowitsch, S. M., and T. V. McAdams. 1994. Distribution and requirements of plants on the upper Mississippi River: literature review. Unit Cooperative Agreement 14-1-0009-1560, Work Order 36. Iowa Cooperative Fish and Wildlife Research Unit, Ames, IA.

Gernes, M. C., and J. C. Helgen. 1999. Indexes of biotic integrity (IBI) for wetlands: vegetation and invertebrate IBI's. Final report to U.S. Environmental Protection Agency CD995525.01. Minnesota Pollution Control Agency, St. Paul, MN.

Hanson, M. A., and M. R. Riggs. 1995. Potential effects of fish predation on wetland invertebrates: comparison of wetlands with and without fathead minnows. Wetlands 15(2):167–175.

Hanson, B. A., and G. A. Swanson. 1989. Coleoptera species inhabiting prairie wetlands of the Cottonwood Lake area, Stutsman County North Dakota. Prairie Naturalist 21(1):49–57.

Hilsenhoff, W. L. 1987. An improved biotic index of organic stream pollution. Great Lakes Entomologist 20:31–39.

Huebert, D. B., B. S. Dyck, and J. M. Shay. 1993 The effect of EDTA on the assessment of Cu toxicity in the submerged aquatic macrophyte, *Lemna trisulca* L. Aquatic Toxicology 24:183-194.

Hungerford, H. B. 1948. The Corixidae of the Western Hemisphere (Hemiptera). University of Kansas Scientific Bulletin 32:5-827.

Kantrud, H. A., and W. E. Newton. 1996. A test of vegetation-related indicators of wetland quality in the prairie pothole region. Journal of Aquatic Ecosystem Health 5:177–191.

Karr, J. R., K. D. Fausch, P. L. Angermeier, P. R. Yant, and I. J. Schlosser. 1986. Assessing biological integrity in running waters: a method and its rationale. Illinois Natural History Survey Special Publication 5.

Karr, J. R., and E. W. Chu. 1999. Restoring life in running waters. Island Press, Washington DC.

Metcalfe, J. L., M. E. Fox, and J. H. Carey. 1984. Aquatic leeches (Hirudinea) as bioindicators of organic chemical contaminants in freshwater ecosystems. Chemosphere 13(1):143–150.

Mueller-Dombois, D. and H. Ellenberg. 1974. Aims and methods in vegetation ecology. Wiley, New York.

Omernik, J. M. 1995. Ecoregions: defining acceptable biological status by comparing with reference conditions. Pages 49–62 *in* W. S. Davis and T. P. Simon (eds.), Biological assessment and criteria tools for water resource planning and decision making. Lewis Publishers, London.

Rankin, E. T., and C. O. Yoder. 1999. Adjustments to the index of biotic integrity: a summary of Ohio experiences and some suggested modifications. Pages 625–638 *in* T. P. Simon (ed.), Assessing the sustainability and biological integrity of water resources using fish communities. CRC Press, Boca Raton, FL.

Saether, O. A. 1979. Chironomid communities as water quality indicators. Holarctic Ecology 2:65–74.

Wilcox, D. 1995. Wetland and aquatic macrophytes as indicators of anthropogenic hydrologic disturbance. Natural Areas Journal 15(3):240–258.

Yoder, C. O., and E. T. Rankin. 1998. The role of biological indicators in a state water quality management process. Environmental Monitoring and Assessment 51:61–88.

9 Restoration of Wetland Plant Communities

JULIA BOHNEN and SUSAN GALATOWITSCH

Wetlands restored to reestablish biotic communities that existed historically often require planting to attain a composition of species resembling a natural system. Revegetation is important for restored wetlands that lack on-site sources of propagules from refugial populations, seedbanks, or dispersal from nearby existing wetlands. Detailed site plans, including maps of existing environmental conditions, intended modifications to landform and hydrology, and desired vegetation zones, transform general objectives into a spatially explicit vision for implementation. Planting specifications developed with the site plan should include the type and amount of propagules needed for each species, acceptable conditions for planting, and restrictions on acquiring plants (i.e., geographic range, substitutions). Planting success for restored wetlands with an existing vegetation and/or seedbank of perennial species often depends on minimizing weed populations prior to reflooding. Repeated mowing, burning, and/or chemical treatments over a year may be needed for adequate site preparation. In addition, before reflooding, a year is often needed for seed collection and propagation of species not commercially available in adequate quantities. In temperate climates, seeds collected in one growing season overwinter in cold storage. Seed needed for plant production is stratified for 30 to 90 days to optimize germination prior to greenhouse production, which requires an additional 60 to 90 days. Wetland plants survive transplanting best if they are installed at the shallow end of their water depth tolerance range and if they have adequate time to establish sufficient root reserves prior to overwintering. Manipulating water levels with water control structures and irrigation can be critical to planting success. Newly planted sites will often require additional perennial weed control and protection from herbivores to ensure vegetation establishment.

Bioassessment and Management of North American Freshwater Wetlands, edited by Russell B. Rader, Darold P. Batzer, and Scott A. Wissinger.
0-471-35234-9

Wetland restoration is done for one of two general reasons: (1) to provide ecosystem services such as stormwater storage, recreation, or water quality improvement (e.g., Mitsch et al. 1998), or (2) to reestablish biotic communities that existed historically (e.g., Zedler 1993, Galatowitsch and van der Valk 1994). In most cases, managers assume that when they design a wetland to be optimal for one of these two objectives, the other will be only partially fulfilled. A design focus on either ecosystem services or community reconstruction is adopted whether or not the restoration is mitigating the loss of an existing wetland. Success is usually assessed based on achieving the primary objective with less critique of the secondary objective. Consequently, revegetation of restored wetlands has followed two distinct approaches. Wetland restorations designed for the primary purpose of providing ecosystem services tend to rely on natural recolonization of vegetation since the specific community composition is unimportant for providing stormwater storage or improving water quality (Mitsch et al. 1998). In contrast, wetlands restored to replace the biotic communities of a specific kind of wetland will probably be planted with specific plants unless there is a large reserve of on-site propagules (Galatowitsch and van der Valk 1995, 1996). In this chapter we focus on revegetation of wetlands where project success depends on exactly what kind of vegetation reestablishes.

SITE DESIGN

Designing a wetland restoration is the process of transforming general objectives (i.e., to restore a floodplain forest) to a spatially resolved plan for implementation. Often, a site design consists of a series of maps, rendered accurately to scale, that depicts tract boundaries, topography and anticipated landform modifications, predicted water levels, the desired vegetation zones in the wetland and surrounding upland, and infrastructure to accommodate human use, such as boardwalks and paths. A site design needs to be sufficiently detailed and accurate to serve as a preliminary assessment of the relationship between anticipated water levels and vegetation zones, as well as property boundaries, boardwalks, and paths.

The design should be based on information obtained from recent field assessments as well as existing data sources such as topographic maps, soil surveys, aerial photographs, and historical accounts. Typically, it is most effective to compile the available published data for the site prior to field surveys. A 7.5-minute topographic map is often adequate for delineating catchment boundaries. Land uses within the watershed should be mapped from aerial photographs and from field observations. A watershed map should also identify likely sources of pollutants to the wetland (e.g., road salt) and their probable flow path. These watershed data can be used to develop watershed budgets of the restored site (e.g., Galatowitsch and van der Valk 1994, Brooks et al. 1991). A more detailed topographic site map (1 inch = 100 feet) should be developed for on-site assessment and design. This map should be annotated with field verification of soil surveys; existing vegetation (especially invasive species populations); drainage infrastructure such as ditches and tiles; and other infrastructure, such as utility lines, roads, and adjacent buildings. If the site has been farmed, herbicide application records should be compiled to determine if herbicide residual may limit vegetation reestablishment.

Site designs that depict hydrologic predictions of both minimum and maximum flooding depth and duration provide more meaningful information for configuring planting zones than do predictions of normal pool (water) level. Many types of wetlands exhibit clear vegetation patterns that correspond to flooding depth and/or duration of saturation (e.g., Kantrud et al. 1989). For example, prairie glacial marshes of midcontinental North America tend to have zones of wet prairie, sedge meadow, emergent marsh, and open water as flooding trends change from ephemeral to permanent and from saturated to deep (Stewart and Kantrud 1971). Determining the types of vegetation to target for a particular wetland restoration site should be based on plant communities typical of the ecoregion, on hydrogeomorphic setting (i.e., riverine, depression) and on specific site conditions (e.g., basin size, soils). Sources of information to determine suitable target communities include published natural history guides, state agency databases on natural features, original land survey notes, and historic accounts (Galatowitsch and van der Valk 1990, 1994).

Each vegetative zone appears distinctive because of shifts in abundance of some dominant species, although adjacent zones have many species in common (Kantrud et al. 1989). Consequently, the planting list for each zone should not be unique, but should include many species common to adjacent zones. It often makes sense to plant each species across a wide elevational range, both above and below their anticipated optimal elevation for growth, since hydrologic predictions are often only approximate for wetland restorations.

DEVELOPING PLANTING SPECIFICATIONS

The maps produced during site design for wetland restorations usually characterize the desired patterns of plant communities and topography; however, these alone are not specific enough to implement a revegetation. Restorationists need to tabulate the type and quantity of propagules for each species indicated on the site design by providing detailed planting specifications. Acceptable conditions for planting, including time of year and water level, should also be noted. Specific planting treatments to minimize mortality during establishment may also be included.

Planting specifications for external contractor bids should also indicate acceptable substitutions for species that typically have low availability. Some vendors of wetland plants market species not native to a region or perhaps not even the continent. Several nonnative species are so commonly used for conservation or erosion control (e.g., reed canary grass, *Phalaris arundinacea*) that they may be perceived to be native. Cultivars of some native species, selected to be showy or exhibit improved vigor, are genetically distinct from their wild parent populations. Comprehensive guides to the plants of a geographic region (floras) will often indicate if a species is introduced or native.

Whether seeds or plants are more efficient propagules for revegetation depends on their relative cost and likelihood of successful establishment. Seeds should be used for wetland revegetation when their establishment success is higher than the number of plants that can be acquired and established for the same cost. Species with relatively high establishment success from seed are generally those that germinate under a broad range of environmental conditions in the field and that are not competition sensitive (Betz 1980). Species with relatively low cost seed are those that produce large seed crops with high viability. Wetland seeding has been prone to failure for several reasons: (1) native seed vendors often do not assess viability and so do not base their weights on "pure live seed" (PLS), (2) site hydrology or soil conditions may be unsuitable for germination, (3) wetland seed floats and may be redistributed to the windward side of the site, and (4) newly emerging seedlings may compete poorly amidst faster-growing weeds. Consequently, seeding will be a more reliable revegetation approach for restorations where reported seed quality is verified, water level is carefully managed, and invasive species are managed for several years post-planting.

Wetland plants can be obtained as greenhouse-grown seedlings in plug trays or larger plants in containers, as nursery-grown or wild-harvested adult divisions, bare-root stock, detached rhizomes or tubers, or pole cuttings. Seedlings that have been established for 1 to 3 months in 48- or 72-cell plug trays are commonly used for revegetation of forbs and graminoids. The additional expense of using larger plants in containers or bare-root stock is generally only warranted for those species that are competition sensitive or uncommon, so that maximizing the survival of each individual is critical (e.g., some orchid species). Detached rhizomes and tubers are often used for submersed and floating, and occasionally emergent, aquatic plants that cannot be propagated under typical greenhouse conditions. Because detached rhizomes and tubers lack shoots and stems, they can experience high mortality (presumably from anoxia) if they cannot regenerate new shoots immediately after transplanting (e.g., Gallagher 1983, Wijte and Gallagher 1991, Yetka and Galatowitsch 1999). Revegetation success probably depends on whether detached propagules are large enough to promote regrowth and if they are transplanted at a time of year when new roots and shoots would normally be initiated (often, spring in temperate regions).

Wetland revegetation projects that require more than a few woody plants usually use bare-root stock or pole cuttings. Bare-root plants can be purchased from many sources including government nurseries. Bare-root trees and shrubs are inexpensive and easier to handle in large numbers than potted plants. They are available in spring as dormant stock and must be planted in early spring while still dormant. Pole cuttings are dormant stem cuttings of woody species that root and sprout readily when they are pounded or dug into the ground (Henderson et al. 1999). Willows (*Salix* spp.), dogwood (*Cornus* spp.), and viburnums (*Viburnum* spp.) are commonly used because they readily develop adventitious roots. Pole cuttings can be planted upright or laid horizontally in trenches. Horizontal plantings (also called wattles, fascines, brush mattresses) are often used as bioengineering approaches to stream bank or lakeshore erosion control.

Quantities of seed and/or plants needed for revegetation are estimated from the area of each planting zone, desired initial densities, expected viability, and the potential for rapid vegetative expansion. For most wetland ecosystems, seeding rates and plant spacing are not developed. One way to establish a suitable seeding rate is to use published seedbank data as a guide. For example, temperate, freshwater herbaceous wetlands have been reported to have between 1000 and 70,000 seeds per square meter (Galatowitsch and Biederman 1998), depending on wetland type. The highest values are from emergent wetlands with dominant species that have long-lived seed (e.g., common burreed, *Sparganium eurycarpum*). Consequently, they result in a very generous estimate of what is actually required to establish a stand of vegetation.

The quantity of seed needed for each species is estimated by multiplying relative abundances desired by the total density and the area to be planted. This estimate assumes that pure live seed is to be planted. Pure live seed (PLS) can be estimated using the following equation:

$$\text{PLS} = \%\ \text{weight as seed (not other parts)} \times \%\ \text{germination)} \times \text{seeds/unit weight}$$

The needed mass of seed is then determined by dividing the total number of seeds needed by the PLS weight per gram (or pound) of seed. The quantities of seeds planted should also be adjusted to downweight species that are likely to spread rapidly vegetatively (Diboll 1997). The amount of adjustment will generally be a matter of best judgment and past experience. Although upward adjustments to account for poor germination from seed may also be warranted for some species (e.g., *Carex lacustris*), using seedlings or divisions may be a more efficient approach.

Rates established for seeding prairie restorations may be appropriate for temperate wetlands. Tallgrass prairies are drilled at a rate of 11 kg PLS-grasses/ha (10 lb PLS/acre) and 2 kg PLS-forbs/ha (2 lb PLS/acre), for a total seeding rate of 13 kg PLS/ha. This rate is doubled for broadcast seeding (Morgan et al. 1995). Planting density for seedlings, rhizomes, and divisions is also not well established for most wetland types. Denser plantings are often presumed to provide some protection against invasive species establishment and reduce the length of time to closure of the vegetative stand. In addition, dense plantings can compensate for anticipated mortality from transplanting. On the other hand, since many wetland species spread clonally, planting them too densely is probably expensive and inefficient. Many herbaceous wetland species planted at intervals of 30 to 60 cm can have significant cover within a few years in persistently saturated or flooded wetlands of temperate North America. Strongly rhizomatous species (i.e., lake sedge, *Carex lacustris*) can be planted farther apart, up to 90 cm. Woody plants can be planted at a density corresponding to their ultimate size at maturity, or more densely to allow for self-thinning (attrition due to competition). Dense initial plantings have an advantage of mimicking the kinds of conditions that may be expected to occur in a wooded wetland during secondary succession, where dense thickets of small trees eventually give way to sparser configurations of larger trees.

The detailed planting specifications for wetland revegetation can be presented as a table or as a computer-generated spreadsheet. If a combination of seeding and planting is to be used, specifications for each can be summarized in separate tables. In addition to noting the quantity of seeds or plants for each species, the table can include annotations of special planting requirements, such as time of planting or sowing, special treatments such as fertilizers or water-retaining acrylimide, and acceptable types of propagules.

SITE PREPARATION

The vegetation planned for a wetland restoration may fail to establish successfully when introduced to a site where species are at a competitive disadvantage from the existing vegetation and/or seedbank. The least successful revegetation efforts are typically those where seeds and seedlings are planted amidst a dense cover of perennial species. Some of these species, like reed canary grass, are both drought and flooding tolerant, so can thrive in drained and reflooded wetlands (Marten and Heath 1985). Many of these invasive perennial weeds spread aggressively by rhizomes and spread rapidly into cleared areas. Even if existing weed species (e.g., many agronomic species) are not flood tolerant, the litter layer produced as the stand of vegetation dies may persist for several years, precluding seed germination or seedling establishment of wetland plants (van der Valk 1986). Seedbanks may contain seeds of invasive species (such as purple loosestrife, *Lythrum salicaria*) that will not germinate until the site is reflooded. To ensure the success of seeding or planting efforts, it is necessary to eliminate the existing perennial weed population prior to planting.

Vegetation management to minimize perennial weed populations is usually more efficient prior to reflooding (i.e., site preparation) because many control techniques that are least labor intensive, including mowing, burning, and chemical treatment, are difficult or even impossible to employ in standing water. An assessment of weeds present at a site should be done during the earliest planning phases because effective site preparation may require a year or even two of management in advance of the intended seeding or planting (see Table 9.1 for a list of invasive aquatic weeds). Map-

TABLE 9.1. Species Invasive in North America That Probably Limit Wetland Revegetation Success if Not Managed

Botanical Name	Common Name	Region[a]	Life Span/Growth Form[b]
Ailanthus altissima	tree of heaven	1,2,6	PW
Arundo donax	giant reed	1,7,10	PHR
Butomus umbellatus	flowering rush	1,3	PHR
Cortaderia selloana	pampas grass		PH
Eichornia crassipes	water hyacinth		PHF
Eleagnus angustifolia	Russian olive		PW
Eucalyptus globus	blue gum		PW
Ficus carica	common fig		PW

Botanical Name	Common Name	Region[a]	Life Span/Growth Form[b]
Hedera helix	English ivy	1–3,9,10	PWC
Hydrilla verticillata	hydrilla	1,2,6,7,9,10	A/PHS
Iris pseudacorus	yellow iris	1,3	PH
Lepidium latifolium	perennial pepperweed		B/PH
Lygodium japonicum	Japanese climbing fern		PHR
Lythrum salicaria	purple loosestrife	1–6,8–10	PH
Melaleuca quinquinervia	paperbark tree		PW
Myriophyllum spicatum	Eurasian water milfoil	1–3,6–10	P/AH
Phalaris arundinacea	reed canary grass	1,3,9	PHR
Phragmites australis	common reed	1,10	PHR
Pistia stratiotes	water lettuce	2,10	PHF
Rhamnus frangula	glossy buckthorn		PW
Rubus discolor	Himalaya berry		PW
Salvinia molesta	giant salvinia		HF
Sapium sebiferum	Chinese tallow tree	2,6	PW
Schinus terebinthifolius	Brazilian pepper	2,10	PW
Senecio mikanioides	German ivy		PH
Sesbania punicea	rattlebox		PW
Solanum tampicense	soda apple		H
Spartina alterniflora	cordgrass		PH
Tamarix chinensis (also *T. parviflora, T. ramosissima*)	saltcedar	4–10	PW
Typha angustifolia	narrow-leaf cattail		PHR
Typha x *glauca*	hybrid cattail		PHR
Viburnum opulus var. *opulus*	European cranberry		PW

Source: Data from Randall and Marinelli (1996), Austin et al. (1999), Colorado Weed Management Association (1999), Howald et al. (1999), the Nature Conservancy (2000), NBII (2000), USDA (2000), and USFWS (2000).

[a]1, Delaware, Connecticut, Kentucky, Maine, Maryland, Massachusetts, New Hampshire, New Jersey, New York, Ohio, Pennsylvania, Rhode Island, Vermont, West Virginia; 2, Alabama, Arkansas, Florida, Georgia, Louisiana, North Carolina, South Carolina, Mississippi, Tennesee; 3, Illinois, Indiana, Iowa, Michigan, Minnesota, Missouri, Wisconsin; 4, Montana, North Dakota, South Dakota, Wyoming; 5, Colorado, Kansas, Nebraska; 6, Oklahoma, Texas; 7, Arizona, New Mexico; 8, Colorado, Nevada, Utah; 9, Idaho, Montana, Oregon, Washington, Wyoming; 9, Idaho, Montana, Oregon, Washington, Wyoming; 10, California.

[b]A, annual; B, biennial; P, perennial; H, herbaceous; W, woody; C, climbing; F, floating; R, rhizomatous; S, submersed.

ping the existing vegetation of the site, noting the dominant species and overall cover of each stand of vegetation, and identifying stands of remnant natural vegetation will often generate adequate information to develop a site preparation approach. An emergence assay of the seedbank (see Galatowitsch and Biederman 1998 for methods) should also be done if additional problematic species not observed in the existing vegetation could be present. For example, sites adjacent to land with weed populations or to a watercourse that serves as dispersal corridor for seeds should be presumed to have a contaminated seedbank and be assayed.

Large areas dominated by native vegetation may be conserved during site preparation. Scattered populations of desirable plants, however, may not be feasible to save in situ. Populations that are uncommon or of special concern may be salvaged and later replanted into the restoration. Prior to reflooding, annual weeds can be controlled by mowing or cultivating prior to anthesis. Since standing water will kill most annuals, weed control is usually most important in areas that will be saturated but not flooded. Repeated shallow cultivation using a harrow will reduce the annual weed seed bank.

A combination of techniques is often required for effective control of perennial weed populations. An initial herbicide application followed by prescribed burning removes the plant canopy and encourages resprouting of rhizomes and seeds. Subsequent applications of glyphosate (Rodeo for wet sites or Roundup for drained sites) can further reduce populations of herbaceous invasive species such as reed canary grass. For many sites, more than four iterations of chemical control and burning over a year may be required to reduce weed populations adequately. At least 7 days should elapse from the last glyphosate application to tillage or mowing and planting; other herbicides used on drained sites may require waiting from 2 weeks to several months before planting. Repeated cultivation can also be used to control perennial weeds. Initially, weed populations may increase because chopping rhizomes creates more propagules, but eventually (i.e., more than 1 year), plants exhaust their carbohydrate reserves. Flooding after mowing or cultivating can reduce weed populations further, because plants may not have a conduit of air from stems to roots. Cutting and flooding will be most effective to control species restricted to portions of the site that will be covered for long periods by deep water. If weed populations extend to the saturated periphery, plants not flooded may serve as recolonists.

Scraping, the removal of a layer of topsoil and associated seeds, roots, and rhizomes, is a more resource-intensive means of controlling perennial weeds. The depth of soil to be scraped depends on the rooting depth of the species to be controlled but is often less than 15 cm in poorly drained soils typical of wetland restoration sites. Limitations on the use of this technique include cost, disposal of the scraped soil, and the desirability of planting into the exposed subsoil.

After weed populations have been reduced, additional soil preparation prior to seeding is generally minimal. Deeper cultivation prior to seeding can bring deeply buried weed seeds to the surface and can cause new weed problems. Since seeds of most wetland species should be sown on the soil surface, the soil can be lightly scratched with a springtooth prior to sowing to create a good seedbed. A harrow or drag can be used to increase seed-to-soil contact after sowing. Organic topsoil may need to be top-dressed on sites that have been graded or excavated to subsoil to improve soil tilth and fertility. Sites should not be enriched with fertilizer because growth of weed populations and algae can be enhanced disproportionately.

ACQUIRING PLANTS FOR REVEGETATION PROJECTS

Plants can sometimes be salvaged from a wetland slated for destruction or even from the restoration site itself if extensive site preparation is needed. The transplant site should be similar to the donor site in most regards (i.e., soils, water depth, hydrologic regime). If feasible, the entire plant (including intact root system) should be removed; otherwise, rhizomes, tillers, tubers, or seeds can be gathered. It is possible to hold the plants indefinitely if they can be heeled into temporary nursery areas. Rhizomes and tubers may be stored in coolers, depending on what time of year they are collected. Species underrepresented in the native nursery industry, due to such difficulties as finding seed sources or poor germinability, are especially important to salvage. Opportunities to salvage may present themselves in time to plan ahead but are often unanticipated. Plants should not be salvaged from areas infested with invasive species because it is difficult to clean roots sufficiently to ensure that no weed seeds are transmitted to the restored site.

The demand for commercially available native plants often greatly exceeds local supplies. Therefore, obtaining a large quantity of a diverse selection of plants in a timely manner can be problematic. To obtain the species and quantities desired for large projects, it may be necessary to work with nurseries a year prior to planting. If a native plant producer has a full year to fill an order, they can collect or purchase seed from other vendors and grow the quantities of seedlings or divisions needed. Because native plant populations often do not reliably set seed every year, it may take several years to complete a planting according to the original specification.

Nursery orders need to be detailed adequately to avoid confusion. Request plants by their botanical names. Common names may be applied to more than one species or to different species in different regions. If local ecotype plants are preferred, request information from the nursery on the source of their seeds or plants. Some vendors sell seed or plants from sources far removed from the nursery location. Plant quantities and identifications should be checked upon receipt to ensure that orders were properly filled. It is not unusual to receive plants that have been misidentified at some stage during production. For example, it is possible that a grass growing as a weed may be mistaken for a sedge. The best defense against this problem is to work with a reputable vendor and have a knowledgeable person check plants when they are received.

Local sources of seed or plants may not be commercially available for some species. In these cases it may be necessary to harvest seed from nearby wetlands. Ideally, wetlands similar to the proposed restoration may provide seed for restoration projects. Regulations controlling seed collecting on public lands will vary from state to state. Check with local conservation agencies for detailed information. Permits to collect seed on public lands can be obtained by organizations with a specific need. Wildlife refuges, state parks, wildlife management areas, and local or regional parks are examples of public lands that may allow collecting. The permit will generally establish collecting guidelines such as off-limit areas, species that may or may not be collected, amount or percentage of seed of a species to be collected from an area, techniques to be used, and the period of time during which collecting is allowed. Wetlands on private lands may also be a source of seeds. It may be desirable to negotiate a written agreement or lease with the landowner.

The implications of repeatedly collecting seeds from intact wetlands are not well understood. Continuous selective collection from an area may have long-term impacts on the wetland seed bank and plant population dynamics. It is wise to collect conservatively. To avoid negatively affecting the population, it has been suggested that only 10% of annual species seed and between 25 and 50% of perennial species be collected (Apfelbaum et al. 1997). When hand collecting, it is easy to move through an area in a random manner, leaving an appropriate amount of seed. Although seed is lost to shattering during hand and mechanical harvest, entire areas within populations should be left unharvested, to avoid overcollecting.

Timing of seed collecting takes practice and vigilance. Some species can be collected before they are fully ripened, to prevent loss of the seed due to shattering or explosive dispersion. Some should be left on the plant until completely ripe. A few species will hold their seeds in persistent seedheads indefinitely. Timing of ripeness will vary from year to year depending on the growing season, but can be determined with practice by using one or more of the following cues: color, firmness, and the "tug test." Seeds of the grasses and sedges are ripe when they are firm, not "doughy." Squeezing a seed between your fingernails will determine if it is sufficiently firm. If so, the other cues will help determine if it is ready to harvest. The color of seed for many species will turn from green to brown or golden when ripe. The tug test involves attempting to strip the seed gently from the inflorescence with your hand. If seed doesn't strip readily, it is not ripe.

When mechanical harvesting techniques such as combining are used in native plant communities, many species are collected at once. These seed mixes, however, are unlikely to include significant quantities of species that ripen earlier or later than the harvest date. It may be desirable to supplement such a mix with hand-collected seed or seed combined at other times of the year. Combining and other methods of mechanical harvesting are not practical for most wetlands, due to the wet or uneven terrain. When ripe, hand-collected seedheads can either be snipped off with garden shears or snapped off by hand and stuffed into bags or buckets. Small quantities of seed can be cleaned by hand. Sets of soil sieves or homemade frames with different grades of hardware cloth or screen stapled to one side are effective for separating and cleaning seed. Most wetland forbs and grasses can be broadcast on the surface and so do not need to be well cleaned and sorted. Large volumes of mixed seeds and stems from the field need only be hammer-milled to a uniform size to pass through a broadcaster. In contrast, using a seed drill requires that seed be thoroughly cleaned and some grasses debearded. Cleaning large quantities of seed of different sizes and weights can be expensive and time consuming. Native plant nurseries and agricultural seed-cleaning companies can be contracted to hammermill, sort, clean, and/or debeard large volumes of seed.

Commercial producers of native species dry seed prior to storage to prevent fungal contamination. Grain drying wagons, or bins with an air supply along the bottom, are used to force air through the seed. For smaller volumes, seed can be piled on a tarp in an area protected from rodents, wind, and rain. The floor of an open shed works well. Even smaller amounts can be piled on newspapers on benches or tabletops in a sheltered location. The piles should be turned periodically to ensure that all material is uniformly dry. Small volumes of seed can be stored temporarily in paper bags. For longer storage, place the seed in plastic buckets or bins with secure lids. For long-term storage, temperatures should be maintained at 2 to 7°C (35 to 45°F) at a humidity of 20 to 25% (Hartmann and Kester 1983). Small volumes of seed can be stored in a refrigerator. Large volumes of seed are generally stored in a barn or shed at ambient temperatures. The longevity of the seed of some species is greater than that of others. Legumes typically have longer viability; *Carex* species appear to have short viability. The grasses require an after-ripening period in which the embryo continues to develop. Therefore, grasses germinate best the spring after they were harvested, whereas *Carex* species appear to be most viable immediately after harvest. *Carex* species continue to lose viability over time in dry storage (Budelsky and Galatowitsch 1999).

Seed that will be used to produce plants rather than be sown directly into a wetland should be stratified to optimize germination (Hartmann and Kester 1983). Stratification, a cold, moist chilling period, consists of sowing the seeds on or mixing them into a moist medium (sand, peat, or even a paper towel), drenching with a fungicide (e.g., Ridomil, Truban, Domain, Rovral) and placing them in a cooler, refrigerator, or cold shed for a period of 30 to 90 days at temperatures between 1 and 4°C (33 to 40°F). If space allows, seed can be stratified directly in flats that can subsequently be placed in the greenhouse for germination. An alternative is to sow the flats and leave them in an unheated greenhouse in winter. In the spring when the greenhouse is heated, germination will occur. Sowing seeds on the surface of the media is best for species that require light to germinate (many small-seeded wetland species and the *Carex* species); whereas others, with hard seed coats that must deteriorate (e.g., *Sparganium, Iris*), germinate best if covered with a germination medium. Some species, such as wild roses, require alternating cold and warm cycles to promote germination, while the seed of golden alexanders (*Zizia* spp.) requires leaching to remove germination inhibitors. Some species with impermeable seed coats require scarification (scarring of the seed coat) to allow imbibition (uptake of water). For some of these species, stratification will promote further germination. Baskin and Baskin (1998) provide extensive information on specific factors that enhance germination in many wetland species.

PROPAGATING WETLAND PLANTS

Many wetland species, the sedges in particular, can be grown using traditional horticultural techniques (Bohnen et al. 1999). Choosing whether to produce plants in a greenhouse versus an outdoor nursery will depend on the facilities available and the products desired. Large numbers of seedling plugs can be grown efficiently in greenhouses. Smaller plants or plugs are cheaper to produce, easier to ship and plant, and may suffer less transplant shock than do larger plants. For example, tussock sedge (*Carex stricta*) plugs need 8 to 10 weeks of growing time in a greenhouse. Older plugs become root-bound, which reduces transplant success because encircling roots do not readily grow into the soil. Production on raised benches (versus ground level) is most functional for seedling production because flats are easier to maintain, seedlings have fewer disease problems, and plants will not root into the ground.

Most wetland forbs and graminoids can be grown in a peat-based mix in plugs or bands on greenhouse benches. Potting and germination media can be purchased from commercial sources or mixed on site from available soils and additives (Nelson 1978). Although commercial mixes are convenient, sterile, uniform, and lightweight, they can be expensive. In addition, they require storage space prior to use and are usually not reusable. Custom growing media can be produced from sand and soils from nearby sources, along with bark and wood chip products. Custom media need to be sterilized and can be very heavy. In addition, unless extensive, uniform supplies of the raw materials are available, it may be difficult to plan for uniform production from year to year. A modified commercial mix (i.e., 5 parts peat-based mix to 1 part coarse vermiculite) constitutes a light, sterile media with adequate water-holding capacity. For seed germination, a fine-textured medium is important so that emergence of the seedlings is not inhibited.

Seedlings can be produced in shallow flats containing a choice of several liners with cells of different sizes. The size is reported as the number of cells per tray (i.e., 72 cells per tray). Open cell packs with no dividers are useful for initial seeding. Seventy-two's are an appropriate size for the final seedling product. Bands are systems of taller containers that fit into trays. Tall, narrow containers provide more rooting volume and may save bench space if well configured. Plants produced in cells slip easily out of the containers and do not require a large hole for planting. Because the roots are undisturbed, they suffer little transplant shock. Seedlings can also be produced in peat pots, which are biodegradable, resulting in less trash at the site. Peat pots have several disadvantages: They have a tendency to fall apart during production or in transit, and plants may begin to grow together through the pots if not planted in a timely manner.

Newly germinated seedlings need to be thinned by transplanting or by removing and discarding seedlings. Seedlings grown too close together will be thin and straggly and will not stand up to transplanting into the field. One to three forb seedlings to a cell and 5 to 10 graminoid seedlings to a cell are suitable growing densities. Forbs may need an additional thinning to reduce competition further. Initial transplanting should occur when the seedlings are just an inch or two tall. The smaller they are, the less damage will be done to their roots when they are teased apart. Rare, competition-sensitive, or slow-growing plants (e.g., orchids and lilies) may be further transplanted to larger pots and maintained in the greenhouse prior to planting. Containers range from 4-inch (diameter) to 1 gallon or more. Some slower-growing species, such as gentians, may require as many as two years in containers (in a greenhouse) before being planted into the restoration site.

Although wetland plants are relatively pest free in most greenhouse settings, some organisms, including thrips (family Thripidae), fungi (especially powdery mildew), and algae can severely reduce seedling survival and growth. Pest problems tend to be more common for forbs than graminoids; they are prone to infestation by thrips, fungus gnat (*Mycetophila* spp.), spider mite (Tetranychidae), scale (Coccidae), and mealy bug (Pseudococcidae). Biological control can be used to minimize these pest problems (Hussey and Scopes 1985). Many pest populations are reduced when the plants are placed outdoors to harden off in the cooler temperatures. Keeping a clean house and beginning with clean supplies is the best defense against insect and disease problems. If trays and pots are reused, they should be rinsed in a solution of disinfectant (i.e., 1 part bleach to 9 parts water).

Large plants are desirable for areas where seedlings would be completely inundated. These plants are best grown in outdoor, in-ground beds that can be flooded. Either bare-root or container stock can be harvested from flooded beds. Lumber or brick can be used to frame-in a bed that is lined with polyurethane pond liners, heavy ultraviolet-resistant plastic, or clay to contain water. Using soil to create berms is a short-term alternative for flooded beds. Stock tanks are sometimes used to grow emergent aquatic plants, such as arrowheads (*Sagittaria* spp.). Any locally available sandy or loamy soil (not clayey) can be used as a growing medium. Systems such as drip or trickle irrigation, overhead irrigation by risers, and flooding via canals can be used to deliver water to the beds. The use of mulch, landscape fabric, or mechanical control between beds is usually needed to minimize weed populations.

Soil tests should be done on a regular basis to determine nutrient needs. Slow-release fertilizers can be incorporated in flooded nursery beds or containers at planting time. Beds or containers may also be top-dressed with slow-release fertilizer when necessary. A soluble, all-purpose fertilizer such as Peters can be metered through an irrigation system. Cattails (*Typha* spp.), bulrush (*Scirpus* spp.), willow (*Salix* spp.), cottonwoods (*Populus* spp.), and many other weedy plants will quickly populate flooded beds. Hand weeding is usually adequate for controlling unwanted species on a small scale.

Plants produced in flooded nurseries are ready to transplant to wetland sites in the second growing season in northern temperate climates. Plants have been overwintered successfully in drained but saturated beds. Frost heaving may damage wooden or brick bed frames, but saturation is not likely to reduce overwintering survival in wetland plants. Rodents (e.g., mice, voles) can cause significant overwintering mortality if they graze to the crowns of the plants. Drained but moist beds are easiest to harvest. Draining beds and harvesting plants without destroying the liner can be problematic in flooded beds. Plants are excavated by hand, and soil is either completely washed or shaken from the roots to make storage or transportation more efficient. If it is necessary to dig more plants than can be transplanted on the same day, transplants need to be stored bare-root, in cool conditions. Many native herbaceous perennial species can be stored dormant for 1 to 3 months in plastic bags in moist peat at 2°C (Testor 1999).

Larger propagules have a greater likelihood of transplanting success. Propagules should have well-developed root systems or, for dormant tubers, several healthy root initials. For example, lake sedge (*Carex lacustris*) may have a vigorous shoot every 10 to 20 cm along a rhizome; each main shoot has several roots as well as thick white rhizomes. Each shoot with roots attached is a viable propagule. Plants that will not be completely inundated can have shoots cut to about 20 cm to facilitate planting and regrowth. Plants for relatively deep water should not have their shoots clipped, due to increased mortality from anoxia.

INSTALLING VEGETATION

Plants should be kept moist and sheltered from wind and sun during the transition from greenhouse or nursery to wetland. Plants need to be well watered before they are shipped to the restoration site, particularly if they will not be planted into standing water. They should be transported in a covered vehicle or covered with a tarp when in an open bed of a truck. A staging area should be selected at the restoration site that is shaded and protected from the wind. If plants will be held for more than a day before planting, they need to be irrigated.

Planting time affects the survival of rhizomes (Yetka and Galatowitsch 1999) and presumably that of other propagule types. Best results occur when emergent plants are planted in spring or early summer. Divisions, rhizomes, and bare-root stock are also best planted in spring. Bare-root stock is sold dormant and should be planted before it begins to grow. Divisions need time to overcome the shock of being divided. Plugs or container plants can be planted in saturated soil or wet prairie from spring to early fall. Fall plantings are generally less successful, possibly due to the lack of time to store sufficient reserves to overwinter. In addition, root development in the fall may not be adequate to prevent heaving by high waves or freezing and thawing.

When water levels can be manipulated, emergent aquatic plants can be planted in standing water that is no deeper than emergent shoots. To prevent the drowning of emergent propagules when water cannot be manipulated, plant them on the high end of their elevational range to compensate for water-level changes during establishment. Emergent species, such as tussock sedge (*Carex stricta*) and lake sedge (*Carex lacustris*), can be planted in shallow water and allowed to propagate vegetatively or seed down into the deeper water. Larger plants will be better able to respire in emergent situations. Emergent vegetation transports gases from the air through its leaves and stems in specialized cells called *aerenchyma*. If enough leaf tissue is not exposed to air, the plant may not be able to support its developing root system and thus drown.

Successful revegetation depends on plant preparation and care after planting. Plants grown in containers with roots circling should be loosened with fingers, by cutting the outside layer of roots with a pocket knife, or by rolling the root ball in your hands or on the ground to promote new root growth. The plant should be placed into ground having a firm base that will not allow it to settle below the ground level when watered. Propagules planted in water should be anchored into a firm substrate (below the loose muck) in which roots can develop to hold the plant in place. The plant's root ball should not be buried deeper than it grew in the container. Plants placed in dry soils should be watered. Cutting back some very leafy plants will make handling them easier and will reduce stress to the plant, allowing it to establish faster. Sedges and grasses can be cut back to 15 to 30 cm. Flowering plants should have flowers removed so that resources are put toward root establishment instead of seed development.

It is recommended that wetland revegetation efforts combine the use of seeds and vegetative propagules to compensate for the lack of success of some species by a single method. If revegetation efforts do use plants only, they should be spaced 30 to 40 cm apart for more rapid canopy closure and exclusion of weeds. When planting is used in combination with seeding, plants can be planted as much as 60 to 90 cm apart. Seeds of wetland species and prairie forbs should be broadcast, while prairie grasses should be drilled (or broadcast at a higher rate). Seeds of emergent species such as softstem bulrush, water plantain, and arrowhead will germinate in still water 15 to 30 cm deep. Other species, such as emergent sedges (e.g., *Carex lacustris, C. stricta*), do not appear to germinate readily when broadcast on-site.

Due to the uncontrolled nature of restoration sites, there are many threats to survival and establishment after planting. Wave action, flowing water, freeze–thaw, erosion, sedimentation, and herbivory all provide challenges for restoration managers. Some sites, such as windward shores with large fetch or high-flow riparian wetlands, may be unsuitable areas for planting. In other sites, more moderate conditions may require some precautions. Anchoring is necessary in areas where moderate wave action or flowing water are factors. Anchoring can be done in a number of ways, such as placing dormant tubers or rhizomes in mesh bags weighted with a stone. U-shaped pins pushed in over the plant crown can help hold it in place. Wave action can be moderated by the use of wave breaks or fiber rolls anchored under water in the planting zone. Overland flow of water into wetlands can result in erosion of soils away from planting areas or the deposition of soils over a planting. Soil erosion can be abated through the use of silt fences, erosion blankets, mulch, cover crops, and rapidly establishing rhizomatous plants such as prairie cord grass (*Spartina pectinata*).

Plants should be watered immediately after planting and throughout the first growing season wherever plants are not in saturated or flooded situations. On larger sites, overhead irrigation systems, which pump water from flooded portions of the wetland to newly planted areas, may deliver water efficiently. Water tanks mounted on trucks or trailers may be useful on some sites. Use care with high-pressure hoses so that plants and seeds are not washed away. On sites where postplanting irrigation is not feasible, water-absorbing acrylimide beads (e.g., Terrasorb, SoilMoist) can be used in the planting hole to help retain water and release it slowly to the plant, as needed.

MONITORING REVEGETATION SUCCESS

The species assemblages of natural wetlands are the consequence of hundreds to thousands of years of propagule dispersal and seed production, varying in response to regional and local conditions. Yet most restored wetlands are planted in one or two events in a single growing season. If climatic conditions during the establishment year are abnormally wet or dry, hot or cold, initial vegetation survivorship may be limited. Plants that do establish initially may experience severe mortality from competition with invasive species and from herbivores (i.e., Canada geese and muskrats), which could cause them to be lost from the site. In many kinds of wetlands, revegetation remains experimental, so lack of survivorship can result from unsuitable growing conditions. Since most wetlands experience high initial mortality of at least some species planted, it is important to monitor revegetation success several times a year during the first few years of establishment and annually thereafter. Monitoring information should be adequate to discern any midcourse corrections needed, such as replanting and weed control.

Compiling a spreadsheet with plant species, numbers, size, and planting location will enable follow-up surveys and documentation. This information should be recorded during planting. Packing slips, production records, and order forms will provide some details needed for documentation, but quantities should be verified in the field. For large projects, one person may need to be assigned to planting documentation. This spreadsheet can be annotated during postplanting surveys with comments on whether the plants were observed, and whether they were thriving or appeared stressed. It may be possible to count plants and calculate mortality rates for each species on small projects. If the monitoring goal is to discover whether some plants completely failed to survive and require replanting or whether water levels should be adjusted to reduce stress, qualitative data are often adequate.

Annual weed scouting is essential for many wetland restorations so that populations can be identified and controlled when small. Often, it is helpful to have a base map to record any infestations, which will be controlled during a later site visit. Weed scouting and control should occur prior to seed set of the most common invasive species in a locale, to prevent further contamination of the restoration seedbank. Quantitative monitoring of vegetation establishment may be desirable if project success is contingent on a specific aspect of plant growth. For example, a wetland creation at Sweetwater Marsh in San Diego, California was undertaken to provide nesting habitat for the light-footed clapper rail (*Rallus longirostris levipes*). Suitable habitat requires *Spartina foliosa* shoots of a minimum height (Zedler 1993). For this wetland, monitoring needed to include measurements of shoot densities and heights. In other cases, it may be important to estimate aerial cover of some plant species to discern whether suitable habitat exists for other wildlife, such as secretive birds (Delphey and Dinsmore 1993).

Monitoring data should also be evaluated to ascertain whether further research is needed to improve the practice of wetland restoration. Failure of certain plants to reestablish or to grow sufficiently may indicate a need for a systematic study that could explore underlying causes of the problem. Many of these failures may be expected to reoccur on other restorations if not resolved. In addition, continued monitoring of species composition beyond establishment could yield some important long-term insights into restoration success which are rarely available.

ACKNOWLEDGMENT

This is a publication of the University of Minnesota Experiment Station.

REFERENCES

Apfelbaum, S. I., B. J. Bader, F. Faessler, and D. Mahler. 1997. Obtaining and processing seeds. Pages 99–126 *in* S. Packard and C. F. Mutel (eds.) The tallgrass restoration handbook. Island Press, Covelo, CA.

Austin, D., K. Craddock Burks, N. Coile, J. Duquesnel, D. Hall, K. Langeland, J. Maguire, M. McMahon, R. Pemberton, D. Ward, and R. Wunderlin. 1999. Florida Exotic Pest Plant Council, 1999 list of Florida's most invasive species, http://www.fleppc.org/99list.htm (accessed Jan. 6, 2000).

Baskin, C. C., and J. M. Baskin. 1998. Seeds: ecology, biogeography and evolution of dormancy and germination. Academic Press, New York.

Betz, R. 1980. One decade of research in prairie restoration at the Fermi National Acceleration Laboratory, Batavia, Illinois. Pages 179–185 *in* G. K. Clambey and R. H. Pemble (eds.), The prairie: past, present and future. Proceedings of the 9th North American Prairie Conference, Moorhead State University, Moorhead, MI.

Bohnen, J., S. Galatowitsch, and P. Olin. 1999. Horticultural practices in sedge meadow restoration. Pages 177–184 *in* Proceedings of the 16th North American Prairie Conference, University of Nebraska, Kearney, NE.

Brooks, K. N., P. F. Ffolliott, H. Gregersen, and J. L. Thames. 1991. Hydrology and the management of watersheds. Iowa State University, Ames, IA.

Budelsky, R. A., and S. M. Galatowitsch. 1999. Effects of moisture, temperature and time on seed germination of five wetland *Carices*: implications for restoration. Restoration Ecology 7:86–97.

Colorado Weed Management Association. 1999. http://www.cwma.org/2_bad_weed.html (accessed Jan. 6, 2000).

Delphey, P. J., and J. J. Dinsmore. 1993. Breeding bird communities of recently restored and natural prairie potholes. Wetlands 13:200–206.

Diboll, N. 1997. Designing seed mixes. Pages 135–150 *in* S. Packard and C. F. Mutel (eds.), The tallgrass restoration handbook. Island Press, Covelo, CA.

Galatowitsch, S. M., and L. A. Biederman. 1998. Vegetation and seedbank composition of temporarily flooded *Carex* meadows and implications for restoration. International Journal of Ecology and Environmental Sciences 24:253–270.

Galatowitsch, S. M., and A. G. van der Valk. 1990. Using the original land survey notes to reconstruct presettlement landscapes in the American West. Great Basin Naturalist 50: 181–191.

Galatowitsch, S. M., and A. G. van der Valk. 1994. Restoring prairie wetlands: an ecological approach. Iowa State University Press, Ames, IA.

Galatowitsch, S. M., and A. G. van der Valk. 1995. Natural revegetation during restoration of wetlands in the southern prairie pothole region of North America. Pages 129–141 *in* B. D. Wheeler, S. C. Shaw, W. J. Fojt, and R. A. Robertson (eds.), Restoration of temperate wetlands. Wiley, Chichester, West Sussex, England.

Galatowitsch, S. M., and A. G. van der Valk. 1996. The vegetation of restored wetlands. Ecological Applications 6:102–112.

Gallagher, J. L. 1983. Seasonal patterns in recoverable underground reserves in *Spartina alterniflora* Loisel. American Journal of Botany 70:212–215.

Hartmann, H. T., and D. E. Kester. 1983. Plant propagation: principles and practices. Prentice Hall, Englewood Cliffs, NJ.

Henderson, C. L., C. J. Dindorf, and F. J. Rozumalski. 1999. Lakescaping for wildlife and water quality. Minnesota Department of Natural Resources, St. Paul, MN.

Howald, A., J. Randall, J. Sigg, E. Wagner, and P. Warner. 1999. California Exotic Pest Plant Council, http://www.caleppc.org/info/plantlist.html (accessed Jan. 6, 2000).

Hussey, N. W., and N. Scopes (eds.). 1985. Biological pest control: the glasshouse experience. Cornell University Press, Ithaca, NY.

Kantrud, H. A., J. B. Millar, and A. G. van der Valk. 1989. Vegetation of wetlands of the prairie pothole region. Pages 132–187 *in* A. G. van der Valk, (ed.) Northern prairie wetlands. Iowa State University Press, Ames, IA.

Marten, G. C., and M. E. Heath. 1985. Reed canarygrass. Pages 207–216 *in* M. E. Heath, R. F. Barnes, and D. S. Metcalfe (eds.), Forages: the science of grassland agriculture. Iowa State University Press, Ames, IA.

Mitsch, W. J., X. Wu, R. W. Nairn, P. E. Weihe, N. Wang, R. Deal, and C. E. Boucher. 1998. Creating and restoring wetlands. Bioscience 48:1019–1030.

Morgan, J. P., D. R. Collicutt, and J. D. Thompson. 1995. Restoring Canada's native prairies: a practical manual. Manitoba Naturalists Society, Winnipeg, Manitoba, Canada.

Nature Conservancy. 2000. http://tncweeds.ucdavis.edu/worst.html (accessed Jan. 6, 2000).

NBII (National Biological Information Infrastructure). 2000. http://www.nbii.gov/invasive/spp.html (accessed Jan. 6, 2000).

Nelson, P. V. 1978. Greenhouse operation and management. Reston Publishing Co., Reston, VA.

Randall, J. and J. Marinelli (eds.). 1996. Invasive plants: weeds of the global garden. Science Press, Brooklyn, NY.

Stewart, R. E., and H. A. Kantrud. 1971. Classification of natural ponds and lakes in the glaciated prairie region. Resource Publication 92 U.S. Fish and Wildlife Service, Washington, DC.

Testor, G. 1999. Personal communication, horticulturist, Prairie Moon Nursery, Winona, MN.

USDA. 2000. U.S. Department of Agriculture APHIS plant protection and quarantine. http://aphisweb.aphis.usda.gov/ppq/weeds/weedshome.html (accessed Jan. 6, 2000).

USFWS. 2000. U.S. Fish and Wildlife Service Invasive Species Program. http://invasives.fws.gov.html (Accessed Jan. 6, 2000).

van der Valk, A. G. 1986. The impact of litter and annual plants on recruitment from the seedbank of a lacustrine wetland. Aquatic Botany 24:13–26.

Wijte, A. H. B. M., and J. L. Gallagher. 1991. The importance of dead and young live shoots of *Spartina alterniflora* (Poaceae) in a mid-latitude salt marsh for overwintering and recoverability of underground reserves. Botanical Gazette 152:509–513.

Yetka, L. A., and S. M. Galatowitsch. 1999. Factors affecting revegetation of *Carex lacustris* and *Carex stricta* from rhizomes. Restoration Ecology 7:162–171.

Zedler, J. B. 1993. Canopy architecture of natural and planted cordgrass marshes: selecting habitat evaluation criteria. Ecological Applications 3:123–138.

10 Plant and Invertebrate Communities as Indicators of Success for Wetlands Restored for Wildlife

SCOTT A. WISSINGER, SCOTT G. INGMIRE,
and JENNIFER L. BOGO

The U.S. Fish and Wildlife Service's Partners for Fish and Wildlife Program (PFW) assists private landowners with the restoration of wetlands to enhance their value as wildlife habitat. To assess the success of these restoration projects, we compared invertebrate and plant communities in restored wetlands to those in hydrogeomorphically similar reference habitats. We found that percent cover of emergent vegetation, plant species, and plant diversity were similar to that in reference wetlands within 5 to 6 years after construction. Many plant species in these habitats are valuable as food and cover to both game and nongame birds. Cattails, which tend to form monotypic stands in constructed wetlands, comprised less than 10% of emergent vegetation at 10 of the 12 restoration sites. We attributed the rapid assembly of diverse plant communities at PFW restoration sites to (1) the presence of remnant populations of wetland plants at the restoration sites; (2) construction design features that include minimal disturbance, shallow depths (<1 m), and seasonal drying, at least along the basin margins; and (3) passive dispersal of plant propagules by waterfowl that use these wetlands. Invertebrate communities in 3-year-old PFW marshes had similar densities, species diversity, and composition as in reference wetlands, and were dominated by taxa important for game and nongame water birds (e.g., chironomids, odonates, crustaceans, caddisflies, water bugs, beetles). The rapid assembly of invertebrate communities in these wetlands is probably less related to remnant populations and more to rapid aerial colonization by immigrants from nearby source populations. Plant and invertebrate communities at wooded PFW sites were depauparate compared to reference wetlands. The absence of remnant plant populations is

Bioassessment and Management of North American Freshwater Wetlands, edited by Russell B. Rader, Darold P. Batzer, and Scott A. Wissinger.
0-471-35234-9

probably responsible for the relatively slow development of vegetation at wooded sites, which in turn is likely to reduce invertebrate diversity. Our results suggest that the presence of diverse remnant communities of wetland plants at open-marsh restoration sites should preclude the need for inoculation with plant propagules. We also concluded that because many wetland invertebrates are excellent dispersers, self-assembly of invertebrate communities should be rapid, especially in regions with an abundance of source populations. Several studies that report high rates of use by game and nongame birds at PFW sites are consistent with the idea that plant and animal communities can be used as a biomonitoring tool to assess the success of wetland restoration for enhancing wildlife habitat.

During the past 30 years there has been a wetland construction boom. Created and restored wetlands are increasingly used to (1) mitigate for jurisdictional encroachments, (2) serve as coastal and riparian storm and flood buffers, (3) improve surface water and groundwater quality, and (4) provide wildlife habitat (e.g., Kentula et al. 1992, van der Valk and Jolly 1992, Wheeler et al. 1995, Kadlec and Knight 1996, Mitsch et al. 1998). Although the increased pace of wetland construction is encouraging, there has been considerable debate about whether constructed wetlands are successful in reaching their goals. This debate has been especially vigorous for mitigation wetlands created to replace ecosystem functions and services lost when jurisdictional wetlands are filled or drained (Roberts 1993, Zedler 1996, Malakoff 1998, Mitsch et al. 1998). Consequently, there has been considerable interest in developing performance standards for comparing wetland functions at the affected site to those at the mitigation site (e.g., Brinson and Rheinhardt 1996, Wilson and Mitsch 1996, Rheinhardt et al. 1997). Less attention has been paid to assessing whether nonmitigation restoration projects have been successful in repairing damaged wetlands (Young 1996, Palmer et al. 1997).

In this chapter we report on the development of plant and invertebrate communities in wetlands restored by the U.S. Fish and Wildlife's Partners for Fish and Wildlife Program (PFW). Since 1987, PFW has assisted private landowners with the voluntary restoration of nearly 500,000 acres of degraded wetlands across the United States (see *http://partners.fws.gov*). In Pennsylvania the program is easily the single largest contributor to wetland restoration, with 300+ sites and over 3500 acres restored to date (Brown 1994). Our study sites are located in the glacial drift zone of northwestern Pennsylvania, where many depressional wetlands were historically drained and filled for agriculture (Tiner 1989).

For a site to be eligible for PFW restoration, there must be evidence of the historical presence of hydric soils. Restoration in this region typically involves the removal or blockage of drain tiles and the construction of low berms downslope from depressions or sources of groundwater discharge (Brown 1994). The wetlands are designed to be shallow (maximum depth typically < 1 m), with seasonal drying at least along the wetland–upland ecotone. Upland and remnant wetland topsoils are set aside and reapplied to surfaces after construction. Some sites include control structures to facilitate water-level regulation by landowners (Snyder 1998).

The primary goal for these PFW restoration projects is to reestablish natural habitats for game and nongame wildlife, particularly wetland-associated birds. To assess the suitability of these habitats for wildlife, we compared their invertebrate and plant communities to those at hydrogeomorphically similar reference sites. Invertebrates and plants are the two most important biological criteria used by managers of wetlands designed as wildlife habitat (Fredrickson and Reid 1988, Payne 1992, Baldassarre and Bolen 1994, Galatowitsch and van der Valk 1994, Euliss et al. 1999). Vegetation is important both as food and as structural habitat for nesting, foraging, and protection from predators (see reviews by Martin et al. 1951, Bellrose 1980, Burger 1985, Weller 1999). Vegetation is also one of the most important factors that influences the abundance and diversity of wetland invertebrates (Batzer and Wissinger 1996). Aquatic and adult stages of invertebrates are the primary trophic links between primary production and most wetland vertebrates (amphibians, birds, mammals; see reviews by Weller 1999, Wissinger 1999). Even waterfowl species that are mainly herbivorous as adults rely heavily on invertebrates during critical early-life stages of growth and development (Bellrose 1980, Sedinger 1992, Cox et al. 1998).

METHODS

Vegetation

We used a synoptic, rapid bioassessment approach to compare the plant and animal communities at restoration sites to those at nearby reference wetlands (see Abbruzzese et al. 1988, Abbruzzese and Leibowitz 1997). The diversity and species composition of plant communities in 12 PFW sites were compared to six established (>25 years old) reference wetlands (Ingmire 1998). We chose four PFW sites in each of three age categories (1 to 2, 3 to 4, and 5 to 6 years since restoration) to obtain information on the early stages of plant community assembly. All of the wetlands were small (0.2 to 2.5 ha) shallow (maximum depth < 1 m) depressional marshes surrounded by upland habitat. Surface water chemistry in restored and reference wetlands was circumneutral (pH 6.5 to 7.5) with high alkalinity (100 to 150 mg/L, high total dissolved solids (40 to 80 ppm), and high levels of nutrients (total nitrogen 1.5 to 3.0 mg/L, total reactive phosphorus 0.1 to 0.3 mg/L; Paulovich 1998). The average area of the reference wetlands was slightly larger (1.3 ha) than that of the PFW sites (1 to 2 years, 1.2 ha; 3 to 4 years, 0.8 ha; 5 to 6 years, 0.8 ha), although mean areas among categories did not differ statistically ($F_{3,14} = 0.58$, $p = 0.64$). Thus, diversity comparisons among the various ages of PFW sites and between restored and reference wetlands were not confounded by species–area effects.

We sampled each wetland twice in fall 1998. We first conducted a qualitative census to establish a species list for all emergent (herbaceous and woody) plants located within the jurisdictional boundary of each wetland (delineation criteria based on U.S. Army Corps of Engineers 1987). Plant species were identified (Gleason and Cronquist

1963) and categorized by jurisdictional wetland status (OBL, obligate hydrophyte; FACW, facultative wetland; FAC, facultative; FACU, facultative upland; Reed 1997). We then conducted a quantitative study at each site to determine the relative abundances of the dominant taxa (based on percent cover) and to estimate overall percent vegetation cover. Quadrats (1 m^2) were sampled at five elevations along three transects from the wetland/upland boundary to the center of each basin.

Plant species richness, Shannon–Wiener diversity, and percent cover were compared among groups using one-way ANOVA and Scheffé's method for a posteriori contrasts (after Day and Quinn 1989). We also calculated Jaccard's index of community similarity as $CC_j = c/ (s_1 + s_2 - c)$ where s_x represents the number of species in each community and c represents the number common to both communities. Distance (or dissimilarity = 1 - similarity) values were depicted graphically in a dendrogram to indicate clustering patterns among sites (after Sneath and Sokal 1973, Legendre and Legendre 1983).

Invertebrates

Invertebrate diversity, species composition, and densities at six (three wooded and three open) PFW sites were compared to six hydrogeomorphically similar reference wetlands (established > 25 years; Bogo 1997). The PFW sites were all 3 years old and the open-field marshes were a subset of those in the 3- to 4-year-old age class in the vegetation study described above. The wooded sites were included in this study because postconstruction observations by PFW personnel suggested that plant communities were less developed at these sites than at open-field restoration sites. The early successional woodlands at these sites were dominated by red maple (*Acer rubrum*), elm (mainly *Ulmus americana*), and black ash (*Fraxinus nigra*) on abandoned farmland. Although the trees in the constructed basins were removed, they were not in the surrounding upland areas.

The four types of wetlands in this study also did not differ significantly in mean area (PFW wooded = 1.21 ha; PFW open = 1.0; reference wooded = 1.2; reference open-1.9; ANOVA $F_{3,8} = 1.8$, $p = 0.22$). An initial vegetation census indicated that the combined area covered by emergent and/or submergent vegetation in restored wooded wetlands was consistently lower (all < 20%) than that at open-field restoration sites (all > 70%). The substrate at the wooded sites was covered by a relatively thin organic layer of leaves from the adjacent woodlands compared to the relatively thick organic layer of herbaceous detritus at the open-field sites.

Invertebrates were first surveyed qualitatively using D-frame net sweeps in all subhabitats to establish taxonomic lists for each wetland (after Cheal et al. 1993). We then returned to each wetland to collect four quantitative samples using a 0.125-m^2 drop box. The drop box was swept repeatedly with a D-frame net until all emergent and submergent vegetation and detritus were removed. Because invertebrate community species composition and densities change rapidly from early to late fall (e.g., Wissinger 1989, Wissinger and Gallagher 1999), we stratified our sampling so there was no seasonal bias by wetland category. Total time from first qualitative to last quantitative samples was 6 weeks (mid-September to late-October).

Invertebrate taxa were separated immediately from the detritus and later identified to level of genus or species. We compared the overall diversity and densities of invertebrates between the sites using two-way ANOVA (restored versus reference and wooded versus open), subsequent protected one-way ANOVA, and Scheffé's posthoc tests (after Day and Quinn 1989).

RESULTS

Vegetation

All the PFW sites including those that were only 1 to 2 years old, had surprisingly diverse plant communities (Table 10.1 and Appendix 10.1). Plant species richness did not differ between restored and reference wetlands, or among the age groups of the restored wetlands (ANOVA $F_{3,14} = 2.16$, $p = 0.14$). Shannon–Weiner diversity was also similar among sites (ANOVA $F_{3,14} = 2.15$, $p = 0.15$). Species richness was positively affected by wetland area ($F_{1,16} = 15.2$, $p = 0.012$; Fig. 10.1). The strength of this species–area relationship, even over a relatively small range of wetland sizes, emphasizes the importance of choosing sites of similar size for comparative floristic surveys in wetlands.

TABLE 10.1. Summary of Vegetation Characteristics of Restored and Reference Wetlands[a]

	Restored			
Characteristic[b]	1–2 (*n* = 4)	3–4 (*n* = 4)	5–6 (*n* = 4)	Reference (*n* = 6)
Species richness	29 + 2	33 + 3	31 + 3	35 + 4
Vegetative cover (%)[b]	48 + 9	57 + 5	72 + 4	71 + 3
Category[c]				
Obligate (%)	58 + 6	36 + 6	48 + 4	63 + 7
Faculative wetland (%)	39 + 6	58 + 11	47 + 5	34 + 6
Facultative (%)	2 + 0.4	5 + 0.8	3 + 1	3 + 0.8
Faculative upland (%)	1 + 0.1	1 + 0.2	2 + 0.6	0
Cattails (%)	11 + 7	6 + 2	8 + 3	6 + 1

[a]All values are means + 1 S.E.

[b]Vegetation cover is percent coverage of emergent hydrophytes within the jurisdictional boundary.

[c]Wetland indicator categories (obligate, facultative wetland, facultative, facultative upland; after Reed 1997) indicate percent coverage of hydrophytic (i.e., not including upland) taxa. Cattails include *Typha latifolia* and *T. angustifolia*.

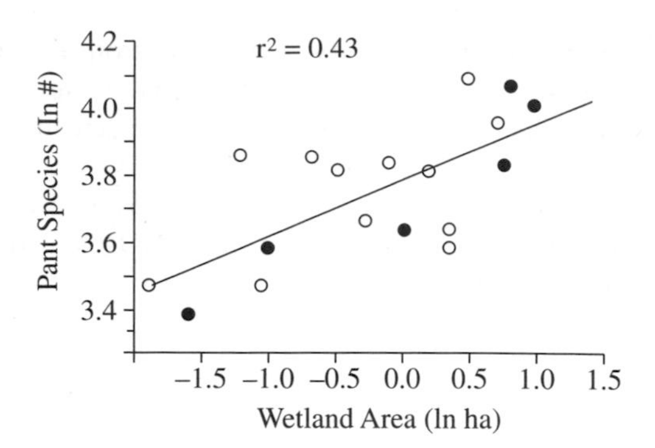

FIGURE 10.1. Relationship between wetland basin area and number of plant species. All values are log transformed. Open circles, restored wetlands; filled circles, reference wetlands.

Although species composition varied considerably among sites, a number of taxa were typically dominant (percent cover) at both reference and restored wetlands [*Bidens cernua, Cornus amomum, Eupatorium perfoliatum, Hypericum mutilum, Juncus effusus, Leersia oryzoides, Ludwigia palustris, Onoclea sensibilis, Phalaris arundinacea, Scirpus cyperinus, Scirpus validus (Schoenoplectus tabernaemontani), Sparganium emersum, Typha latifolia,* and *Verbena hastata*]. Several plant species were more abundant in the reference wetlands than at the restoration sites (*Eleocharis palustris, Scutelleria laterifolia, Sparganium eurycarpum, Triadenum virginicum, Vaccinium corymbosum, Dulichium arundinaceum, Galium trifidum, Sambucus canadensis, Viburnum dentatum,* and *Carex pseudocyperus*). Conversely, species more abundant at restoration than reference sites included *Alisma plantago-aquatica, Eleocharis intermedia, Epilobium coloratum, Hypericum mutilum, Polygonum pennsylvanicum,* and *Scirpus atrovirens.*

Cattails (*Typha latifolia, T. angustifolia*) were no more abundant at the restored sites than at the reference sites ($F_{3,14} = 0.94$, $p = 0.45$; Table 10.1). Cattails represented > 10% of the total vegetation coverage at only two of the 12 restoration sites, and the relatively large average value (11%) for 1- to 2-year-old restored wetlands was due primarily to abundance at one site (26.7%).

There was not a clear overall trend in terms of the degree of similarity in community composition among the different groups of wetlands (Fig. 10.2). Jaccard distances suggest that several of the reference wetlands were more similar in community composition to each other than to the restored sites, but there was not a strong successional pattern of clustering between reference and restored, or among the different ages of restored habitats (Fig. 10.2).

Percent vegetation cover differed significantly among wetland categories ($F_{3,14} = 3.98$, $p = 0.03$; Table 10.1). Although percent vegetation cover at 1- to 2-year-old and 3 to 4-year-old PFW sites was surprisingly high (50%), it was significantly lower that than at 5- to 6-year-old PFW sites and reference wetlands (Scheffé's contrast $p < 0.05$). Percent cover at 5- to 6-year-old restoration sites was nearly identical to that at reference sites (Table 10.1).

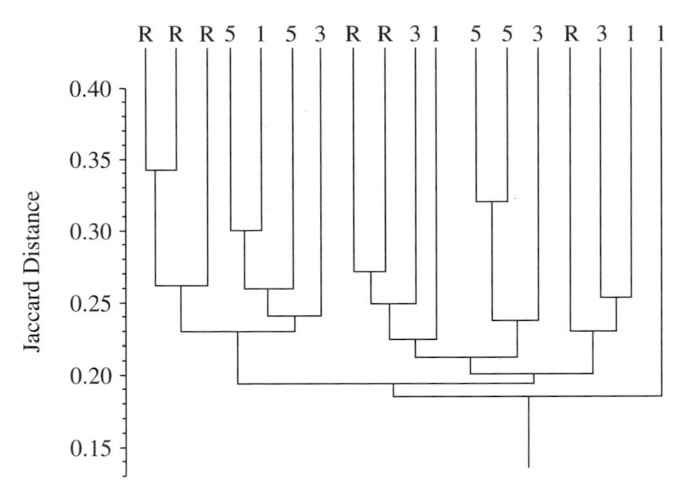

FIGURE 10.2. Jaccard distances (1-similarity) for three age groups of PFW wetlands and reference wetlands. 1, 1-2 yr. old restored; 3, 3-4 yr. old restored; 5, 5-6 yr. old restored; R, reference wetlands.

Invertebrates

Open-field restoration sites and reference sites had similar invertebrate communities (Table 10.2 and Appendix 10.2). The most diverse taxonomic groups were true flies (21 species, mostly chironomids), beetles (20 species), odonates (dragonflies and damselflies, 22 species), mollusks (15 species), water bugs (12 species), and a variety of benthic crustaceans (amphipods, isopods, ostracods, crayfish) (details in Bogo 1997). The numerically dominant taxa at all sites were chironomid dipterans, water bugs, amphipods, odonates, and mayflies (*Caenis hilaris*).

Reference sites had more invertebrates species than restored wetlands (two-way ANOVA, habitat main effect, $F_{1,7} = 8.7$, $p = 0.021$; Table 10.2). However, this effect was due primarily to differences at the wooded sites where restored wetlands were much less diverse than reference wetlands (one-way ANOVA $F_{1,3}= 51.84$, $p = 0.005$). In contrast, restored and reference open sites did not differ in invertebrate diversity (one-way ANOVA $F_{1,3}= 0.75$, $p = 0.43$; Table 10.2). Across both restored and references habitats, invertebrate diversity was not significantly different between wooded and open sites (two-way ANOVA, $F_{1,7} = 2.55$, $p = 0.15$). Total invertebrate densities ranged from about 1000 to over 5000 animals/per square meter and were generally lower in wooded in than open wetlands (Table 10.2). However, because of extremely high variation among replicates, average densities were not different between habitat types (two-way ANOVA, $F_{1,7}= 0.85$, $p = 0.38$) nor between restored and reference wetlands (two-way ANOVA $F_{1,7}= 2.24$, $p = 0.12$).

The most dramatic taxon-specific trend in this study was the dominance of chironomid dipterans at wooded restoration sites. Wooded restoration sites had much higher densities than those in wooded reference wetlands (density $F_{1,7}= 16.9$, $p =$ 0.004; Fig. 10.3). The high chironomid densities at the wooded restored sites were

TABLE 10.2. Summary of Benthic Invertebrate Communities in Restored and Reference Wetlands[a]

	Open Marsh			Wooded		
Characteristic	Restored ($n = 3$)	Reference ($n = 3$)		Restored ($n = 3$)		Reference ($n = 3$)
Species richness	40 ± 4	42 ± 5		24 ± 2		41 ± 3
Total density	3924 ± 1643	2725 ± 833	*	1037 ± 439		1102 ± 558
Taxa						
Diptera (%)	8 ± 1	6 ± 2	*	80 ± 15	*	24 ± 13
Crustaceans (%)	26 ± 13	29 ± 16	*	0	*	14 ± 8
Mollusks (%)	20 ± 7	24 ± 9	*	0	*	10 ± 5
Ephemeroptera (%)	23 ± 12	17 ± 9		3 ± 1	*	22 ± 14
Odonata (%)	10 ± 4	9 ± 3		12 ± 4		17 ± 4
Trichoptera (%)	2 ± 1	3 ± 1		0		2 ± 1
Hemiptera (%)	2 ± 1	8 ± 4		1 ± 1		2 ± 1
Other taxa[b] (%)	8 ± 6	4 ± 2		4 ± 3		9 ± 5

[a]Values are means ± 1 S.E. Asterisks indicate significant differences (Scheffé's a posteriori test < 0.5) between habitat types or between restored and reference wetlands within habitat type. Center column asterisks indicate significant differences between wooded and open marshes (see text for ANOVA statistics).

[b]Mainly beetles, leeches, and oligochaetes.

due primarily to an abundance of *Chironomus* and *Microtendipes*, two genera that were not dominant at the other sites. Together, these two chironomids comprised over 75% of the invertebrates collected at the wooded restoration sites and over 50% of the biomass. Beetles, mayflies, odonates, and caddisflies were all less abundant and less diverse at the wooded restoration wetlands than in the other reference and restoration sites (Table 10.2; for more details, see Bogo 1997).

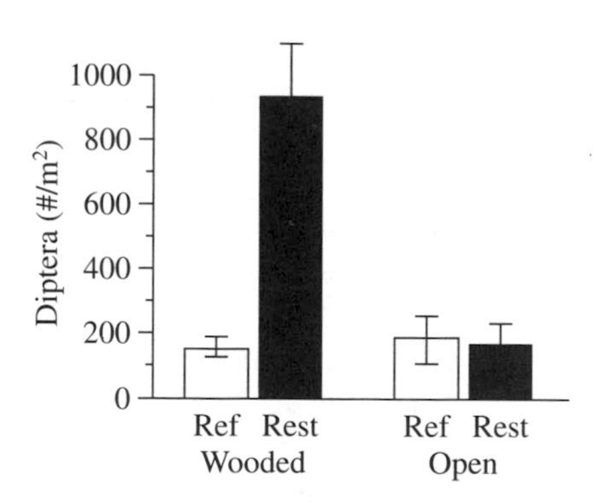

FIGURE 10.3. Diptera densities in restored and reference wetlands. Values are means ± 1 S.E.

DISCUSSION

Plant Community Development at PFW Restoration Sites

Our results suggest that wetland plant community assembly is rapid at PFW restoration sites. Within a few years after construction, PFW plant communities were taxonomically similar to and as diverse as the reference wetlands. Vegetation coverage at 5- to 6-year-old PFW sites was nearly identical to that of reference sites. Although we do not know much about the preagricultural composition of these plant communities, we do know that many of the plant species at PFW wetlands are the same as those at established reference sites in the region. Many of these species (see Appendix 10.1) are important as food (mainly seeds) and structural habitat (nesting, cover) for waterfowl and other wetland-associated birds (Martin et al. 1951, Bellrose 1980). The diversity of sedges (nine *Carex* spp., *Dulichium arundinaceum*) at PFW sites was particularly interesting given that they are considered to be slow colonizers of restored wetlands in other regions (Budelsky and Galawowitsch 1999). We did not sample the seedbanks in these restored wetlands and therefore do not know whether these newly restored sites would be more vulnerable to interannual variations in climate than wetlands with more established seedbanks (Galatowitsch and van der Valk 1996a,b).

Most studies that report similarly rapid plant community assembly have been at constructed wetlands that were inoculated with one or more types of plant propagules (seeds, seedbanks, seedlings, tubers) (e.g., van der Valk et al. 1992, Reinartz and Warne 1993, Brown and Bedford 1997), or were riparian and hence could be colonized during flooding by waterborne propagules (hydrochory; Mitsch et al. 1998). The rapid assembly of plant communities in this study is surprising given that propagules were not introduced and that these are depressional, nonecotonal (Tiner 1993) wetlands surrounded by upland habitat. We suggest three explanations for the rapid assembly at PFW wetlands. First, many of these sites probably had remnant populations of hydrophytes that had persisted in the wettest areas in or along fields. Such remnant populations are common on the poorly drained glacial soils in this region and should often provide an on-site source of propagules for restoration. Many of the species that were common at nearly all the PFW sites can frequently be observed in active and abandoned farm fields in this area (e.g., *Carex scoparia, Eupatorium perfoliatum, Juncus effusus, Leersia oryzoides, Onoclea sensibilis, Polygonum pennsylvanicum, Polygonum sagittatum, Spirea alba, Typha latifolia*).

Farming has steadily decreased in the region during the past 30 years (Rumney 1992), and many PFW sites are fallow at the time of restoration. This should enhance the importance of refugial hydrophyte populations and perhaps increase the importance of active seedbanks (Madsen 1986, LaGrange and Dinsmore 1989). Long-term viability of seeds in well-drained cultivated areas of these fields are a less likely source of propagules. Wienhold and van der Valk (1989) found that over 75% of all wetland species had disappeared from seed banks 40 years after they had been drained and cultivated. Most of our sites had probably been drained for > 50 years (D. Brown, personal communication).

A second potential explanation for this rapid "self-assembly" (Mitsch et al. 1998) is that birds are reported to colonize PFW restoration sites rapidly (Dick 1993, Cashen 1998, Snyder 1998) and should therefore act as vectors for passive seed dispersal (see deVlaming and Proctor 1968, Powers et al. 1978). Censuses at six of the 12 PFW sites in our study indicated that several species of waterfowl were observed as often (e.g., black ducks, green-winged teal, wood ducks) or more often (e.g., mallards) than at reference wetlands during fall migration (Repko 1998). A 3-year study across all seasons of 18 PFW sites in central Pennsylvania provides evidence for high usage rates not only by waterfowl but also by shorebirds, and songbirds including several species of special concern (Cashen 1998). It seems likely that if the initial stages of community development at restoration sites exceed some threshold in diversity or plant cover that is attractive to wildlife, subsequent colonization should be enhanced by a positive feedback loop of passive immigration associated with bird use.

Finally, several features of PFW project design should encourage rapid revegetation. PFW design specifications explicitly acknowledge the "too deep and too permanent" syndrome that has historically plagued wetland construction in North America (see reviews by Mitsch and Wilson 1996; Mitsch et al. 1998). To reduce the tendency for creating pondlike wetlands, the fill material for berms is often taken from outside and adjacent to, rather than inside, the restored site. The main basins are therefore shallow (< 1 m maximum) and surrounded by other types of wet meadow and moist soil habitats. This encourages plant species that require partially or completely exposed substrates for germination as well as those that can tolerate permanent inundation (see reviews by Gallinato and van der Valk 1986, Budelsky and Galatowitsch 1999).

Cattail Dominance

Early colonization by cattails can preempt the establishment of a diverse plant community in constructed wetlands (Reinartz and Warne 1993; Brown and Bedford 1997), and once established, cattails are difficult to control (e.g., Ball 1990). Although cattails are beneficial to some wildlife and extremely important for establishing rapid cover in some types of pollution treatment wetlands (e.g., McKinstry and Anderson 1994, Horstman et al. 1998, O'Sullivan et al. 1999), the tendency to form monocultures reduces their overall value for wildlife conservation (Sojda and Solberg 1993).

Cattail dominance was not a problem at most of our study sites, perhaps because of the minimal disturbance associated with PFW construction techniques and/or to their lack of dominance in remnant communities. At our study sites, and in this region in general, draining for cultivation was typically achieved with subsurface tiles rather than ditches. Although in-field depressions and wet margins along fields with drainage tiles support remnant communities of wet-meadow and moist-soil hydrophytes, they are not typically dominated by cattails. In contrast, drainage ditches are often dominated by cattails, and restoration projects at sites with ditches should be more vulnerable to early cattail dominance than those without (e.g., Brown and Bedford 1997). We suggest that a preconstruction survey of the composition of remnant hydrophyte communities should be conducted at restoration sites to determine (1) the presence/extent of such communities, (2) their diversity, and (3) the degree to which they are dominated by species that tend to form monotypic stands if they are among the earliest colonists (cattails or invasives such as common reed, purple loosestrife, etc.). The results of such a survey could be used to determine whether some form of inoculation (seeds, seed banks, seedlings, other propagules) would increase the chances for early establishment of a diverse plant community.

Invertebrate Community Development at Open PFW Sites

The results of this study provide evidence that invertebrate community assembly is also rapid at open-field PFW restoration sites. This observation is consistent with several other studies that report rapid invertebrate community assembly in constructed marshes that have well-established plant communities (Streever and Portier 1994, Streever et al. 1995, 1996, Brown et al. 1997, Mitsch et al. 1998). In contrast with the plants (see the discussion above), invertebrate community assembly is unlikely to be strongly tied to on-site remnant populations or egg banks. Most of the taxa at PFW marshes are characteristic of inundated wetland habitats in this region (see Wissinger and Gallagher 1999) and differ from the suite of moist-soil and upland-soil invertebrates that are likely to inhabit these fields before restoration.

A parsimonious explanation for the rapid assembly of a diverse and abundant community of invertebrates in PFW wetlands is immediate and frequent immigration by aerial dispersal. Unlike many aquatic invertebrates, wetland invertebrates, exhibit a wide variety of adaptations for colonizing temporary habitats, including several aerial colonization strategies (reviewed by Wissinger 1999). For example, many chironomids and other dipterans (e.g., mosquitoes) are "*r-strategists*" that can be observed ovipositing in wetlands within a few days of inundation (Batzer and Wissinger 1996, Wrubeleski 1999). Beetles and water bugs exhibit a different aerial colonization strategy that involves cyclic migrations between temporary and permanent habitats. Cyclic colonizers (see Wissinger 1997) such as corixids and predaceous diving beetles depend on permanent habitats (ponds and lakes) as refugia in winter or when temporary habitats dry. Cyclic colonizers are more likely to colonize a restoration site rapidly in this region, where source populations are ubiquitous than in relatively arid regions (e.g., Anderson et al. 1999, Hall et al. 1999). A third group of aerial colonizers are long-lived, strong-flying aquatic insects such as the odonates that can readily disperse among habitats on a local, regional, and even latitudinal scale (Wissinger 1999).

In a follow-up study to the work presented here, Cramer (1998) compared the invertebrate communities at PFW sites that ranged in age from less than 6 months to more than 6 years. Only 6 months after restoration, he found taxa representing all three types of aerial colonization including several genera of chironomids and other diptera (especially *Chironomus, Microtendipes, Paratanytarsus, Culex*), dragonflies (*Pantala hymenea, Libellula lydia, Anax junius*), water bugs (*Hesperocorixa, Sigara*), predaceous diving beetles (*Hydroporus, Agabetes*), and even several taxa that are typically considered poor dispersers (e.g., the mayfly *Callibaetis*, and the caddisflies *Ptilostomis* and *Platycentropus*). In landscapes with an abundance of aquatic and wetland habitats, the assembly of invertebrate communities is less likely to be constrained by dispersal than has been argued for plants (Reinartz and Warne 1993).

Nonwinged invertebrate taxa that rely on passive dispersal (e.g., mollusks, annelids, crustaceans) were less abundant at our PFW sites that at the reference wetlands. In a separate study, Bolden et al. (in review) compared the odonate and mollusk faunas of 12 PFW sites to those of 12 reference wetlands. These taxa were chosen because (1) they represent the end members of a dispersal continuum from highly mobile taxa to relatively sedentary, passive dispersers (see Boag 1986, Brown 1991); (2) they are relatively large and easy to identify to species without long-term life history studies, and (3) both are important prey to a variety of game and nongame birds (e.g., Bellrose 1980, Gammonly and Heitmeyer 1990, Turner and McCarty 1998). As we found in this study, species composition, diversity, and density of odonates in the PFW restored wetlands studied by Bolden et al. were all similar to that at reference wetlands after only 3 years. In contrast, mollusks colonized less rapidly and did not reach reference levels of diversity and abundance until 5 years after restoration. This suggests that passively dispersing species such as molluscs are perhaps a better indicator of the maturity of invertebrate communities in restored wetlands than are aerial dispersers.

Because these restorations are so shallow, they do not support fish populations, which can inhibit the development of a diverse wetland invertebrate fauna. Fish reduce invertebrate densities (e.g., Mallory et al. 1994, Fairchild et al. 1999) and therefore are often considered unwanted predators by managers of waterfowl habitats (e.g., Hunter et al. 1986). In addition, many nongame birds depend on large invertebrate taxa which are the most likely to be eliminated by fish predators (see reviews by Batzer and Wissinger 1996, Wellborn et al. 1996). In many regions, fish are excluded from restoration sites only if they dry, whereas at our study sites, even shallow permanent wetlands are often fishless because of winterkill (Wissinger and Gallagher 1999). There is a abundant evidence that shallow and temporary restoration sites are more likely to support assemblages of wetland invertebrates rather than the aquatic taxa that are likely to be found when sites are made too deep and too permanent (Wellborn et al. 1996, Wissinger 1999).

Wooded PFW Sites

In contrast to restored wetlands at open-field sites, the PFW restorations at wooded sites were not as diverse as reference wetlands and were typically dominated by chironomids, especially *Chironomus*. Dominance by this early-colonizing species and the relative paucity of other taxa including other chironomids, suggests that these restoration sites have remained in a relatively early stage of invertebrate community succession (e.g., see Wrubeleski 1999 and references therein). Dominance by chironomids can be viewed as a positive attribute of wetlands managed for migrating waterfowl, especially if those chironomids are extremely abundant (Baldassarre and Bolen 1994, Weller 1999). However, at these restoration sites, the overall abundance of invertebrates is not exceptionally high (Table 10.2), and several other groups of taxa important for breeding waterfowl and nongame birds are rare or absent. One explanation for the depauparate invertebrate communities at these sites is that plant communities are poorly developed compared to those observed at the open marsh sites. It is well established that wetland invertebrate diversity is positively correlated with plant diversity, vegetation cover (Fairchild et al. 1999, Wissinger 1999), and amount of vascular plant detritus (Evans et al. 1999). Although wetland ecologists do not understand all of the cause-and-effect relationships that underlie correlations between vegetation and invertebrates, it is clear that a diverse and extensive community of wetland plants is an important correlate of a diverse assemblage of invertebrates.

Self Assembly Versus Ecological Engineering

There is considerable debate among wetland ecologists about the need for inoculating constructed wetland sites with plant (and/or invertebrate) propagules (see the review by Mitsch et al. 1998). One concern about the "ecological engineering" approach to constructed wetlands is that inoculation can interfere with natural sequences of immigration and establishment associated with "self-assembly." For wetland plants, such inoculations are probably more often necessary for created than for restored wetlands because the latter have on-site sources of colonists from remnant populations and associated seed banks (Reinartz and Warne 1993, Stauffer and Brooks 1997, Parikh and Gale 1998, Sistani et al. 1999). However, at restoration sites without such remnant populations (such as the wooded sites in our study), the introduction of plant propagules should enhance the development of a diverse plant community (van der Valk et al. 1992, Galatowitsch and van der Valk 1996a,b). Introducing propagules of desirable plants species might also be necessary at sites when preconstruction populations are dominated by invasive taxa (Brown and Bedford 1997, Galatowitsch et al. 1999).

A second criterion that should affect the need for plant propagule inoculation is whether constructed wetlands are nonecotonal (surrounded by terrestrial habitat) versus ecotonal (adjacent to lakes, rivers, estuaries) (see Tiner 1993, Wissinger 1999). Mitsch et al. (1998) found that seed dispersal by floodwaters at created riparian wetlands resulted in the self-assembly of a vegetation community at nearly the same rates as in wetlands that were inoculated. Such self-assembly of plant communities is less likely in depressional wetlands because of dispersal constraints related to distance to the nearest seed sources (Reinartz and Warne 1993). It is important to distinguish between nonecotonal and ecotonal wetlands and between restored and created sites when considering the need for plant propagule inoculation. Data from plant community assembly that support or refute the ecological engineering versus self-assembly arguments collected from one wetland category should not be generalized to other categories.

Because of the rapid dispersal abilities of wetland invertebrates, the ecotonal versus nonecotonal or created versus restored distinctions are probably less important for predicting invertebrate than plant community assembly (Streever et al. 1995, 1996, Brown et al. 1997). Instead, invertebrate community assembly is more likely to depend on the abundance of local sources of aerial and passive dispersers in the surrounding landscape (Anderson et al. 1999, Wissinger and Gallagher 1999). Whether immigrants are successful in becoming established will depend in turn on the presence of a well-established plant community and the absence of fish predation.

Plants and Invertebrates as Indicators of Success in Wildlife Restorations

For restoration efforts that are designed specifically to enhance the value of wetlands for wildlife, the single best indicator of success will be some measure of usage and production of wildlife (e.g., VanRees-Siewert and Dinsmore 1996, Cashen 1998). Although surrogate measures of habitat suitability (plants and invertebrates) cannot replace bird censuses, they do offer the advantage of being less seasonally constrained and more easily included in rapid, synoptic bioassessment studies. There is an abundance of evidence that suggests the three easily measured vegetation attributes for such a study should include (1) percent basin coverage by emergent species; (2) species composition, including dominance by undesirable species, and (3) species diversity. Percent cover is important because (1) of the obvious benefits to waterfowl and non-game birds as nesting habitat and for protection (see the references in Chapter 1), (2) it is highly correlated with invertebrate secondary production (Wissinger 1999), and (3) it provides bird foraging habitat on both the aquatic (e.g., Delphy and Dinsmore 1993, Batzer and Wissinger 1996, VanRees-Siewert and Dinsmore 1996), and aerial stages of wetland invertebrates (King and Wrubeleski 1998, Turner and McCarty 1998). Percent cover by hydrophytes should be low both when restoration efforts fail to recreate hydric soil conditions and when sites are too deep and permanent (Tiner 1991, 1993). Many wetland plants that are important to waterfowl germinate only on exposed soils (Kadlec 1962; Harris and Marshall 1963, Gallinato and van der Valk 1986); thus, successful restoration sites must be shallow, nonpermanent, or at least have exposed margins in some years.

Plant species composition is an important indicator of restoration success for wildlife for several reasons. It is important to establish first, that restoration sites actually contain a preponderance of wetland vegetation (i.e., hydrophytes) (Tiner 1991), and second, that disturbed habitat species (e.g., cattails, common reed) do not form monotypic stands and hence preclude the development of a diverse plant community of other species that are important as food and invertebrate habitat for waterfowl (see the reviews by Reinartz and Warne 1993, Brown and Bedford 1997).

Finally, plant diversity is an important attribute of restoration success for wildlife because (1) different species of plants create different types of structural habitat for nesting and protection from predators (e.g., Burger 1985, Greenwood et al. 1987, de Szalay and Resh 1997), (2) seeds of different wetland plants are important for different species of waterbirds (e.g., Martin et al. 1951, Baldassarre and Bolen 1994, Lillie and Evard 1994), (3) different species of invertebrates are associated with different wetland plants (Krull 1970, Olson et al. 1995, Batzer et al. 1999), and (4) diverse plant communities create diverse seed banks that can respond to interannual fluctuations in water levels (van der Valk and Davis 1978, van der Valk et al. 1992), thus enhancing ecosystem stability. Unless restoration efforts are designed for a particular species, a diverse plant community should always be included as a goal for wildlife restoration projects.

Invertebrate community attributes that indicate restoration success for wildlife should include some measure of overall diversity, abundance, and community composition. Total abundance is correlated with habitat choice by game and nongame wetland birds (e.g., Colwell and Oring 1988, Safran et al. 1997, Turner and McCarty 1998). Taxonomic diversity is an indicator of the potential for supporting a variety of game and non-game birds (Bellrose 1980, Weller 1999), and species composition indicates the presence of those particular taxa that are important in wetland bird diets (e.g., chironomids, molluscs, crustaceans, odonates) (Bellrose 1980, Kaminski and Prince 1981, Nudds and Bowlby 1984, Euliss and Grohaus 1987, Fredrickson and Reid 1988, Eldridge 1992, Nummi 1993, Baldassarre and Bolen 1994, de Szalay and Resh 1997, Safran et al. 1997, Wrubeleski 1999, Weller 1999). Because particular species of wildlife consume only certain invertebrate taxa, classifying invertebrates to only order or family rather than to genus or species might be inadequate for assessing the success of restoration for supporting a diverse assemblage of wildlife (Batzer and Wissinger 1996).

There are at least three caveats to the conclusion that invertebrates and plants can be used to assess the success of wildlife-centered restoration projects. First, all else being equal, overall wetland size can be extremely important for habitat choice by wetland birds (Erwin et al. 1991, Cashen 1998). Small sites are unlikely to attract as many individuals or species of waterbirds as large sites. Second, water depth can limit the availability of food to waterfowl and shorebirds (Safran et al. 1997), as will the timing of drawdown or drying (Hands et al. 1991). Third, assessing the degree to which plant and animal communities at restored wetlands sites actually approaches the desired goals of repairing damaged ecosystems may ultimately depend on assessing ecosystem function as well as species composition (Palmer et al. 1997). One important research agenda for the future will be to establish the relationship between community assembly at restored wetland sites (which can be quickly determined) and levels of ecosystem function.

APPENDIX 10.1. Hydrophytic Plant Species Found in Wetlands Restored by Partners for Fish and Wildlife and in Nearby Reference Wetlands

Species	Status[b]	Restored			Reference
		1–2 years	3–4 years	5–6 years	> 25 years
Alisma plantago-aquatica	OBL	×	×	×	×
Asclepias incarnata	OBL	×	×	×	
Aster borealis	OBL	×			
A. novae-angliae	FACW	×	×		
A. umbrellatus	FACW	×			
A. lateriflorus	FACW	×	×	×	×
Bidens cernue	OBL	×	×	×	×
B. connata	FACW	×	×		
Boehmeria cylindrica	FACW	×			
Brasenia schreberi	OBL	×	×	×	
Calamagrostis canadensis	FACW	×			
Cardamine pratensis	OBL	×			
Carex asa-grei	FACW	×			
C. bebbi	OBL	×	×		
C. comosa	OBL	×	×		
C. crinita	OBL	×			
C. grayi	FACW	×			
C. gynadra	OBL	×			
C. lupuliformis	FACW	×	×	×	
C. lupulina	OBL	×	×		
C. psuedocyperus	OBL	×			
C. scoparia	FACW	×	×	×	×
C. stricta	OBL	×	×		
C. tuckermanii	OBL	×			
C. vulpinoidea	OBL	×	×	×	×
Cephalanthus occidentalis	OBL	×	×		
Chelone glabra	OBL	×			
Cicuta bulbifera	OBL	×	×	×	

Species	Status[b]	Restored			Reference
		1–2 years	3–4 years	5–6 years	> 25 years
Conium maculatum	FACW	×			
Coptis trifolia	FACW	×			
Cornus amomum	FACW	×	×	×	×
C. stolonifera	FACW	×			
Cyperus engelmanni	FACW	×			
C. strigosus	FACW	×	×	×	
Dryopteris cristata	FACW	×			
Dulichium arundinaceum	OBL	×	×		
Eleocharis acicularis	OBL	×	×		
E. intermedia	FACW	×	×	×	×
E. obtusa	OBL	×	×		
E. ovata	OBL	×	×		
E. palustris	OBL	×			
Epilobium coloratum	OBL	×	×	×	×
Eupatorium maculatum	FACW	×	×		
E. perfoliatum	FACW	×	×	×	×
Galium claytoni	OBL	×			
G. palustre	OBL	×			
G. tinctorium	OBL	×	×		
G. trifidum	FACW	×	×	×	×
Geranium maculatum	FACU	×			
Geum macrophyllum	FACU	×			
Hypericum ellipticum	OBL	×	×	×	
H. mutilum	FACW	×	×	×	×
Impatiens biflora	FACW	×			
I. capensis	FACW	×	×	×	×
Juncus acuminatus	OBL	×	×	×	
J. brachycarpus	FACW	×	×	×	
J. canadensis	OBL	×	×	×	×
J. effusus	FACW	×	×	×	×

APPENDIX 10.1. (*continued*)

Species	Status[b]	Restored 1–2 years	Restored 3–4 years	Restored 5–6 years	Reference > 25 years
J. interior		×			
Kalmia polifolia	OBL	×			
Leersia oryzoides	OBL	×	×	×	×
Lemna minor	OBL	×	×	×	
Lophotocarpus depauperatus	OBL	×			
Ludwigia palustris	OBL	×	×	×	×
Lycopus americanus	OBL	×	×		
L. rubellus	OBL	×	×		
L. uniflorus	OBL	×	×	×	×
L. virginicus	OBL	×			
Mimulus ringens	OBL	×	×		
Myosotis scorpioides	OBL	×			
Nasturtium officinale	OBL	×			
Nuphar variegatum	OBL	×	×		
Onoclea sensibilis	FACW	×	×	×	×
Osmunda cinnamomea	FACW	×			
Panicularia grandis	OBL	×			
Panicum clandestinum	FAC	×	×		
P. dichotomiflorum	FACW	×			
Penthorum sedoides	OBL	×	×	×	
Phalaris arundinacea	FACW	×	×	×	×
Pilea pumila	FACW	×	×	×	×
Poa palustris	FACW	×			
P. sylvestris	FACW	×			
Polygala viridescens	FACU	×			
Polygonum hydropiper	OBL	×	×	×	
P. hydropiperoides	OBL	×	×		
P. pennsylvanicum	FACW	×	×	×	×
P. punctatum	OBL	×	×		

Species	Status[b]	Restored 1–2 years	Restored 3–4 years	Restored 5–6 years	Reference > 25 years
P. sagittatum	OBL	×	×	×	×
Potamogeton americanus	OBL	×			
P. vaseyi	OBL	×	×		
Rumex britannica	OBL	×			
R. orbiculatus	OBL	×			
R. verticillatus	OBL	×			
Sagittaria latifolia	OBL	×	×	×	
Salix discolor	FACW	×	×	×	×
S. interior	OBL	×	×		
S. petiolaris	OBL	×	×		
Salvinia natans	OBL	×			
Sambucus canadensis	FACW	×	×		
Scirpus atrovirens	OBL	×	×	×	×
S. cyperinus	FACW	×	×	×	×
S. microcarpus	OBL	×			
S. polyphyllus	OBL	×	×		
S. validus	OBL	×	×	×	×
Scutellaria laterifolia	FACW	×			
Solidago graminifolia	OBL	×	×		
S. neglectica	OBL	×	×		
S. patula	OBL	×	×	×	×
Sparganium emersum	OBL	×	×	×	×
S. eurycarpum	OBL	×			
S. fluctuans	OBL	×	×	×	
Sphagnum spp.	OBL	×			
Spiranthes cernua	FACW	×			
Spirea alba	FACW	×	×	×	
S. tomentosa	FACW	×			
Spirodela polyrrhiza	OBL	×			
Symplocarpus foetidus	OBL	×	×		

APPENDIX 10.1. (*continued*)

Species	Status[b]	Restored 1–2 years	Restored 3–4 years	Restored 5–6 years	Reference > 25 years
Thelypteris palustris	FACW	×	×		
Tracaulon arifolium	OBL	×			
T. saggittatum	OBL	×	×	×	×
Triadenum fraseri	OBL	×			
T. virginicum	OBL	×			
Typha angustifolia	OBL	×	×		
T. latifolia	OBL	×	×	×	×
Vaccinium corymbosum	FACW	×			
Vallisneria americana	OBL	×			
Verbena hastata	FACW	×	×	×	×
Viburnum dentatum	FACW	×	×	×	
V. trilobum	FACW	×			
Wolffia arrhiza	OBL	×	×		

[a]The crosses indicate a species was present in either the qualitative survey or in one of the quantitative samples.

[b]FAC, facultative; FACU, facultative upload; FACW, facultative wetland; OBL, obligate hydrophyte. Indicator status is based on Reed (1997).

APPENDIX 10.2. Invertebrates Found in Wetlands Restored by Partners for Fish and Wildlife and in Nearby Reference Wetlandsa

	Restored (3 yr)		Reference (> 25 yr)	
Invertebrate Taxa	Open	Wooded	Open	Wooded
Ephemeroptera				
Caenis hilaris	×	×	×	×
Callibaetis ferrugineus	×	×	×	×
Odonata				
Anax junius	×	×	×	
Aeshna tuberculifera	×			
Enallagma basidens	×	×	×	×
E. boreale	×	×	×	×
E. civile	×			
E. signatum	×			
Epitheca cynosura	×	×	×	×
Erythemis simpliciollis	×	×		
Gomphus cornutus	×	×	×	
Ischnura posita	×	×		
I. verticalis	×	×	×	×
Lestes disjunctus	×	×	×	
L. eurinus	×	×	×	
Leucorrhinia glacialis	×	×	×	
L. intacta	×	×	×	×
Libellula luctuosa	×	×		
L. lydia	×	×	×	
L. pulchella	×	×	×	×
Pachydiplax longipennis	×	×	×	×
Perithemis tenera	×	×	×	×
Symptrum vicinum	×	×	×	×
Tramea lacerata	×			
Hemiptera				
Belostoma	×			
Callicorixa	×	×		
Hesperocorixa	×	×	×	×

APPENDIX 10.2. (*continued*)

	Restored (3 yr)		Reference (> 25 yr)	
Invertebrate Taxa	Open	Wooded	Open	Wooded
Hemiptera (*cont.*)				
Notonecta	×	×	×	×
N. 2	×	×		
Mesovelia	×			
Microvelia	×			
Neoplea	×	×		
Pelocoris	×	×		
Ranata	×	×		
Sigara	×	×	×	
Coleoptera				
Acilius	×	×		
Agabetes	×	×		
Agabus	×	×	×	×
Coptotomus	×	×	×	×
Dineutus	×			
Donacia	×	×		
Gyrinus	×			
Haliplus	×	×	×	×
Hydrocanthus	×	×		
Hydroporus	×	×	×	×
H. 2	×			
Hydrovatus	×		×	
Laccophilus	×	×	×	×
Liodessus	×			
Peltodytes	×	×	×	×
Potamonectes	×			
P. 2	×	×		
Rhantus	×			
Rhysodidae	×			
Tropisternus	×	×		

Invertebrate Taxa	Restored (3 yr) Open	Restored (3 yr) Wooded	Reference (> 25 yr) Open	Reference (> 25 yr) Wooded
Diptera				
Ablebesmyia	×	×	×	×
Acricotopus	×			
Anopholes	×	×		
Bezzia	×	×	×	×
Chaoborus	×	×		
Chironomus	×	×	×	×
Chrysops	×			
Clinotanypus	×	×	×	
Cryptochironomous	×			
Dicrotendipes	×	×	×	
Dixella	×	×		
Dolichopodidae	×			
Endochironomus	×	×	×	×
Microtendipes	×	×	×	×
Natarsia	×			
Polypedilum	×			
Procladius	×	×	×	×
Psectrotanypus	×	×		
Rheotanytarsus	×			
Stratiomys	×	×		
Tanypus	×			
Trichoptera				
Anabolia	×			
Fabria	×	×	×	×
Limnephilus	×	×		
L. 2	×	×		
Nemotaulius	×			
Oecetis	×			
Phryganea			×	
Platycentropus	×	×	×	

APPENDIX 10.2. (*continued*)

Invertebrate Taxa	Restored (3 yr)		Reference (> 25 yr)	
	Open	Wooded	Open	Wooded
Trichoptera (*cont.*)				
Pychnopsyche	×			
Ptilostomis	×	×	×	×
Mollusca				
Fossaria	×			
Gyraulis circumstriatus	×	×		
G. pavris	×	×		
Musculium	×	×	×	
Physella heterostropha	×	×	×	×
Planorbella campanulata	×			
P. trivolis	×	×	×	
Pseudosuccina columella	×	×	×	×
Crustacea (benthic)				
Caecidotea	×			
Cypridodopsis	×	×	×	×
Hyallela	×	×	×	
Procambarus	×	×	×	×
Oligochaeta				
Enchytraeidae	×			
Lumbriculidae	×	×		
Naididae	×			
Hirudinea				
Erpobdella punctata	×	×		
Helobdella stagnallis	×			
H. fusca	×			
Mooreobdella	×			
Placobdella ornata	×	×		
Total taxa	73	51	67	66

[a]The crosses indicate the presence of species in either the qualitative surveys or in one of the quantitative samples. "Open" PFW wetlands were restored in fields dominated by herbaceous vegetation, whereas "wooded" refers to those restored in early successional woodlands that had previously been farmed. Open reference wetlands are marshes in open fields whereas "wooded" reference habitats are wetlands adjacent to woodlands.

ACKNOWLEDGMENTS

We are especially indebted to Dennis Brown of the Pennsylvania office of the U.S. Fish and Wildlife Partners for Fish and Wildlife Program and to the landowners who permitted access to their wetlands. Comments by Chris Orr, Ben Mason, Nicky Mason, Kristy Swanson, Jesse Laux, Christie Knight, Susan Bolden, and Tim Grahl greatly improved an earlier version of the manuscript. Thanks to Russ Rader for insightful recommendations and editing. This research was supported by Allegheny College Fund of '39 grants to SGI and JLB and by a National Science Foundation grant (DEB-9407856) to SAW.

REFERENCES

Abbruzzese, B., and S. G. Leibowitz. 1997. A synoptic approach for assessing cumulative impacts to wetlands. Environmental Management 21:457–475.

Abbruzzese, B., A. B. Allen, S. Henderson, and M. E. Kentula. 1988. Selecting sites for comparison with created wetlands. Pages 291–297 *in* C. D. A. Rubec and R. P. Overend (eds.), Wetlands/Peatlands Symposium '87, Edmonton, Alberta, Canada.

Anderson, C. R., B. L. Peckarsky, S. A. Wissinger. 1999. Tinajas of southeastern Utah: Invertebrate reproductive strategies and the habitat templet. Pages 741–810 *in* D. P. Batzer, R. B. Rader, and S. A. Wissinger (eds), Invertebrates in freshwater wetlands of North America: Ecology and Management. Wiley, New York.

Ball, J. P. 1990. Influence of subsequent flooding depth on cattail control by burning and mowing. Journal of Aquatic Plant Management 28:28–32

Batzer, D. P., and S. A. Wissinger. 1996. Ecology of insect communities in nontidal wetlands. Annual Review of Entomology 41:75–100.

Batzer, D. P., R. B. Rader, and S. A. Wissinger. 1999. Invertebrates in freshwater wetlands of North America: ecology and management. Wiley, New York.

Bellrose, F. C. 1980. Ducks, geese, and swans of North America, 3rd ed. Stackpole Books, Harrisburg, PA.

Boag, D. A. 1986. Dispersal in pond snails: potential role of waterfowl. Canadian Journal of Zoology 64:904–909.

Bogo, J. L. 1997. Invertebrate communities in natural vs. constructed wetlands in northwestern Pennsylvania. B.S. thesis, Allegheny College, Meadville, PA.

Bolden, S. R., S. A. Wissinger, and K. M. Brown. In review. Colonization rates by active and passive dispersing invertebrates in restored depressional wetlands. Journal of the North American Benthological Society.

Brinson, M. M., and R. Rheinhardt. 1996. The role of reference wetlands in functional assessment and mitigation. Ecological Applications 6:69–76.

Brown, K. M. 1991. Mollusca: Gastropoda. Pages 285–214 *in* J. H. Thorp and A. P. Covich (eds.), Ecology and classification of North American freshwater invertebrates. Academic Press, New York.

Brown, D. F. 1994. A partnership for wildlife. Pennsylvania Game News 65:28–32.

Brown, S. C., and B. L. Bedford. 1997. Restoration of wetland vegetation with transplanted wetland soil: an experimental study. Wetlands 17:424–437.

Brown, S. C., K. Smith, and D. Batzer. 1997. Macroinvertebrate responses to wetland restoration in northern New York. Environmental Entomology 26:1016–1024.

Budelsky, R. A., and S. M. Galatowitsch. 1999. Effects of moisture, temperature, and time on seed germination of five wetland *Carices*: implications for restoration. Restoration Ecology 7:86–97.

Burger, J. 1985. Habitat selection in marsh-nesting birds. Pages 253–281 *in* Habitat selection in birds. M. L. Cody (ed.), Academic Press, New York.

Cashen, S. T. 1998. Avian sue of restored wetlands in Pennsylvania. M.S. thesis, Pennsylvania State University, University Park, PA.

Cheal, F., J. A. Davis, J. E. Growns, J. S. Bradley, and F. H. Whittles. 1993. The influences of sampling method on the classification of wetland macroinvertebrate communities. Hydrobiologia 257:47–56.

Colwell, M. A., and L. W. Oring. 1988. Nonrandom shorebird distribution and fine-scale variation in prey abundance. Condor 95:94–103.

Cox, R. R., M. A. Hanson, C. C. Roy, N. H. Euliss, D. H. Johnson, and M. G. Butler. 1998. Mallard duckling growth and survival in relation in aquatic invertebrates. Journal of Wildlife Management 62:124–133.

Cramer, B. J. 1998. Using a wetlands index of biotic integrity for management. B.S. thesis, Allegheny College, Meadville, PA.

Day, R. W., and G. P. Quinn. 1989. Comparisons of treatments after an analysis of variance in ecology. Ecological Monographs 59:433–463.

Delphey, P. J., and J. J. Dinsmore. 1993. Breeding bird communities of recently restored and natural prairie potholes. Wetlands 13:200–206.

de Szalay, F. W., and V. H. Resh. 1997. Responses of wetland invertebrates and plants in important in waterfowl diets to burning and mowing of emergent vegetation. Wetlands 17:149–156.

deVlaming, V., and V. W. Proctor. 1968. Dispersal of aquatic organisms: viability of seeds recovered from the droppings of captive killdeer and mallard ducks. American Journal of Botany 55:20–26.

Dick, T. M. 1993. Restored wetlands as management tools for wetland-dependent birds. Pennsylvania Birds 7:4–6.

Eldridge, J. 1992. Management of habitat for breeding and migrating shorebirds in the midwest. U.S. Fish and Wildlife Service, Washington, DC.

Erwin, R. M., D. K. Dawson, D. B. Stotts, L. S. McAllister, and P. H. Geissler. 1991. Open marsh water management in the mid-Atlantic region: aerial surveys of waterbird use. Wetlands 11:209–228.

Euliss, N. H., and G. Grodhaus. 1987. Management of midges and other invertebrates for waterfowl wintering in California. California Fish and Game 73: 238–243.

Euliss, N. H., D. A. Wrubleski, and D. M. Mushet. 1999. Wetlands of the Prairie Pothole Region: invertebrate species composition, ecology, and management. Pages 471–514 *in* D. P. Batzer, R. B. Rader, and S. A. Wissinger (eds.), Invertebrates in freshwater wetlands of North America: ecology and management. Wiley, New York.

Evans, D. L., W. J. Streever, and T. L. Crisman. 1999. Natural flatwoods marshes and created freshwater marshes of Florida: factors influencing invertebrate distribution and comparisons between natural and created marsh communities. Pages 81–104 *in* D. B. Batzer, R. B. Rader, and S. A. Wissinger (eds.), Invertebrates in freshwater wetlands of North America: ecology and management. Wiley, New York.

Fairchild, G. W., A. M. Faulds, and L. L. Saunders. 1999. Constructed marshes in southeast Pennsylvania: invertebrate foodweb structure. Pages 423–446 *in* D. P. Batzer, R. B. Rader, and S. A. Wissinger (eds.), Invertebrates in freshwater wetlands of North America: ecology and management. Wiley, New York.

Fredrickson, L. H., and F. A. Reid. 1988. Invertebrate responses to wetland management. Leaflet 13.2.1. U.S. Fish and Wildlife Service, Washington, DC.

Galatowitsch, S. M., and A. G. van der Valk. 1994. Restoring prairie wetlands. Iowa University Press, Ames, IA.

Galatowitsch, S. M., and A G. van der Valk. 1996a. Characteristics of recently restored wetlands in the prairie pothole region. Wetlands 16:75–83.

Galatowitsch, S. M., and A G. van der Valk. 1996b. The vegetation of restored and, natural prairie wetlands. Ecological Applications 6:102–112.

Galatowitsch, S. M., N. O. Anderson, and P. D. Ascher. 1999. Invasiveness in wetland plants in temperate North America. Wetlands 19:733–755.

Gallinato, M. I., and A. G. van der Valk. 1986. Seed germination traits of annuals and emergent recruited during drawdowns in the Delta Marsh. Ecology 44:331–343.

Gammonly, J. H., and M. E. Heitmeyer. 1990. Behavior, body condition, and foods of bufflehead and lesser scaups during spring migration. Wilson Bulletin 10:672–683.

Gleason, H. A., and A. Cronquist. 1963. Manual of vascular plants of Northeastern United States and adjacent Canada. Willard Grant Press, Boston.

Greenwood, R. J., A. B. Sargeant, L. M. Cowardin, and T. L. Shaffer. 1987. Factors associated with duck nest success in the prairie pothole region of Canada. Transactions of the North American Wildlife and Natural Resources Conference 52:298–309.

Hall, D. L., R. W. Sites, E. B. Fish, T. R. Mollhagen, D. L. Moorhead, and M. R. Willig. 1999. Playas of the Southern high plains. Pages 635–636 *in* D. P. Batzer, R. B. Rader, and S. A. Wissinger (eds.), Invertebrates in freshwater wetlands of North America: Ecology and management. Wiley, New York.

Hands, H. M., M. R. Ryan, and J. W. Smith. 1991. Migrant shorebird use of marsh, moist soil, and flooded agricultural habitats. Wildlife Society Bulletin 19:457–464.

Harris, S. W., and W. H. Marshall. 1963. Ecology of water-level manipulations on a northern marsh. Ecology 44:331–343.

Horstman, A. J., J. R. Nawrot, and A. Woolf. 1998. Mine-associated wetlands as avian habitat. Wetlands 18:298–304.

Hunter, M. L., J. J. Jones, K. E. Gibbs, and J. R. Moring. 1986. Duckling responses to lake acidification: do black ducks and fish compete? Oikos 47:26–32.

Ingmire, S. G. 1998. Plant communities in restored wetlands in northwestern Pennsylvania. B.S. thesis, Allegheny College, Meadville, PA.

Kadlec, J. A. 1962. Effects of a drawdown on a waterfowl impoundment. Ecology 43:267–281.

Kadlec, R. H., and R. L. Knight. 1996. Treatment wetlands. Lewis Publishers, New York.

Kaminski, R. M., and H. H. Prince. 1981. Dabbling duck and aquatic macroinvertebrate responses to manipulated wetland habitat. Journal of Wildlife Management 45:1–15.

Kentula, M. E., R. P. Brooks, S. E. Gwin, C. C. Holland, A. D. Sherman, and J. C. Sifneos. 1992. An approach to improving decision making in wetland restoration and creation. Island Press, Washington, DC.

King, R. S., and D. A. Wrubeleski. 1998. Spatial and diel availability of flying insects as potential duckling food in prairie wetlands. Wetlands 18:100–114.

Krull, J. N. 1970. Aquatic plant–invertebrate associations and waterfowl. Journal of Wildlife Management 34:707–718.

LaGrange, T. G., and J. J. Dinsmore. 1989. Plant and animal community responses to restored Iowa Wetlands. Prairie Naturalist 21: 39–48.

Legendre, L., and P. Legendre. 1983. Numerical ecology. Elsevier, New York.

Lillie, R. A., and J. O. Evrard. 1994. Influence of macroinvertebrates and macrophytes on waterfowl utilization of wetlands in the Prairie Pothole Region of northwestern Wisconsin. Hydrobiologia 279/280:235–246.

Madsen, C. 1986. Wetland restoration: a pilot project. Journal of Soil and Water Conservation 41:159–160.

Malakaff, D. 1998. Restored wetlands flunk real-world test. Science 280:371–372.

Mallory, M. L., P. J. Blancher, P. J. Weatherhead, and D. K. McNichol. 1994. Presence or absence of fish as a cue to macroinvertebrate abundance in boreal wetlands. Hydrobiologia 279/280:345–351.

Martin, A. C., H. S. Zim, and A. L. Nelson. 1951. American wildlife and plants: a guide to wildlife food habits. McGraw-Hill, New York.

McKinstry, M. C., and S. H. Anderson. 1994. Evaluation of wetland creation and waterfowl use with abandoned mine lands in northeast Wyoming. Wetlands 14:284–292.

Mitsch, W. J., and R. F. Wilson. 1996. Improving the success of wetland creation and restoration with know-how, time, and self-design. Ecological Applications 6:77–83.

Mitsch, W. J., X. Wu, R. Nairn, P. E. Weihe, N. Want, R. Deal, and C. E. Boucher. 1998. Creating and restoring wetlands. Bioscience 48:1019–1030.

Nudds, T. D., and J. N. Bowlby. 1984. Predator–prey size relationships in North American dabbling ducks. Canadian Journal of Zoology 62:2002-2008.

Nummi, P. 1993. Food–niche relationship of sympatric mallards and green- winged teals. Canadian Journal of Zoology 71:49–55.

Olson, E. J., E. S. Engstrom, M. R. Doeringsfeld, and R. Bellig. 1995. Abundance and distribution of macroinvertebrates in relation to macrophyte communities. Journal of Freshwater Ecology 10:325–335.

O'Sullivan, A. D., O. M. McCabe, D. A. Murray, and M. L. Otte. 1999. Wetlands for rehabilitation of metal mine wastes. Proceedings of the Royal Irish Academy 99B:11–17.

Palmer, M. A., R. F. Ambrose, and N. L. Poff. 1997. Ecological theory and community restoration ecology. Restoration Ecology 5:291–300.

Parikh, A., and N. Gale. 1998. Vegetation monitoring of created dune swale wetlands, Vandenberg Air Force Base, California. Restoration Ecology 6:83–93.

Paulovich, J. P. 1998. Water chemistry of constructed and natural wetlands in northwestern Pennsylvania. B.S. thesis, Allegheny College, Meadville, PA.

Payne, N. F. 1992. Techniques for wildlife habitat management of wetlands. McGraw-Hill. New York.

Powers, K. D., R. E. Noble, and R. H. Chabreck. 1978. Seed distribution by waterfowl in southwestern Louisiana. Journal of Wildlife Management 42:598–605.

Reed, P. B. 1997. National list of vascular plant species that occur in wetlands: update of national summary. U.S. Fish and Wildlife Service. Washington, DC.

Reinartz, J. A., and E. L. Warne. 1993. Development of vegetation in small created wetlands in southeastern Wisconsin. Wetlands 13:153–164.

Repko, M. 1998. Habitat preference of migratory waterfowl: a comparison of natural and constructed wetlands. B.S. thesis, Allegheny College, Meadville, PA.

Rheinhardt, R. D., M. M. Brinson, and P. M. Farley. 1997. Applying wetland reference data to functional assessment, mitigation, and restoration. Wetlands 17:195–215.

Roberts, L. 1993. Wetlands trading is a losers game, say ecologists. Science 260:1890–1892

Rumney, T. A. 1992. Patterns in Pennsylvania's agricultural land on the eve of the twenty-first century: changes and processes. Pennsylvania Geographer 30:55–67.

Safran, R. J., C. R. Isola, M. A. Colwell, and O. E. Williams. 1997. Benthic invertebrates at foraging locations of nine waterbird species in managed wetlands of the northern San Joaquin Valley, California. Wetlands 17:407–415.

Sedinger, J. S. 1992. Ecology of prefledgling waterfowl. Pages 109–127 *in* B. D. J. Batt, A. D. Afton, M. G. Anderson, C. D. Ankeyn, D. H. Johnson, J. A. Kadlec, and G. L. Krapu (eds.), Ecology and management of breeding waterfowl. University of Minnesota Press, Minneapolis, MN.

Sistani, K. R., D. A. Myas, and R. W. Taylor. 1999. Development of natural conditions in constructed wetlands: biological and chemical changes. Ecological Engineering 12:125–131.

Sneath, P. H., and R. R. Sokal. 1973. Numerical taxonomy: the principles and practice of numerical classification. W. H. Freeman, New York.

Snyder, R. 1998. Restoring wetlands. Birder's World 12:72–76.

Sojda, R. S., and K. L. Solberg. 1993. Management and control of cattails. Leaflet 13.4.13. U.S. Fish and Wildlife Service, Washington, DC.

Stauffer, A. L., and R. P. Brooks. 1997. Plant and soil responses to salvaged marsh surface and organic matter amendments at a created wetland in central Pennsylvania. Wetlands 17:90–105.

Streever, W. J., and K. M. Portier. 1994. A comparison of dipterans from ten created and ten natural wetlands. Wetlands 16:416–428.

Streever, W. J., D. L. Evans, C. Keenan, and T. L. Crisman. 1995. Chironomidae (Diptera) and vegetation in a created wetland and implications for sampling. Wetlands 14:285–289.

Streever, W. J., T. L. Crisman, and J. H. Kiefer. 1996. Constructing freshwater wetlands to replace impacted natural wetlands: a subtropical perspective. Pages 295–303 *in* F. Schiemer and K. T. Boland (eds.), Perspectives in tropical limnology. SPB Academic Publishing, Amsterdam, The Netherlands.

Tiner, R. W. 1989. Current status and recent trends in Pennsylvania's wetlands. Pages 368–387 *in* S. K. Majumdar, R. P. Brooks, F. J. Brenner, and R. W. Tiner, Wetlands Ecology and Conservation: Emphasis in Pennsylvania. Pennsylvania Academy of Science Press, Easton, PA.

Tiner, R. W. 1991. The concept of a hydrophyte for wetland identification. BioScience 41: 236–247.

Tiner, R. W. 1993. Wetlands are ecotones: reality or myth? Pages 1–15 *in* B. Gopal, A. Hillbrichet-Iklowska, and R. G. Wetzel (eds.), Wetlands and Ecotones: studies on land–water interactions. National Institute of Ecology, New Delhi, India.

Turner, A. M., and J. P. McCarty. 1998. Resource availability, breeding site selection, and reproductive success of red-winged blackbirds. Oecologia 113:140–146.

U.S. Army Corps of Engineers. 1987. USACE wetlands delineation manual. Technical Report Y-87-1. Environmental Laboratory, U.S. Army Crops of Engineers Waterway Experiment Station, Vicksburg, Mississippi.

van der Valk, A. G., and C. B. Davis. 1978. The role of seed banks in the vegetation dynamics of prairie glacial marshes. Ecology 59:322–335.

van der Valk, A. G., and R. W. Jolly. 1992. Recommendations for research to develop guidelines for the use of wetlands to control rural nonpoint source pollution. Ecological Engineering 1:115–134.

van der Valk, A. G. , R. L. Pederson, and C. B. Davis. 1992. Restoration and creation of freshwater wetlands using seed banks. Wetlands Ecology and Management 1:191–197.

VanRees-Siewert, K. L., and J. J. Dinsmore. 1996. Influence of wetland age on bird use of restored wetlands in Iowa. Wetlands 16:577–582.

Wellborn, G. A., D. K. Skelly, and E. E. Werner. 1996. Mechanisms creating community structure across a freshwater habitat gradient. Annual Review of Ecology and Systematics. 27:337–363.

Weller, M. W. 1999. Wetland birds: habitat resources and conservation implications. Cambridge University Press, New York.

Wheeler, B. D., S. C. Shaw, W. J. Fojt, and R. A. Robertson. 1995. Restoration of temperate wetlands. Wiley, New York.

Wienhold, C. E., and A. G. van der Valk. 1989. The impact of duration of drainage on the seed banks of northern prairie wetlands. Canadian Journal of Botany 67:1878–1884.

Wilson, R. F., and W. J. Mitsch. 1996. Functional assessment of five wetlands constructed to mitigate wetlands loss in Ohio, USA. Wetlands 16:436–451.

Wissinger, S. A. 1989. Seasonal variation in the intensity of competition and intraguild predation among larval dragonflies. Ecology 70:1017–1027.

Wissinger, S. A. 1997. Cyclic colonization and predictable disturbance: a template for biological control in ephemeral crop systems. Biological Control 10:1–15.

Wissinger, S. A. 1999. Ecology of wetland invertebrates: synthesis and applications for conservation and management. Pages 1013–1042 *in* D. P. Batzer, R. B. Rader, and S. A. Wissinger (eds.), Invertebrates in freshwater wetlands of North America: Ecology and management. Wiley, New York.

Wissinger, S. A., and L. J. Gallagher. 1999. Beaver pond wetlands in northwestern Pennsylvania: modes of colonization and succession after drought. Pages 333–362 *in* D. P. Batzer, R. B. Rader, and S. A. Wissinger (eds.), Invertebrates in freshwater wetlands of North America: ecology and management. Wiley, New York.

Wrubeleski, D. A. 1999. Northern prairie marshes (Delta Marsh, Manitoba): II. Chironomidae (Diptera) responses to changing plant communities in newly flooded habitat. Pages 571–602 *in* D. P. Batzer, R. B. Rader, and S. A. Wissinger (eds.), Invertebrates in freshwater wetlands of North America: ecology and management. Wiley, New York.

Young, P. 1996. The "new science" of wetland restoration. Environmental Science and Technology 30:292–296.

Zedler, J. B. 1996. Ecological issues in wetland mitigation: an introduction to the forum. Ecological Applications 6:33–37.

11 Birds, Plants, and Macroinvertebrates as Indicators of Restoration Success in New York Marshes

STEPHEN C. BROWN and DAROLD P. BATZER

Ongoing losses of wetlands have resulted in dramatically increased efforts at restoration of wetlands, often in an attempt to increase habitat for wildlife. Assessment of the success of these restoration efforts must often be done with severely limited resources and time. We examined the use of birds, plants, and macroinvertebrates as indicators of the success of wetland restoration in shallow marshes in northern New York. Analysis of the taxonomic richness of macroinvertebrates and species richness of plants and birds indicated that restored and natural reference sites were not significantly different. However, in terms of community composition, birds, plants, and macroinvertebrates each had significant differences between restored and natural wetlands. Richness measures for the different groups did not correlate with each other, so each group must be analyzed separately. Plants were the most sensitive measure, and bird sampling required the least effort. Macroinvertebrates should be analyzed when one of the goals of a restoration project is to restore macroinvertebrate diversity or overall food web structure. In these wetlands, no single group of organisms indicates overall success of restoration.

The loss of wetland habitats has been severe in a majority of the states, and particularly extreme in the Great Lakes Basin (Dahl 1990). The national goal of no net loss of wetlands has dramatically increased the interest in and practice of wetland restoration by state and federal wildlife agencies, and by individuals and businesses as mitigation requirements for permitted wetland losses (World Wildlife Fund 1992). However, recent national estimates predict increased losses of wetland habitats, despite existing restoration programs (Smith and Tome 1992). In addition, efforts of regulatory programs to create wetlands have limited success at replacing wildlife

Bioassessment and Management of North American Freshwater Wetlands, edited by Russell B. Rader, Darold P. Batzer, and Scott A. Wissinger.
0-471-35234-9

habitat (Brown and Veneman 1998), so the success of restoration programs is even more critical in offsetting continuing wetland losses.

Wetlands are being created and restored in large numbers across the country, often in an effort to replace the habitat values lost through destruction of natural wetlands (Hammer 1992). For example, the U.S. Fish and Wildlife Service is dramatically increasing efforts to restore wetlands by adding projects on privately owned lands to its more traditional work on national wildlife refuges (Smith and Tome 1992; Brown 1995). For restoration goals to be met successfully, restored wetlands must provide comparable habitat to natural wetlands. Resource agencies involved in restoration have severely limited resources for assessing the ecological results of restoration projects. Rapid assessment techniques are needed to determine the degree to which restoration projects provide habitat comparable to natural wetlands.

One common goal of wetland restoration is to recreate the diversity of plants and animals typical of natural wetland communities. However, different groups of organisms respond to different environmental characteristics, and this makes assessing the success of restoration for diverse groups of organisms complex. In many aquatic habitats, macroinvertebrate communities are useful indicators of biotic integrity (Plafkin et al. 1989). For other groups, such as wetland birds, the factors affecting diversity in restored systems can be complex (Hemesath and Dinsmore 1993, Brown and Smith 1998). We compared the use of data on macroinvertebrates, birds, and plants as indicators of the success of restoration, and examined the correlations among levels of diversity between each group.

Managers and scientists studying wetland restoration are generally limited by time and resources and must understand advantages and disadvantages of sampling for different groups of organisms. In addition, it can be difficult to predict the outcome of restoration efforts from conditions existing on the site prior to restoration of wetland hydrology, such as remnant vegetation or seed banks (Brown 1998). We describe the methods, protocol, and resources needed to sample macroinvertebrates, plants, and birds for comparisons among restored and natural wetlands. In addition, we compare the results that would be obtained from using each group as the primary indicator of restoration success.

METHODS

Study Sites

The study sites were located in the Lake Ontario and St. Lawrence River plains in Jefferson County, New York. This region is part of the Great Lake Plains as defined by O'Connor and Cole (1986) and the Erie/Ontario Lake Plain as defined by Omernik (1987). The topography is nearly level along the lakeshore and gradually changes to gently rolling hills inland. The soils are predominantly high-clay lacustrine deposits. Each restoration site shared several characteristics, including clear evidence of drainage by construction of ditches, remnant hydric soils, and a history of agricultural use. Each restoration site contained a remnant wetland in the drainage ditch that was constructed at least 40 years ago to remove water from the basin. The original wetland vegetation at higher elevations was replaced by upland field plants as the drained areas became too dry to support hydrophytic vegetation. The drained areas were used primarily for pasture and for forage crops, and none showed any evidence of recent cultivation.

Sites were restored by constructing dikes across the drainage ditches to retain water and began to refill in the fall of 1991 with the onset of heavy seasonal rains. Each site had deepwater habitats in the center (>1 m), grading to shallow marsh along the edges. All restored sites held water continuously over the subsequent 1992–1994 study period. Macroinvertebrates, birds, and plants were sampled in each year. Due to the different amounts of sampling effort involved, different numbers of sites were sampled for each group in each year, as described below.

We chose natural reference wetlands following the selection procedures described by Abbruzzese et al. (1988). This process includes map identification of study regions and wetland types, random selection of sites, and field reconnaissance to verify site characteristics and obtain permission for entry. The criteria for selecting appropriate sites included basin morphometry, hydrogeology, soils, watershed size, watershed land use, and watershed soils that were similar to the restored sites. The reference wetlands were all located in the same county as the restoration sites, and all had soils of the same or a closely related series. All reference sites were permanently flooded. The number of reference sites sampled for each part of the study varied as described below.

Macroinvertebrate Sampling

Macroinvertebrate communities in restored and natural sites were sampled in all study years using a D-frame sweep net. This technique has been widely used for describing general macroinvertebrate communities in wetlands and for discriminating among communities in different wetland types (Cheal et al. 1993; Chapter 15, this volume). Samples consisted of five 2-m-long sweeps through the water and vegetation at randomly selected locations in the shallow marsh zone (<1 m). We collected six weekly samples per site from June 15 through August 10 in 1992, five weekly samples per site from July 10 through August 15 in 1993, and five weekly samples per site from July 5 through August 2 in 1994. Field sampling and identification was completed by one full-time field assistant.

Because removing large numbers of invertebrates from the sites during sweep net sampling could influence subsequent community development, we processed samples on-site and returned live invertebrates to their habitats. Voucher specimens were identified to family or genus using taxonomic keys compiled by Peckarsky et al. (1990) or Merritt and Cummins (1984). In 1994 we also recorded the dominant plant community where the samples were taken. We did not sample to determine if fish were present in restored or natural wetlands. Details of sampling protocols are reported in Brown et al. (1997).

Data were collected at eight restored wetlands and two reference sites in 1992, but one of the reference sites was drained by the landowner during the study period and was deleted from the study. In 1993, one restored site was flooded by beaver (*Castor canadensis*), and we selected two additional reference sites for study, all of which were sampled again in 1994. Thus, we had eight restored sites in year 1 and seven in years 2 and 3, a single reference site in year 1, and three reference sites in years 2 and 3.

Plant Community Sampling

We established three sets of five 1-m^2 permanent plots for vegetation analysis in each site prior to restoration and at each natural wetland. Plot sets were established at each site at five fixed elevations relative to the proposed maximum water level at the restored sites and the observed maximum water level at natural sites: +10 cm, 0 cm, –10 cm, –20 cm, and –30 cm. All plots at restored sites were established outside the area that would be disturbed by dike construction. We recorded the percent cover for all species in 1991 before restoration, and in 1992, 1993, and 1994 following restoration. Percent cover was determined individually for each species, so the totals did not necessarily sum to 100%. For each species that could not be identified conclusively in the field, we pressed a specimen for later identification. Data were collected between July 5 and August 15 in each year by the same full-time field assistant. Details of the plant community sampling methodology are available in Brown (1999). Data were collected at 20 potential restoration sites, but one was never restored, and during the course of the study, four sites had dike failures, and two more were flooded by beaver. This left a total of 13 restoration sites with data on plant development for all three years of the study. We collected comparable data at four natural wetlands in each study year.

Bird Sampling

Birds were counted at restored and natural reference sites during the breeding season in each year between 1992 and 1994. We considered observations at each type of wetland to be independent because the sites are all within the same landscape, and individual birds observed during the study could choose either type of wetland. Although breeding birds at natural wetlands might return to the same location because of site fidelity, we considered habitat selection on the basis of site quality in any particular year to be sufficient to allow the assumption of statistical independence among observations.

A location was chosen at each study site from which the maximum area could be observed and used for observation in each year. We identified the wetland border and conducted an unlimited radius point count within each wetland during each sample period. Birds outside the wetland were not counted. Each sampling period lasted for 20 minutes and was conducted in all weather conditions except thunderstorms. Each site was sampled five times in each year. We recorded each individual bird that was seen or heard, and its type of activity (i.e., singing, nesting, foraging, flying over the site). For active, numerous species such as swallows, we recorded the number observed at the start of the census period and increased this number only if a greater number of individuals was observed at one time later in the census period. We included in the count birds that flushed as we approached the observation point. Observations began at about 0500 hours, and ended by 1100 hours Eastern Standard Time. We rotated our site sampling schedule to ensure that all sites were sampled during all time periods in each year. Data were collected from May 19 to June 25 in 1992, from May 28 to June 30 in 1993, and from June 4 to June 30 in 1994, by one full-time field assistant. Details regarding bird sampling methods are reported in Brown and Smith (1998).

Bird censuses were conducted on 21 sites in 1992. Seventeen of these were restoration sites, and the other four were natural reference wetlands. In 1993 one restored site was added following dike repair, one reference site was drained by the landowner and replaced with a comparable site, and four additional reference wetlands were added. These additions resulted in a total of 25 sites, with 18 restored and 8 reference sites. In 1994, one additional reference site was drained by the landowner, and one new reference site was added to the study, for a total of 25 sites, with 18 restored and 8 reference sites.

Data Analysis

We compared the mean number of taxa of macroinvertebrates, and the mean number of species of birds and plants, at restored and natural sites in each year using one-way ANOVAs. Statistical tests on macroinvertebrate numbers were only performed for data from 1993 and 1994 when multiple reference sites were used. We compared the abundances of macroinvertebrates within taxa between restored and natural wetlands using a two-way ANOVA, with factors of type of site and taxa, and the interaction. We tested the effect of plot elevation on the difference between numbers of wetland plant species among restored and natural sites using two-way ANOVAs with factors of type of site, plot elevation, and the interaction. We also used two-way ANOVAs to compare taxa richness of macroinvertebrates, species richness of birds, and species richness of plants. In each analysis we used the factors of study year, the taxonomic group being analyzed, and the interaction. When the interaction was significant, we repeated the comparison within years. Normal probability plots were used to test assumptions of normality for data and residuals.

RESULTS

Macroinvertebrates

When all samples were combined to estimate the total number of taxa present at each wetland, the number of macroinvertebrate taxa was not significantly different between restored and natural sites in 1993 ($F = 0.9$; df = 1, 8; $p = 0.360$) or in 1994 ($F = 4.0$; df = 1, 8; $p = 0.080$) (Fig. 11.1). However, there were significant differences in the abundances of specific taxa between the two types of sites (Brown et al. 1997). The mean abundance of macroinvertebrates at restored wetlands was significantly

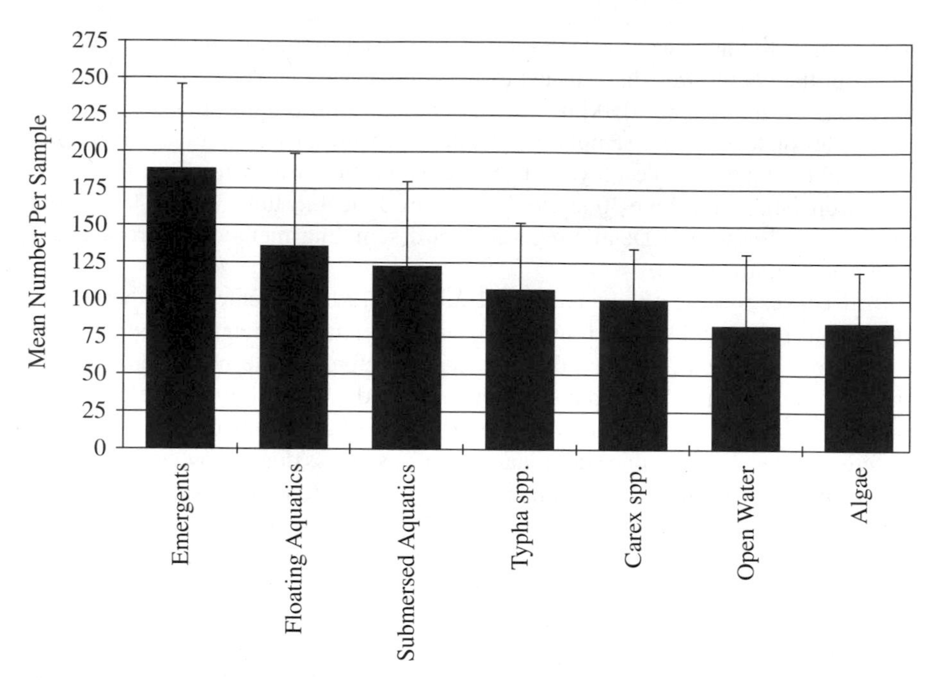

FIGURE 11.1. Mean numbers of macroinvertebrate taxa, wetland plant species, and bird species, at restored and natural wetlands in each study year, with standard deviations.

lower than at natural wetlands in 1993 ($F = 9.0$; df = 1, 360; $p = 0.003$), and taxa was a significant factor ($F = 8.4$; df = 44, 360; $p < 0.001$). There was also a significant interaction between abundances of taxa across wetland type ($F = 3.2$; df = 44, 360; $p < 0.001$), with some taxa more abundant at restored wetlands and others more abundant at natural wetlands. The results were similar for 1994, with significant factors of both wetland type ($F = 12.8$, df = 1, 376; $p < 0.001$) and macroinvertebrate taxa ($F = 12.6$; df = 46, 376; $p < 0.001$), and a significant interaction between the two ($F = 3.0$; df = 46, 376; $p < 0.001$).

Three taxa were significantly more abundant at natural wetlands in both 1993 and 1994, including leeches in two families, *Hirudinidae* ($F = 11.75$; df = 1, 8; $p = 0.009$) and *Erpobdellidae* ($F = 17.39$; df = 1, 8; $p = 0.003$), and also pygmy backswimmers in the family *Pleidae* ($F = 65.44$; df = 1, 8; $p < 0.001$). Three taxa (baetid mayflies, syrinid beetles, and chaoborid phantom midges) had higher mean abundances in the restored sites than in the natural sites, but these differences were not significant. Additional differences in taxonomic composition are discussed in Brown et al. (1997).

Macroinvertebrate abundances varied greatly within specific plant community types (Fig. 11.2). Samples from emergent plant communities had over twice the abundance of macroinvertebrates compared to samples from open water or algae-dominated areas. The greatest abundances were found among tall emergent species such as reed canary grass (*Phalaris arundinacea*). Floating-leaved aquatics, includ-

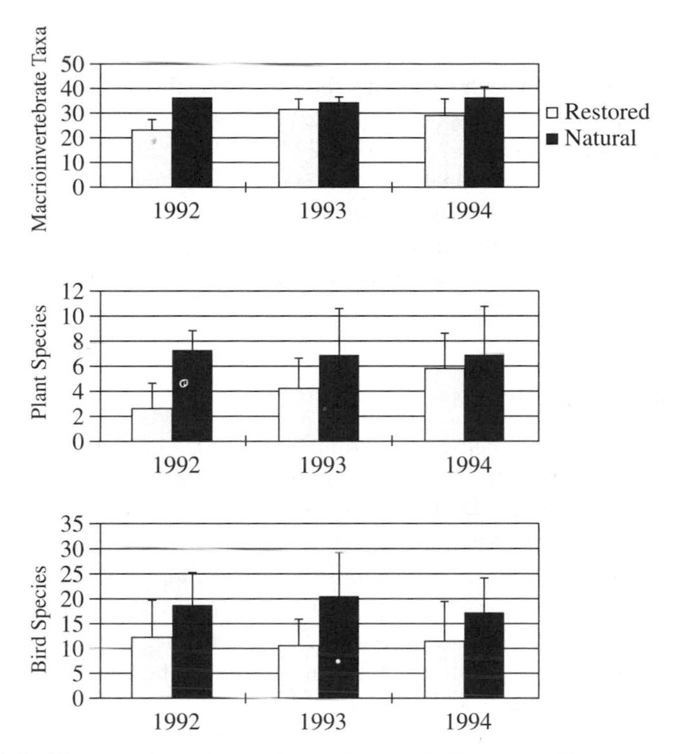

FIGURE 11.2. Mean number of macroinvertebrates of all taxa combined per sample from each plant community type.

ing young water plantain shoots (*Alisma subcordatum*), had higher abundances than shorter emergents such as *Carex* spp. Stands of *Typha* spp., although emergent, had consistently lower abundances than did other emergent species.

Plants

The number of wetland plant species did not differ between restored and natural wetlands in any year when data from all plots were combined into a single species richness estimate for each site (1992, $F = 2.3$; df = 1, 16; $p = 0.146$; 1993, $F = 1.6$; df = 1, 16; $p = 0.219$; 1994, $F = 0.2$; df = 1, 16; $p = 0.664$) (Fig. 11.1). However, there were significant differences between restored and natural sites when plot elevation was included as a factor. In 1992 and 1993 there were more wetland plant species at natural sites (1992, $F = 113.5$; df = 1, 261; $p < 0.001$; 1993, $F = 41.0$; df = 1, 263; $p < 0.001$), and plot elevation was a significant factor (1992, $F = 11.1$; df = 1, 16; $p < 0.001$; 1993 $F = 9.5$; df = 1, 263; $p < 0.001$). By 1994, when wetland plants had become more established at the restored sites, there was no significant difference between restored and natural sites ($F = 2.6$; df = 1, 265; $p = 0.108$), but plot elevation was still significant ($F = 3.6$; df = 1, 265; $p = 0.008$).

Birds

When all samples within each year were combined to estimate bird species richness for each wetland site, there was no difference between restored and natural wetlands in any year (1992 $F = 0.2$; df = 1, 19; $p = 0.697$; 1993 $F = 2.9$; df = 1, 23; $p = 0.101$; 1994 $F = 0.02$; df = 1, 23; $p = 0.897$) (Fig. 11.1). However, there were significant differences between the abundances of particular bird species, with some species more abundant at restored wetlands and others more abundant at natural wetlands (Brown and Smith 1998). In addition, the number of species present per hectare was consistently lower at restored wetlands.

Comparisons Among Groups

There was no relationship between macroinvertebrate taxa richness and the number of plant species ($F = 0.2$, df = 1, 23, $p = 0.650$), but both study year and the interaction were significant ($F = 8.42$, df = 2, 23, $p = 0.002$; $F = 5.73$, df = 2, 23, $p = 0.010$, respectively) (Fig. 11.3). When the analysis was repeated separately for each year, however, there was no significant relationship between macroinvertebrate and plant richness. Similarly, there was no relationship between macroinvertebrate and bird species richness ($F = 2.1$, df = 2, 23, $p = 0.157$), and neither year nor the interaction were significant. The only significant relationship among the three groups is between the species richness of plants and birds ($F = 11.4$, df = 2, 23, $p = 0.003$), with neither year nor the interaction significant. The relationship is reasonably strong, with an r^2 value of 0.542. However, when data are pooled for all years, the relationship between bird and plant species richness is not significant ($r^2 = 0.026$, $p = 0.400$) (Fig. 11.3).

DISCUSSION

For a specific indicator to serve as an estimate of restoration success, it must be sensitive to differences between recently restored and natural wetlands. The patterns of changes during the three years of the study differ for the taxonomic richness of macroinvertebrates, species richness of wetland plants, and species richness of birds. This suggests that even the most general patterns are not consistent among these groups, and that differences among each of the groups in each year cannot be generalized to the other groups. Macroinvertebrate richness was lower at restored sites during the first year, but recovered to natural levels during years 2 and 3. Using macroinvertebrates as the only index would suggest that the restored sites had become similar to the natural sites after only 2 years. The only invertebrate populations slow to recover were some flightless forms (leeches and pleids). Similarly, bird species richness in 1993 and 1994 was not significantly different at restored and natural marshes, although bird species numbers were lower at restored sites in each

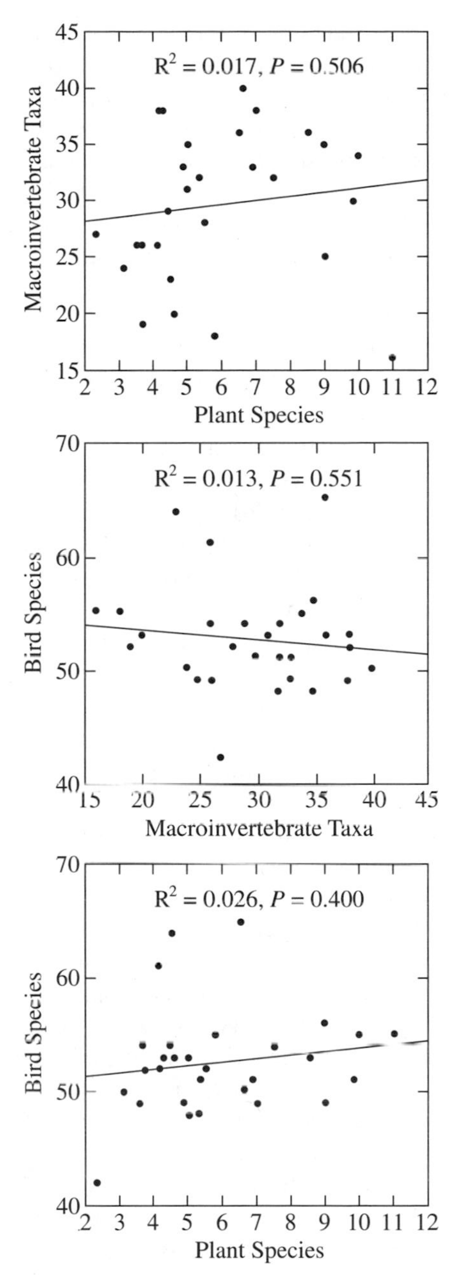

FIGURE 11.3. Relationships between the number of macroinvertebrate taxa, the number of species of wetland plants, and the number of species of birds at restored and natural wetlands in all study years.

year. For macroinvertebrates and birds, differences between restored and natural sites were evident only if abundances of specific taxa were examined. In contrast, plant data suggested that recovery was somewhat slower than for macroinvertebrates and birds. Although the number of plant species at restored wetlands increased gradually during each year of the study, plant species richness did not become similar between restored and natural wetlands until the third year. In addition, the development of a shrub layer at the restored sites will require many years, and the lack of this structural diversity can be expected to affect the bird community using the restored sites (Brown and Smith 1998).

A common goal of wetland restoration is to reestablish ecological functions at degraded sites. Most ecologists agree that the plant base of an ecosystem will have a profound influence on higher trophic levels (Hunter and Price 1992). This would suggest that plants should be prioritized for monitoring when ecological functioning of wetlands is to be assessed. Macroinvertebrates are also useful in that regard because they represent a major link between plants and higher trophic levels (Batzer and Wissinger 1996). In Chapter 5, Cummins and Merritt discuss how invertebrates can be used to describe some attributes of wetland ecological functioning.

Restoration efforts often are undertaken to reestablish populations of select taxa. For example, endangered or threatened birds and plants, or game birds, are often prioritized at specific restoration projects. While macroinvertebrates are vital foods of many birds, and many taxa depend on specific plant species or plant forms for habitat, our results indicate that macroinvertebrate richness poorly predicts the richness of bird or plant species. If the goal of a restoration is reestablishing populations of specific taxa, both initial wetland design and analysis of success should be directed at those species.

Restoration efforts are usually limited by time and financial constraints. Of the three groups monitored in this study, collection of macroinvertebrate data was more time consuming than collection of data on plant and bird distributions. In approximately equal amounts of time during each study year, we collected comparable data on macroinvertebrates at 10 wetland sites, on wetland plants at 17 sites, and on birds at 25 sites. This greater time input for macroinvertebrate sampling was a result of the need for multiple composite samples at each site and the difficult task of identifying large numbers of macroinvertebrate individuals. When the goal of a monitoring program is to assess a large number of sites, this increased field time would be a limiting factor. However, fewer than the five to six collections per year used in this study may be sufficient for many monitoring efforts.

Statistical issues can also complicate the use of macroinvertebrates in wetland restoration assessment. The very high natural variability in numbers of macroinvertebrates among wetlands makes detection of differences between groups of sites difficult. When variability is very high, detection of differences between groups requires large numbers of samples, exacerbating the logistic problems mentioned above. Because the number of macroinvertebrates varied among different plant community types, future studies should consider stratifying sampling among plant communities to explore whether this reduces variability in macroinvertebrate numbers and therefore makes comparisons among sites more practical.

The choice of appropriate indicators for analysis of wetland restoration success depends on the goals of the restoration project. If the goal is to restore macroinvertebrate communities, they must be measured directly. If the goal is to restore specific bird, plant, or macroinvertebrate communities, or overall food web functioning, each group must be measured directly. However, if the goal is to restore habitat for a range of species, it is unlikely that measuring the abundances of one species group will adequately indicate patterns of restoration success for other groups. Furthermore, our study suggests that problems with detection sensitivity (at least for measures of macroinvertebrate taxa richness), logistic problems, and statistical constraints might make it difficult to use macroinvertebrate monitoring to measure the success of some wetland restoration efforts. Fortunately, rapid assessment procedures and biotic indices are now being developed to improve the practicality and efficacy of macroinvertebrate sampling for biological monitoring in wetlands (Hicks 1997, Chapter 8, this volume).

ACKNOWLEDGMENTS

This research was funded by Region 5 of the U.S. Fish and Wildlife Service under the Partners for Fish and Wildlife Program. Additional funding was provided by the National Fish and Wildlife Foundation, Ducks Unlimited, Inc., the New York Cooperative Fish and Wildlife Research Unit, Cornell University Department of Natural Resources, and the Edna Bailey Sussman Fund. Extensive consultation on study design was provided by B. Bedford, C. Smith, M. Richmond, and B. Peckarsky. We are grateful for the commitment and dedication of our field assistants, including Kevin Smith, Bill Yarnell, Kim Claypoole, Metta McGarvey, and Christian Ottke.

REFERENCES

Abbruzzese, B., A. B. Allen, S. Henderson, and M. E. Kentula. 1988. Selecting sites for comparison with created wetlands. Pages 291–297 *in* C. D. A. Rubec and R. P. Overend (compilers), Proceedings of Symposium '87: Wetlands/peatlands, Edmonton, Alberta, Canada.

Batzer, D. P., and S. A. Wissinger. 1996. Ecology of insect communities in nontidal wetlands. Annual Review of Entomology 41:75–100.

Brown, S. C. 1995. Wetland restoration: factors controlling plant community response and avifaunal habitat value. Ph.D. dissertation, Cornell University, Ithaca, NY.

Brown, S. C. 1998. Remnant seed banks and vegetation as predictors of restored marsh vegetation. Canadian Journal of Botany 76:620–629.

Brown, S. C. 1999. Vegetation similarity and avifaunal food value of restored and natural marshes in northern New York. Restoration Ecology 7:56–68.

Brown, S. C., and C. R. Smith. 1998. Breeding season bird use of recently restored versus natural wetlands in New York. Journal of Wildlife Management 62:1480–1491.

Brown, S. C., and P. L. Veneman. 1998. Compensatory wetland mitigation in Massachusetts. University of Massachusetts Agricultural Experiment Station Research Bulletin 746.

Brown, S. C., K. Smith, and D. Batzer. 1997. Macroinvertebrate responses to wetland restoration in northern New York. Environmental Entomology 26:1016–1024.

Cheal, F., J. A. Davis, J. E. Growns, J. S. Bradley, and F. H. Whittles. 1993. The influences of sampling method on the classification of wetland macroinvertebrate communities. Hydrobiologia 257:47–56.

Dahl, T. E. 1990. Wetlands losses in the United States, 1780's to 1980's. U.S. Fish and Wildlife Service, Washington, DC.

Hammer, D. L. 1992. Creating freshwater wetlands. Lewis Publishers, Chelsea, MI.

Hemesath, L. M., and J. J. Dinsmore. 1993. Factors affecting bird colonization of restored wetlands. Prairie Naturalist 25:1–11.

Hicks, A. L. 1997. New England freshwater wetlands invertebrate biomonitoring protocol. Environmental Institute, University of Massachusetts, Amherst, MA.

Hunter, M. D., and P. W. Price. 1992. Playing chutes and ladders: heterogeneity and relative forces of bottom–up and top–down forces in natural communities. Ecology 73:724–732.

Merritt, R. W., and K. W. Cummins. 1984. An introduction to the aquatic insects of North America, 2nd ed. Kendall/Hunt, Dubuque, IA.

O'Connor, S., and N. B. Cole. 1986. Freshwater wetlands inventory data analysis. U.S. Fish and Wildlife Reference Service No. MIN 31/8680555. New York State Department of Environmental Conservation, Albany, NY.

Omernik, J. M. 1987. Ecoregions of the conterminous United States. Annals of the Association of American Geographers 77:118–125.

Peckarsky, B. L., et al. 1990. Freshwater macroinvertebrates of northeastern North America. Cornell University Press, Ithaca, NY.

Plafkin, J. L., M. T. Barbour, K. D. Porter, S. K. Gross, and R. M. Hughes. 1989. Rapid bioassessment protocols for use in streams and rivers: benthic macroinvertebrates and fish. Report EPA/44/4-89-001. U.S. Environmental Protection Agency, Assessment and Watershed Protection Division, Washington, DC.

Smith, D. A., and M. W. Tome. 1992. Wetlands restoration: will we make a difference? Proceedings of the 13th Annual Conference of the Society of Wetland Scientists, New Orleans, LA, South Central Chapter of the Society of Wetland Scientists.

World Wildlife Fund. 1992. Statewide wetlands strategies: a guide to protecting and managing the resource. Island Press, Washington, DC.

12 Bacteria as Biomonitors

J. VAUN McARTHUR

Bacteria are ubiquitous in nature. However, bacteria are used only marginally as indicators of ecosystem status or health. Because of their extremely small size, bacteria have the highest surface-to-volume ratios of any living organisms. Thus, bacteria should be the most environmentally aware organisms and as such, excellent early warning signals. Several methods are described that use bacteria or their genes as biomonitors and require increasing levels of sophistication and experience. Methods described include both phenotypic and genetic approaches. Only the creativity and level of experience of the investigator limit applications of these techniques to wetlands. Bacteria can be used to indicate levels of stress, degree of recovery and as measures of ecosystem functional status.

Bacteria are found in every habitat that researchers have been clever enough to sample. They are found at the lowest depths of the oceans, highest reaches of the atmosphere, and in all terrestrial and aquatic environments. In addition, they can be found under the most extreme environmental conditions. For example, bacteria can be found in hydrothermal vents (Gugliandolo et al. 1999, Jeanthon 2000), Antarctic ice (Staley and Gosink 1999), pH ranging from 1 to 12 (Edwards et al. 2000), and they are able to withstand both natural and anthropogenic radiation (Wise et al. 1996). The seeming cosmopolitan nature of bacteria makes their use as a biomonitoring tool suspect. While we are just beginning to appreciate microbial diversity, it is clear that various associations are found under natural conditions. When these natural conditions are altered by human activity, nuisance species of bacteria increase in density. Thus assemblages of bacteria found under natural conditions can be considered as native bacteria and those assemblages that develop under stress as exotic bacteria.

Bacteria are integral to the functioning of most ecosystems, but much of their ecology is yet to be described. However, details of various nutrient cycles have been described in a variety of habitats, including wetlands (Golterman 1995). The role of bacteria in nitrogen, phosphorus, and carbon cycles has been carefully examined.

Bioassessment and Management of North American Freshwater Wetlands, edited by Russell B. Rader, Darold P. Batzer, and Scott A. Wissinger.
0-471-35234-9

From these and other studies it has been shown that specific bacteria are involved in specific transformations. Nitrification, the transformation of ammonia to nitrite and then to nitrate, is performed by members of the genera *Nitrosomonas* and *Nitrobacter*. Numerous other ecological functions are performed by specific groups of bacteria. The presence or absence of these specific bacteria would be a strong indicator that a particular function would or would not be performed in a habitat. However, such an observation is not necessarily "indicative" of environmental perturbation. The lack of function may be due to naturally occurring environmental conditions that do not favor or promote the function. Unimpacted sites are needed to define reference or natural conditions.

The ability to break down diverse carbon compounds is often found in bacterial assemblages. In fact, certain specific bacteria (i.e., *Burkholderia cepacia*) are capable of using over 200 different organic molecules as their sole source of carbon. These bacteria might be considered "weedy" species and their distribution assumed to be cosmopolitan. As such, these organisms would not be good candidates as biomonitors. In contrast, other bacteria are very restricted in their ability to process carbon. These bacteria would have greatly reduced distributions and abundance compared to the carbon generalists. The presence or absence of specific bacteria capable of using various organic substrates would be indicative of ecosystem status but not necessarily ecosystem stress or perturbation.

Bacteria have been used for decades in biomonitoring programs, but we have only begun to utilize their full potential as a biomonitoring tool. Public health officials use the presence and abundance of various enteric bacteria as indicators of human and agricultural sewage pollution. *Escherichia coli* has become one of the most widely known bacteria because of this monitoring program. The presence of *E. coli* in surface water or groundwater is telling evidence of introduction of sewage whether from domestic or agricultural sources. *E. coli* is not the only bacteria used to indicate sewage pollution. Other bacteria include species from the genera *Salmonella, Streptococci, Citrobacter, Enterobacter,* and *Klebsiella.* These bacteria are characterized as bacteria that ferment lactose to carbonic acid with gas formation (either H_2 or CO_2) and are commonly called *coliform bacteria.* While the importance of coliform bacteria as biomonitors of sewage pollution cannot be underestimated, other bacteria can also indicate stress associated with a variety of human-induced disturbances. *Sphaerotilus natans,* a sheathed bacteria, often grow to such high numbers that the colonies can be seen with the naked eye. These bacteria form thick lawns of "sewage fungus" that can coat most substrata and severely reduce the available oxygen. As such, they are not good indicator organisms but rather, a severe nuisance and biological hazard. However, early detection of *S. natans* and related organisms would be indicative of organic pollution. Unfortunately, *S. natans* has its optimum growth rate at 10°C, a temperature of many natural waters, especially during fall and spring. At these temperatures, *S. natans* can outcompete most native bacteria and lower the oxygen concentration to levels that affect other microbes, micro- and macroinvertebrates, and vertebrates such as fish. Thus their effectiveness as an indicator of ecosystem status is often too much, too late.

Various other bacteria are regularly monitored because of their importance to various industries. Certain species of *Vibrio* affect shellfish (Jones and Summer-Brason 1998). *Pfiesteria* has recently been identified with fish die-off along the Chesapeake Bay and have been traced to chicken and hog production along rivers (Fairey et al. 1999, Boesch 2000, Silbergeld et al. 2000).

WETLAND BACTERIA AS BIOMONITORS

Wetlands are diverse and varied in their locations, chemistry, biology, hydrology, geomorphology, and importance to human activities. In most cases, water is the single unifying aspect among wetlands. Equally diverse and varied are the ecosystem-level functions performed by wetlands. Microbial processes control most of these functions. These processes include decomposition of allochthonous and autochthonous organic matter, nutrient cycling (e.g., nitrogen, phosphorus), and dissimilatory reduction of various natural and anthropogenic organic compounds. Because of the great divergence in wetland types, microbial processes are similarly different among wetlands. These differences can be seen in the ability to break down various compounds, such as lignin or cellulose. Wetlands that can break this material down efficiently do not accumulate organic matter; those that cannot, develop peat deposits. In many cases, the differences are due to pH, which is similarly influenced by the nature and timing of organic matter inputs.

Wetlands have been used as natural processors of human-produced effluents. Wetland bacteria and plants have been effective in removing heavy metals (Lung and Light 1996, Qian et al. 1999), processing sewage (Chaguc-Goff et al. 1999, Gopal 1999), and in the breakdown of various toxic materials. It is this aspect of wetland bacterial ecology that makes them attractive as biomonitors. Identification of specific sentinel or indicator species or genes would provide critical insight into wetland health and condition. For example, the presence and abundance of specific genes that are involved in metal tolerance or resistance should be directly related to the presence or absence of heavy metals in the environment. Bacteria cannot migrate significant distances to avoid contact with toxic or inhibitory compounds. Therefore, natural selection will favor bacteria with genes that tolerate metal contamination, and the proportion of these bacteria should increase relative to bacteria without such genes. Although some baseline level of these genes might be expected, increased abundance of these genes would be a strong indication of metal contamination even before the metals started to accumulate in the tissues of higher organisms.

BACTERIAL DIVERSITY

Most ecologists, environmental scientists, and managers refer to bacteria at the highest taxonomic level possible. In 1969, Margulis suggested that all living organisms could be grouped into five kingdoms. In this scenario, bacteria are part of one group, the Monera. Subsequently, most scientists, with the exception of the health sciences, lumped all bacteria into a single category at the kingdom level. Using 16 S rRNA sequences, Woese (1990), proposed a new scheme for the classification of life. This scheme has three domains: Eukarya, Bacteria, and Archaea. Two of the three domains are bacterial; all of the other known organisms fit into the Eukarya. This construct is based on the genetic divergence among groups as expressed in the 16 S rRNA subunit. It is clear from this scheme that there exists tremendous genetic diversity among the bacteria. In fact, based on 16 S rRNA, fish are more similar to cypress trees than are *E. coli* (Bacteria) and *Methanococcus* sp. (Archaea), two bacterial species that used to be lumped under the single taxon Monera. The lumping continues today.

Given the diversity of bacteria, studies that seek to measure some aspect of overall bacterial physiology, metabolism, or activity are too general. What does "bacterial" activity mean? What is "bacterial" production? In essence, these questions are similar to asking: What is the production of all plants and animals? The answer is a very coarse approximation of living processes. Any efforts to document changes in ecosystem function utilizing all bacteria lumped together will surely fail. Given the incredible functional diversity of bacteria, there would be no way to interpret such information.

PROBLEMS OF SCALE

Bacteria live at the smallest ecological scales. Most native bacteria are smaller than 0.2 μm. As observed from microscopy, small colonies of bacteria may form in association with particles, and complex assemblages often develop in biofilms, but individual bacteria carry out their existence at very small scales. Their small sizes make bacteria particularly effective in monitoring changes in the environment. Bacteria have the highest surface-to-volume ratios of any living organism. This fact makes bacteria the most environmentally aware organisms. Minute changes in the environment are quickly detected by bacteria that subsequently respond by rapid growth, sudden death, or dormancy.

Because of the difficulties associated with identifying bacteria, most scientists simply report total numbers. However, total numbers gives little, if any, information on the condition of a particular habitat. Nuisance or "weedy" species of bacteria may be extremely abundant but provide little information concerning ecosystem processes. The use of bacteria as indicators of ecosystem conditions must take this observation into consideration. Although total bacterial densities may be high ($> 10^6$ mL^{-1} or $>10^8$ g^{-1} soil), detection of indicator bacterial species must be sensitive enough to account for extreme spatial heterogeneity at the microscopic level. In addition, because the very act of sampling will disturb bacterial associations, methods must be able to detect species of interest in slurries resulting from sample collection.

MODERN TECHNIQUES

Modern molecular techniques have made it possible to detect specific bacteria or specific genes without the need to culture the bacteria involved. This is a major breakthrough. Previous to the development of these techniques, researchers had to be able to culture bacteria before they could try to identify or characterize the organisms. Culturing is a labor-intensive and often futile undertaking because all the requirements for growth must be identified correctly and be present in the culture medium. Many of the new techniques can be performed in situ, while others require only the extraction of environmental DNA. The development of these techniques allows detection of various levels of taxonomy or specific gene sequences. Because specific genes code for specific functions, their detection is indicative of the capability of that function in nature but is not necessarily indicative of the occurrence of that function. Estimates of gene abundance is further evidence but not sufficient to state definitively that a function is or is not being performed. Bacteria may harbor certain genes that are not functional under current conditions. These background levels of a gene need to be known to provide meaningful comparisons to impacted or disturbed systems.

Many bacterial genes that aid in tolerance or confer the ability to process various compounds are carried on *plasmids*. Plasmids are small circular pieces of DNA that can be transferred between bacteria. Lateral transfer of plasmid DNA among bacteria is known as *conjugation*. Several studies have shown that conjugation occurs under natural conditions and that transfer can be made between different unrelated taxa (Davison 1999). Under strong selection, both the relative and actual number of copies of a particular plasmid that confers an advantage will increase. Plasmid profiles of bacteria from disturbed and undisturbed samples may be indicative of specific selective pressures on the bacterial assemblages. Bacteria and/or their genes may be particularly attractive as biomonitors. Given that the goals of biomonitoring are the early detection of stress or as indicators of recovery following mitigation or restoration, bacteria are a powerful but underutilized tool for addressing these problems.

SAMPLE COLLECTION

Collection of samples suitable for microbial analyses is basically very simple. A simple coring device is required to sample soft or sandy sediments, whereas sterile bottles may be used to sample bacterioplankton. The coring device can be a piece of PVC pipe (2.54 cm in diameter) that can be shoved into the substratum at least 10 cm. Care must be taken to wash the coring device between sample locations or to have sufficient corers for every sample to prevent contamination between samples. Cores should be placed into sterile plastic bags (Whirlpacs are ideal) labeled and placed on ice. For rocky substrate either of two methods can be used. First, representative rocks can be collected or if the substrate is too large, a standardized area scraped with a sterile scalpel and collected into Whirlpacs with sterile physiologic saline (0.9%

NaCl wt/vol) before placing on ice. Wood, leaves and macrophytes can either be collected whole or scraped similar to rocks. Samples should be kept below 5°C before processing to prevent further bacterial growth. The disturbance created by sampling may alter nutrient availability and thus favor various species or strains over others and affect the results. Therefore, it is extremely important to place the samples on ice to restrict growth.

ANALYZING BACTERIAL SAMPLES

Level of Resolution

The various techniques described in this section are effective in answering questions at different levels of taxonomic or functional resolution (Table 12.1). Some of these techniques require higher levels of experience and specific equipment and instrumentation; others can easily be performed. Managers can apply the first two methods, while the next methods would require the assistance of a suitably trained microbial ecologist. As with any study, using the right tools is essential to answering the question. Furthermore, baseline microbial information on undisturbed wetlands must be available to determine significance of effects. If such data are not available for the wetland of concern, representative reference systems must be sampled concurrently. Baseline information must include adequate temporal and spatial characterization; the full range of the reference condition should be documented. Comparisons should be made to the appropriate season and similar habitat type.

Diversity of Organic Matter Utilization

The number of different organic substrates that bacterial assemblages can metabolize is an indication of the metabolic and functional diversity of a particular habitat. This information can be used to determine changes due to various stresses. Since the diversity of organic substrates utilized by bacteria is habitat specific, the same habitat type should be compared among reference and test sites. Conditions that would lower the metabolic/functional diversity can easily be detected and would be an early signal of ecosystem stress. Various commercial products are available that make this type of survey easy and accessible. Biolog plates have been used in a variety of novel ways to compare the metabolic potential of various bacterial assemblages (Derry et al. 1999, Gamo and Shoji 1999, Rutgers and Breure 1999). Each Biolog plate is a 96-well microtiter plate. Each well contains a single different organic compound. Environmental samples are homogenized through either vortexing (spinning samples at speeds to create a vortex) or mild sonicating (use of ultrahigh sound frequencies) to disperse cells, and then the inoculum is added to each well. If the assemblage is able to metabolize the specific organic substrate of a well, there is a color change. The number and location of these positive responses can be used to give a numerical score to the sample. Comparisons using clustering or other multivariate procedures are then used to determine similarity among samples. For example, you could cluster on

TABLE 12.1. Decision Matrix

	Application	Equipment Required	Ease of Use
Organic matter utilization	Compare between/among sites in terms of microbial community metabolism. Increases or decreases relative to reference conditions indicate ecosystem degradation or recovery.	Biolog or similar plates—can be read with or without automated plate reader	Easy
Selective media	Determine levels of resistance, tolerance, or inhibition of bacteria by various substances or conditions. Increased levelsindicate increased stress.	Autoclave, petri dishes, various media and chemicals	Moderately easy
Immunofluoresence	Detect species and/or roles in the ecosystem.	Bacterial culturing facility, access to lab that produces antibodies, instruments to detect fluorescence (i.e., microscope)	Experience required
Polymerase chain reaction	Detect specific bacteria, genes, or sequences. Allows detection of indicator species or genes even at low concentrations.	Well-equipped molecular biology laboratory	Experience required
Hybridizations	Compare strains, species, or assemblages between sites or times. Detect indicator species, genes, genes, or functions in situ.	Well-equipped molecular biology laboratory	Experience required

percent similarity between test and reference samples using Jaccard's index. This procedure is simple, fairly inexpensive, and relatively quick. One basic assumption of the procedure is that the inoculum introduced into each well is homogeneous among wells. This is difficult to determine without suitable replication. Failure to have a homogeneous mix in each well would result in significant differences among replicate samples caused by initial differences in the suite of bacteria introduced into the well. In this case, replication is both the number of plates inoculated by a single environmental sample and the number of samples collected per habitat.

Selective Media

The ability to culture bacteria on solid media and in batch culture has both helped and hindered microbial ecology. Less than 10% and often less than 1% of the bacteria detected through direct microscopy can be cultured on agar or other solid media plates. Bacterial culturing is extremely important in medical and agricultural applications, for obvious reasons. Many of the known species of bacteria are the direct result of efforts to understand some disease. This technique is the basis of monitoring for coliform and enteric bacteria. The use of solid media plates can be extended to provide meaningful insights into the functioning and or stresses of native bacteria.

Various constituents can be added either to chemically defined or to undefined media and used as selection for certain traits. For example, if metal pollution were suspected, plating bacteria from an environmental sample onto a medium containing heavy metals would show the number of metal-tolerant or metal-resistant individuals and the effect on the total culturable counts. Under metal stress, bacteria have the ability to acquire resistance genes. An increase in the number of bacteria with resistance genes would be a strong indicator of metal pollution.

This method can be extended to include any compound, element, or physical condition (e.g., pH changes). However, the presence of a trait is not necessarily evidence of stress or perturbation, nor is absence of a trait an indication otherwise. Background levels of a trait need to be determined to determine responses to a particular stress. For example, various methanogen bacteria (methane eating) have the ability to cometabolize trichloroethylene (TCE) (Arcangeli and Arvin 1997, Tartakovsky et al. 1998, Bielefeldt and Stensel 1999). TCE is one of the most, if not the most, pervasive toxicant in the environment. The word *cometabolize* is misleading because although these bacteria can break down TCE, they get no energy or structural carbon from the process. It is simply fortuitous. Therefore, the presence of TCE cannot and does not select for an increase in TCE-consuming bacteria. But the presence and abundance of methanogens would be a strong indicator of potential TCE degradation. In other words, if methane-producing genes and genes for methane consumption are present at a site where TCE had been released, there is a greater probability that TCE will be metabolized there then at sites where these genes are absent or are in low abundance. Because of the simplicity of this technique, it is often overlooked as a monitoring tool. But if one understands the constraints and limitations of the method, it can be an effective tool for monitoring the environment. The primary advantage of the procedure is the ability to directly select for bacteria that indicate a specific concern.

Modern Molecular Techniques

The ability to use bacteria as biosensors of the environment has greatly increased with the development of numerous molecular biological techniques. It is not our intent in this chapter to review the available procedures thoroughly but rather, to give a general overview of the possibilities that exist. These procedures require varying levels of instrumentation and sophistication. The sensitivity and power of these new methods make them ideal tools for monitoring various aspects of bacteria without the constraints of culturing.

Immunofluoresence

Although not necessarily a new technique, the use of immunofluoresence for the detection of specific bacteria or bacterial products is poorly known. The principle of the procedure is the specificity of the antigen–antibody response in mammalian systems. Monoclonal antibodies are highly specific and are useful in typing closely related species (Joly and Washington 1984). Antibodies are produced against a specific antigen, which may be bacteria that indicate some environmental condition. For example, certain *Aeromonas* species are responsible for diseases among many freshwater fishes. Cultured *Aeromonas* would be used as an antigen, injected into a host (usually a mouse), and the resulting antibodies cloned. These antibodies are fused to a fluorescent tag and then introduced into a sample from the environment. When observed under a microscope, cells of *Aeromonas* that bind with the fluorescent antibody can easily be observed. The technique is highly specific and can be used to detect individual cells. However, the bacteria needed to make the antigen must be cultured. This is difficult for many environmental bacteria.

Polymerase Chain Reaction

The polymerase chain reaction (PCR) results in the amplification of exact sequences of target DNA from bacteria collected from environmental samples. This is a highly specific reaction, and single copies of the target can be amplified. The procedure requires knowledge about a specific gene sequence or segment of a gene. For example, mercury resistance is carried on a gene complex known as *mer* (Reyes et al. 1999). Knowledge of the sequence of bases that make up *mer* is essential in constructing gene probes specific to *mer* sequence. Clones of a known sequence are tagged with primers and fluorescent or radioactive indicators and used to amplify the specific gene of interest. Because of the potential for high specificity, this technique can be used to detect indicator species or genes. For example, resistance to mercury is due to the interaction of five genes in a complex known as *mer*. Detection of any of these genes would be an indicator of mercury stress in the environment. Numerous other environmentally important genes have been sequenced and used to detect various functions or potential activities. However, the method is not quantitative, so no information on the number of copies of the target gene in the sample can be obtained.

Hybridizations

Hybridization of DNA requires the reannealing of two complementary nucleic acid strands. Hybridization reactions occur between a target and a probe that has been labeled with either a radioactive or a nonradioactive tag. The use of nucleic acids to identify specific bacteria or groups of bacteria or to indicate functional capabilities requires that environmental DNA be available. Therefore, either DNA or whole cells must be extracted or separated from the samples. Extraction of DNA from environmental samples yields much more nucleic acids than separating cells from the sample (Leff et al. 1995). In either case, once the environmental DNA is obtained, it can be used as a template for various nucleic acid probes that are specific to varying levels of taxonomic or functional resolution. For example, DNA/DNA hybridizations were used to identify fish pathogens and determine their phylogenies, and in epidemiology studies (Bernardet 1996). In another study DNA/DNA hybridizations were used to establish the phylogeny and taxonomy of *Methanococcus* spp. (Keswani et al. 1996). *Aeromonas* strains from different sources (dead and live fish, drinking, lake, river and seawater, municipal sewage and aluminum rolling emulsion) were identified using this technique (Kaznowski 1998). These results could then be used to identify potential stresses that may increase specific *Aeromonas* strains. Because this is a fairly new application of a technology, innovations and applications are being developed or discovered almost daily. The power of the technique has not been fully determined nor the multitude of its uses exhausted. Researchers clever enough to modify the technique to their specific problem can expect higher resolution than researchers constrained by other techniques, such as culturing.

In situ hybridizations are a powerful tool for the detection of specific functions or groups of organisms without the need for isolation or culturing. This technique has been used to identify nitrifying bacteria (Okabe et al. 1999), identifying bacteria used for oil recovery (Fujiwara et al. 2000), and to describe the microbial ecology of hydrothermal vents (Jeanthon 2000), to list just a few examples. Fixed or fresh samples are incubated with the probe of interest. After a suitable incubation time any nonannealed probe is washed from the sample and the amount of hybridization determined through autoradiography or through chemoluminescence or fluorescence. Nonisotopic methods are becoming more sensitive and more attractive because they are easy to use and safer than isotopic probes. Colony hybridization is performed on colonies that have been transferred to a nylon membrane and lysed. This technique allows many individuals to be screened but requires that the bacteria can be grown to colonies under laboratory conditions.

Often, differences in community structure are more ecologically telling than changes in the abundance of a specific taxon. Community analysis of bacterial samples using traditional techniques is rudimentary, the primary problems being the disruption of structure during sampling and the inability to observe directly. Changes in bacterial assemblages may be at the strain level (i.e., genetic difference within a species) and not replacement of species. Prior to the development of molecular techniques, it was impossible to assess assemblage level changes effectively. Today, there are several methods that allow characterization and comparison of bacterial communities. Such techniques as denaturing gradient gel electrophoresis (DGGE) and modifications allow direct comparisons between and among communities (Muyzer et al. 1995). This would allow the use of entire communities or subsets to be compared before and after a disturbance. As with some monitoring procedures, it is necessary to have predisturbance samples taken over multiple time intervals to be able to make meaningful comparisons. Microbial assemblages change naturally due to season and inputs of nutrients and resources. Therefore, care must be taken with comparative data to ensure that differences are attributable to the appropriate cause.

The methods and techniques described above are not exhaustive. There are many others, and more are being developed. The purpose of this narrative is to demonstrate the potential of using bacteria as monitors of environmental change. No other groups of organisms are in greater contact with their environment. No other group have had as long of interaction with the environment (3.8 billion years), and thus no other organisms can be expected to show the level of response as the bacteria.

Although bacteria and microbes in general have not been used in most biomonitoring efforts, they offer clear advantages. From sentinel species, to indicators of ecosystem function, to warnings of human health problems, to monitors of ecosystem health, bacteria can be utilized as biomonitors in the management, restoration, and maintenance of wetlands and other ecosystems.

ACKNOWLEDGMENTS

I thank Russell. B. Rader for the opportunity to write this chapter. Manuscript preparation was supported by Financial Assistance Award DE-FC09-96SR18546 from the U.S. Department of Energy to the University of Georgia Research Foundation.

REFERENCES

Arcangeli, J. P., and E. Arvin. 1997. Modeling of the cometabolic biodegradation of trichloroethylene by toluene oxidizing bacteria in a biofilm system. Environmental Science and Technology 31:3044–3052.

Bernardet, J. F. 1996. Bacterial fish pathogens: outcome of molecular studies for taxonomy, epidemiology and identification. Zoological Studies 35:71–77.

Bielefeldt, A. R., and H. D. Stensel. 1999. Biodegradation of aromatic compounds and TCE by a filamentous bacteria-dominated consortium. Biodegradation 10:1–13.

Boesch, D. F. 2000. Measuring the health of the Chesapeake Bay: toward integration and prediction. Environmental Research 82:134–142.

Chague-Goff, C., M. R. Rosen, and M. Roseleur. 1999. Water and sediment chemistry of a wetland treating municipal wastewater. New Zealand Journal of Marine and Freshwater Research 33:649–660.

Davison, J. 1999. Genetic exchange between bacteria in the environment. Plasmid 42(2): 73–91.

Derry, A. M., W. J. Staddon, P. G. Kevan, and J. T. Trevors. 1999. Functional diversity and community structure of micro-organisms in three arctic soils as determined by sole-carbon-source-utilization. Biodiversity and Conservation 8:205–221.

Edwards, K. J., P. L. Bond, T. M. Gihring, and J. F. Banfield. 2000. An archael iron-oxidizing extreme acidophile important in acid mine drainage. Science 287:1796–1799.

Fairey, E. R., J. S. G. Edmunds, N. J. DeamerMelia, H. Glasgow, F. M. Johnson, P. R. Moeller, J. M. Burkholder, and J. S. Ramsdell. 1999. Reporter gene assay for fish-killing activity produced by *Pfiesteria piscicida*. Environmental Health Perspectives 107:711–714.

Fujiwara, K., S. Tanaka, M. Ohtsuka, H. Yonebayashi, and H. Enomoto. 2000. Identification of bacteria used for microbial enhanced oil recovery process by fluorescence in situ hybridization technique. Sekiyu Gakkaishi (Journal of the Japan Petroleum Institute) 43: 43–51.

Gamo, M., and T. Shoji. 1999. A method of profiling microbial communities based on a most-probable-number assay that uses BIOLOG plates and multiple sole carbon sources. Applied and Environmental Microbiology 65:4419–4424.

Golterman, H. L. 1995. The labyrinth of nutrient cycles and buffers in wetlands: results based on research in the Camargue (southern France). Hydrobiologia 315:39–58.

Gopal, B. 1999. Natural and constructed wetlands for wastewater treatment: Potentials and problems. Water Science and Technology 40:27–35.

Gugliandolo, C., F. Italiano, T. L. Maugeri, S. Inguaggiato, D. Caccamo, and J. P. Amend. 1999. Submarine hydrothermal vents of the Aeolian Islands: relationship between microbial communities and thermal fluids. Geomicrobiology Journal 16:105–117.

Jeanthon, C. 2000. Molecular ecology of hydrothermal vent microbial communities. Antoine Van Leeuwenhoek International Journal of General and Molecular Microbiology 77: 117–33.

Joly, J. R., and C. W. Washington. 1984. Correlation of subtypes of *Legionella pneumophila* defined by monoclonal antibodies with epidemiological classification of cases and environmental sources. Journal of Infectious Diseases 1250:667–671.

Jones, S. H., and B. Summer-Brason. 1998. Incidence and detection of pathogenic *Vibrio* sp. in a northern New England estuary, USA. Journal of Shellfish Research 17:1665–1669.

Kaznowski, A. 1998. Identification of Aeromonas strains of different origin to the genomic species level. Journal of Applied Microbiology 84:423–430.

Keswani, J., S. Orkand, U. Premachandran, L. Mandelco, M. J. Franklin, and W. B. Whitman. 1996. Phylogeny and taxonomy of mesophilic *Methanococcus* spp. and comparison of ribosomal-RNA, DNA hybridization, and phenotypic methods. International Journal of Systematic Bacteriology 46:727–735.

Leff, L. G., J. R. Dana, J. V. McArthur, and L. J. Shimkets. 1995. Comparison of methods of DNA extraction from stream sediments. Applied and Environmental Microbiology 61: 1141–1143.

Lung, W. S., and R. N. Light. 1996. Modeling copper removal in wetland ecosystems. Ecological Modelling 93:89–100.

Muyzer, G., A. Teske, C. O. Wirsen, and H. W. Jannasch. 1995. Phlogenetic relationships of *Thiomicrospira* species and their identification in deep-sea hydrothermal vent samples by denaturing gradient gel electrophoresis of 16S rDNA fragments. Archives of Microbiology 164:165–172.

Okabe, S., H. Satoh, and Y. Watanabe. 1999. In situ analysis of nitrifying biofilims as determined by in situ hybridization and the use of microelectrodes. Applied and Environmental Microbiology 65:3182–3191.

Qian, J. H., A. Zayed, Y. L. Zhu, M. Yu, and N. Terry. 1999. Phytoaccumulation of trace elements by wetland plants: III. Uptake and accumulation of ten trace elements by twelve plant species. Journal of Environmental Quality 28:1448–1455.

Reyes, N. S., M. E. Frischer, and P. A. Sobecky. 1999. Characterization of mercury resistance mechanisms in marine sediment microbial communities. FEMS Microbiology Ecology 30:273–284.

Rutgers, M., and A. M. Breure. 1999. Risk assessment, microbial communities, and pollution-induced community tolerance. Human and Ecological Risk Assessment 5:661–670.

Silbergeld, E. K., L. Grattan, D. Oldach, and J. G. Morris. 2000. Pfiesteria: harmful algal blooms as indicators of human: ecosystem interactions. Environmental Research 82:97–105.

Staley, J. T., and J. J. Gosink. 1999. Poles apart: biodiversity and biogeography of sea ice bacteria. Annual Review of Microbiology 53:189–215.

Tartakovsky, B., C. B. Miguez, L. Petti, D. Bourque, D. Groleau, and S. R. Guiot. 1998. Tetrachloroethylene dechlorination using a consortium of coimmobilized methanogenic and methanotrophic bacteria. Enzyme and Microbial Technology 22:255–260.

Wise, M. G., J. V. McArthur, and L. J. Shimkets. 1996. 16S rRNA gene probes for *Deinococcus* species. Systematic and Applied Microbiology 19:365–369.

Woese, C. R., O. Kandler, and M. L. Wheelis 1990. Towards a natural system of organisms: proposal for the domains Archaea, Bacteria, and Eucarya. Proceedings of the National Academy of Sciences of the United States of America 87:4576–4579.

13 Sampling Algae in Wetlands

GORDON GOLDSBOROUGH

Algae are often neglected in wetland monitoring programs, but their contributions as food resources for herbivores and regulators of the physical and chemical environment can be significant. In this chapter I introduce terminology pertaining to planktonic and benthic algal assemblages in wetlands. Some common methods for algal sampling and analysis, and their respective advantages and disadvantages are described. Also discussed are issues affecting the expression and comparability of data from different sites and studies.

Much of the freshwater wetland literature labors under two misconceptions about the primary producer community. One is that detritus from emergent and submersed plants (macrophytes) is the main fuel for wetland food webs. Recent studies of algal production (e.g., Robinson et al. 1997a,b, Steinman et al. 1997) and food web traces (Neill and Cornwell 1992, Keough et al. 1998) indicate that algal cells provide an abundant, rapidly renewed, easily assimilated food resource that could be as or more important than macrophyte detritus (Campeau et al. 1994). A second misconception is that when the role of algae in such food webs is recognized at all, attention is focused on planktonic algae in the water column (e.g. Reeder 1994, Klarer and Millie 1994, Robarts et al. 1995, Westlake et al. 1998, Casamatta et al. 1999) under the assumption it is most important. The abundance of submersed surfaces in shallow systems such as wetlands argues that benthic algae may be at least as common as algae in the water column. The diversity of habitats available for colonization is emphasized by the variety of terms used by wetland ecologists to describe algal assemblages (Table 13.1).

The group referred to collectively as *algae* is difficult to describe because it encompasses diverse phylogenetic origins and, consequently, a wide range of anatomical, morphological, reproductive, and ecological characteristics (Sze 1993). There are five algal divisions that are widely distributed in freshwater wetlands (Table 13.2), but there are usually many more that can occur at specific sites and times. Most

Bioassessment and Management of North American Freshwater Wetlands, edited by Russell B. Rader, Darold P. Batzer, and Scott A. Wissinger.
0-471-35234-9

TABLE 13.1. Terms Pertaining to Wetland Algae

Epilithon	Algae attached to the submersed surfaces of rocks or other inorganic matter.
Epipelon	Algae that migrate through the loose, uppermost layers of well-illuminated sediments; usually composed of diatoms and unicellular flagellates; sometimes referred to as *edaphic algae* (Sullivan and Moncreiff 1990).
Epiphyton	Firmly or loosely attached layers of diatoms, cyanobacteria, and green algae on the submersed surfaces of emergent and submersed macrophytes.
Epixylon	Algae associated with the external surfaces of submersed trees, such as in swamps or riparian wetlands; a specialized form of epiphyton.
Metaphyton	Dense aggregations of filamentous green algae that arise from profuse growth and dislodgement of epiphyton or plocon; may be distributed diffusely in the water column or concentrated in floating mats on the surface; sometimes referred to as *flab* (Hillebrand 1983) (for "floating algal bed") (Scheffer 1998).
Neuston	Concentrated layer of algae occurring at the air–water interface; generally found in quiet bodies of water with limited water movement.
Periphyton	Algae and heterotrophic biota associated with any submersed surface; in general use, it is often used to refer specifically to the algal component.
Phytoplankton	Algae that are entrained in the water column as a result of suspension from benthic habitats (tychoplankton) or in situ growth (euplankton).
Plocon	Nonmigratory algal crust that forms on the surface of well-illuminated sediments; usually composed of mucilaginous cyanobacterial trichomes, among which are motile and nonmotile diatoms; possibly equivalent to *äfja* (Hejny and Segal 1998); contrast with *epipelon*.

algae are photolithotrophic, requiring supplies of light and a suite of inorganic nutrients for growth. The concentrations of two macronutrients, inorganic nitrogen (N) and phosphorus (P), are important determinates, in general, of what algal taxa occur in a specific wetland, although some species may be growth limited by other inorganics, such as silicon (diatoms and chrysophytes). There is evidence that some algae can use organic substrates facultatively (Tuchman 1996) or, in rare cases, obligately. All photosynthetic algae contain one or more types of chlorophyll pigment, but coloration imparted by accessory pigments can mask the green chlorophyll. In some cases this may be a general diagnostic for which algal group predominates at a site. Diatoms tend to be brown, whereas cyanobacteria range from aqua to red shades. Green algae and euglenids are variously green, but in a few instances may be other colors. For instance, euglenid blooms in the neuston of Delta and Oak Hammock Marshes in Manitoba give the water surface a marked brick-red color.

Algae are involved in the function of wetlands in numerous ways. First, as alluded to above, their contribution to annual primary productivity can be high, in some cases

TABLE 13.2. Characteristics of Some Common Algal Groups Occurring in North American Freshwater Wetlands[a]

Group	Classification	Major Habitats	Typical Abiotic Preferences	Major Pigment Groups and Color[b]
Cryptomonads	Division Cryptophyta	Phytoplankton	High organic C	Chlorophylls, phycobilins
Cyanobacteria	Division Cyanophyta	Epipelon, epiphyton, phytoplankton	High N and P but May occur at low N if capable of N-fixation	Chlorophylls, phycobilins
Diatoms	Division Chrysophyta, class Bacillariophyceae	Epilithon, epipelon, epiphyton, phytoplankton	Low to moderate N and P, high silicon	Chlorophylls, carotenoids
Euglenids	Division Euglenophyta	Neuston, phytoplankton	High organic C	Chlorophylls
Green algae	Division Chlorophyta	Epiphyton metaphyton, phytoplankton	high N and P, high light	Chlorophylls

[a]Classification in the kingdoms Monera (cyanobacteria) and Protista follows Sze (1993).

[b]The color of an alga is determined by its major pigments: chlorophylls, green; phycobilins, red or blue-green; carotenoids, yellow, orange, red, or brown.

exceeding 1 kg C/m^2 (Table 13.3), providing food for a wide range of wetland consumers (e.g., Mason and Bryant 1975, Hann 1991, Campeau et al. 1994). Algal biomass can equal that of submersed macrophytes, especially if metaphyton is present (Table 13.4). Proliferation of filamentous green algae in floating metaphyton beds is

TABLE 13.3. Reported Range of Algal Chlorophyll Concentration and Annual Primary Productivity in North American Freshwater Wetlands, Where Each of the Specific Assemblages Occurred

Assemblage	Chlorophyll Concentration	Primary Productivity g C/m^2 per year
Phytoplankton	0–358 μg/L	1–381
Epipelon and plocon	0.4–435 mg/m^2	0–29
Epiphyton	2–650 mg/m^2	2–548
Metaphyton	17–838 mg/m^2	13–1119

Source: Modified from data tabulated in Goldsborough and Robinson (1996), and Robinson et al. (2000).

TABLE 13.4. Comparative Biomass of Algae and Macrophytes at Oak Hammock Marsh, Manitoba, Canada in July 1997

	Biomass[a] (g/m^2)	Percent of total
Phytoplankton	6.1	0.9
Epiphyton	12.2	1.8
Metaphyton	179.6	26.8
Submersed macrophytes	100.3	15.0
Emergent macrophytes (aboveground only)	372.6	55.5
Total algae and macrophytes	670.8	

Source: R. L. McDougal and L. G. Goldsborough, unpublished data.

[a]Algal biomass was estimated using a chlorophyll conversion factor of 0.25% (Fig. 13.3).

a fairly common occurrence following wetland flooding (Hosseini and van der Valk 1989, Robinson et al. 1997a, Wu and Mitsch 1998). The high rate of metabolism has resulting effects on water chemistry, by increasing dissolved oxygen, increasing pH (by depleting the carbonate buffering system), and decreasing supplies of dissolved inorganic carbon (DIC) and other macronutrients. Some algae that are capable of fixing atmospheric nitrogen (free-living cyanobacteria and those occurring as endosymbionts in certain diatom taxa) enrich the environment with inorganic and organic N. Epipelic algae and plocon at the sediment–water interface mediate the flux of nutrients from the sediments to the comparatively nutrient-poor water column (e.g., Jansson 1980, Woodruff et al. 1999). Gelatinous plocon crusts on the sediments stabilize them against resuspension (Grant et al. 1986) and in stimulating precipitation of calcite and other minerals, may aid in sediment formation (Browder et al. 1994). When metaphyton mats occur in abundance, they can reduce subsurface irradiance substantially (Robinson et al. 2000). A thick layer of epiphyton shades the underlying macrophyte and thereby reduces its growth (e.g., Phillips et al. 1978, Sand-Jensen 1990). Finally, there is some indication that the epiphyton matrix on macrophytes harbors bacteria that foster decomposition (Neely 1994). Because of the many ways in which algae can affect ecosystem structure and function, they are of obvious importance for wetlands biomonitoring (e.g., John 1993, Mayer and Galatowitsch 2000, Slate and Stevenson 2000).

Although algae can clearly play a major role in many wetland ecosystems, a broad appreciation for this role has been hampered by a general lack of standardized protocols for sampling and analysis. The manager or nonspecialist wishing to collect data on algal abundance in a wetland is faced with a daunting set of terminology, sampling, and analytical methods. There have been a few reviews of sampling protocols (e.g., Sládečková 1962, Weitzel 1979, Aloi 1990, APHA 1998, Stevenson and Bahls 1999) but, as yet, no comprehensive manual. The purpose of this chapter is to acquaint the reader with commonly used methods for collecting and measuring planktonic and benthic algal assemblages in freshwater wetlands. A simple flow chart (Fig. 13.1) is offered as a way of focusing on the pertinent issues and some of the

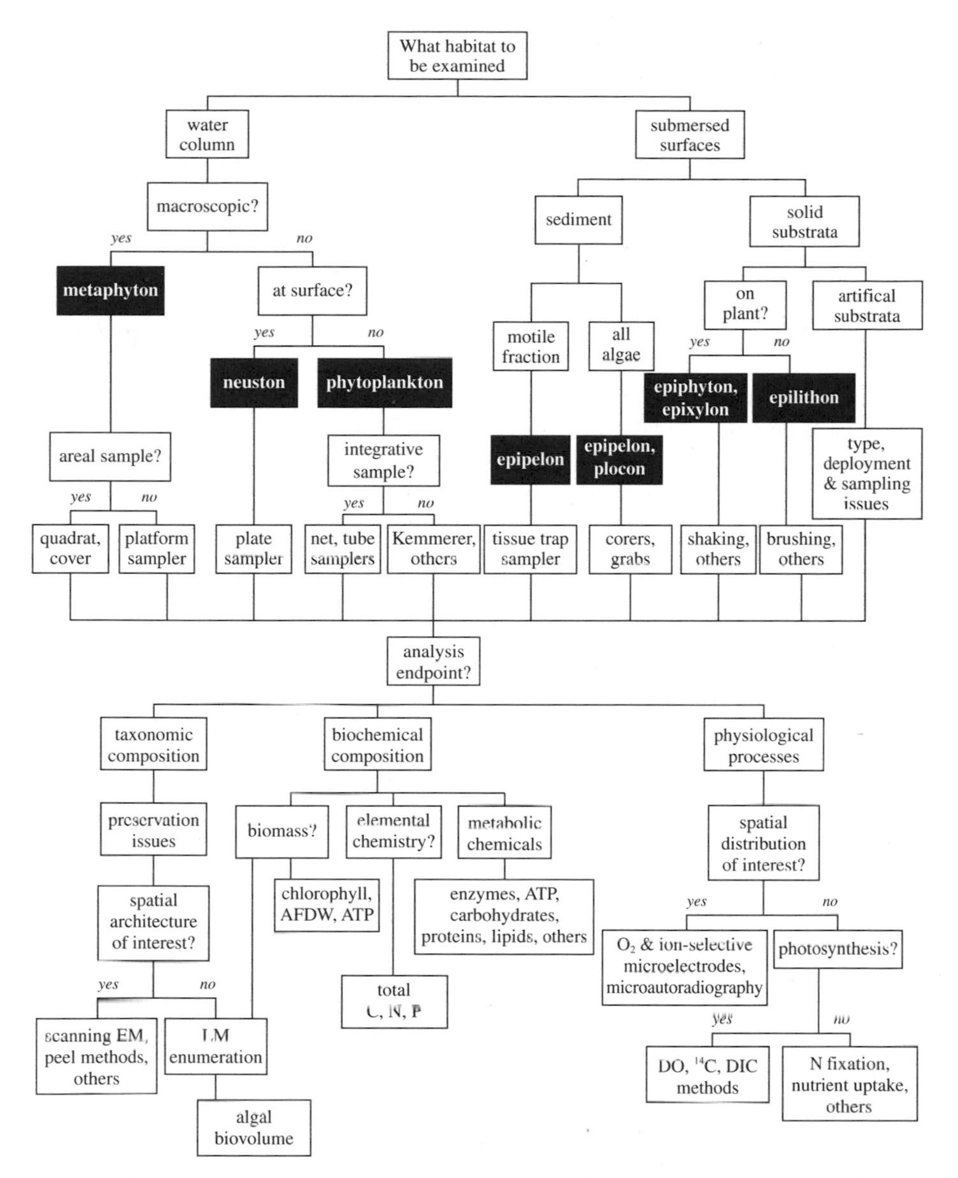

FIGURE 13.1. Decision tree for the sampling and analysis of planktonic and benthic algal assemblages in freshwater wetlands (black boxes), based on issues and methods described in the text. AFDW, ash-free dry weight; DIC, dissolved inorganic C; DO, dissolved oxygen; EM, electron microscopy; LM, light microscopy.

major methods that may be considered during design of a specific study. It is neither exhaustive nor complete so the interested reader is referred to references cited here for more specific details. Stevenson describes the use of these data for wetland monitoring and risk assessment in Chapter 6.

SAMPLE COLLECTION METHODS

The method used to collect wetland algae depends, in large part, on the specific assemblage to be examined (Table 13.1). Phytoplankton are sampled using different equipment than that used for benthic algae, which, in turn, are sampled in a manner appropriate with their substrata (*substratum* is used here to distinguish from *substrate,* the chemical upon which an enzyme acts, consistent with international convention). In most cases, collected samples of planktonic and benthic algae can be processed using the same methods, sometimes making small modifications to account for differences in sample origin. Some of the questions to ask when planning a collection program are addressed below. In many cases the answers will depend on the objectives of the study, circumstances of the sampling site and time, the number and type of analyses, and the spatial scale at which results are desired.

How often should samples be collected? Algal turnover time, the time needed for a cell to complete one cycle of growth and reproduction, is typically measured in hours to days. Consequently, an assemblage growing at optimal light and nutrient conditions can undergo dramatic changes in biomass, productivity, and taxonomic composition within a week. On the other hand, an assemblage growing under stress (e.g., low nutrients or light, presence of toxicants, high herbivory or hydrodynamic shear) may change very little over the same time interval. The sampling frequency in most studies ranges between 2 or 3 days and 1 month, the latter being less desirable; weekly or biweekly sampling is a common compromise. Ideally, a preliminary investigation to consider the sampling frequency at which changes in algal assemblage can be detected at the specific site should precede more extensive sampling. Logistic factors such as the remoteness or accessibility of the sampling site, and the cost (in money and time) of sample analysis, will necessarily affect the decision.

Must samples be collected quantitatively? In some studies it is necessary to characterize algal abundance on an exact volumetric (e.g., cell density per liter), areal (cells per square centimeter), or mass (cells per gram of host tissue) basis to enable comparisons between samples from different sites or times. This requires precise (low replicate variability) and accurate (representative of real abundance) measurements of both the algal assemblage and the volume, area, or mass of the environment from which it was collected, which can significantly inflate the time and cost of sampling. For some routine biomonitoring purposes, or rapid contrasts of several sites, it may be sufficient to characterize the assemblage qualitatively in terms of its relative abundance (e.g., percent cover by metaphyton of a 1-m^2 quadrat) or some proxy variable that can be measured more readily (e.g., Secchi depth as a measure of phytoplankton abundance).

How many samples should be collected from a site? The degree of spatial heterogeneity in the biotic and abiotic environment at a site bears on the number of replicate samples that may be needed to characterize its algal assemblage adequately. Replicate variability tends to increase as a direct function of algal production (Morin and Cattaneo 1992). It is a routine practice to collect triplicate samples so single analytical outliers can be identified more readily, or to collect several samples and pool them prior to analysis. But this may not be required in all cases. For example, the shallow water column (< 1 m) of prairie wetlands is typically well mixed by frequent winds (Carper and Bachman 1984), so vertical patchiness in phytoplankton abundance rarely develops. A single water sample collected at the surface is probably representative of the entire column. On the other hand, sharp chemical, physical, and biological gradients across the sediment–water interface (Carlton and Wetzel 1988, Jorgensen et al. 1988), under thick metaphyton mats (Robinson et al. 2000), and around the submersed surfaces of macrophytes may promote heterogeneous patterns of benthic algal distribution, at scales as small as a few micrometers (Goldsborough 1994). Methods by which algal samples are collected from specific habitats are discussed below.

Phytoplankton

A dip sample collected by filling a bottle at the water surface may be adequate to characterize the phytoplankton in a well-mixed water column (Ytow et al. 1994, Pan and Stevenson 1996). However, the absence of vertical gradients in algal abundance should be verified before intensive sampling begins. In pelagic systems, this would normally be accomplished by collecting samples at discrete depths using a Van Dorn or Kemmerer sampler or any of several equivalents (Wetzel and Likens 1991, APHA 1998). Although these devices have been used in wetlands (e.g., Hanson and Butler 1994), the characteristic shallowness (by definition, < 2 m depth; Mitsch and Gosselink 2000) makes them difficult to operate effectively and the coarse scale at which they collect (normally, at least 1 m increments) may provide inadequate resolution. An alternative is to position tubing connected to a vacuum pump or syringe at the desired sampling depth and withdraw water from that stratum (e.g., Merks 1979). If no information on vertical gradients is available, or they exist but are not pertinent to the study, collection of a sample that integrates the entire water column is strongly recommended. This can be done by drawing a traditional plankton net from the sediments to the water surface (e.g., Kasanof 1973, John 1993). However, unless this is done with extreme care, a substantial amount of resuspended sediment (and associated algae) may be included in the sample. A better solution is to lower to near the bottom a 6- to 10-cm diameter vertical tube of length that is about the same as the site depth, plug the top end, and then transfer an intact water column into a waiting basin or large jar (Schoenberg and Oliver 1988, Murkin et al. 1991, King et al. 1997). Commercial bailer devices that incorporate an automatic check valve at the tube bottom also work well (Swanson 1995). In deeper water, a small pump can be lowered to the bottom, then drawn up slowly, collecting water from each stratum that it passes (Hwang et al. 1998). The resulting sample is usually well mixed during the transfer, so a subsample can simply be withdrawn for phytoplankton analysis. For quantitative analyses, the volume of water collected must be recorded. Whatever the collection method, it may be desirable to remove herbivorous zooplankton or phytophilous invertebrates (when sampling is done in a water column with profuse submersed macrophytes) by passing the sample through a 50- to 200-μm mesh net or bucket, depending on the size distribution of organisms present (Wetzel and Likens 1991).

Neuston

Algal taxa comprising neuston may be found in low abundance throughout the water surface, but they are much more concentrated in the uppermost millimeter or less of the column, along with bacteria and a suite of organic compounds (Round 1981). A qualitative sample can be collected easily by skimming the surface with a small bottle or bag. However, quantitative sampling at this small spatial scale is difficult because surficial films, which tend to be hydrophobic, adhere persistently to sampling gear. This adhesion is the basis of an effective sampling method, wherein an acid-washed glass or Teflon plate (Harvey and Burzell 1972, Hillbricht-Ilkowska et al. 1997) or stainless steel screen (Danos et al. 1983) is dipped into the water or laid briefly on its surface. A glass microscope slide may also be used. The sample is transferred from the plate to a sample container quantitatively using a squeegee or, depending on the intended analyses, using an organic solvent wash. Alternatively, the plate can be kept moist in a plastic bag until it can be analyzed directly. For quantitative analyses, the area of wetland surface collected by the sampler, or the volume of water associated with it, must be determined.

Metaphyton

In the early stages of metaphyton development, mats can be sampled effectively using the same methods as for epipelon or epiphyton (see below). However, when macroscopic metaphyton mats are more fully developed, methods for sampling macrophytes are more appropriate. At the descriptive level it may be sufficient to describe the percentage of the wetland surface that is occupied by metaphyton (Murkin et al. 1994, Wu and Mitsch 1998) or the frequency of metaphyton occurrence in randomly positioned quadrats (Schoenberg and Oliver 1988) or along transects (Gurney and Robinson 1988). A hooked pole can be used to take qualitative grab samples (Hosseini and van der Valk 1989). When the mat is concentrated discretely at the water surface, quantitative samples from a delimited area can be collected readily (Atchue et al. 1983). However, metaphyton that is distributed more diffusely through the water column is more difficult to sample. Integrative water column samplers, such as can be used for phytoplankton, do not typically work well, especially when densely intertangled filamentous green algae are abundant. In such cases, a floatable platform, such as a piece of rigid foam, can be positioned below the metaphyton. Bringing it slowly to the water surface, metaphyton is accumulated on top for collection (Gurney and Robinson 1988). A sharpened "cookie cutter" cylinder can be used to take subsamples of the metaphyton.

Epipelon/Plocon

Epipelon can occur wherever the sediments are well illuminated. In a wetland environment, surface microtopography and patchy submersed macrophyte distribution can result in considerable spatial variation in algal abundance. In shallow clear water, cyanobacteria can flourish, producing mucilaginous compounds that consolidate sediment particles, forming a discrete plocon crust. Accumulation of sediment gases (methane, CO_2) and photosynthetic oxygen in these crusts provide buoyancy, which, in abundance, causes them to detach from the sediment and float to the surface. Floating fragments can be sampled readily by transferring them onto a piece of rigid foam placed below them. The surface area of the sampled fragment must be recorded for quantitative analyses. It is worth noting that such fragments probably detach in areas where algal abundance (and resulting oxygen production) is especially high, so such samples should not be interpreted as representative of sediment algal abundance generally. A more reliable estimate is obtained by in situ sampling, which can be done using either of two basic methods.

The standard method for sampling epipelic algae, the migratory component of the sediment assemblage, is based on their phototactic response (Eaton and Moss 1966). The uppermost few centimeters of sediment are aspirated using a tube connected to a vacuum source and transferred to a blackened beaker. The sediment is allowed to settle and the overlying water is decanted. A piece of tissue is placed on the exposed, moist sediment and the "trap" is exposed to natural sunlight for a period of several hours. Algal cells migrating toward the light become entangled in the cellulose fibers of the tissue and are removed from the sample when the tissue is collected Algae are dislodged from the tissue by shaking it vigorously in water (e.g., Murkin et al. 1991, Kassim and Al-Saadi 1994). The primary disadvantages of this method are that it is laborious and time consuming, there is a lengthy time in sample processing during which algal growth and death can occur, and it collects only some proportion of the motile fraction of the sediment assemblage. Nonmigratory algae and plocon crusts are not sampled by this method, and it is unclear whether all migratory algae migrate into the trap. A recent study at Delta Marsh demonstrated that the trap method seriously and consistently underestimated algal abundance in this system, compared to data obtained from analysis of the whole, aspirated sediment or intact sediment cores (Table 13.5). The extreme represented by floating plocon fragments was also shown clearly.

Given the logistic difficulty of the trap method and, indeed, the artificial conditions imposed by collection of sediment using an aspiration method (including the slurry method), collection of an intact sediment core is recommended. The core can be collected in situ using a plain tube pushed into the bottom, plugged at the top, and

TABLE 13.5. Total Chlorophyll *a* ($mg/m^2 \pm$ SE in parentheses) of Sediment-Associated Algae Measured Using Four Sampling Methods[a]

Sampling Method	Blind Channel		Crescent Pond		Saline Pond	
Tissue trap	0.8	(0.1)	1.4	(0.2)	2.1	(0.2)
Sediment slurry	22.3	(4.4)	34.0	(5.5)	122.1	(17.5)
Intact sediment core	8.3	(1.4)	15.6	(2.0)	21.3	(2.2)
Floating plocon fragments	151.7	(14.3)	127.6	(20.4)	No data	

Source: A. L. E. Bourne and L. G. Goldsborough, unpublished data.

[a]Each value is the mean of 8 to 10 samples collected over a 4-month period in 1998 at three sites in Delta Marsh, Manitoba, Canada.

withdrawn carefully to minimize disturbance caused by dislodgement of enclosed sediment gases (e.g., Shamess et al. 1985). Sediment coring apparatus developed for paleolimnological studies of surficial sediments, such as the mini-KB (Glew 1991), may also be used (Hwang et al. 1998), as can modified plastic syringes (Robertson et al. 1997). Alternatively, an intact sediment sample can be retrieved by any of several grab or dredge samplers (APHA 1998). Core processing should be carried out in the field to minimize disturbance, as soon after collection as possible, with sectioning of discrete depth strata done at whatever resolution is permitted by the fluidity of sediments and the objectives of the study. The volume of sediment collected or area delimited by the core tube must be known for quantitative analyses.

For some analyses, it may be desirable or necessary to obtain pure algae from heterogeneous bulk sediments. Siliceous algal parts, such as diatom frustules, are usually isolated readily from organic sediments by digesting the sample with a mixture of hydrogen peroxide or concentrated nitric acid and potassium dichromate (e.g., Slate and Stevenson 2000), or combusting it at about 600°C in a muffle furnace; treatment with hydrochloric acid can help to remove carbonates remaining in the mineralized sample. Algal biochemistry (Hellebust and Craigie 1978) and stable isotopic composition (Lajtha and Michener 1994) involve analyses where removal of organic (e.g., fauna, macrophytes, detritus) and inorganic (e.g., clays, carbonates) contaminants is required. Density gradient centrifugation is one method for purifying algae (Pertoft et al. 1978, de Jonge 1979). A whole sample is added to a column of silica particles (10 to 30 μm in diameter) of continuously increasing density, then centrifuged, after which algal cells and other constituents are distributed on the column according to their density.

Epiphyton/Epixylon

Macrophytes are often the visually dominant element of wetlands, so it should follow that epiphytic algae on their submersed (and, in a few rare cases, exposed) surfaces can be abundant. As for metaphyton, qualitative estimates of abundance, based on visual assessment of plant surface coverage by the algal film, can be obtained. It is usually necessary to collect plants and remove algae for quantitative analysis. Differences in the morphology between macrophyte species mean that there is no single method that samples all species adequately. Emergent plants within a quadrat can be cut at the sediment surface using shears and transferred carefully into bags. Whole submersed plants can be harvested using submersible samplers that shear the plants at the sediment interface (e.g., Pip and Stewart 1976), but their effectiveness at collecting a known bottom area tends to decrease in dense plant beds. A known mass of plant tissue can be collected using devices such as the Downing box (Downing 1986), essentially a hinged plastic briefcase that encloses and cuts plant tissue when it is closed. Epixylic algae on detached tree snags are collected easily, whereas those on submersed tree roots or trunks requires a piece to be cut from the tree. Floating macrophytes delimited by a quadrat can be sampled using a screen, such as a kitchen colander, that is brought up from below (Goldsborough 1993).

It is possible to study sparse epiphyton while it remains attached to plant surfaces using bleaching and light microscopy or epifluorescent or scanning electron microscopy (e.g., Carter 1982, Delbecque 1985). Most chemical and physiological analyses require them to be removed to prevent contamination by macrophyte tissue. In so doing, physiological effects associated with gradients of light, nutrients, and waste products that develop in a three-dimensional matrix on the intact plant are lost. Numerous methods are used to remove algal cells from plant surfaces, most often by vigorous shaking of plants in filtered water (e.g., Gough and Woelkerling 1976, Delbecque 1985, Steinman et al. 1997), or scraping their surfaces with rubber, glass, or steel blades (Roos et al. 1981, Cronk and Mitsch 1994, Vymazal et al. 1994, Hwang et al. 1998) or brushes (Pan and Stevenson 1996). Algae are not always removed completely by these treatments, so the plant should be examined closely after processing. The degree to which plant tissue is degraded by the removal method should be evaluated, too, particularly when the intended analysis may be compromised by contaminating plant parts. The end result is generally an algal suspension in water that can be processed like phytoplankton samples.

Epilithon

Epilithic algae can be sampled in the same general ways as epiphyton. Rocks to be sampled are usually sufficiently small that the entire rock can be collected. Otherwise, various in situ brush-type samplers must be used to scrub algae from a known surface area of the rock surface (reviewed by Aloi 1990). For whole rocks, it is feasible to wrap them with paper or metal foil and determine the area of material needed to cover the rock completely.

Periphyton on Artificial Substrata

Logistical difficulties in sampling algae from solid, natural substrata in wetlands, such as plants or rocks, can be circumvented through the use of artificial substrata (e.g., Cronk and Mitsch 1994, McCormick et al. 1996, Robertson et al. 1997, Mayer and Galatowitsch 2000). For lack of a more specific term, *periphyton* will be used here to denote the algae that colonize such surfaces, although many use it more generally for all benthic algae. There are numerous materials that have been used for this purpose (see Sládečková 1962 for many examples), of two basic types. *Point substrata* such as glass microscope slides, clay tiles, bricks, foam blocks, wood blocks, rock slices, or scanning electron microscopy (SEM) stubs allow for sampling of discretely localized communities. Spatial sampling with this technique requires the use of several individual substrata. *Linear substrata,* from which spatially distinct subsamples may be collected from the same substratum, include nylon monofilaments, glass and plastic rods, ropes, or plastic strips. Some substrata are deployed in a separate support frame, whereas others are freestanding. All should be wiped with alcohol or acetone prior to deployment to remove contamination arising from hand contact.

In most cases, artificial substrata sacrifice morphological similarity to natural substrata for the sake of simpler deployment and collection, higher algal removal efficiency, lower variability between replicate samples, and elimination of confounding effects of environmental factors on natural substrata such as macrophytes. They also permit experimental study of the controls on benthic algal growth; for example, artificial substrata that release nutrients at a controlled rate (Fairchild et al. 1985) are used widely to diagnose algal nutrient deficiency. Chemically inert artificial substrata facilitate biochemical or physiological measurements of intact assemblages that would not be possible using algae on natural substrata. There are a few examples where plastic mimics resembling real macrophytes, at least at a gross morphological level, have been used successfully (Cattaneo and Kalff 1979). However, the validity of doing so probably varies with the importance of inorganic and organic nutrients, growth regulators, and toxic waste products leaked by the host. Being chemically inert and topographically smooth, artificial surfaces tend to support fewer niches and therefore less algae than nearby natural surfaces. A striking contrary example was seen in markedly higher algal growth on smooth acrylic rods deployed among cylindrical *Scirpus validus* (bulrush) stems (Figure 13.2). The difference was maintained throughout the growing season except after plant senescence, when slippery cuticular waxes were eroded, permitting algal attachment and growth to increase biomass to levels similar to that on rods. The lack of consensus on the degree to which algal colonization of artificial surfaces resembles that on natural surfaces (e.g., Cattaneo and Amireault 1992) argues that the use of artificial substrata as a substitute for natural wetland substrata must be evaluated carefully on a site-by-site basis.

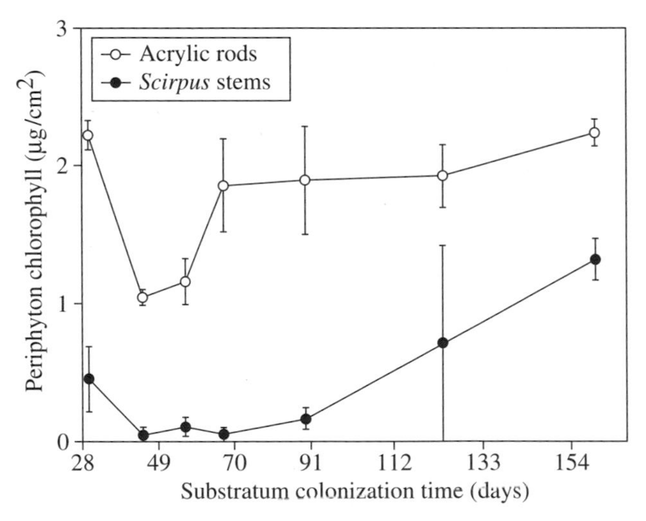

FIGURE 13.2. Comparison of periphytic algal chlorophyll concentration (μg/cm^2) on the surfaces of smooth, acrylic rods deployed among morphologically similar stems of *Scirpus validus* (bulrush) in the littoral zone of Hastings Lake, Alberta, Canada. (Data from Goldsborough and Hickman 1991).

SAMPLE ANALYSIS

Structure and function are the two fundamental metrics that are used in the assessment of wetland algal assemblages. *Structure* refers to community composition, measured by species richness and evenness, numbers, biomass, and spatial distribution. *Function* refers to processes involving energy flow, such as rates of production and respiration and nutrient cycling. On the surface, the difference between them would appear to be the role of taxonomy, being present in the former and absent in the latter. However, this is not always so. It is possible, for instance, to measure algal function in terms of the rates of immigration and emigration, growth, and death of one or more taxa within a mixed assemblage. Parameters used for both metrics are inherently interrelated and provide useful information, depending on the questions being asked and the resources that are available.

Another way to classify the analyses that can be performed on wetland algal samples is to assign them to one of three broad groups: taxonomic composition, biochemical composition, and physiological processes. The information provided by each class of analysis and their respective pros and cons are described below.

Taxonomic Composition

The short generation time of algal cells means that changes in taxonomic composition of an assemblage due to, for instance, toxicant exposure can occur within a few days. As such, changes in the proportions of rare or common taxa, local extinction, or appearance of new species or taxonomic groups, may be sensitive indicators of autogenic or allogenic stress. Algal groups vary in their use by herbivores (e.g., Underwood and Thomas 1990, Botts and Cowell 1992), so taxonomic description may give a better indication of energy flow to higher trophic levels than measures of total dry weight. For benthic algae, species composition of an assemblage can be used to infer its physiognomy, because the growth habits of constituent taxa, whether sessile or in various upright positions, are mostly known. An assemblage composed entirely of sessile taxa, for example, may be inferred to represent an early successional state, occurring as a result of short colonization time, or stress such as hydrodynamic shear, that prevents development of the three-dimensionality typical of more mature stages. Determination of changes in benthic algal composition over time may be used to examine species-specific rates of immigration, accrual, and emigration as affected by environmental factors (e.g., Stevenson and Peterson 1989).

The typical method to analyze algal taxonomic composition by light microscopy is to transfer a small sample of an algal suspension onto a glass slide or counting chamber, and to identify and enumerate algal taxa in several fields of view at random positions or on diametric transects. It might be acceptable, as a faster alternative to counting, to visually estimate the proportion of field area covered by each taxon, analogous to the widely used method for terrestrial macrophytes (Gerbeaux and Lowe 2000). Digested diatom samples are often mounted in highly refractive media such as Naphrax, which accentuates frustular details used in taxonomy. Algal identification usually requires at least 400 × magnification. Specialized optics, such as phase contrast or differential interference, improve the ability to resolve diagnostic features. When cell density is low, it may be necessary to preconcentrate the sample by centrifugation (APHA 1998) or, more usually, by allowing the cells to settle in a columnar chamber, after which most of the liquid is decanted and the material deposited on the chamber bottom is examined.

There is no definitive global reference for the taxonomy of all algal groups. There are many good keys for specific groups and geographic regions, but these are generally not up to date with nomenclatural revisions. Therefore, the analyst is required to consult a large collection of books and journal reprints for species identification. Measurement of species richness, a count of all unique taxa in a sample, may be sufficient for some purposes. More detailed information is obtained from separate enumeration of each species, which can be used to calculate total algal density per unit volume or area. Species diversity indices are calculated using these data. To ascertain that enough cells are counted to detect all (or most) of the species in a sample, cumulative species richness should be plotted against abundance data as the count progresses. The resulting plot asymptote indicates the minimum cell count needed for the sample. In practice, this is often not done because of time constraints and, instead, a standard minimum count (sometimes as few as 100 but more typically 500 or more cells) is done for all samples.

If a representative subsample of each species is measured, usually by matching it to the most similar geometric solid and measuring its major axes, density (typically, cells/L or cells/mm^2) can be multiplied by the volume of a single cell. This is summed for all species to calculate total biovolume. Total algal biovolume usually correlates well with other biomass indicators, such as chlorophyll concentration (LaBaugh 1995).

It may be important to distinguish live from dead cells when the proportion of the latter is high (e.g., Owen et al. 1978). A simple way to do this is to count only cells with intact chloroplast(s), on the assumption that these break down upon death. However, such cells may not be active physiologically. A more rigorous method is to apply a "vital" stain such as tetrazolium violet to fresh, unpreserved samples (APHA 1998). These stains bond to cell contents only through the mediation of a cellular process, so they readily differentiate inactive and dead cells.

Light microscopy provides excellent taxonomic information, but for benthic algal assemblages that are detached from their substratum during sampling, potentially useful information on spatial architecture is lost. Scanning electron microscopy (SEM) of intact, critical-point-dried specimens is an alternative (e.g., Higashi et al. 1981, Delbecque 1985, Burkholder et al. 1990). However, it is often difficult to obtain quantitative counts of species abundance using SEM, especially when cells are obscured by other components of the matrix, and the high equipment cost and time for sample preparation usually makes it unsuited for routine enumeration. In cases where the algal assemblage is essentially two-dimensional, algae can be removed quantitatively from their substrata, with their positions intact, for light microscopy. These methods involve transferring the algae from the substratum to an adhesive layer, then to glass slides, after which the adhesive is removed (e.g., Margalef 1949, Cattaneo 1978, Goldsborough 1989).

Information about the taxonomic composition of an algal assemblage can be inferred from the suite of accessory pigments that it contains (Eloranta 1986), especially those that are specific to certain algal groups (Table 13.2). Efforts to use pigment chromatography as a faster, simpler substitute for light microscopy find considerable discrepancy between the two methods (e.g., Havens et al. 1999b), so it is likely that biochemical methods should be used only for coarse screening purposes.

Despite the clear advantages of taxonomic enumeration of wetland algae, it is less often done than other analyses. The primary reason is that for the results to be meaningful requires considerable resources. Species-level identification generally requires advanced analyst training and an expensive, high-quality microscope. There have been several attempts to develop simple macroscopic measures of algal composition based on predominant coloration, texture, or structure (e.g., Stevenson and Bahls 1999). Enumeration is often done at the aggregate level, with only the proportions of major divisions being reported (e.g., Table 13.2). Such data may provide sufficient information for gross characterization of food availability for herbivores. However, identification at the genus level or lower may provide misleading or incorrect information, particularly when species within a genus have markedly different ecological requirements. Species resolution may require observation of reproductive stages that are difficult to detect, or absent from the sample at the time of collection. In addition, sample analysis can rarely be done in real time. Instead, samples are preserved and stored for later examination. It is also desirable to maintain a collection of preserved reference specimens for reexamination and quality assurance (although high-quality photographs or sketches may suffice in some applications). Lugol's Iodine, neutralized formalin, and glutaraldehyde are the most commonly used algal preservatives (APHA 1998). Preservation usually causes varying degrees of cell discoloration, so a potentially useful diagnostic characteristic is lost. Distortion of cell features and decomposition of the preservative are issues whose severity increases with longer periods of sample storage. On an operational level, accurate identification and enumeration may be difficult when cell clusters of varying size are abundant or when samples contain detritus and other nonalgal material that obscures algae. This is especially true for epipelon samples, which if not collected carefully, can contain substantial amounts of sediment, although epifluorescent microscopy may be used to differentiate algal cells from detrital matter (Stanley 1976). Cell damage, especially during sampling of algae that are tightly adherent to substrata, may complicate analysis. A more fundamental problem is that there may be no taxonomic response to the environmental factor of interest. Instead, changes may be manifest at the biochemical or physiological levels. Consequently, depending on the goals of the study, the ideal sampling program should combine taxonomy with other analyses.

Biochemical Composition

Chemical preservation of algal samples may cause analytical interference or artifacts so those intended for biochemical analysis are typically processed immediately after collection, or other methods of preservation, such as freezing, are used in some cases. The chemistry of unpreserved samples may change rapidly, so analysis, especially for metabolic products, should be done as soon as possible. Similarly, the method of sample storage should be chosen carefully to limit contamination.

It is appropriate and desirable to perform some analyses on the algal suspension. For example, measures of benthic algal N or P in the whole aqueous sample will include nutrients released from cells that were damaged during detachment from the substratum. On the other hand, some analyses, such as dry weight determination, require the algae to be isolated from the medium. The simplest and most common method for this is to collect algal cells onto a filter under weak vacuum. Glass-fiber or cellulose membrane filters with a pore size of 1 μm or less are preferred; larger pore sizes may not retain small algal cells. Filter composition should be selected in view of the intended analysis; glass filters, for instance, will survive incineration up to about 600°C, whereas cellulose will not. Samples can also be centrifuged to collect an algal pellet for analysis.

Biomass is the most readily understood and most often collected data on wetland algae, and there are several ways by which it is measured (Table 13.6). Macrophyte biomass samples are typically dried (at temperatures ranging from 60 to 100°C) to remove associated water and the dry residue is weighed. A problem in using this method for algae is that inorganic material, such as carbonate precipitate, may be abundant at some sites, giving an erroneously high measure of algal biomass. A solution is to oxidize organic matter in the dried sample, collected onto glass filters, at high temperature (typically, 400 to 600°C), then calculate the change in weight,

TABLE 13.6. Pros and Cons of Various Metrics of Algal Biomass

Measure	Pros	Cons
Dry weight, ash-free dry weight	Inexpensive; large body of existing data for comparison; directly comparable to metrics for other organisms	Not specific to algae; can include variable contributions by plant, heterotroph, dead, and abiotic constituents
Chlorophyll concentration	May be inexpensive, depending on method; large body of existing data for comparison; unique to algae and plants	Cellular chlorophyll content varies with growth conditions
ATP concentration	Method sensitive to low levels; useful in examining trophic structure of an assemblage (trophic index)	Not specific to algae; includes variable contributions by heterotrophs; cellular ATP content varies with growth conditions
Species counts, biovolume	Detailed information on algal assemblage structure	Time consuming; accuracy depends on analyst skill (species identification) and estimates of cell volume

Source: Modified from Stevenson (1996).

termed the *ash-free dry weight* (AFDW). The results may vary with duration and temperature of the oxidation and the sample size (Ridley-Thomas et al. 1989). An important assumption is that the sample contains pure algae, when in reality it may contain substantial amounts of nonalgal matter, such as bacteria, fungi, invertebrates, macrophyte parts, and detritus. When nonalgal contributions are suspected to be high, chlorophyll concentration is a superior indicator of algal biomass in the sample.

There are several methods to quantify the chlorophyll concentration in an algal sample, ranging in cost, time requirement, and information provided. A filter containing sampled algae is immersed in one of several organic solvents (acetone, methanol, ethanol, dimethylsulfoxide) to extract chlorophylls quantitatively. Periphyton samples on artificial (and some natural) substrata can be immersed directly into the solvent as long as the substratum is not affected; for instance, many plastics are soluble in acetone. Similarly, solvent can be added directly to whole sediment samples collected for epipelon (Stanley 1976), although there is some evidence that pigments are extracted less efficiently in wetter sediments (Hansson 1988). Pigment extraction efficiency is usually enhanced by grinding or sonicating the filter or freezing it overnight to rupture cell membranes (samples may be stored frozen for weeks to months as long as the filter is buffered against acidification using $MgCO_3$). The elution should be done in the dark over a period of up to 24 hours to prevent chlorophyll degradation by light. Chlorophyll concentration is measured by spectrophotometry (Marker et al. 1980) or fluorometry (APHA 1998). A correction that accounts for interference by pheophytin, a chlorophyll metabolite resulting from herbivory or acid exposure (Leavitt 1993), can be made by reading samples before and after treatment with dilute hydrochloric acid. Some have found the "corrected" results to be unreliable, in which case the sum of calculated chlorophyll and pheophytin concentrations may be reported as "total chlorophyll" (Robinson et al. 1997a). Information on other solvent-soluble pigments that may be present in the extract can be obtained by chromatographic methods; high-performance liquid chromatography is the most common (Millie et al. 1993, Havens et al. 1999b) but, regrettably, the most expensive. Thin-layer and paper chromatographic methods are also used (Eloranta 1986).

The cellular concentration of ATP is another commonly used indicator of algal biomass (Stephens 1987). ATP is measured by hydrolyzing it with firefly luciferin and luciferase in the presence of magnesium and oxygen, causing chemoluminescence, whose intensity can be measured (one photon of light for one molecule of ATP hydrolyzed).

A ratio of the foregoing variables can be instructive (Clark et al. 1979). A trophic index, sometimes called an *autotrophic index,* is calculated by the ratio of chlorophyll concentration to ash-free dry weight, with higher values indicating greater relative dominance of an assemblage by algae as compared to heterotrophic or abiotic constituents. A functional trophic index excludes abiotic components by replacing ATP concentration in the denominator, which is assumed to occur only in living (or recently dead) tissue. A viability index of the ratio of ATP concentration to ash-free dry weight is least specific to algae but considers the proportion of a sample that is metabolically active, or at least biotic in origin.

Measurements of the elemental composition of sampled algae (most commonly, N, P, or C) are done by slight modification of methods for natural water or soil (APHA 1998). These are generally done to assess the nutritional status of algae, in that deficiency in a growth-limiting nutrient is demonstrated by depleted levels of that nutrient, or a metabolic product made from it, with respect to another cellular constituent. Therefore, it is possible to diagnose algal nutrient deficiency (Healey and Hendzel 1980, Murkin et al. 1991), or relative nutritive value to grazers, based on ratios of cellular biochemicals (Table 13.7) and analyses of algal lipid, carbohydrate, and protein content (Hooper-Reid and Robinson 1978a). Similarly, the presence of certain enzymes are also taken as nutrient deficiency indicators. Nitrogenase, an enzyme found in some cyanobacteria, diatoms with endosymbiotic cyanobacteria, and bacteria, catalyzes the reduction of atmospheric nitrogen to ammonia under conditions of limiting N and its concentration over time is used to monitor algal N deficiency (Hooper-Reid and Robinson 1978b). Phosphatases, most commonly the alkaline form, are produced by P-limited algae to hydrolyze organic P molecules in the environment (Hooper-Reid and Robinson 1978b, Goulder 1990, Scholz and Boon 1993).

There are many reasons to perform biochemical analyses of wetland algae. These are often the least consumptive of time and resources, especially if one is interested

TABLE 13.7. Commonly Used Metrics of Nutrient Deficiency in Algae

Parameter[a]	Extreme Deficiency	Moderate Deficiency	No Deficiency
Nitrogen/general			
Cellular N (μg/mg DW)	< 40	40–70	> 70
Protein (μg/mg DW)	< 300	300–400	> 400
Carbohydrate (μg/mg DW)	> 400	300–400	< 300
Chlorophyll (μg/mg DW)	< 5	5–10	> 10
N/C	< 0.10	0.10–0.13	> 0.13
Protein/CH_2O	< 0.7	0.7–1.2	> 1.2
Protein/CH_2O + lipid	< 0.5	0.5–0.6	> 0.6
NH_4^+ uptake (μm/mg per hour)	> 0.7	> 0.7	< 0.7
N debt (μm/mg DW)	> 1	> 1	< 1
Phosphorus/general			
Cellular P (μg/mg DW)	< 5	5–10	> 10
N/P	> 10		< 10
P/C	< 10	10–20	> 20
PO_4 uptake (μm/mg per hour)	> 0.2	> 0.2	< 0.2
P debt (μm/mg DW)	> 0.5	> 0.5	< 0.5
Alkaline phosphatase (μm/mg per hour)	> 2	> 2	< 2

Source: Modified from Healey (1975) and Healey and Hendzel (1980).

[a]DW, dry weight.

only in the total abundance of the assemblage. Algal biomass, measured as chlorophyll, ATP, or AFDW, may represent the availability of the primary resource to the wetland food web; these variables are generally simple to measure under primitive lab conditions. Analyses can provide a "snapshot" of assemblage characteristics under a specified set of environmental conditions: for example, algal production at a certain place and time. In addition, the effects of stress may, depending on the stressor, be detected in the chemical composition of algae within minutes of exposure before species change in the scale of days can occur. Cell biochemistry, especially the content of metabolically active substances such as ATP or chlorophyll, can serve as a proxy for physiological changes. For example, cellular nutrient quotas are good indicators of metabolic deficiency (Droop 1973). Indeed, repeated biochemical analyses can be used to assess function. Differences in oxygen or carbon concentration at two successive sample times can be used to calculate photosynthesis rate.

There can be problems in relating algal biochemistry to function, especially when a specific biochemical parameter varies as a function of numerous environmental influences. For example, when chlorophyll concentration is used as a proxy for biomass, it is assumed, that cellular pigment content is a fixed proportion of total dry weight, generally about 1 or 2% (APHA 1998). Physiological studies done under controlled laboratory conditions, however, have demonstrated that this is not always the case. It is well known that algal cells contain less chlorophyll when growing at high irradiance (e.g., Reynolds 1984). Consequently, in a mixed-species assemblage of species, especially in benthic habitats where light microgradients can be pronounced (Dodds 1992), there is inevitably going to be variation in a linear plot of algal chlorophyll and dry weight (Fig. 13.3). Similarly, comparisons of the chlorophyll/ATP ratio have shown that it is not constant (e.g., Paerl et al. 1976), due perhaps to variation in the chlorophyll concentration with differing irradiance.

In some cases the chemistry of an algal assemblage may not respond to the environmental factor of interest. Communities under stress may be functionally homeostatic if sensitive taxa are replaced by more tolerant taxa having equivalent biomass. Physiological changes may not be detected by gross biochemical analyses. As noted above, for instance, cellular chlorophyll concentration may not vary directly with photosynthetic activity.

Other potential drawbacks of biochemical analyses are the requirements by some methods for specialized equipment and training and the fact that some variables, such as AFDW or ATP concentration, are not specific to algae and may therefore provide misleading information. With respect to interpretation of data from diverse assemblages, it is sometimes tempting to consider results as if pertaining to a single species. For example, deficiency assays using nutrient-diffusing substrata enriched with N or P may show that algal biomass is stimulated more by both nutrients together than by either alone, suggesting that algae are co-limited by these nutrients. This contravenes the principle that growth can be limited solely by one nutrient at a given time. It is important to be mindful that constituents of an assemblage experience differing conditions depending on their spatial position and physiological requirements, so simultaneous limitation by differing resources is possible.

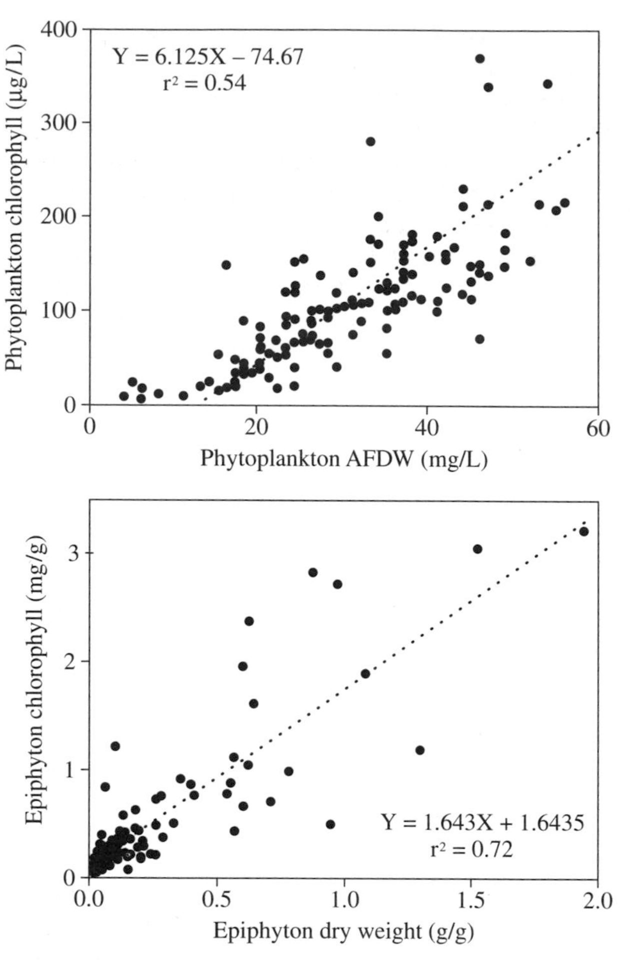

FIGURE 13.3. Positive linear relationships between the chlorophyll concentration of phytoplankton (µg/L) and epiphyton (mg/g) from Delta Marsh, Manitoba, Canada and their corresponding dry weights (mg/L and g/g, respectively; G. Goldsborough, unpublished data).

Physiological Processes

The most commonly measured physiological parameters of wetland algae are rates of photosynthesis, respiration, inorganic N fixation, and N or P uptake. In most cases it is usual to deal with the intact algal assemblage, that is, a whole water sample for phytoplankton or a colonized, undisturbed substratum for benthic algae. This avoids artifacts that may arise due to chemical or physical changes during processing of the water sample or detachment of algae from substrata. The method of containment for the sample during physiological analysis therefore depends on the specific assemblage being examined. Typically, phytoplankton samples are contained in clear glass or plastic bottles and tubes, especially if they require exposure to light (although it is worth remembering that containers affect the spectral quality of irradiance, absorbing nearly all ultraviolet wavelengths). Benthic algae on small substrata may be similarly contained (usually, with filtered water from the collection site), or if the substratum is not amenable for subsampling (large rocks, entire macrophytes), chambers that fasten and seal onto or around the substratum may be used (e.g., Loeb 1981, Jones 1984).

The two most common methods for measuring wetland algal photosynthesis are to quantify oxygen evolution (e.g., Reeder 1994) or carbon assimilation (Hooper and Robinson 1976), each of which has inherent advantages and disadvantages (Table 13.8). Oxygen methods involve collection of three samples, one of which has its dis-

TABLE 13.8. Pros and Cons of Dissolved Oxygen Production and ^{14}C-Uptake Methods for Measuring Wetland Algal Photosynthesis

Method	Pros	Cons
Oxygen production	Provides estimates of gross photosynthesis, net photosynthesis, and respiration from one set of measurements; inexpensive apparatus with few hazardous chemicals	Accuracy of oxygen titration is critical—under ideal conditions, an error of ±0.02 mg/L can be obtained, but this is rarely achieved under field conditions; incubation time is dependent on wetland trophic status; method falsely assumes that respiration (R) is due entirely to photosynthetic organisms; R is falsely assumed to be unaffected by light intensity; photosynthetic quotient (PQ) varies temporally and with species composition of algal assemblage; artifacts due to bacterial wall growth during enclosure; net photosynthesis may decrease in midday due to photoinhibition by high light intensities, oxygen poisoning, nutrient deficiency, high pH due to C depletion, etc.; assumption that photosynthesis is not light saturated may be violated in wetlands
^{14}C uptake	Very sensitive; capable of measuring low levels of photosynthesis; provides direct measurement of C assimilation without need of PQ estimate	Result is neither gross nor net photosynthesis; no estimate of R is provided; method assumes that the only assimilation of ^{14}C is due to photosynthesis; method assumes incorrectly that algal cells do not discriminate ^{14}C versus ^{12}C; some of DOC produced by algae may be reexcreted; delicate cells may rupture under vacuum pressure during filtration; method assumes that addition of ^{14}C radiotracer does not increase significantly the DIC of medium

solved oxygen concentration determined immediately ("initial bottle"; IB). Another sample is contained in a blackened container to exclude all light ("dark bottle"; DB), and the third sample is in a clear bottle ("light bottle"; LB). The latter two samples are incubated under conditions similar to those in situ for periods of up to about 4 hours, after which their oxygen concentration is determined by titration (APHA 1998). These oxygen measurements are used to calculate gross photosynthesis (LB minus DB), net photosynthesis (LB minus IB), and respiration (IB minus DB). Carbon assimilation rate measurements most often use an inorganic C radiotracer such as $NaH^{14}CO_3$ (e.g., Robarts et al. 1995, Robinson et al. 1997b) added to paired clear and darkened bottles. After being incubated for up to 4 hours, the difference in radioactivity of algae from each bottle is used to calculate a value lying between net and gross photosynthesis. No estimate of respiration is provided by this method. In dilute wetlands where DIC concentration is low, it is possible to measure the rate of DIC consumption in paired clear and darkened bottles (Turner et al. 1991).

Typically, photosynthesis methods are applied at a macroscopic scale dictated by the resolution of conventional laboratory equipment. However, there are several methods for measuring it at a more appropriate microscopic scale, in some cases enabling data to be obtained for individual species or cells within the assemblage. One is the use of oxygen-specific microelectrodes to measure static differences in oxygen concentration in benthic films or, if measured at successive times, to calculate oxygen evolution rates (e.g., Jorgensen et al. 1988). Because the method requires a micromanipulator to deploy the electrode and sensitive recording equipment, use is generally limited to samples collected in the field and transported carefully to the lab. Oxygen electrodes are now available commercially, along with others for specific ions. Phosphorus-specific microelectrodes have been used to monitor diurnal changes in P efflux from sediment as mediated by algal crusts at the sediment, interface (Carlton and Wetzel 1988).

Photosynthetic measurements for individual cells can be made using microautoradiographic methods (e.g., Burkholder et al. 1990) wherein a live algal sample is allowed to assimilate radiolabeled inorganic C for a short time interval, then is deposited on a glass slide, coated with a photographic emulsion, and reduction of silver grains in the emulsion is caused by beta particles emitted by assimilated ^{14}C. Depending on the depth of the emulsion and the duration of emulsion exposure, photosynthetic cells acquire tracks of grains extending outward from the cell (track autoradiography) or a halo of exposed silver grains (grain density autoradiography) that is visible using light microscopy. Measuring the density of these grains, in some cases by image analysis software, gives a quantitative measure of photosynthetic rate; by their absence, dead cells are clearly detected. Using the same method, substituting other radiolabeled chemicals, such as ^{14}C-glucose or ^{32}P, it is possible to measure species-specific rates of heterotrophy (Pip and Robinson 1982) or nutrient uptake (Burkholder et al. 1990). Information on species-specific uptake of radioisotopes can be useful for distinguishing viable from nonviable cells (Wasmund 1989) and for measuring differences in metabolic rates and sensitivity to toxicants. Unfortunately, the method has seen limited application in wetlands (Robinson and Pip

1983) because it requires extreme care during all aspects of sample processing, and quantification of silver grains often must be done manually, so it tends to be very time consuming.

A direct, quantitative measure of algal N fixation is obtained using ^{15}N as a tracer. This stable isotope normally comprises about 0.368% of total N; the "excess" is any quantity greater than this natural amount. After a period of exposure of an algal sample to N_2 enriched with ^{15}N, the "excess" in the remaining gas phase and in the sample are measured via mass spectrometry, together with the total N content of the sample. A simpler, less expensive method for measuring algal N fixation is based on the fact that nitrogenase has multiple substrate affinity such that it is capable of reducing several other materials in addition to N_2. This means that, in theory, it is possible to measure the reduction of alternative materials to measure indirectly the potential for N_2 fixation (Jones 1974). Most typically, an algal sample is exposed for a period of hours to acetylene gas (C_2H_2), which is oxidized to ethylene (C_2H_4), then the head-space gas over the sample is analyzed for acetylene and ethylene by gas chromatography. The lack of nitrogenase substrate specificity means that the affinity of nitrogenase for acetylene may differ from that for nitrogen, so the results cannot be expressed in absolute units unless calibrated against coincident N-fixation measurements by the ^{15}N method. The two methods have respective advantages and disadvantages (Table 13.9).

Nutrient debt experiments (e.g., Murkin et al. 1991) assume that the uptake rate of N or P spiked into a natural algal sample will vary directly with the severity of deficiency in that nutrient. Typically, samples of intact phytoplankton or benthic algal substratum in filtered water are taken into the lab, spiked with a known quantity of inorganic N or P, then allowed to assimilate it for a period of about a day. The loss of

TABLE 13.9. Pros and Cons of Acetylene Reduction and ^{15}N-Uptake Methods for Measuring Wetland Algal Nitrogen Fixation

Method	Pros	Cons
Acetylene reduction	Analytically simple; very sensitive to small amounts of N fixation	Lack of nitrogenase substrate specificity; method does not measure absolute N fixation unless calibrated against coincident ^{15}N-fixation measurements; ethylene may be released from organic matter and it may be absorbed by rubber serum stoppers and polyethylene syringes; ethylene may be oxidized by methane-using bacteria
^{15}N uptake	Gives absolute measurements of N fixation; can trace passage of fixed ^{15}N to other organisms	Analytically complex, time consuming, and expensive; about 1000 times less sensitive than acetylene reduction method

N or P from the aqueous phase, compared to an initial analysis, is used to calculate the nutrient uptake rate. These data are usually normalized to algal biomass, indicated by chlorophyll concentration (Healey 1975). An alternative to examining the uptake of spiked P is to measure uptake of the radiotracer ^{32}P (Hwang et al. 1998, Havens et al. 1999a).

As dynamic measures, physiological analyses may reflect algal function more directly than biochemical analyses. They can provide an indication of algal growth that may be easier and faster to measure than information obtained, for instance, by light microscopy of changes in species composition. Responses to environmental changes can be detected quickly, sometimes within seconds. Without physiological analyses such changes may not be recorded until expressed as changes in biomass or taxonomic composition.

Expression of Results

It has been a long-standing practice to express taxonomic, biochemical, and physiological data in units that are pertinent for a specific algal assemblage but which confound comparison with other assemblages or other ecosystems. For instance, phytoplankton data are typically expressed on a volumetric basis, and values for benthic algae are relative to the surface area of the substratum. In many cases, however, with a little more effort it is possible to extrapolate these measurements to an areal basis (e.g., Table 13.4). Phytoplankton chlorophyll concentration (μg/L) can be converted to mg/m^2 simply by multiplying it by the site depth in meters. Epipelon chlorophyll data calculated on a sediment wet weight basis (mg/g) are multiplied by sediment bulk density (g/m^3) and the sampled depth (m) to yield mg/m^2. Metaphyton chlorophyll data are usually expressed initially on an areal basis (mg/m^2), so no conversion should be needed.

Epiphyton chlorophyll data expressed in relation to substratum area (mg/cm^2) must be multiplied by macrophyte surface area available for colonization (cm^2/m^2). The latter measurements are difficult to obtain, especially for submersed macrophyte species with complex morphology. Several methods are available, ranging from direct measurements (either manually or using automated area measurers) to extrapolations based on weight measurements and predetermined weight–area relationships (McDougal et al. 1997), to photometric measurements based on adsorption of colored dyes to plant tissue (Cattaneo and Carignan 1983, Watala and Watala 1994). These methods are ultimately inferior because the irregular microtopography on plant surfaces mean that measurements at a gross scale probably underestimate the total area available to algae, in extreme cases by orders of magnitude. An alternative extrapolation that avoids measurement of surface area is to express algal chlorophyll in units of sampled plant weight (mg/g) multiplied by measurements of plant weight per unit area (g/m^2).

A chlorophyll measurement, which is the de facto standard for assessing algal biomass, is unsatisfying as a basis for comparison with macrophyte data, which are generally based on dry weight. As discussed above, the relationship between algal chlorophyll and dry weight probably varies between places, times, and species, so any conversion of a chlorophyll value to its equivalent dry weight is, at best, an estimate. A data set collected for phytoplankton and epiphyton from Delta Marsh (Figure 13.3) shows, however, a relatively high correlation between the two variables, with the median ratio of chlorophyll to dry weight being 0.25% for phytoplankton and 0.24% for epiphyton. Therefore, for a first-order approximation, it seems reasonable to use a general 0.25% conversion factor for other algal data from this site (Table 13.4). Similarly, we have used a chlorophyll-to-carbon conversion factor of 130 (Robinson et al. 2000), based on a data set including both variables collected at Delta Marsh (Hosseini 1986).

CONCLUSIONS

Recent studies using stable C, N, and S isotopes to establish trophic linkages between primary producers and their prospective consumers have demonstrated that the roles of algae and macrophytes can often be differentiated readily (e.g., Keough et al. 1998). Based on these studies, it is increasingly evident that the contributions by algae can be significant, so they should be considered in any thorough examination of wetland food webs. The knowledge base for wetland algae has flourished in the past decade or so, with more people becoming interested in sampling algal assemblages, more special conference sessions being organized, and more papers being published. The greatest impediment to synthesizing this information, in my view, is the lack of standardized sampling protocols and units of data expression, so that data from studies at different places and times are difficult or impossible to compare and contrast. As a consequence, there are many unresolved questions on such issues as the importance of algae in North American wetlands. Most of the research has focused on shallow lakes, constructed wetlands, and prairie marshes, including clarification of biotic and abiotic controls on primary producers in degraded, turbid wetlands and factors leading to metaphyton proliferation in managed wetlands; and determination of wetland algal responses to exotic species introductions, inputs of nutrients and other anthropogenic chemicals, and climate change. The need for a comprehensive procedures manual for sampling algae in wetlands is imperative. In the meantime, I have tried to show here that methods for collecting and analyzing algae from all habitats in wetlands are generally available and uncomplicated. With only a little additional effort, planktonic and benthic algae can and should be incorporated into routine wetland monitoring programs.

ACKNOWLEDGMENTS

I thank Dale Wrubleski for a preview copy of a forthcoming paper, and my students Rhonda McDougal and Alexandra Bourne for providing data. This is contribution 300 of the University of Manitoba Field Station (Delta Marsh).

REFERENCES

Aloi, J. E. 1990. A critical review of recent freshwater periphyton field methods. Canadian Journal of Fisheries and Aquatic Sciences 47:656-670.

APHA. 1998. Standard methods for the examination of water and wastewater, 20th ed., L. S. Clesceri, A. E. Greenberg, and A. D. Eaton, (eds.). American Public Health Association, Washington, DC.

Atchue, J. A., III, F. P. Day, Jr. and H. G. Marshall. 1983. Algal dynamics and nitrogen and phosphorus cycling in a cypress stand in the seasonally flooded Great Dismal Swamp. Hydrobiologia 106:115–122.

Botts, P. S., and B. C. Cowell. 1992. Feeding electivity of two epiphytic chironomids in a subtropical lake. Oecologia 89:331–337.

Browder, J. A., P. J. Gleason, and D. R. Swift. 1994. Periphyton in the Everglades: spatial variation, environmental correlates, and ecological implications. Page 379–418 *in* S. M. Davis and J. C. Ogden (eds.), Everglades: the ecosystem and its restoration. St. Lucie Press, Delray Beach, FL.

Burkholder, J. M., R. G. Wetzel, and K. L. Klomparens. 1990. Direct comparison of phosphate uptake by adnate and loosely attached microalgae within an intact biofilm matrix. Applied and Environmental Microbiology 56:2882–2890.

Campeau, S., H. R. Murkin and R. D. Titman. 1994. Relative importance of algae and emergent plant litter to freshwater marsh invertebrates. Canadian Journal of Fisheries and Aquatic Science 51:681–692.

Carlton, R. G. and R. G. Wetzel. 1988. Phosphorus flux from lake sediments: effect of epipelic algal oxygen production. Limnology and Oceanography 33:562–570.

Carper, G. L., and R. W. Bachmann. 1984. Wind resuspension of sediments in a prairie lake. Canadian Journal of Fisheries and Aquatic Sciences 41:1763–1767.

Carter, C. C. 1982. A technique for direct microscopic observation of periphyton assemblages on aquatic macrophytes. Journal of Aquatic Plant Management 20:53–56.

Casamatta, D. A., J. R. Beaver, and D. J. Fleischman. 1999. A survey of phytoplankton taxa from three types of wetlands in Ohio. Ohio Journal of Science 99:53–56.

Cattaneo, A., and J. Kalff. 1979. Primary production of algae growing on natural and artificial aquatic plants: a study of interactions between epiphytes and their substrate. Limnology and Oceanography 24:1031–1037.

Cattaneo, A., and M. C. Amireault. 1992. How artificial are artificial substrata for periphyton? Journal of the North American Benthological Society 11:244–256.

Cattaneo, A., and R. Carignan. 1983. A colorimetric method for measuring the surface area of aquatic plants. Aquatic Botany 17:291–294.

Cattaneo, A. 1978. The microdistribution of epiphytes on the leaves of natural and artificial macrophytes. British Phycological Journal 13:183–188.

Clark, J. R., K. L. Dickson, and J. Cairns, Jr. 1979. Estimating aufwuchs biomass. Pages 116–141 *in* R. L. Weitzel (ed.), Methods and measurements of periphyton communities. American Society for Testing and Materials, Philadelphia.

Cronk, J. K., and W. J. Mitsch. 1994. Periphyton productivity on artificial and natural surfaces in constructed freshwater wetlands under different hydrologic regimes. Aquatic Botany 48:325–341.

Danos, S. C., J. S. Maki, and C. C. Remsen. 1983. Stratification of microorganisms and nutrients in the surface microlayer of small freshwater ponds. Hydrobiologia 98:193–202.

de Jonge, V. N. 1979. Quantitative separation of benthic diatoms from sediments using density gradient centrifugation in the colloidal silica Ludox™. Marine Biology 51: 267–278.

Delbecque, E. J. P. 1985. Periphyton on nymphaeids: an evaluation of methods and separation techniques. Hydrobiologia 124:85–93.

Dodds, W. K. 1992. A modified fiber-optic light microprobe to measure spherically integrated photosynthetic photon flux density: characterization of periphyton photosynthesis–irradiance patterns. Limnology and Oceanography 37:871–878.

Downing, J. A. 1986. A regression technique for the estimation of epiphytic invertebrate populations. Freshwater Biology 16:161–173.

Droop, M. R. 1973. Some thoughts on nutrient limitation in algae. Journal of Phycology 9: 264–272.

Eaton, J. W., and B. Moss. 1966. The estimation of numbers and pigment content in epipelic algal populations. Limnology and Oceanography 11:584–595.

Eloranta, P. 1986. Paper chromatography as a method of phytoplankton community analysis. Annales Botanici Fennici 23:153–159.

Fairchild, G. W., R. L. Lowe, and W. B. Richardson. 1985. Algal periphyton growth on nutrient-diffusing substrates: an in situ bioassay. Ecology 66:465–472.

Gerbeaux, P., and R. L. Lowe. 2000. Diatom community patterns in wetlands of the South Island west coast (New Zealand). Pages 290–291 *in* Millennium Wetland Event Abstract, Québec, Canada.

Glew, J. R. 1991. Miniature gravity corer for recovering short sediment cores. Journal of Paleolimnology 5:285–287.

Goldsborough, L. G. 1989. Examination of two dimensional spatial pattern of periphytic diatoms using an adhesive surficial peel technique. Journal of Phycology 25:133–143.

Goldsborough, L. G. 1993. Diatom ecology in the phyllosphere of the common duckweed (*Lemna minor* L.). Hydrobiologia 269/270:463–471.

Goldsborough, L. G. 1994. Heterogeneous spatial distribution of periphytic diatoms on vertical artificial substrata. Journal of the North American Benthological Society 13: 223–236.

Goldsborough, L. G., and M. Hickman. 1991. A comparison of periphytic algal biomass and community structure on *Scirpus validus* and on a morphologically similar artificial substratum. Journal of Phycology 27:196–206.

Goldsborough, L. G., and G. G. C. Robinson. 1996. Patterns in wetlands. Pages 77–117 *in* R. J. Stevenson, M. L. Bothwell, and R. L. Lowe (eds.), Algal ecology: freshwater benthic ecosystems. Academic Press, San Diego, CA.

Gough, S. B., and W. J. Woelkerling. 1976. On the removal and quantification of algal aufwuchs from macrophyte hosts. Hydrobiologia 48:203–207.

Goulder, R. 1990. Extracellular enzyme activities associated with epiphytic microbiota on submerged stems of the reed *Phragmites australis*. FEMS Microbiology Ecology 73:323–330.

Grant, J., E. L. Mills, and C. M. Hopper. 1986. A chlorophyll budget of the sediment–water interface and the effect of stabilizing biofilms on particle fluxes. Ophelia 26:207–219.

Gurney, S. E., and G. G. C. Robinson. 1988. The influence of water level manipulation on metaphyton production in a temperate freshwater marsh. Proceedings of the International Association of Theoretical and Applied Limnology 23:1032–1040.

Hann, B. J. 1991. Invertebrate grazer–periphyton interactions in a eutrophic marsh pond. Freshwater Biology 26:87–96.

Hanson, M. A., and M. G. Butler. 1994. Responses to food web manipulation in a shallow waterfowl lake. Hydrobiologia 279/280:457–466.

Hansson, L.-A. 1988. Chlorophyll *a* determination of periphyton on sediments: identification of problems and recommendation of method. Freshwater Biology 20:347–352.

Harvey, G. W., and L. A. Burzell. 1972. A simple microlayer method for small samples. Limnology and Oceanography 17:156–157.

Havens, K. E., T. L. East, S.-J. Hwang, A. J. Rodusky, B. Sharfstein, and A. D. Steinman. 1999a. Algal responses to experimental nutrient addition in the littoral community of a subtropical lake. Freshwater Biology 42:329–344.

Havens, K. E., A. D. Steinman, H. J. Carrick, J. W. Louda, N. M. Winfree, and E. W. Baker. 1999b. Comparative analysis of lake periphyton communities using high performance liquid chromatography (HPLC) and light microscope counts. Aquatic Sciences 61:307–322.

Healey, F. P. 1975. Physiological indicators of nutrient deficiency in algae. Fisheries Marine Service Research Development Technical Report 585. Canada Department of Fisheries and Oceans, Ottawa, Ontario, Canada.

Healey, F. P., and L. L. Hendzel. 1980. Physiological indicators of nutrient deficiency in lake phytoplankton. Canadian Journal of Fisheries and Aquatic Sciences 37:442–453.

Hejný, S., and S. Segal. 1998. General ecology of wetlands. Pages 1–77 *in* D. F. Westlake, J. Kv́et and A. Szczepański, (eds.), The production ecology of wetlands. Cambridge University Press, Cambridge.

Hellebust, J. A., and J. S. Craigie (eds.). 1978. Handbook of phycological methods, Vol. 2, Physiological and biochemical methods. Cambridge University Press, New York.

Higashi, M., T. Miura, K. Tanimizu, and Y. Iwasa. 1981. Effect of the feeding activity of snails on the biomass and productivity of an algal community attached to a reed stem. Proceedings of the International Association of Theoretical and Applied Limnology 21:590–595.

Hillbricht-Ilkowska, A., I. Jasser, and I. Kostrzewska-Szlakowska. 1997. Air–water interface: dynamics of nutrients and picoplankton in the surface microlayer of a humic lake. Proceedings of the International Association of Theoretical and Applied Limnology 26:319–322.

Hillebrand, H. 1983. Development and dynamics of floating clusters of filamentous algae. Pages 31–39 *in* R. G. Wetzel (ed.), Periphyton of freshwater ecosystems. Dr. W. Junk Publishers, The Hague, The Netherlands.

Hooper, N. M., and G. G. C. Robinson. 1976. Primary production of epiphytic algae in a marsh pond. Canadian Journal of Botany 54:2810–2815.

Hooper-Reid, N. M., and G. G. C. Robinson. 1978a. Seasonal dynamics of epiphytic algal growth in a marsh pond: productivity, standing crop, and community composition. Canadian Journal of Botany 56:2434–2440.

Hooper-Reid, N. M., and G. G. C. Robinson. 1978b. Seasonal dynamics of epiphytic algal growth in a marsh pond: composition, metabolism, and nutrient availability. Canadian Journal of Botany 56:2441–2448.

Hosseini, S. M. 1986. The effects of water level fluctuations on algal communities of freshwater marshes. Ph.D. dissertation, Iowa State University, Ames, IA.

Hosseini, S. M., and A. G. van der Valk. 1989. The impact of prolonged, above-normal flooding on metaphyton in a freshwater marsh. Pages 317–324 *in* R. R. Sharitz and J. W. Gibbons (eds.), Freshwater wetlands and wildlife. DOE Symposium Series 61, U.S. Department of Energy, Office of Scientific and Technical Information, Oak Ridge, TN.

Hwang, S.-J., K. E. Havens, and A. D. Steinman. 1998. Phosphorus kinetics of planktonic and benthic assemblages in a shallow subtropical lake. Freshwater Biology 40:729–745.

Jansson, M. 1980. Role of benthic algae in transport of nitrogen from sediment to lake water in a shallow clearwater lake. Archiv fuer Hydrobiologie 89:101-109.

John, J. 1993. The use of diatoms in monitoring the development of created wetlands at a sand-mining site in Western Australia. Hydrobiologia 269/270:427–436.

Jones, K. 1974. Nitrogen fixation in a salt marsh. Journal of Ecology 62:553–565.

Jones, R. C. 1984. Application of a primary production model to epiphytic algae in a shallow, eutrophic lake. Ecology 65:1895–1903.

Jorgensen, B. B., Y. Cohen, and N. P. Revsbech. 1988. Photosynthetic potential and light-dependent oxygen consumption in a benthic cyanobacterial mat. Applied and Environmental Microbiology 54:176 182.

Kasanof, N. Van M. 1973. Ecology of the micro-algae of the Florida Everglades: 1. Environment and some aspects of freshwater periphyton, 1959 to 1963. Nova Hedwigia 24:619–664.

Kassim, T. I., and H. A. Al-Saadi. 1994. On the seasonal variation of the epipelic algae in marsh areas (southern Iraq). Acta Hydrobiologia 36:191–200.

Keough, J. R., C. A. Hagley, E. Ruzycki, and M. Sierszen. 1998. ^{13}C composition of primary producers and role of detritus in a freshwater coastal ecosystem. Limnology and Oceanography 43:734–740.

King, A. J., A. I. Robertson, and M. R. Healey. 1997. Experimental manipulations of the biomass of introduced carp (*Cyprinus carpio*) in billabongs. I. Impacts on water-column properties. Marine and Freshwater Research 48:435–443.

Klarer, D. M., and D. F. Millie. 1994. Regulation of phytoplankton dynamics in a Laurentian Great Lakes estuary. Hydrobiologia 286:97–108.

LaBaugh, J. W. 1995. Relation of algal biovolume to chlorophyll *a* in selected lakes and wetlands in the north-central United States. Canadian Journal of Fisheries and Aquatic Sciences 52:416–424.

Lajtha, K., and R. H. Michener. 1994. Stable isotopes in ecology and environmental science. Blackwell, Oxford.

Leavitt, P. R. 1993. A review of factors that regulate carotenoid and chlorophyll deposition and fossil pigment abundance. Journal of Paleolimnology 9:109–127.

Loeb, S. L. 1981. An in situ method for measuring the primary productivity and standing crop of the epilithic periphyton community in lentic systems. Limnology and Oceanography 26:394–399.

Margalef, R. 1949. A new limnological method for the investigation of thin-layered epilithic communities. Hydrobiologia 1:215–216.

Marker, A. F. H., C. A. Crowther, and R. J. M. Gunn. 1980. Methanol and acetone solvents for estimating chlorophyll *a* and phaeopigments by spectrophotometry. Archiv fuer Hydrobiologie Beihefte 14:52–69.

Mason, C. F., and R. J. Bryant. 1975. Periphyton production and grazing by chironomids in Alderfen Broad, Norfolk. Freshwater Biology 5:271–277.

Mayer, P. M., and S. M. Galatowitsch. 2000. Diatom communities as ecological indicators of recovery in restored prairie wetlands. Wetlands 19:765–774.

McCormick, P. V., P. S. Rawlik, K. Lurding, E. P. Smith, and F. H. Sklar. 1996. Periphyton–water quality relationships along a nutrient gradient in the northern Florida Everglades. Journal of the North American Benthological Society 15:433–449.

McDougal, R. L., L. G. Goldsborough, and B. J. Hann. 1997. Responses of a prairie wetland to press and pulse additions of inorganic nitrogen and phosphorus: production by planktonic and benthic algae. Archiv fuer Hydrobiologie 140:145–167.

Merks, A. G. A. 1979. A microgradient sampler for shallow waters. Hydrobiological Bulletin 13:61–67.

Millie, D. F., H. W. Paerl, and J. P. Hurley. 1993. Microalgal pigment assessments using high-performance liquid chromatography: a synopsis of organismal and ecological application. Canadian Journal of Fisheries and Aquatic Sciences 50:2513–2527.

Mitsch, W. J., and J. G. Gosselink. 2000. Wetlands, 3rd ed. Wiley, New York.

Morin, A., and A. Cattaneo. 1992. Factors affecting sampling variability of freshwater periphyton and the power of periphyton studies. Canadian Journal of Fisheries and Aquatic Sciences 49:1695–1703.

Murkin, H. R., M. P. Stainton, J. A. Boughen, J. B. Pollard, and R. D. Titman. 1991. Nutrient status in the Interlake region of Manitoba, Canada. Wetlands 11:105–122.

Murkin, H. R., J. B. Pollard, M. P. Stainton, J. A. Boughen, and R. D. Titman. 1994. Nutrient additions to wetlands in the Interlake region of Manitoba, Canada: effects of periodic additions throughout the growing season. Hydrobiologia 279/280:483–495.

Neely, R. K. 1994. Evidence for positive interactions between epiphytic algae and heterotrophic decomposers during the decomposition of *Typha latifolia*. Archiv fuer Hydrobiologie 129:443–457.

Neill, C., and J. C. Cornwell. 1992. Stable carbon, nitrogen, and sulfur isotopes in a prairie marsh food web. Wetlands 12:217–224.

Owen, B. B., M. Afzal, and W. R. Cody. 1978. Staining preparations for phytoplankton and periphyton. British Phycological Journal 13:155–160.

Paerl, H. W., M. M. Tilzer, and C. R. Goldman. 1976. Chlorophyll *a* versus adenosine triphosphate as algal biomass indicators in lakes. Journal of Phycology 12:242–246.

Pan, Y., and R. J. Stevenson. 1996. Gradient analysis of diatom communities in western Kentucky wetlands. Journal of Phycology 32:222–232.

Pertoft, H., T. C. Laurent, T. Laas, and L. Kagedal. 1978. Density gradients prepared from colloidal silica particles coated by polyvinylpyrrolidone (Percoll). Analytical Biochemistry 88:271–282.

Phillips, G. L., D. Eminson, and B. Moss. 1978. A mechanism to account for macrophyte decline in progressively eutrophicated freshwaters. Aquatic Botany 4:103–126.

Pip, E., and G. G. C. Robinson. 1982. A study of the seasonal dynamics of three phycoperiphytic communities using nuclear track autoradiography. II. Organic carbon uptake. Archiv fuer Hydrobiologie 96:47–64.

Pip, E., and J. M. Stewart. 1976. The dynamics of aquatic plant–snail associations. Canadian Journal of Zoology 54:1192–1205.

Reeder, B. C. 1994. Estimating the role of autotrophs in nonpoint source phosphorus retention in a Laurentian Great Lakes coastal wetland. Ecological Engineering 3:161–169.

Reynolds, C. S. 1984. The ecology of freshwater phytoplankton. Cambridge University Press, New York.

Ridley-Thomas, C. I., A. Austin, W. P. Lucey, and M. J. R. Clark. 1989. Variability in the determination of ash free dry weight for periphyton communities: a call for a standard method. Water Research 23:667–670.

Robarts, R. D., D. B. Donald, and M. T. Arts. 1995. Phytoplankton primary production of three temporary northern prairie wetlands. Canadian Journal of Fisheries and Aquatic Sciences 52:897–902.

Robertson, A. I., M. R. Healey, and A. J. King. 1997. Experimental manipulations of the biomass of introduced carp (*Cyprinus carpio*) in billabongs. II. Impacts on benthic properties and processes. Marine and Freshwater Research 48:445–454.

Robinson, G. G. C., and E. Pip. 1983. The application of a nuclear track autoradiographic technique to the study of periphyton photosynthesis. Pages 267–273 *in* R. G. Wetzel (ed.), Periphyton of freshwater ecosystems. Dr. W. Junk Publishers, The Hague, The Netherlands.

Robinson, G. G. C., S. E. Gurney, and L. G. Goldsborough. 1997a. Response of benthic and planktonic algal biomass to water level manipulation in a prairie wetland. Wetlands 17:167–181.

Robinson, G. G. C., S. E. Gurney, and L. G. Goldsborough. 1997b. The primary productivity of benthic and planktonic algae in a prairie wetland under controlled water-level regimes. Wetlands 17:182–194.

Robinson, G. G. C., S. E. Gurney, and L. G. Goldsborough. 2000. Algae in prairie wetlands. Pages 163–199 *in* H. Murkin, J. Kadlec, and A. van der Valk (eds.), Prairie wetland ecology: the contribution of the marsh ecology research program. Iowa State University Press, Ames, IA.

Roos, P. J., A. F. Post, and J. M. Reiver. 1981. Dynamics and architecture of reed periphyton. Proceedings of the International Association of Theoretical and Applied Limnology 21:948–953.

Round, F. E. 1981. The ecology of algae. Cambridge University Press, Cambridge.

Sand-Jensen, K. 1990. Epiphyte shading: its role in resulting depth distribution of submerged aquatic macrophytes. Folla Geobotanica et Phytotaxonomica 25:315–320.

Scheffer, M. 1998. Ecology of shallow lakes. Chapman & Hall, London.

Schoenberg, S. A., and J. D. Oliver. 1988. Temporal dynamics and spatial variation of algae in relation to hydrology and sediment characteristics in the Okefenokee Swamp, Georgia. Hydrobiologia 162:123–133.

Scholz, O., and P. I. Boon. 1993. Alkaline phosphatase, aminopeptidase and ß-D glucosidase activities associated with billabong periphyton. Archiv fuer Hydrobiologie 126:429–443.

Shamess, J. J., G. G. C. Robinson, and L. G. Goldsborough. 1985. The structure and comparison of periphytic and planktonic algal communities in two eutrophic prairie lakes. Archiv fuer Hydrobiologie 103:99–116.

Sládečková, A. 1962. Limnological investigation methods for the periphyton ("Aufwuchs") community. Botanical Review 28:287–350.

Slate, J. E., and R. J. Stevenson. 2000. Recent and abrupt environmental change in the Florida Everglades indicated from siliceous microfossils. Wetlands 20:346–356.

Stanley, D. W. 1976. Productivity of epipelic algae in tundra ponds and a lake near Barrow, Alaska. Ecology 57:1015–1024.

Steinman, A. D., R. H. Meeker, A. J. Rodusky, W. P. Davis, and C. D. McIntire. 1997. Spatial and temporal distribution of algal biomass in a large, subtropical lake. Archiv fuer Hydrobiologie 139:29–50.

Stephens, D. W. 1987. Extraction of periphyton adenosine triphosphate and variability in periphyton-biomass estimates. Archiv fuer Hydrobiologie 108:325-335.

Stevenson, R. J. 1996. An introduction to algal ecology in freshwater benthic habitats. Pages 3–30 *in* R. J. Stevenson, M. L. Bothwell, and R. L. Lowe (eds.), Algal ecology: freshwater benthic ecosystems. Academic Press, San Diego.

Stevenson, R. J., and L. L. Bahls. 1999. Periphyton protocols. Pages 6.1–6.22 *in* M. T. Barbour, J. Gerritsen, B. D. Snyder, and J. B. Stribling (eds.), Rapid bioassessment protocols for use in streams and wadeable rivers: periphyton, benthic macroinvertebrates and fish, 2nd ed. Report EPA 841-B-99-002. U.S. Environmental Protection Agency, Office of Water, Washington, DC.

Stevenson, R. J., and C. G. Peterson. 1989. Variation in benthic diatom (Bacillariophyceae) immigration with habitat characteristics and cell morphology. Journal of Phycology 25:120–129.

Sullivan, M. J., and C. A. Moncreiff. 1990. Edaphic algae are an important component of salt marsh food-webs: evidence from multiple stable isotope analyses. Marine Ecology Progress Series 62:149–159.

Swanson, R. B. 1995. Methods for monitoring physical, chemical, and biological characteristics at selected wetlands in the Platte River Basin, Nebraska. Pages 515–524 *in* K. L. Campbell (ed.), Versatility of wetlands in the agricultural landscape. American Society of Agricultural Engineers, St. Joseph, MI.

Sze, P. 1993. A biology of the algae, 2nd ed. Wm. C. Brown, Dubuque, IA.

Tuchman, N. C. 1996. The role of heterotrophy in algae. Pages 299–319 *in* R. J. Stevenson, M. L. Bothwell, and R. L. Lowe (eds.), Algal ecology: freshwater benthic ecosystems. Academic Press, San Diego, CA.

Turner, M. A., E. T. Howell, M. Summerby, R. H. Hesslein, D. L. Findlay, and M. B. Jackson. 1991. Changes in epilithon and epiphyton associated with experimental acidification of a lake to pH 5. Limnology and Oceanography 36:1390–1405.

Underwood, G. J. C., and J. D. Thomas. 1990. Grazing interactions between pulmonate snails and epiphytic algae and bacteria. Freshwater Biology 23:505–522.

Vymazal, J., C. B. Craft, and C. J. Richardson. 1994. Periphyton response to nitrogen and phosphorus additions in Florida Everglades. Algological Studies 73:75–97.

Wasmund, N. 1989. Micro-autoradiographic determination of the viability of algae inhabiting deep sediment layers. Estuarine, Coastal and Shelf Science 28:651–656.

Watala, K. B., and C. Watala. 1994. A photometric technique for the measurement of plant surface area: the adsorption of Brilliant Blue dye on to plant surfaces. Freshwater Biology 31:175–181.

Weitzel, R. L. (ed.). 1979. Methods and measurements of periphyton communities. American Society for Testing and Materials, Philadelphia. STP 690.

Westlake, D. F., J. Kv́et and A. Szczepański (eds.). 1998. The production ecology of wetlands. Cambridge University Press, Cambridge.

Wetzel, R. G., and G. E. Likens. 1991. Limnological analyses, 2nd ed. Springer-Verlag, New York.

Woodruff, S. L., W. A. House, M. E. Callow, and B. S. C. Leadbeater. 1999. The effects of biofilms on chemical processes in surficial sediments. Freshwater Biology 41:73–89.

Wu, X., and W. J. Mitsch. 1998. Spatial and temporal patterns of algae in newly constructed freshwater wetlands. Wetlands 18:9–20.

Ytow, N., H. Seki, T. Mizusaki, S. Kojima, and A. E. Batomalaque. 1994. Trophodynamic structure of a swampy bog at the climax stage of limnological succession. I. Primary production dynamics. Water, Air, and Soil Pollution 76:467–479.

14 Sampling Macrophytes in Wetlands

CURTIS J. RICHARDSON and JAN VYMAZAL

The use of macrophytes as biomonitors in wetland ecosystems are presented in terms of assessment of plant population and community responses to disturbance or anthropogenic inputs. The life forms of macrophytes are reviewed and the sampling procedures for estimating changes in population size as well as community structure are presented for herbaceous plants, shrubs, and trees. The methods and formulas for determining abundance and cover as well as frequency, density and dominance are given for both plot and plot less sampling techniques. Additional sampling methods are assessed for plant biomass and productivity measurements. We also outline procedures for establishing macrophyte growth rates and nutrient status in wetland plants. Procedures for determining both above- and belowground biomass, nitrogen and phosphorus as well as ash-free dry matter are presented along with representative data for typical wetland species. In this chapter we provide a comprehensive plan for sampling and monitoring of plant populations and communities in wetlands.

Plants have often been used as a biomonitoring tool in terrestrial ecotoxicological studies but not in wetland assessments (Wang et al. 1990, Lewis et al. 1999). In this chapter we outline a wetlands biomonitoring program using macrophytes. Our approach first requires a decision on whether the objectives of the study are monitoring at a landscape scale or site-specific testing. Next, it is important to recognize that macrophyte biomonitoring can occur at the biochemical, physiological, organism, population, community, or ecosystem levels, and it can be focused on either structural changes (e.g., plant cover, density) or functional aspects (e.g., productivity). In this chapter, we focus on population, community, and ecosystem level changes at both the structural and functional levels. Assessment of biochemical and physiological plant responses as well as landscape-level changes are outside the scope of this chapter. Moreover, few examples of plant responses at these levels exist in the literature for wetland macrophytes.

Bioassessment and Management of North American Freshwater Wetlands, edited by Russell B. Rader, Darold P. Batzer, and Scott A. Wissinger.
0-471-35234-9

In a comparison of algal and vascular plant responses to chemical toxicity, Fletcher (1990) suggests that 20% of the time, algal screening tests may not detect a response unique to the growth and development of vascular plants. He also concluded that no single plant type (algal, monocotyledon, or dicotyledon) can be viewed as the least or most sensitive test system in biomonitoring. This suggests a great need to include macrophytes in any biomonitoring assessment plan. Three general categories of macrophyte response to stressors in the environment are given in Table 14.1, along with a cross-listing of which metabolic process are common to both algae and macrophytes. Although assessment of all these features is beyond the scope of this chapter, it is important to recognize their usefulness in vascular plant studies, especially for responses to xenobiotic inputs into wetlands. An excellent review of some of these responses is given in Wang et al. (1990).

Our objectives with this chapter are to (1) provide guidelines to determine the structural characteristics of plant populations and communities so that they can be monitored to determine change over environmental gradients as well as assess the effects of anthropogenic inputs into wetlands; (2) outline procedures to assess macrophyte growth responses and nutrient status to determine and compare the status of disturbed, created, or restored wetlands; and (3) provide some ranges of plant community structure, plant biomass, and nutrient responses that are typically found in wetlands.

INTRODUCTION TO WETLAND PLANTS

Life Forms of Wetland Macrophytes

Numerous lines of evidence indicate that aquatic angiosperms originated on the land. Adaptation and specialization to the aquatic habitat have been achieved by only a few angiosperms (<1%) and pteridophytes (<2%). Consequently, the richness of plant species in aquatic and wetland habitats is relatively low compared with most terrestrial communities. Most are rooted, but a few species float freely in the water (Wetzel 1983). A brief description of the morphological adaptations of macrophytes to wetlands is important when considering plant-sampling techniques for biomonitoring.

The primary groups of rooted and nonrooted aquatic angiosperms have been subdivided according to foliage type, inflorescence type, and whether the leaves and flowering organs are emergent, floating on the water surface, or submersed (Arber 1920). The most frequently used simple classification is based on attachment, which has also proven useful in morphological, physiological, and ecological studies (Arber 1920, Sculthorpe 1967): (1) emergent macrophytes, (2) floating-leafed macrophytes both (a) rooted and (b) freely floating, and (3) submerged macrophytes. We have added to this classification (4) woody plants, which includes both shrubs and trees.

Emergent Plants. Emergent macrophytes are the dominating life form in wetlands and marshes, growing within a water table range from 0.5 m below the soil surface to a water depth of 1.5 m or more (Fig. 14.1*a*–*c*). In general, they produce aerial

TABLE 14.1. Three Categories of Plant Response Used in Biomonitoring

(I) Unique Features[a]		(II) Common Features[b]		(III) Growth Parameters	
Symptom	Records	Symptom	Records	Symptom	Records
Abscission	869[c]	DNA synthesis	105	Cover range	267
Cold hardiness	38	Ion uptake	86	Dry weight[d]	2,965
Stem curvature	530	Photosynthesis	448	Fresh weight[d]	3,526
Deformed organs	704	Protein synthesis	340	Harvest yield	2,055
Desiccation	78	Respiration	721	Kill	5,292
Dormancy	266	RNA synthesis	139	Plant number	5,590
Floral development	188			Plant or organ size[d]	9,062
Seed germination	4,316				
Maturation	535				
Nastic movements	160				
Parthenocarpy	212				
Secondary metabolism	200				
Seed development	115				
Seed number change	223				
Senescence	67				
Flower sex change	130				
Sterility	81				
Stomatal closure	155				
Transpiration	508				
Vascular disruption	31				
Wilt	154				
Total symptom records	11,879		1,839		28,757

Source: Adapted from Fletcher (1990).

[a]Physiological features that are characteristic of vascular plants but not algae.

[b]Metabolic processes that are common to both vascular plants and algae.

[c]Each record in Phytotox pertains to the effect of a single dose of one chemical applied to one plant species.

[d]The numbers of records shown for changes in dry weight, fresh weight, and plant size include data pertaining to plant organs as well as for whole plants. For example, root elongation records are included among the 9062 records dealing with changes in plant size.

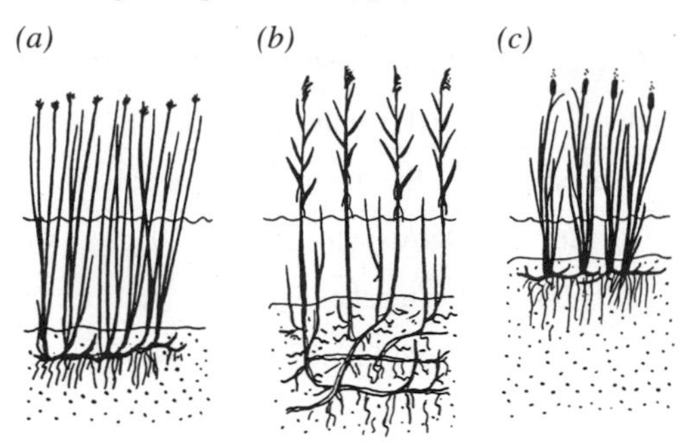

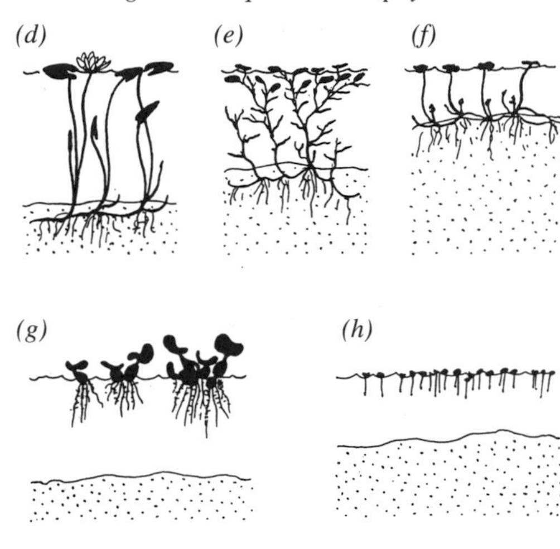

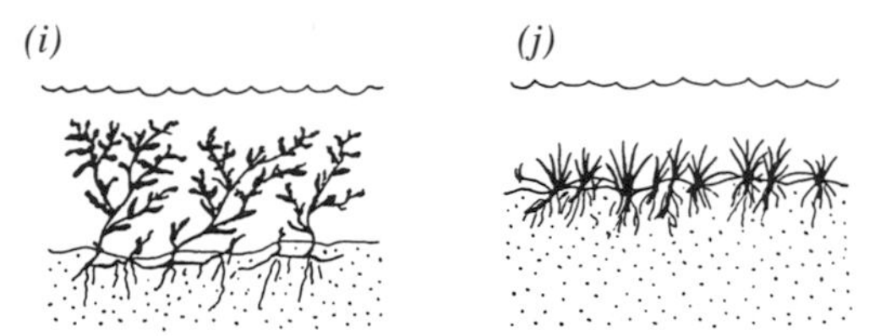

FIGURE 14.1. Life forms of aquatic and wetland macrophytes. I. Emergent macrophytes: *(a) Scirpus lacustris* (bulrush), *(b) Phragmites australis* (common reed), *(c) Typha latifolia* (common cattail). II. Floating-leaved macrophytes: *(d) Nymphaea alba* (waterlily), *(e) Potamogeton gramineus* (variable pondweed)), *(f) Hydrocotyle vulgaris* (pennywort), *(g) Eichhornia crassipes* (water hyacinth), *(h) Lemna minor* (common duckweed). III. Submerged macrophytes: *(i) Potamogeton crispus* (curly pondweed), *(j) Littorella uniflora* (shoreweed). (From Brix and Schierup 1989. Reprinted with permission.)

stems and leaves and an extensive root and rhizome system (Brix and Schierup 1989). These plants are morphologically adapted to growing in a waterlogged or submersed substrate by virtue of large internal air spaces for transportation of oxygen to roots and rhizomes. Part of the oxygen may leak into the surrounding rhizosphere, creating oxidized conditions in an otherwise anoxic environment and stimulating both decomposition of organic matter and growth of nitrifying bacteria (Brix and Schierup 1989). Aerial stems and leaves of emergent macrophytes possess many similarities in both morphology and physiology to related terrestrial plants. Emergent monocotyledons such as *Phragmites* and *Typha* produce erect, approximately linear leaves from an extensive anchoring system of rhizomes. Epidermal cells are elongated parallel to the long axis of the leaf, which allows flexibility for bending. The cell walls are heavily thickened with cellulose, which provides the necessary rigidity. Emergent dicots produce erect, leafy stems, which show greater anatomical differentiation (Wetzel 1983).

Emergent macrophytes assimilate nutrients from sediments. Emergent macrophytes also act as nutrient pumps and play a key role in seasonal changes in available N, P, and K. The net effect of rooted emergent vegetation is to transfer nutrients from the soil to the surface water via leaching and litterfall, especially at the end of the growing season (e.g., Klopatek 1975, Richardson et al. 1978, Richardson and Marshall 1986).

Examples of emergent macrophytes: *Phragmites australis* (Common reed), *Phalaris arundinacea* (reed canary grass), *Typha* spp. (cattails), Scirpus spp. (bulrushes), *Carex* spp. (sedges), *Eleocharis* spp. (spikureshes), *Juncus* spp. (rushes), *Sagittaria* spp. (arrowheads), *Spartina* spp. (cordgrasses), *Alternanthera philoxeroides* (aligatorweed), *Cladium jamaicense* (sawgrass), *Glyceria maxima* (sweet mannagrass), *Panicum hemitomon* (maidencane), *Pontederia cordata* (pickerelweed), *Zizania aquatica* (wild rice).

Floating-Leafed Macrophytes. The floating-leafed macrophytes include species that are rooted in the substrate and species that are freely floating on the water surface.

Rooted Floating-Leafed Macrophytes. The rooted floating-leafed macrophytes are primarily angiosperms that occur attached to submersed sediments at water depths from about 0.5 to 3.0 m (Fig. 14.1*d–f*). In heterophyllous species, submersed leaves precede or accompany the floating leaves. Reproductive organs are floating or aerial and floating leaves are on long, flexible petioles (e.g., *Nuphar* or *Nymphaea*), or on short petioles from long ascending stems (e.g., *Brasenia, Potamogeton natans*) (Wetzel 1983).

The surface of the water is a habitat subject to severe mechanical stresses from wind and water movements. Adaptations to these stresses by floating-leafed macrophytes include the tendency toward peltate leaves that are strong, leathery, and circular in shape with an entire margin. The leaves usually have hydrophobic surfaces and long, pliable petioles. Despite these adaptations, strong winds and water movements restrict these macrophytes to relatively sheltered habitats in which there is little water movement (Wetzel 1983).

Positioning of leaf surfaces parallel to the water surface creates a vigorous competition for space to expose maximum leaf area to incident light. Leaf growth some distance from the root or rhizome system is accommodated by long, very pliable petioles. A uniform proportionality is found between water depth ranging between 10 cm and 4 m and the length of the leaf petioles of waterlilies. Petioles are about 20 cm longer than the water depth, which permits leaves to remain on the surface among undulating waves (Wetzel 1983).

Examples of floating-leafed rooted macrophytes: *Brasenia schreberi* (watershields), *Hydrocotyle* spp. (pennyworts), *Limnobium spongia* (frog's-bit), *Nelumbo* spp. (lotuses), *Nuphar* spp. (spatterdocks, cowlilies), *Nymphaea* spp. (waterlilies), *Nymphoides* spp. (floatinghearts), *Potamogeton* spp. (pondweeds).

Free Floating-Leafed Macrophytes. Free-floating macrophytes (Fig. 14.1*g, h*) are highly diverse in form and habit, ranging from large plants with rosettes of aerial and/or floating leaves and well-developed submerged roots (e.g., *Eichhornia, Pistia*) to minute surface-floating plants with few or no roots (e.g., *Lemna, Spirodela, Wolffia*) (Brix and Schierup 1989). *Eichhornia* is one of the fastest-growing plants in the world, causing nuisance in the tropics and the southern United States by blocking waterways.

Freely floating macrophytes generally are restricted to sheltered habitats and slow-flowing rivers. Their nutrient absorption is completely from the water, and most of these macrophytes are found in waters rich in dissolved salts (Wetzel 1983). Most free-floating plants possess little lignified tissue. Rigidity and buoyancy of the leaves are maintained by turgor of living cells and extensively developed lacunate mesophyll tissue (often, > 70% gas by volume). All freely floating rosette plants form well-developed adventitious roots, lateral roots, and epidermal hairs. The root system of the water hyacinth, for example, represents 20 to 50% of the plant biomass (Wetzel 1983).

Examples of free-floating macrophytes: *Azolla caroliniana* (mosquito fern), *Eichhornia crassipes* (water hyacinth), *Lemna* spp. (duckweeds), *Pistia stratiotes* (water lettuce), *Salvinia rotundifolia* (water fern), *Spirodela* spp. (giant duckweeds), *Wolffia* spp. (watermeals).

Submerged Macrophytes. These macrophytes have their photosynthetic tissue entirely submerged, but the flowers are usually exposed to the atmosphere (Brix and Schierup 1989). They occur at all depths within the photic zone, but vascular angiosperms occur only to about 10 m (1 atm hydrostatic pressure) (Fig. 14.1*i, j*). Leaf morphology is highly variable, from finely divided to broad (Wetzel 1983). Among the vascular submerged macrophytes, numerous morphological and physiological modifications are found that allow existence in a totally aqueous environment. Stems, petioles, and leaves usually contain little or no lignin, even in vascular tissues. Conditions of reduced illumination under the water are reflected in numerous characteristics: an extremely thin cuticle, leaves only a few cells in thickness, and an increase in number of chloroplasts in epidermal tissue. Leaves tend to be much more divided and reticulated than in terrestrial or other aquatic plants. The vascular system is greatly reduced, and all major conducting vessels have been lost from the stem (Wetzel 1983).

Various experiments have proven that minerals can be taken up directly by shoot tissues of submerged plants. However, there is also no question regarding the uptake capability of nutrients by the roots of these plants. Investigations on quantitative contribution of either the shoots or the roots to the overall nutrition of submerged plants are numerous but not decisive (Vymazal 1995).

Examples of submerged macrophytes: *Cabomba caroliniana* (fanwort), *Ceratophyllum* spp. (coontails), *Elodea* spp. (waterweeds), *Isoetes* spp. (quillworts), *Myriophyllum* spp. (watermilfoils), *Najas* spp. (naiads, water nymphs), *Utricularia* spp. (bladderworts)

Woody Species. Woody species vary from low-growing shrubs to towering *Taxodium* (cypress), *Picea* (spruce), and *Thuja* (cedar). Upper portions of stems and leaves are generally similar to terrestrial forms and they may be deciduous or evergreen. The difference lies in the lower portion of the stem or trunk and in root structures. Most woody wetland plants possess specialized structures—knees, adventitious roots, prop roots, lenticles, and butt swellings—to increase gas exchange between the roots and the atmosphere or to provide support (Hammer 1992). Measurements of aboveground biomass and nutrient uptake have been made for many of the commercial species like *Taxodium* (Schlesinger 1978) but few other species. Studies of belowground biomass or nutrients storage are rare. Recent studies using root rhizotrons have been used as a biomonitor to assess root growth responses under varying stress conditions (Megonigal and Day 1992).

Examples of wetland trees: *Acer rubrum* (red maple), *Nyssa aquatica* (water tupelo), *Nyssa sylvatica* (swamp tupelo, blackgum), *Taxodium distichum* (baldcypress), *Salix nigra* (black willow), *Fraxinus* spp. (ashes), *Liquidambar styraciflua* (sweetgum), *Pinus serotina* (pond pine), *Chamaecyparis thyoides* (Atlantic white cedar), *Thuja occidentalis* (northern white cedar), *Quercus nigra* (water oak), *Avicennia germinans* (black mangrove).

Examples of shrub species: *Ludwigia* spp. (primrose), *Vaccinium macrocarpon* (cranberry), *Ledum groenlandicum* (Labrador tea), *Hibiscus* spp. (mallow), *Myrica cerifera* (wax myrtle).

SAMPLING METHODS FOR PLANT POPULATIONS AND COMMUNITIES

A qualitative description of the wetland (e.g., sedge marsh or forested wetland) or complete flora list of all species present at a site provides only a superficial concept of the structure of a community. How many trees are present per hectare? Which are more abundant, and which are larger? Are the plant species evenly dispersed in the wetland, or are there populations shifts along some nutrient, water, or elevational gradient? What is the frequency, density, or abundance of the dominant species? How do plant densities in a disturbed area compare to those in a natural area? To quantify these measurements, a variety of sampling methods have been developed over the past 100 years (Vollenweider 1969). A complete and through review of vegetation sampling techniques are given in books by Phillips (1959), Kershaw (1964), Mueller-Dombois and Ellenberg (1974), Barbour et al. (1980), and Gauch (1982). Our chapter is intended to highlight some of the key features from these texts and adapt them for biomonitoring wetland surveys of macrophytes. Each of these methods has advantages and disadvantages in terms of accuracy, cost, and sampling time.

The selection of the stand sampling locations and the method used for site selection are of prime importance. In this case, we use the term *stand* interchangeably with *plant community* to represent an assemblage of plants of the same floristic composition. Consideration of both the overstory and understory must be included in forested and shrub areas. In a recent volume, Tiner (1999) suggests that random selection of stands and random sampling within them are inefficient. He indicates that a better approach would be subjective location of stands and random sampling within locations (including regular placement of samples) for research studies. Thus, the most preferable method is a stratified random system in which quadrats or sample points are located at random within a definite stand. A detailed analysis of vegetation sampling methods and survey protocols is covered in the excellent volume by Mueller-Dombois and Ellenberg (1974). Locating samples subjectively in representative plant stands (determined by trained researchers) is typically used for delineation of wetland boundaries and is not recommended for a complete macrophyte survey.

To aid in the location of representative stands, it is recommended that aerial photos or low-level aerial flights be used to determine the number of communities in the study area, especially if the wetland is greater than 100 ha. Next, the purpose of the study determines if detailed vegetation information is need (e.g., density and frequency of all species) or if relative information (e.g., relevé or percentage vegetation cover) can be used. A general listing of the steps taken at both the landscape scale of biomonitoring or site specific assessment is shown in Figure 14.2. Specific information on the procedures used at each step is given in the sections that follow. Once the scale is determined, the researcher must select the size, number, and shape of plots to sample within the study area. Plotless sampling techniques can also be employed to cover more area in a shorter time (Mueller-Dombois and Ellenberg 1974). Most biomonitoring is done at the site-specific level, where key species (e.g., keystone species), population dynamics, or changes in community structure can be compared among sites. Numerous reference or undisturbed areas (e.g., 5 to 10) must be included in all biomonitoring analyses if changes in communities are to be assessed accurately.

Sample Size

Regardless of the method used, it should be of sufficient area to contain almost all (>95%) the species within the plant community. It is recommended that a *minimal area* be determined for the community. This is defined as the smallest area on which the species composition of the community in question is adequately measured. This is usually determined by first applying the relevé technique of the Swiss plant ecolo-

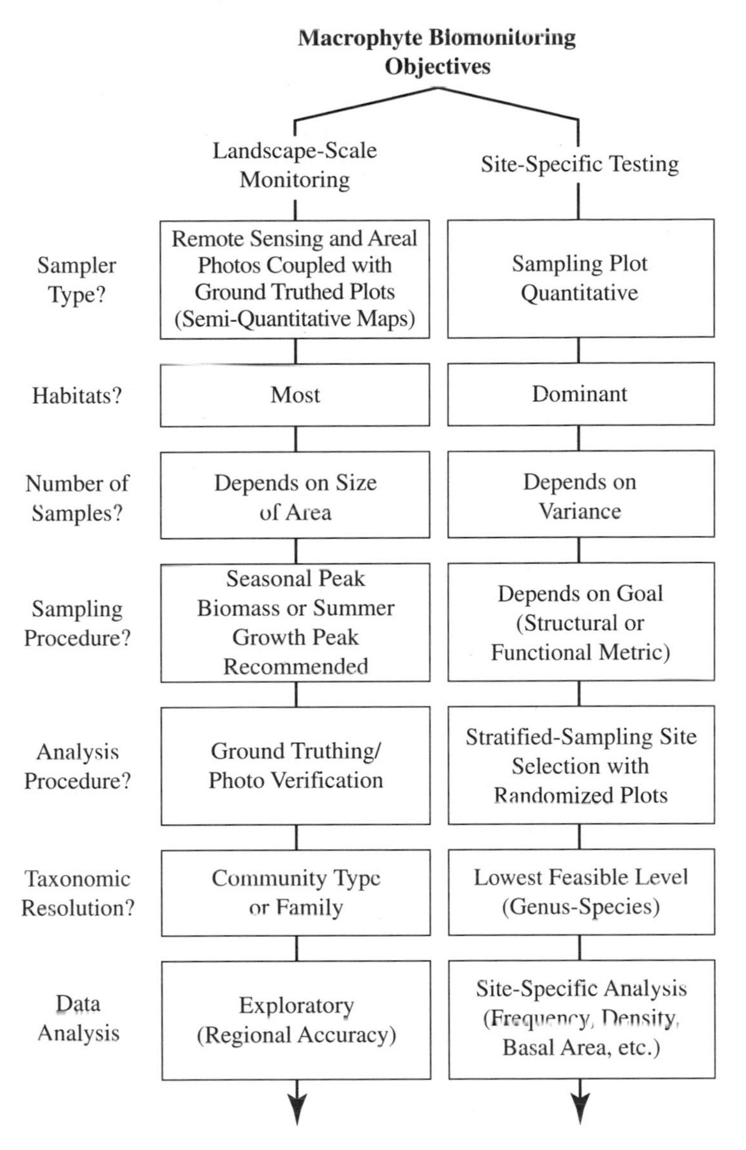

FIGURE 14.2. General guideline for macrophyte sampling in wetland communities.

gist Braun-Blanquet (1965). In this technique a scientist familiar with the local flora makes a list of the species on plots of an ever-increasing size (e.g., 1 m^2, 2 m^2, 4 m^2, etc.) until 90 to 95% of the species observed in the entire community have been observed in one plot (Fig. 14.3*a*). The sampling area should always be a little larger than the minimal area shown on a species–area curve (e.g., 13 to 15 plots near the asymptote on the curve), where the *y*-axis represents the number of species and the *x*-axis is the area or cumulative number of sample plots. It should be noted that if the

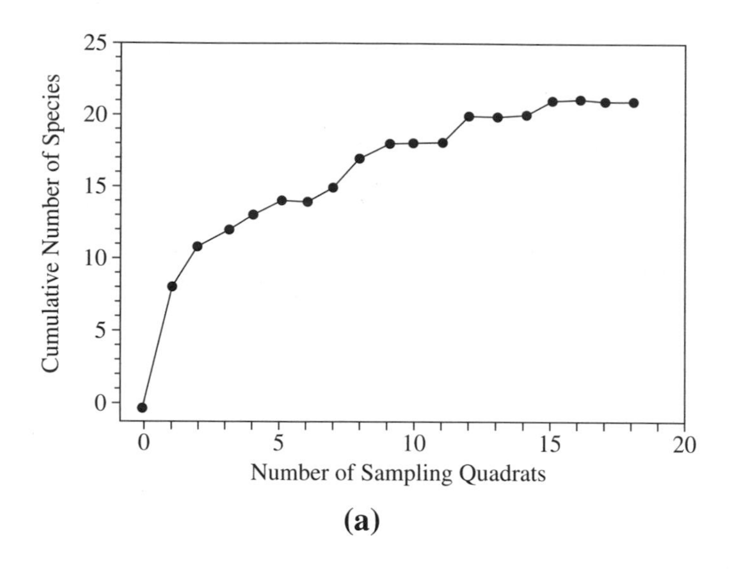

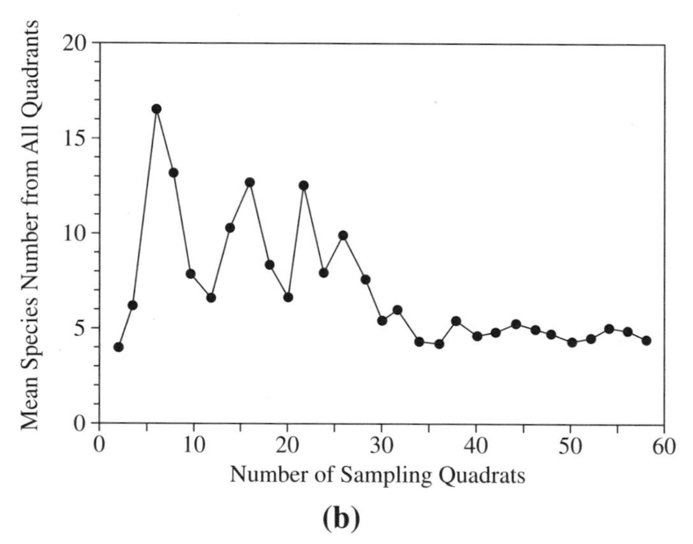

FIGURE 14.3. *(a)* Cumulative number of species in relation to increasing number of sample quadrats in the Everglades (C. J. Richardson, unpublished data); *(b)* Variation in the number of species in relation to an increasing number of quadrats.

sampling emphasis is on determining the quantitative variation of dominant species within large communities or across gradients, it is better to use an increased number of smaller plots spread over each vegetation segment instead of a few large minimal area quadrats or plots (Mueller-Dombois and Ellenberg 1974). The minimal area depends on the community type and is quite variable in size. Early research on this by Cain (1938), Oosting (1956), and Daubenmire (1968) suggest the following minimal area for various plant types: lichen communities (0.1 to 1.0 m^2), moss communities (1.0 to 4.0 m^2), grass meadows (10.0 to 25.0 m^2), dwarf-shrub heath (10.0 to 25.0 m^2), forest understory (50 to 200 m^2), and forest overstory (200 to 500 m^2). These total plot sizes give a general rule for the area required for each community type, but for an accurate estimate of the number of sampling plots or measurements required, it is suggested that a minimum of five quadrats be taken and a mean value for species number or other variables be calculated (Kershaw 1964). The sample size is then increased to 10, 15, or 20 and a new mean is calculated each time. A plot of the value of the mean versus sample size gives an estimate of the reduced variance. For example, data for plant species numbers fluctuate up to about 30 quadrats, and a plot number of about 40 would provide an accurate estimate of the true number of plants per quadrat (Fig. 14.3*b*). Generally, one should use the smallest plot sizes possible with the size of the vegetation dictating the size of the individual plots.

Quadrats or sample plots are historically the oldest method for obtaining quantitative data on vegetation but are sometimes hard to establish in dense vegetation. Square plots are the least efficient and most difficult to establish since right angles must be established. The usual size of forest square quadrats is 10×10 m while the 1×1 m or 0.5×0.5 m quadrat is commonly used in sedge or grass-type vegetation. The size of sampling plots for low shrubs and tall herbs is 4 m^2, tall shrubs 16 m^2, and a low herb layer 1 to 2 m^2 (Mueller-Dombois and Ellenberg 1974). Circular and strip quadrats are often used because they are easy to establish. Circular quadrats have the advantage of minimal edge effect (5.65 m in radius = 100 m^2 = 0.01 ha) and can easily be established from a central stake using a tape or marked line to circumscribe the area. The total area of all samples is related to the number of times a quadrat should be repeated or the given density of plots. This is often decided arbitrarily, but in forested timber volume surveys it is often common practice to set a percent sampling intensity of 5 to 10% (Mueller-Dombois and Ellenberg 1974). For example, a 6-ha area sampled at 5% intensity would result in an area of $0.05 \times 60{,}000$ m^2 = 3000 m^2. This could be done with thirty 10×10 m^2 quadrats. A sampling effort of less than 1% is usually not acceptable unless the vegetation is nearly a monoculture (e.g., sedge meadow).

Transects

Transects are important in the determination of vegetation changes along an environmental gradients, successional sequences in relation to topographic changes, or sampling along a known pollution gradient. Transects are formed by laying plots (varying in area for each study as determined by the size of the vegetation) along a sampling line across the landscape at known distances from each other. With modern GPS (global positioning system) methodology, permanent plots can be established (latitude and longitude). The length of the transects and number of plots along the transect are dictated by the size of the entire study area. A stratified-random sampling scheme is usually employed to establish plot positions on the transect. However, plots are often established at set equidistant locations along the transect. The number of transect lines is usually dictated by the need to cover a portion or all of the wetland cover types. Once the transects are established, various measurements of density, frequency, or cover can be measured along the line.

Estimation of Abundance and Cover

The Braun–Blanquet method (1965) for cover abundance provides a rapid assessment method for all plant community types that uses little time per relevé or plot. It is based on a cover–abundance scale that is used to estimate the percentage of cover or abundance of plant species in a plot. The scale is fixed within the size of the plot and ranges from 5 to 1 based on the following cover percentages:

5: Any number of plants, with cover more than ¾ of the plot area (>75%)
4: Any number of plants, with ½ to ¾ cover (50 to 75%)
3: Any number of plants, with ¼ to ½ cover (25 to 50%)
2: Any number of plants, with 1/20 to ¼ cover (5 to 25%)
1: Numerous, but less than 1/20 cover or scattered, with cover up to 1/20 (5%)
x: Few with small cover
r: Solitary, with small cover (this has been added to the scale by some investigators)

The upper four scale categories (5, 4, 3, 2) refer only to cover of the crown or shoot area of the species in the plot. The lower categories are used as an estimate of the abundance or number of individuals in the plot. The scale is quantitatively crude, but if applied properly, it allows for a quick and reasonably accurate estimate of the vegetation cover and plant abundance. An example of the use of this approach is shown for a recolonization experiment in the Everglades (three replicate plots per treatment; Table 14.2). The estimates were done at the peak of the growing season on 2 × 2 m plots to determine what species would invade a series of disturbed plots. All 12 plots

TABLE 14.2. Braun–Blanquet Estimated Cover After First Year of Treatment in Hydrology–Disturbance Plots in WCA-2B in the Northern Everglades

	Chara sp.	*Utricularia* spp.	*Typha* sp.	*Cladium jamaicense*	Algal Mat	Filamentous Algae
Control	—	—	—	5	—	—
Vegetation removed	3 ± 1	2 ± 1	×	×	1 ± 1	0
Disturbed soil	1 ± 2	2 ± 2	×	0	1 ± 1	×
Phosphorus	3 ± 3	2 ± 1	×	×	1 ± 2	×
Water depth (60 cm)	×	2 ± 2	×	0	×	×

Source: C. J. Richardson, unpublished data.

could be analyzed within one day. Cover estimates showed that the controls remained 75 to 100% *Cladium jamaicense,* while this species was sparse in the phosphorus and removal plots. *Chara* sp. dominated the phosphorus and removal plots. The water depth plots had some covering (5 to 25%) of *Utricularia* spp. after 1 year but little else. This approach can estimate successional change in communities after disturbance, with limited cost and time. More complex categories (1 to 10) have been developed by Daubenmire (1959, 1968), but these take more training to reduce the level of subjectivity. Training is important any time that visual estimates are used to assign species subjectively to cover classes, especially when more than one person is involved in the sampling process.

Plant Sampling Methods and Procedures

Quadrat Methods. To obtain data to describe the structure of the tree, shrub, or herbaceous components of a wetland community by the quadrat method requires that the species of each plant be noted and tree and shrub diameters measured. Trees are generally classed as plants greater than 10 cm (4 in.) in diameter at breast height [DBH, taken at 1.37 m (4½ feet)]. Trees are usually measured in the summer when it is easier to identify species. The time of sampling is critical for herbaceous species since many taxa are often found only in early spring or late fall. Several seasonal measurements may have to be done to capture the entire population for some wetlands dominated by herbaceous plants. Using these data, one may estimate the following for each species (see techniques by Curtis and Cottam 1962, Cox 1967).

1. *Frequency.* The percentage of sample plots in which a given species occurs. This parameter is a measure of the evenness of distribution of the species within the study area. Comparing this value to density allows for contrasts between distribution and patterns of abundance (see Table 14.3.). Frequency = total number of quadrats in which a species occurs/total number of quadrats. Relative frequency = (frequency of the species/sum of the frequencies of all species) × 100.
2. *Density.* The average number of individuals per unit area. Density = number of individuals sampled per quadrat/total number of quadrats (or area sampled). Relative density = (density for a species/sum of densities for all species) × 100.
3. *Dominance (basal area).* For an individual tree, basal area is determined as the area of a circle whose diameter is equal to that of the tree. Since tree diameters measured at several different heights would result in divergent values, the procedure has been standardized. Diameters are measured at breast height (DBH at 137 cm or 4½ feet). The trees of largest diameter in a community are usually also greatest in height and tend to play a dominant role in the structure and foundation of an ecosystem. Thus, basal area is often used as an index of dominance. Dominance = total basal area of the species/sum of the areas of quadrats sampled. Relative dominance = (dominance of the

TABLE 14.3. Calculations for Quadrat Method Data: Frequency, Density, and Dominance of Four Tree Species in 20 Quadrats

Species	Number of Quadrats of Occurrence	Number of Trees	Total Basal Area (cm)	Relative Frequency, F	Relative Density, D	Relative Dominance, DO	Importance Value, $F + D +$ DO
Cypress	17	75	3921	17/41 = 41.5%	75/126 = 59.5%	3921/9484 = 41.3%	142.3
Red maple	10	22	1928	10/41 = 24.4%	22/126 = 17.4%	1928/9484 = 20.3%	62.1
Ash	8	23	1320	8/41 = 19.5%	23/126 = 18.3%	1320/9484 = 13.9%	51.7
Gum	6	6	2315	6/41 = 14.6%	6/126 = 4.8%	2315/9484 = 24.5%	43.9
Total	41	126	9484	100.0%	100.0%	100.0%	300.0

Trees per quadrat (126/20) = 6.3; basal area per quadrat (9484/20) = 472.2
Trees per hectare (6.3 × 100) = 630; basal area per hectare (474.2 × 100) = 47,420

Source: Modified from Curtis and Cottam (1962).

species/sum of the dominance values of all species) × 100, or relative dominance = (basal area of a species/basal area of all species) × 100.

4. *Importance value.* It has become common practice in quantitative descriptions of plant communities to use the importance values (IVs) of Curtis (1959) as an index of the importance of a species in a community. IV is defined as the sum of relative frequency, relative density, and relative dominance.

See Table 14.3 for an example of these calculations and stand information that can be developed from a quadrat analysis.

Distance Methods or Plotless Sampling. These techniques are based on the concept that the number of trees or stands per unit area can be calculated from the average distance between trees. Distance methods attempt to determine the mean area of the population. Mean area is defined as the amount of space occupied by a single plant. It is, therefore, the reciprocal of density. That is, if there are 100 trees per hectare in a stand, the mean area of the trees is: 10,000 $m^2 \div 100$. All distance methods attempt to measure the square root of the mean area. This figure, when squared and divided into the number of square meters per hectare, gives density per hectare. In addition to obtaining distances, which may be converted to mean area, distance methods can also include the species and size of trees or shrubs. This is done by recording the species and basal areas of the plants to which distance is measured. All distance methods are sampled by establishing sampling points within the stand. These are selected before sampling or by a system of paced compass lines on a transect, with distance between points selected by a random numbers table. The plots should not overlap. The number of compass lines, their direction, and the number of paces between them will depend on the size and nature of the stand. After a sampling point has been located, the criteria for selecting trees and shrubs are applied.

The quarter method (or more appropriately the point-centered quarter method) is the most useful of the plotless methods. The area around a point is divided into four equal sectors, or *quarters,* resulting from the intersection of the paced compass line and its perpendicular. The distance to the midpoint of the nearest tree or shrub from the sampling point is measured in each quarter. In addition, the species and its diameter can be recorded. It is important to include analysis of both living and dead trees in a biomonitoring survey. It is also easy to measure only the nearest four live trees so that both calculations (live and dead tree density, etc.) can be determined (Richardson and Cares 1976).

ANALYSIS OF DATA

The data may consist of two parts (distance or plant species) and can be analyzed as follows:

1. The average distance is obtained. This is simply the sum of all distances divided by the number of distances measured. (See Table 14.4.)

2. The average distance is squared to give the mean area.
3. The mean area is divided into the unit area to give density per unit area. The unit area is of your choice; 1 hectare is often used. (*Note:* This is not the same as relative density.)

Plant species data are analyzed as follows:

1. Number of points of occurrence, number of plants (usually, trees or shrubs in this method), and total basal area are computed for each species. The selection of a given species in one or more of the four quarters of a sampling point is a single occurrence.
2. From the above, relative frequency, relative density, and relative dominance are obtained. (see Table 14.4).
3. In addition, the size of the average tree or shrub is obtained. This is the total basal area divided by the total number of plants.

From these calculations, the importance value (IV) may be determined. In addition, the plant data and distance may be combined as follows:

1. Total plants per unit area multiplied by the relative density gives density per unit area for each species.
2. Total plants per unit area multiplied by the size of the average gives basal area per unit area.
3. Total basal area per unit area multiplied by relative dominance gives basal area per unit area per species. (See Table 14.4 for sample calculations.)

Line Intercept Method

A great deal of information about the composition of a stand can be obtained from data on the numbers, linear extent, and frequency of occurrence of different species intercepted by a series of line transects through the stand. All standard measurements except absolute density may be obtained by this technique. This technique is especially useful in sampling nonforested wetlands and is excellent for herbaceous, sedge, or grass stands.

The most satisfactory device for laying out a line transect is a measuring tape 20 to 30 m in length. The transect line can be subdivided into intervals of any length (e.g., 10 to 100 cm) for the determination of frequency. The tape-measuring scale also provides a convenient means of measuring the length of the segments intercepted by individual plants. Only those plants that are touched by the transect or that underlie or overlie it should be recorded. Measurements of the length of the transect line intercepted by individual plants should be recorded. The length of transect segments overlying bare ground should also be measured and recorded. In summary, the total number of individuals, the total intercept length, and the number of transect intervals in which each species occurred should first be determined. From these values,

TABLE 14.4. Data on the Number of Quarters Occupied, Total Number of Trees, and Total Basal Area

Species	Number Points of Occurrence	Number of Trees	Total Basal Area (cm)	Relative Frequency, F	Relative Density, D	Relative dominance, DO	Importance Value, $F + D + \mathrm{DO}$
Cypress	17	47	2490	17/41 = 41.5%	47/80 = 58.8%	2490/6022 = 41.3%	141.6
Red maple	10	14	1224	10/41 = 24.4%	14/80 = 17.5%	1224/6022 = 20.3%	62.2
Ash	8	13	838	8/41 = 19.5%	13/80 = 16.2%	838/6022 = 13.9%	49.6
Gum	6	6	1470	6/41 = 14.6%	6/80 = 7.5%	1470/6022 = 24.4%	46.5
Total	41	80	6022	100.0%	100.0%	100.0%	299.9

Total distance = 320.51 m; trees per hectare ($10{,}000/4.01^2$) = 6.22; average basal area per tree (6022/80) = 75.3.
Average distance = (320.51/80) = 4.01 m; total basal area = 6022; basal area per hectare (75.3×621.89) = 46,828.

Source: Modified from Curtis and Cottam (1962).

various vegetational measurements can be calculated according to the following formulas.

1. Relative density = (total individuals of a species/total individuals of all species) × 100.
2. Dominance or cover (as a percent of ground surface) = total of intercept lengths for a species × 100/total transect length. Relative dominance = (total of intercept lengths for a species/total of intercepts for all species) × 100.
3. Frequency = (Total number of intervals in which species occurs/total number of transect intervals) × 100. Relative frequency = (frequency value for a species/total frequency value for all species) × 100. Importance value = relative density + relative dominance + relative frequency.

An estimate of the total percentage of the ground surface covered by vegetation may be obtained by totaling cover percentages if measurements of intercept distances were taken in a nonoverlapping manner. If overlap on intercept measurements did occur, owing to the sampling of individuals belonging to different strata, the total plant coverage must be obtained by the formula: Total coverage = (total transect length – total bare ground)/total transect length.

Comparison of Plant Communities

Once sampling of plant communities is complete, it is possible to compare plant population or community structural characteristics among sites. Calculations of relative frequency, relative density, and relative dominance can be used individually to assess and compare the communities or they can be combined to calculate an importance value (IV) for each site as shown in Tables 14.3 and 14.4. In addition, it is possible to calculate the number of plants or trees per hectare, tree basal area per hectare, and even the total percentage of ground surface covered or bare if the line-intercept method is used. Basal area for shrubs and herbs is too time consuming and is seldom measured. An example of a vegetation survey taken in the Everglades utilizing the line-intercept method in June through October of 1990, 1992, 1994 and 1996 along a permanent 10-km transect in WCA-2A shows plant species composition changes along a nutrient enrichment gradient (Fig. 14.4; Richardson and Qian 1999). Cattail *(Typha)* was found primarily in the P-enriched areas (>1250 mg TP/kg in the top 10 cm of soil) along the first 3.5 km of the gradient. Sawgrass had its highest frequency in areas more than 5 km from P input sources (<500 mg TP/kg in the top 10 cm of soil). Collectively, this sampling approach shows the species composition and shifts in plant populations along the nutrient gradient. The amount of bare area as well as other species (>10% frequency) are also shown.

Another way to compare communities is to compare their similarity or dissimilarity in terms of species composition. Various indices of similarity have been developed based on Jaccard's index (1928). He developed a very simple mathematical expression based on the presence–absence of species (Mueller-Dombois and Ellenberg 1974). This coefficient represents the ratio of common species to all species

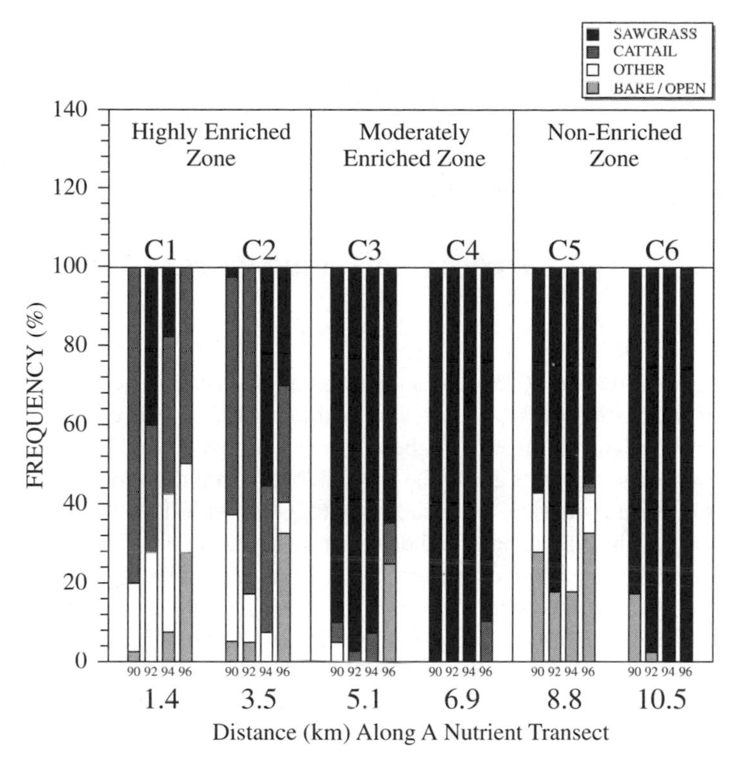

FIGURE 14.4. Percent frequency of macrophytes along a nutrient gradient in WCA-2A of the Everglades. All species comprising less than 10% plus bare ground are grouped as other category. The enriched zone is found at distances of 1.4 to 5.1 km. Unenriched or nonimpacted >5.1 km. (Adapted with permission from Richardson and Qian 1999. Copyright 1999 American Chemical Society.

found in two sites. This index can be used to compare large areas, sample stands, or relevés. *Jaccard's index* can be calculated as follows: $ISJ = (c/a + b + c) \times 100$, where c is the number of common species, b is the number of unique species to the second stand, and a is the number of species unique to the first stand. Many other similarity indexes have been developed and are reviewed in (Mueller-Dombois and Ellenberg 1974).

Sørensen (1948) also developed a similarity index that is often used to compare one stand to another in plant biomonitoring. Sorensen's index is: $ISS = [c/\frac{1}{2}(A + B)] \times 100$ or $[2c/(A + B)] \times 100$, where c is the number of species common to two stands (relevés), A is the total number of species in relevé A, and B is the total number of species in relevé B. When applied to the same data, Sorensen's community coefficient results in a greater similarity than Jaccard's, because the definition of $A + B$ is different in each index. Sorensen's reasoning for modifying Jaccard's index was that both the numerator and the denominator change simultaneously, whereas in Sorensen's formula the denominator is independent of the numerator (Mueller-Dombois and

Ellenberg 1974). Mueller-Dombois and Ellenberg (1974) suggested that Sorensen's index may be better since it includes a statistical probability term.

In addition to these estimators of population or community analysis, it may be important to assess some functional aspects of the plant community. One of the simplest but most useful biomonitoring metrics is the use of biomass or productivity data, especially when compared to reference or controls.

BIOMASS AND PRODUCTIVITY MEASUREMENTS

Biomass

Biomass is most frequently defined as the mass of all living tissue at a given time in a given unit of Earth's surface (Lieth and Whittaker 1975). It is commonly divided into belowground (roots, rhizomes, tubers, etc.) and aboveground biomass (all vegetative and reproductive parts above the ground level). In plant ecology the term *biomass* usually includes all live and dead parts together with litter (Janetschek 1982). *Standing crop* includes live parts and dead parts of live plants that are still attached. These dead parts of plants together with still standing dead plants are called *standing litter*. *Litter* refers to those dead parts of the plant that have fallen on the ground or sediment. *Peak standing crop* is the single largest value of plant material present during a year's growth. In tropical communities, with an almost constant biomass, it is not profitable to search for an annual maximum (Westlake 1969). However, in all other climatic regions the biomass fluctuates widely throughout the year (e.g., Dykyjova and Kvet 1978, Richardson 1978, Shew et al. 1981, Kaswadji et al. 1990). In Figure 14.5, the seasonal variation in various forms of biomass of *Spartina alterniflora* is shown. The range of standing crop of wetland plants is quite large (Table 14.5).

Biomass is most commonly expressed as dry matter. However, other units are also possible: ash-free dry matter, amount of carbon, amount of assimilated CO_2, amount of released O_2, or amount of energy bound in the mass. The approximate relationships between dry mass and other units were given by Jakrlova (1989): 1 g dry mass (ash <10%) = 0.9 to 1.0 g AFDM = 0.4 g C = 1.5 g CO_2 = 1.07 g O_2 = 17.6 kJ. Wetzel (1983) gives the following values of energy content for various groups of wetland macrophytes (per gram of dry matter): emergent, 18.8 kJ (17.6 to 20.9 kJ); floating-leafed, 20.0 kJ (19.1 to 21.5 kJ); submerged, 19.2 kJ (17.4 to 21.8 kJ); average for all species, 19.1 kJ.

Frequent sampling is necessary to determine the seasonal maximum biomass. If the approximate time of the maximum is known, samples at 2-week intervals during the month before and after will suffice. If the seasonal maximum is unknown or variable, this samplinjg frequency may need to be continued for longer periods. Sampling at monthly or even bimonthly intervals over the growing season will often suffice; and in general, numerous samples taken at intervals of 1 or 2 months are a better use of labor than a few samples taken over shorter intervals (Westlake 1969). The ratio between below- and aboveground macrophyte biomass varies considerably, but Wetzel (1983) and Vymazal (1995) reported similar ranges for belowground biomass (as

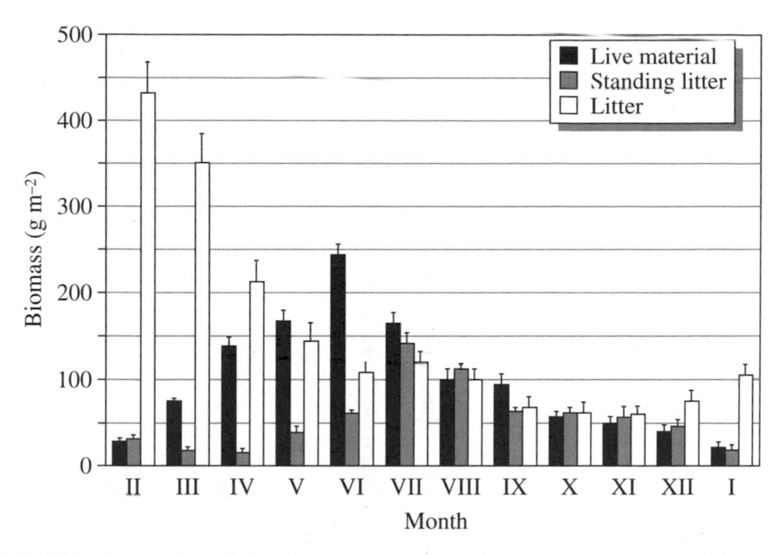

FIGURE 14.5. Seasonal variation in compartments of *Spartina alterniflora* biomass in a southeastern North Carolina marsh. (From Shew et al. 1981.)

percentage of total biomass) for freshwater species: emergent = 7 to 96% and 12 to 89%, submerged = 3 to 55% and 16 to 54%, and floating-leafed = 48 to 80% and 46 to 77%, respectively. Sampling of biomass for biomonitoring purposes should take into account not only the live biomass but also the amount of standing dead and litter accumulation. These measurements quite often differ along disturbance gradients and are excellent indicators of impacts on ecosystem function.

Productivity

Gross primary production is normally defined as the assimilation of organic matter by a plant community during a specified period, including the amount used by plant respiration. Net primary production is the biomass that is incorporated into a plant community during a specified time interval, less that respired. This is the quantity that is measured by harvest methods and which has also been called *net assimilation* or *apparent photosynthesis*. *Net aerial* (or *aboveground*) *primary production* (NAPP) is the biomass incorporated into the aerial parts (leaf, stem, seed, and associated organs) of the plant community (Milner and Hughes 1968).

Biomass accumulation, a conspicuous result of the growth of larger plants, is used to estimate primary production in many ecosystems. Accuracy of primary production estimates using changes in plant biomass, however, can be seriously compromised unless appropriate corrections are made for mortality or loss of plant parts during the growing season (Dickerman et al. 1986). The types and magnitudes of corrections in net production depend in turn on both plant growth characteristics and sampling frequency (Bradbury and Hofstra 1976, Whigham et al. 1978).

TABLE 14.5. Maximum Standing Crop of Wetland and Macrophytes and Woody Species (g dry mass/m^{-2}) Reported from Localities in North America

Species	Aboveground	Belowground	Locality
Emergent			
Acorus calamus	605–819		New Jersey
Bidens leavis	17–282		Virginia, South Carolina, New Jersey, Pennsylvania
Carex spp.[a]	101–1145	433–1430	Alberta, Quebec, Minnesota, Michigan, New York, Iowa, Northwest Territories, California, New Jersey
Distichlis spicata	280–970	12,400–1399	Florida, Delaware, New Jersey, Louisiana, North Carolina
Juncus roemerianus	480–2100	4060–12,400	North Carolina, Louisiana, Florida, Mississippi
Peltandra virginica	84–386		New Jersey, South Carolina
Phragmites australis	934–8147	1121–1565	Florida, Iowa, Maryland, New Jersey, Louisiana
Polygonum arifolium	12–200		South Carolina, Virginia, New Jersey
Pontederia cordata	257–716		South Carolina
Sagittaria latifolia	460–12563		Florida, Iowa
Scirpus spp.[b]	28–1381	40–211	Iowa, South Carolina, Quebec
Sparganium eurycarpum	638–1950	681–1945	Iowa, Michigan, Wisconsin
Spartina spp.[c]	640–2410	100–17,500	Virginia, Louisiana, North Carolina, New Jersey, Georgia, Louisiana, Maryland, Mississippi, Oregon, Nebraska
Typha spp.[d]	48–2895	393–11,280	Texas, Florida, North Carolina, South Carolina, Iowa, New York, Michigan, Oklahoma, Wisconsin, Alberta, New Jersey, SE U.S.A., North Dakota, South Dakota
Zizania aquatica	560–1390		New Jersey, Virginia

Submerged			
Myriophyllum spicatum	60–220		Quebec, Vermont, Wisconsin
Potamogeton natans	295		Alberta
Potamogeton pectinatu	297		Alberta
Utricularia sp.	143		Florida
Rooted floating-leafed			
Nelumbo lutea	184		South Carolina
Nuphar advena	84–1125	1146–4799	South Carolina, Virginia, New Jersey, Pennsylvania
Nymphaea odorata	110–256		Minnesota, South Carolina
Free floating-leafed (total)			
Eichhornia crassipes	440–3000		Louisiana, Alabama, Florida, Iowa
Woody species—trees[e]			
Bog forest[1]	4650	2280	Manitoba
Marginal fen[2]	9880		Minnesota
Bottomland swamp[3]	17,300		Louisiana
Mangrove systems	900–20,800		Florida
Cedar swamp[4]	15,960–22,100		Minnesota, Virginia
Cypress swamp[5]	31,300–45,200		Florida, Georgia, Virginia, Louisiana, Illinois

Source: Data from Richardson (1978), Brown (1978), and Vymazal 1995.

[a]*C.aquatilis, C. atherodes, C. lacustris, C. lasiocarpa, C. nebraskensis, C. rostrata, C. stricta.*

[b]*S. americanus, S. validus.*

[c]*S. alterniflora, S. cynosuroides, S. patens.*

[d]*T. angustifolia, T.domingensis, T. glauca, T. latifolia.*

[e]Dominant species: [1]*Picea mariana;* [2]*Fraxinus nigra, Acer rubrum;* [3](hardwood) *Acer rubrum, Nyssa aquatica;* [4]*Thuja occidentalis, Chamaecyparis thyoides, Nyssa sylvatica;* [5]*Taxodium distichum* (+ tupelo, Nyssa aquatica).

NAPP is an important variable for the analysis of energy and nutrient flows in wetlands. Estimation of NAPP has been a difficult task because it can only be assessed indirectly from the storage and flows of biomass in an ecosystem (Hsieh 1996). Various methods have been devised for estimating primary production based on changes in biomass. Some methods assume negligible losses of plant material, at least until the seasonal maximum biomass occurs (Milner and Hughes 1968). Other methods incorporate one or more terms to correct for losses due to early mortality, grazing, and leaf turnover (e.g., Smalley 1959, Wiegert and Evans 1964, Lomnicki et al. 1968). Peak standing crop is simply the single largest value of live material present during a year's growth. The method provides a severe underestimate of primary production because it does not account for mortality, decomposition, or growth occurring after the peak (Bradbury and Hofstra 1976, Shew et al. 1981). The method introduced by Milner and Hughes (1968) involves a summation of positive changes in live material through sampling intervals spanning one year. Although the method considers the time, it does not account for dead material or decomposition.

The method devised by Smalley (1959) compensates for changes in both living and dead components, but decomposition is not estimated. The method suggested by Valiela et al. (1975) estimates production in steady-state habitats by summing losses of dead plant material over a growing season. The Wiegert–Evans method calculates loss of dead material each month and adds that to the change in dead standing stock to estimate mortality. Mortality, in turn, is added to the change in live biomass, to estimate production (Wiegert and Evans 1964). Lomnicki et al. (1968) presented a modification of the Wiegert–Evans method, basing production calculations on growth of green material and production of dead material. In this method, mortality is estimated directly by the appearance of dead plant material 1 month after removal of all dead culls. A thorough review of various techniques for estimating NAPP has been given by Singh et al. (1975). The range of productivity rates of vegetation in North American wetlands is shown in Table 14.6.

Dickerman et al. (1986) pointed out that because different methods for calculating production from changes in biomass use different correction terms and depend on different assumptions, production estimates for the same sites using different techniques can be quite disparate (Bradbury and Hofstra 1976, Kirby and Gosselink 1976, Kruczynski et al. 1978, Linthurst and Reimold 1978, Shew et al. 1981, Dickerman et al. 1986, Kaswadji et al. 1990, de Leeuw et al. 1996). Selection of the appropriate method for an individual study depends on the amount of accuracy required and time the researcher can allocate to the study. The most accurate methods follow Wiegert and Evens (1964). The important feature of this method is the accounting for plant mortality by comparing both green and dead plant material over the growing season. Milner and Hughes (1968) carefully document the procedures used in estimating plant production.

Tagging and measuring individual plants or plant parts is a basic technique in studying the dynamics of a plant population; using additional phytometric studies, plant demographic information can be used to estimate NAPP. This method is nondestructive (Dai and Wiegert 1996). Although harvest methods measure biomass change directly, they are destructive, labor intensive, and affected by the spatial

TABLE 14.6. Maximum Net Productivity of Wetland Macrophytes and Woody Species (g dry mass/m^2 per year) Reported from Localities in North America

Species	Aboveground	Belowground	Locality
Emergent			
Acorus calamus	712–940		New Jersey
Carex spp.[a]	32–2858	130–548	Alberta, Quebec, Northwest Territories, British Columbia, Manitoba, Minnesota, Virginia, New Jersey, Iowa, New York, Wisconsin
Distichlis spicata	1291–3366		Louisiana, Mississippi
Juncus roemerianus	796–3794		North Carolina, Louisiana, Mississippi, Georgia
Peltandra virginica	144–800		Virginia, Pennsylvania, New Jersey
Phragmites australis	1825–2811		Louisiana
Spartina spp.[b]	63–6163	460–3500	New Jersey, Alabama, Mississippi, Maine, Georgia, Maryland, North Carolina, Virginia, Louisiana, Delaware
Typha spp.[c]	330–3035	1050–2505	Texas, Florida, South Carolina, Minnesota, Iowa, Oregon, New York, New Jersey, Virginia, North Dakota, Oklahoma, Pennsylvania
Zizania aquatica	605–1547		New Jersey, Pennsylvania
Submerged			
Rooted floating-leafed			
Nuphar spp.[d]	222–1188		Virginia, New Jersey, Pennsylvania, North Carolina
Free floating-leafed (total)			
Eichhornia crassipes	1473–6520		Alabama, Louisiana

(*continues*)

TABLE 14.6. (*continued*)

Species	Aboveground	Belowground	Locality
Woody species—shrubs[e]			
Muskeg and bog[1]	86–308		Alberta, Manitoba
Fir[2]	108–650	130–548	Alberta, Manitoba, Michigan
Woody species—trees[f]			
Bog forest[1]	480	190	Manitoba
Marginal fen[2]	710		Minnesota
Cedar swamp[3]	1030		Minnesota
Cypress swamp[4]	595–1607		Florida, Georgia, Louisiana
Bottomland swamp[5]	1374–1570		Louisiana
Mangrove systems	380–2740		Florida

Source: Data from Reader and Stewart (1972), Richardson (1978), Vymazal (1995), and Szumigalski and Bayley (1996).

[a]*Carex aquatilis, C. atherodes, C. lacustris, C. lyngbyei, C. rostrata, C. stricta, C. subspathacea.*

[b]*S. alterniflora, S. cynosuroides, S. patens.*

[c]*T. angustifolia, T.domingensis, T. glauca, T. latifolia, Typha* sp.

[d]*N. advena, N. lutea.*

[e]Dominant shrub species: [1]*Ledum groenlandicum, Chamaedaphne calyculata;* [2]*Betula pumila, Chamaedaphne calyculata, Salix predicellaris.*

[f]Dominant tree species: [1]*Picea mariana;* [2]*Fraxinus nigra, Acer rubrum;* [3]*Thuja occidentalis;* [4]*Taxodium distichum;* [5](hardwood) *Acer rubrum, Nyssa aquatica.*

variation between sampling sites. Whittaker (1962, 1965) outlines the nondestructive allometric technique of developing regression equations for predicting biomass from representative plant equations comparing diameter to dry weight biomass or diameter times stem length versus biomass for each species. For example, a universal set of equations developed for all eastern U.S. forest species follows the linear form $\ln Y = a + b \ln X$, where Y = biomass (bole, branch, foliage), X is the independent variable (diameter at breast height), and a and b are regression constants. Example equations are shown in Table 14.7. From these equations it is possible to obtain a reasonable estimate (10 to 15%) of tree biomass of each component (Harris et al. 1972). In contrast to harvest methods, nondestructive methods have the following advantages: (1) losses of aerial biomass (leaves and stems) can be easily tracked, permitting an accurate assessment of biomass turnover and a consistent NAPP estimation; (2) they require only small sampling sites, are sensitive to temporal change, and allow long-term continuous monitoring; (3) they give detailed data about individuals or plant parts, permitting estimates of variation within a stand; (4) information on leaf dynamics can be combined with leaf-level physiology to study canopy physiological processes using simulation models; and (5) they are laborsaving when only simple censuses (e.g., measurements of stem diameter) are required (Dai and Wiegert 1996). Three disadvantages are usually associated with nondestructive methods: (1) they do not measure belowground productivity, (2) they are timeconsuming if many parameters are to be collected for large samples, and (3) allometric equations have to be developed

TABLE 14.7. Summary of Uncorrected Regressions (1n–1n) of Tree Component Weight (kg) on Tree Diameter Breast Height (cm)

Dependent Variable	Intercept, (a)	Slope, (b)	Coefficient of Determination, r^2	Sample Size, n
Leaf				
All species	–3.498	1.695	0.86	302
Hardwoods	–3.862	1.740	0.88	178
Conifers	–2.907	1.674	0.91	65
Branch				
All species	–3.188	2.226	0.91	298
Hardwoods	–3.173	2.224	0.89	231
Conifers	–3.461	2.292	0.95	51
Bole				
All species	–2.437	2.418	0.97	298
Hardwoods	–2.270	2.385	0.98	231
Conifers	–3.787	2.767	0.96	51
Branch–bole				
All species	–2.126	2.393	0.96	371
Hardwoods				
Conifers				

Source: Data from Harris et al. (1972).

for each component from at least 15 harvested plants covering the range of plant sizes in the community. Nondestructive methods have been used in estimating NAPP in only a few marshes (Williams and Murdoch 1972, Hopkinson et al. 1980, Dickerman et al. 1986, Cranford et al. 1989, Morris and Haskin 1990, Daoust and Childers 1998).

HARVESTING

The harvesting technique depends on the purpose of harvesting. The general aim is to remove and weight the vegetation from enough plots to obtain a mean biomass sufficiently accurate to show significant differences between different sampling periods and sites. Removal of the plants may be direct, by hand from a quadrate, or indirect, by a sampling apparatus. This choice is influenced by the method of approach and the nature of the substratum (Westlake 1969). Plant samples are then dried to a constant temperature (60 to 80° C) prior to comparing biomass values among sites.

Positioning of Sampling Quadrats

It is often necessary to decide prior to sampling what plants or plant parts are to be included in the quadrat (e.g., all plants rooted within the quadrat, or all plants projecting skyward through the quadrat). Large areas may be marked with stakes and strings, whereas quadrat frames are sufficient for smaller areas (Westlake 1969). It is especially important to determine if the plants will be harvested at the water surface or at the sediment surface when submerged and floating-leafed rooted macrophytes are sampled. The most accurate sampling for aboveground biomass of rooted plants requires sampling at the sediment surface. The placing of quadrats (plastic, metal, or wood) into dense vegetation as a cutting guide is often difficult and not a completely objective operation. The investigator must often decide if a plant should be included within the quadrat. This problem may be partially overcome by making the quadrat out of very thin material or by using a rigid, two-piece quadrat. A two-piece quadrat can be pushed into the tangled base of the vegetation and subsequently fitted together (e.g., PVC pipe) allowing efficient separation of the leaves within the sampling area (Milner and Hughes 1968, Westlake 1968). Another difficulty is positioning a quadrat in deep water. Sinking frames with legs that can be pushed into the sediment can solve this problem. This problem always arises when aboveground biomass of submerged or floating-leafed rooted macrophytes is harvested. Quadrats for sampling free-floating macrophytes must be made of floating material, and very thin floating quadrats are necessary for sampling small species (e.g., *Lemna*).

Sampling of woody species such as shrubs, requires larger plots (2 × 2 m to 5 × 5 m). Multiple stems of shrubs make it quite difficult to determine individual plants, so biomass estimates are usually based on harvesting the entire plot. Trees are tallied in plots or quadrats (10 m × 10 m), as mentioned earlier. Trees and shrubs are not often harvested to determine biomass directly since it is difficult and expensive. Indirect means of estimating biomass have been developed using allometric regression techniques for each species (Whittaker 1962, 1965). This method is called *dimension analysis* of woody plants (Whittaker, 1961, 1962; Whittaker and Woodwell 1968). It involves three steps: (1) trees are tallied by species, and diameters at breast height (DBH) are measured within a quadrat (usually several 10 × 10 m plots as noted earlier); (2) sample trees are harvested and regressions developed (Table 14.7) to relate to biomass and growth parameters of interest (Whitttaker and Woodwell 1968); and (3) conversion to a land area basis is done by applying the allometric equations developed for DBH and biomass estimates for each species to the tree tallies and sizes per plot. Productivity estimates require more intensive analysis and are not usually part of routine biomonitoring analysis for trees or shrubs.

Aboveground Biomass

Rooted Macrophytes. To harvest the aboveground biomass it is necessary to clip plants at the ground or sediment level with knives, scythes, or similar tools inside the quadrat perimeter. Pruning shears are also useful. Harvested material is then placed in plastic bags and labeled with a waterproof marker or small aluminum tags. When sampling vegetation containing large amounts of standing litter, it is useful to place old litter (already detached from stems) in a separate bag during harvesting. Standing litter may break during transportation and it is difficult to separate this biomass fraction from old litter in the laboratory.

Floating Macrophytes. Larger species (e.g., *Eichhornia crassipes, Pistia stratiotes*) may be grabbed by hand and put into plastic bags. Smaller species (e.g., *Lemna* spp., *Salvinia*) may be harvested by hand, scoops, or strainers of appropriate size. Plants can be placed in labeled plastic bags or buckets (for small species).

Woody Species. Litterfall is usually collected in litter traps. The size varies between 0.25 and 1 m^2 depending on the density of the trees. The box sides (ca. 20 cm) may be made of wood or fiberglass; fiberglass screen (1-mm^2 mesh) is the best material for the bottom. The boxes are positioned about 1 m above the ground level. Falling litter should be collected regularly, depending on the season, with shorter intervals (1 to 2 weeks) for deciduous trees in autumn (Conner and Day 1976).

Belowground Biomass

Herbaceous Macrophytes. If the plants are rooted in flooded soft mud or sediment, whole plants may often be pulled out by loosening the substrate. If the shoots are cut off at a ground level or if the soil is compact, the roots must be sampled separately. The sampling method is also influenced by the purpose of sampling. It is important to decide (1) if the sample should maintain the root system structure in natural three-dimensional space (this maintains the belowground biomass in discrete layers of the rhizosphere), or (2) if it is enough to determine the total belowground biomass regardless of the spatial distribution. Various types of core samplers can be used when it is necessary to maintain the vertical distribution of the belowground biomass. If the vertical distribution is not important, the belowground biomass can be dug out from inside the perimeter of a quadrat or root cores can be taken.

The most commonly used samplers for belowground biomass are either coring tubes (PVC or fiberglass) or rectangular box corers. The diameter of coring tubes and the edge of the box corers should be at least about 7.5 cm in diameter. In peat soils, box corers with a removable side are very useful. Pushing the three-sided corer into soil or sediment and then sliding the fourth side to cut the soil profile minimizes compaction (Cuttle 1983). The sample may be sectioned into several increments in the field and placed in labeled plastic bags or boxes. Roots and rhizomes are then separated from the soil in the laboratory.

Quadrats are best, especially for sampling in sites with sparse or multispecies vegetation cover. After positioning the quadrat, it is necessary to cut the soil or sediment along the quadrat perimeter to the depth of the belowground biomass sample. For most plants, the majority of belowground organs are found within the top 30 cm. However, roots and rhizomes of some plants penetrate as deep as 80 cm (e.g., *Scirpus, Phragmites*). A long-bladed root-pruning shovel is best for digging the belowground biomass. It is usually necessary to collect the small remains of the sediment or soil containing belowground biomass by hand. In flooded sites, it is useful to carefully wash as much soil as possible at the site to prevent the transportation of excessive soil. In dry sites, it is possible to separate the biomass from the soil by sieving through a mesh screen. The diameter of the screen depends on the grain size of the substrate and length and thickness of roots and rhizomes. Washed or sieved belowground plant material is then placed in labeled plastic bags or boxes, depending on the sample size, as samples may be very heavy. The fine separation of roots and rhizomes from the soil matrix is done in the laboratory.

SAMPLE ANALYSIS

Separation

If only biomass of the entire community is needed, all plant species can be combined for a total dry weight. Separation is necessary in any study where species information is required. The following categories of aboveground biomass may also be determined: live parts, standing litter (standing dead plants and dead parts that are still attached to live parts), and litter (dead parts that have already fallen on the ground). Live parts can be further distinguished into stems, leaves, and inflorescence. Separation should be accomplished as soon as possible after transporting the plants to the laboratory and before drying. This is especially true for the separation of dead parts that are still attached to live parts, as some macrophyte species dry and change their color very quickly. In addition, trapped litter from woody ecosystems may be sorted into woody (twigs, branches, bark) and nonwoody (leaves, flowers, seeds) fractions before drying.

Belowground biomass may be separated into roots and rhizomes. Roots and rhizomes may be very difficult to separate from heavy soils, and the process is always time consuming. The separation technique includes an initial wash with a jet of water over a fine mesh screen to collect small rootlets and pieces of broken root and rhizomes. The sieved material can be floated to recover as many rootlets as possible.

It is sometimes difficult to distinguish belowground parts of different macrophyte species. Moreover, it is always difficult to distinguish between dead and live roots and rhizomes. Rhizomes of different species might be distinguished based on color and consistency, but to distinguish between live versus dead parts of the roots without special treatment is nearly impossible. Jakrlova (1989) reported that live and dead root fractions can be distinguished by means of various dyes or labeled isotopes ^{14}C, ^{32}P, or ^{89}Sr.

Hand separation of different species requires the least skill and is probably the most accurate of the available techniques. However, it is always time consuming, difficult among some species, and costly. If plant species are morphologically distinct, less skilled labor may be used (Milner and Hughes 1968). If the material is to be dried, it is useful to cut plant material into short pieces and place them into weighed paper bags. Each bag should display identification information. Most commonly, the information includes the date of sampling, location, and biomass fraction.

Drying Process

Biomass, standing crop, or productivity in wetlands is frequently expressed on a dry matter basis. In the older literature, the term *dry weight* has often been used instead of *dry matter*. However, this term is not correct, as *weight* is not a mass unit. Dry matter is determined by drying the biomass to a constant weight in an oven at temperatures of 60 to 80° C. Some authors recommend higher temperatures up to 105° C, but lower temperatures are preferred to prevent the loss of volatile constituents. For most aboveground parts of herbaceous macrophytes, 24 hours is sufficient to reach a constant weight. For old litter soaked with water, the drying time may be longer. For huge belowground organs (e.g., rhizomes of *Typha* spp. or *Nymphaea odorata,* tussocks of *Cladium jamaicense,* or roots of *Acrostichum danaeifolium*) this period may take a week or more. Drying time for woody materials may be even longer. If the drying time is longer, it is necessary to check the weight of the sample at regular intervals until the weight is constant. Samples should be dried in paper bags to prevent biomass losses. Remember to include the weight of the empty bag. Values of air-dried matter or freshly cut material are not used because the water content varies widely.

Biomass Analysis

Before nutrient or ash analyses of the macrophyte biomass can be performed, it is necessary to obtain a homogenized sample. This is achieved by grinding the plant material in a mill. To increase sample homogeneity, it is necessary to sieve it through a fine mesh screen (e.g., 2-mm diameter). At this stage, it is possible to store the samples for an extended period of time, but it is always necessary to dry samples to a constant weight prior to further analysis.

Ash-Free Dry Matter. The ash content of wetland plants varies among groups (emergent, submerged, floating-leafed), plant parts (roots, rhizomes, stems, leaves, seeds) as well as seasons. Ash-free dry matter is the difference between the weight of oven-dried matter and the weight of ash after combusting. Wetzel (1983) gives the following values of ash as a percentage of the dry matter for various groups of wetland macrophytes: emergent species 12% (5 to 25%), floating-leaved 16% (10 to 25%), submerged 21% (9 to 25%), with an average of 18% for all species. Boyd (1978) noted a 6 to 40% range of ash content in 40 wetland plant species from various localities in the United States. However, this study included a calcium precipitating macrophytic algae *(Chara)* and other species that usually have a high ash content (e.g., *Cladophora;* Westlake 1963, Vymazal 1995).

The ash content of a particular plant fraction is determined gravimetrically after combusting a milled sample at 475 to 550° C in a muffle furnace for 2 hours. The amount of dry matter for combusting may be small. One gram is enough to provide reliable data, but when material is in short supply, 50 mg is sufficient.

Carbon. Compared to nutrients, there is much less information on the organic carbon content of wetland plants. Commonly, organic carbon is assumed to be 45% of the dry matter, but this is rarely correct as it can vary from about 5% in highly calcareous or siliceous organisms to nearly 70% in very fatty plants. Carbon measurements are much less variable when divided by the amount of organic matter rather than the amount of total dry matter. For example, if 20% of the dry matter is ash and the true carbon content is 47% of the organic (ash-free) matter, the carbon will only be 38% of the dry matter. Most whole plants contain 46 to 48% carbon in their organic matter under natural growth conditions (Westlake 1963).

Nitrogen and Phosphorus. Nitrogen and phosphorus are taken up and assimilated by growing plants throughout the growing season. However, the highest concentrations in live plant tissue occur early in the growing season and decrease as plants mature and senesce (Johnston 1991). This trend is documented in Fig. 14.6. The content of mineral nutrients in macrophytes is controlled by both habitat and interspecific variations of plants. Also, the developmental rate of various species must be take into consideration. In some species, such as *Acorus calamus* (sweet flag), the vegetative phase of aboveground parts is very short. In temperate regions, its reproductive phenophase is over in July. After having attained its maximum biomass, the shoots become senescent. On the other hand, species such as *Glyceria maxima* (mannagrass) begin to sprout in very early spring and the formation and growth of new tillers continues during the entire growing season, sometimes even during the mild winter periods (Dykyjova 1973). Concentrations of nitrogen and phosphorus in biomass of various species of wetland macrophytes and trees are presented in Tables 14.8 and 14.9. The results indicate broad variation among various groups of wetland plants as well as among plant parts. The methods to analyze for nutrients are outside the scope of this volume. Excellent books by Allen (1974) and Page et al. (1982) provide detailed analysis methods for both plants and soil nutrient analysis.

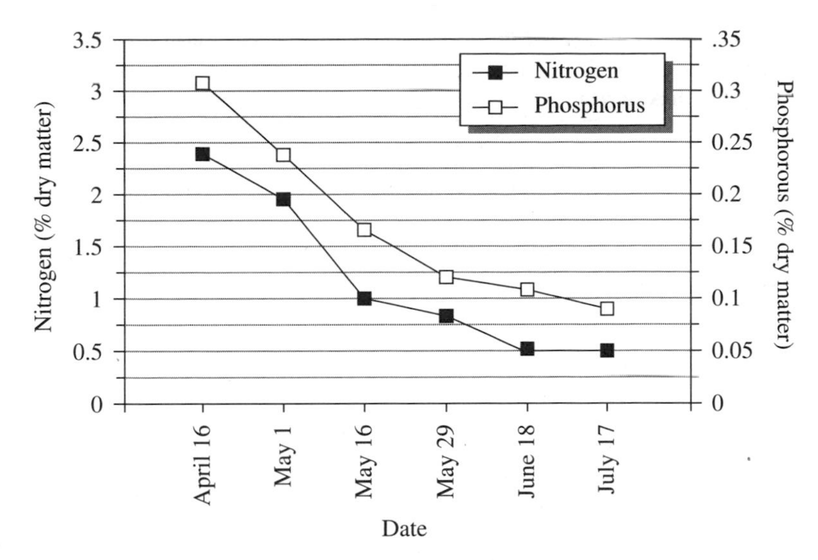

FIGURE 14.6. Variation in nitrogen and phosphorus content in *Typha latifolia* aboveground live shoots during the growing season. (From Boyd 1970.)

TABLE 14.8. Average Ranges of Nitrogen Concentrations (mass in percent dry matter) and Standing Stocks (g/m²) in Wetland Macrophytes, Shrubs, and Trees from Localities in North America

	Concentration		Standing Stock	
	Aboveground[a]	Belowground	Aboveground	Belowground
Herbaceous macrophytes				
Emergent	0.18–3.77*		1.03–222	0.92–49
Submerged	1.48–4.43		1.98–8.45	
Floating-leafed rooted	1.80–2.66		1.90–9.86	
Free-floating	1.30–4.30** (total)		0.40–90	
Litter	0.30–1.30			
Woody species				
Leaves	0.70–4.50		4.70–6.20	
Shrub stems	0.35–0.91			
Tree boles and roots	0.08–0.68			
Leaf litter	0.60–1.80		4.50–24.1	

Source: Macrophyte data from Vymazal (1995) and Vymazal and Richardson (unpublished data); woody species and litter data from Johnson (1991).

[a]In systems designed for wastewater treatment up to *9.50%, **7.28%.

TABLE 14.9. Average Ranges of Phosphorus Concentrations (mass in percent dry matter) and Standing Stocks (g/m²) in Wetland Macrophytes, Shrubs and Trees from Localities in North America

	Concentration		Standing-Stock	
	Aboveground[a]	Belowground	Aboveground	Belowground
Herbaceous macrophytes				
Emergent	0.01–0.64*	0.02–0.72	0.15–19.0	0.24–6.87
Submerged	0.08–1.05		0.07–1.29	
Floating-leafed rooted	0.14–0.42		0.16–4.48	0.45–3.54
Free-floating	0.14–0.87** (total)		0.10–34.5	
Litter	0.01–0.24			
Woody species				
Leaves	0.04–0.60		0.08–1.70	
Shrub stems	0.02–0.08			
Wood			0.18–4.28	
Tree boles and roots	0.003–0.044			
Leaf litter	0.03–1.13		0.21–1.94	

Source: Macrophyte data from Vymazal (1995) and Vymazal and Richardson (unpublished data); woody species and litter data from Johnston (1991).

[a]In systems designed for wastewater treatment up to *1.20%, **2.84%.

Nutrient Ratios and Limitations. Bridgham et al. (1995) suggested that some minimum level of nutrients is required to sustain an ecosystem, and that a balanced carbon/nitrogen/phosphorus (C/N/P) stoichiometry is essential to sustain ecosystem productivity. The nutrient limitation criteria by Koerselman and Meuleman (1996) and Verhoeven et al. (1996) suggest that wetland plants with mass N/P ratios above 16 (>36 molar ratio) are P limited, those below 14 (<31 molar ratio) are N limited, and between 14 (31 molar ratio) and 16 (36 molar ratio) they could be either N or P limited. The use of plant nutrient ratios C/N/P and N/P provide a basis for determining the nutrient status as well as the growth conditions for wetland plants. A recent study by Richardson et al. (1999) clearly demonstrates that plant nutrient ratios were useful in determining areas of P limitation in the Everglades as well as areas that were eutrophic due to nutrient additions (Table 14.10). The N/P and C/P ratios of sawgrass shoot tissue increased about twofold and fivefold, respectively, from the highly enriched zone to the nonenriched zone, suggesting increased P limitations (Table 14.10). Utilizing the N/P criteria of Koerselman and Meuleman (1996) and Verhoeven et al. (1996), the molar ratios for sites C4, C5, and C6 are definitely P limited. Surprisingly, C2 is N limited, and C1 and C3 are either N or P limited. Trends in C:N leaf ratios along thc C nutrient gradient were more variable than N/P and C/P ratios, but all ratios were generally higher in the nonenriched zones. Indices like these can also be used to determine the effects of nutrient inputs or other environmental factors on plants.

TABLE 14.10. Molar Ratios of Leaf Nutrients, Soil P Content (0 to 10 cm), and Incidence of Herbivory and Fungal Infection for Sawgrass Plants Along a P Nutrient Gradient in Water Conservation Area 2A of the North-Central Everglades (Fig. 14.1)

Site	Soil P Content (mg/kg)	Ratio[a] C/N	Ratio[a] C/P	Ratio[a] N/P	Observation[b] (February 1996) Herbivory	Observation[b] (February 1996) Fungal Infection
C1 (most enriched)	1600	48:1 (0.4)	1705:1 (47.8)	36:1 (0.95)	+	+
C2	1050	74:1 (6.0)	1883:1 (117)	26:1 (3.8)	+	+
C3	800	60:1 (2.4)	2058:1 (176)	34:1 (1.6)	–	+
C4	600	74:1 (4.5)	3150:1 (328)	43:1 (3.8)	–	–
C5	450	66:1 (4.9)	5460:1 (134)	84:1 (4.0)	–	–
C6	400	113:1(10.0)	8355:1(1147)	77:1(17.5)	–	–

Source: Adapted from Richardson et al. (1999).

[a]Values shown in parentheses are ±1 SE of the ratio (n = 3).

[b]+, Observed; –, not observed.

Standing Stock. Total storage of a substance in a particular compartment is called standing stock. Nutrient standing stocks in the vegetation are commonly computed by multiplying nutrient concentrations in the plant tissue by biomass per unit area and are expressed as mass per unit area (usually g/m^2 or kg/ha; Johnston 1991). It implies that nutrient standing stocks in live biomass depend on both nutrient concentrations in the plant tissue as well as on the amount of live plant biomass. However, it is well known that peak nutrient concentration and peak live biomass do not occur at the same time of the growing season in northern temperate latitudes. Maximum nutrient concentrations are found early in the growing season, while peak live aboveground biomass occurs later in the growing season (usually July–August in temperate climates and earlier in the spring at lower latitudes). The major factor influencing nutrient standing stocks is biomass. Although the nutrient concentrations at the time of peak biomass may be 50% lower compared to the beginning of the season (e.g., April, May), the live biomass dry matter during the peak biomass may be 10 to 30 times higher for some species. Therefore, the peak nutrient standing stock in northern temperate latitudes usually occurs at the time of peak live biomass or shortly before that, and will then decline during autumn senescence. Results presented by Boyd (1970) for *Typha latifolia* from a South Carolina marsh are a good example of this trend (Fig. 14.7).

Measurements of standing stocks gives a snapshot view of the total amount of nutrients contained in a storage compartment at a particular time, Some standing stocks (e.g., nutrients stored in tree boles) change very little over time, while the standing stocks in dynamic storage compartments (e.g., leaves, herbaceous plants) change

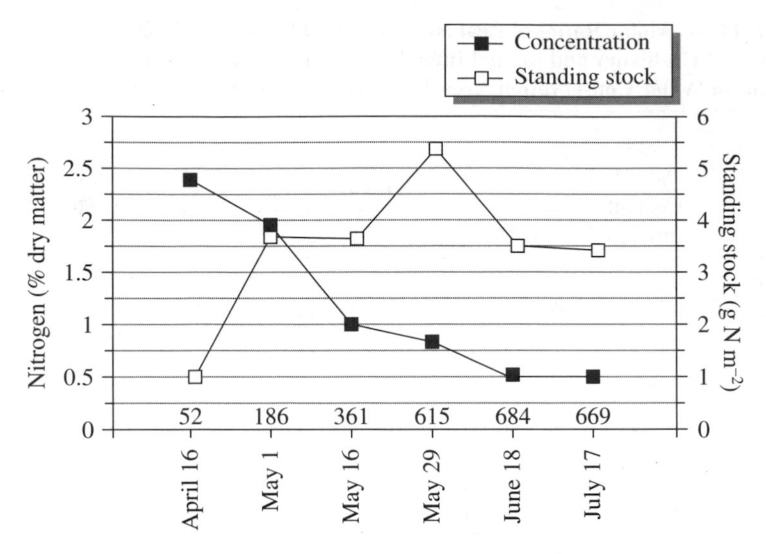

FIGURE 14.7. Variation in nitrogen concentration and standing stock in *Typha latifolia* aboveground shoots during the growing season. The numbers denote biomass dry matter. (From Boyd 1970.)

substantially during the course of the year (Johnston 1991). Nutrient standing stocks vary among species as well as among plant parts (Tables 14.8 and 14.9).

FUTURE OF WETLAND BIOMONITORING

Assessment of plant populations and community structure as well as biomass, nutrient status, and productivity estimates will be the basis for most assessments. However, recent studies have focused on more physiological or biochemical approaches to biomonitoring (Table 14.1; Wang et al. 1990, Richardson et al. 1999). Little of this work has been done in wetland systems, but new techniques that show considerable promise include chlorophyll fluorescence as an important tool for monitoring the electron transport system of photosynthesis (Judy et al. 1990). This nondestructive technique can be done on intact leaves in the field with modern portable equipment. Changes in leaf chlorophyll content, secondary compounds, ion accumulation patterns, and delayed fluorescence of plant leaves have proven useful as indicators of toxicity due to chemical stress (Richardson et al. in press). Selection of the appropriate plant species and prior determination of natural versus stress patterns are essential to the successful use of these techniques.

Finally, it is essential that any plant biomonitoring program be aware of seasonal differences in the presence or absence of plants in the community as well as the need for the proper scale of sampling procedures. The trends in recent years are to develop rapid assessment metrics for assessing plant communities. This suggests that newer and simpler techniques need to be developed for macrophytes. To date, the revelé method and use of similarity indices is still the most rapid tool available for assessing plant communities. This approach, coupled with quadrat analysis and biomass estimates at peak season, has provided the basis for most biomonitoring in wetlands. The interest in key species responses suggests that more growth and physiological indices will be developed in the future.

REFERENCES

Allen, S. E. 1974. Chemical analysis of ecological materials. Blackwell, Oxford.

Arber, A. 1920. Water plants. a study of aquatic angiosperms. Cambridge University Press, Cambridge.

Barbour, M. G., J. H. Burk, and W. D. Pitts. 1980. Terrestrial plant ecology. Benjamin/Cummings, Menlo Park, CA.

Boyd, C. E. 1970. Production, mineral accumulation and pigment concentrations in *Typha latifolia* and *Scirpus americanus*. Ecology 51:285 --290.

Boyd, C. E. 1978. Chemical composition of wetland plants. Pages 155–167 *in* R. E. Good, D. F. Whigham, and R. L. Simpson (eds.), Freshwater wetlands: ecological processes and management potential. Academic Press, New York.

Bradbury, I. K., and G. Hofstra. 1976. Vegetation death and its importance in primary production measurements. Ecology 57:209–211.

Braun-Blanquet, J. 1965. Plant sociology: the study of plant communities. Translation of first ed. of Pflanzensoziologie (1928) by G. D. Fuller and H. S. Conard. McGraw-Hill, New York.

Bridgham, S. D., J. Pastor, C. A. McClaugherty, and C. J. Richardson. 1995. Nutrient-use efficiency: a litterfall index, a model and a test along a nutrient-availability gradient in North Carolina peatlands. American Naturalist 145(1):1–21.

Brix, H., and H.-H. Schierup. 1989. The use of aquatic macrophytes in water-pollution control. Ambio 18:100–107.

Brown, S. 1981. A comparison of the structure, primary productivity, and transpiration of cypress ecosystems in Florida. Ecological Monographs 5:403–427.

Cain, S. A. 1938. The species-area curve. American Midland Naturalist 19:573–581.

Conner, W. H., and J. W. Day, Jr. 1976. Productivity and composition of a bald cypress-water tupelo site and a bottomland hardwood site in a Louisiana swamp. American Journal of Botany 63:1354–1364.

Cottam, G., and J. T. Curtis. 1956. The use of distance measures in phyto-sociological sampling. Ecology 37:451–460.

Cox, G. W. 1967. Laboratory manual of general ecology. Wm. C. Brown, Dubuque, IA.

Cranford, P. J., D. C. Gordon, and C. M. Jarvis. 1989. Measurements of cordgrass, *Spartina alterniflora,* production in a macrotidal estuary, Bay of Fundy. Estuaries 12:27–34.

Curtis, J. T. 1959. The vegetation of Wisconsin: an ordination of plant communities. University of Wisconsin Press, Madison, WI.

Curtis, J. T., and G. Cottam. 1962. Plant ecology workbook: laboratory, field and reference manual. Burgess Publishing, Minneapolis, MN.

Cuttle, S. P. 1983. Chemical properties of upland peats influencing the retention of phosphate and potassium ions. Journal of Soil Science 34:75–82.

Dai, T., and R. G. Wiegert. 1996. Ramet population dynamics and net aerial production of *Spartina alterniflora*. Ecology 77(1):276–288.

Daoust, R. J., and D. L. Childers. 1998. Quantifying aboveground biomass and estimating net aboveground primary production for wetland macrophytes using a non-destructive phenometric technique. Aquatic Botany 62:115–133.

Daubenmire, R. F. 1959. Canopy coverage method of vegetation analysis. Northwest Science 33:43–64.

Daubenmire, R. F. 1968. Plant communities: a textbook of plant synecology. Harper & Row, New York.

de Leeuw, J., A. Wielemaker, W. de Munck, and P. M. J. Herman. 1996. Net aerial primary production (NAPP) of the marsh macrophyte *Scirpus maritimus* estimated by a combination of destructive and non-destructive sampling methods. Vegetatio 123:101–108.

Dickerman, J. A., A. J. Stewart, and R. G. Wetzel. 1986. Estimates of net aerial aboveground production to sampling frequency. Ecology 67:650–659.

Dykyjova, D. 1973. Content of mineral macronutrients in emergent macrophytes during their seasonal growth and decomposition. Pages 163–172 *in* S. Hejny (ed.), Ecosystem study on wetland biome in Czechoslovakia. Czechoslovak IBP/PT-PP Report 3. Trebon, Czech Republic.

Dykyjova, D., and J. Kvet (eds.). 1978. Pond littoral ecosystems: structure and functioning. Springer-Verlag, Berlin.

Fletcher, J. S. 1990. Use of algae versus vascular plants to test for chemical toxicity. Pages 33–39 *in* W. Wang, J. W. Gorsuch, and W. R. Lower (eds.), Plants for toxicity assessment. American Society for Testing and Materials, Philadelphia.

Gauch, H. G. 1982. Multivariate analysis in community ecology. Cambridge University Press, New York.

Hammer, D. A. 1992. Creating freshwater wetlands. Lewis Publishers, Chelsea, MI.

Harris, F. W., R. M. Anderson, T. L. Cox, B. E. Dinger, P. Sollins, L. A. Stephens, F. G. Taylor, and D. E. Todd. 1972. Terrestrial primary production. Pages 56–68 *in* S. I. Auerbach and D. J. Nelson (eds.), Ecological Sciences Division annual progress report for period ending Sept. 30, 1971. Publication 430. Ecological Sciences Division, Oak Ridge, TN.

Hopkinson, C. S., J. G. Gosselink, and R. T. Parrondo. 1980. Production of Louisiana marsh plants calculated from phenometric techniques. Ecology 61:1091–1098.

Hsieh, Y. P. 1996. Assessing aboveground net primary production of vascular plants in marshes. Estuaries 19:82–85.

Jaccard, P. 1928. Die statistisch-floristiche Methode als Grundlage der Pflanzensoziologie. Handbuch der Biologischen Arbeitsmethoden 11:165–202.

Jakrlova, J. 1989. Primary production. Pages 304–330 *in* D. Dykyjov (ed.), Methods for ecosystem studies. Academia Praha, Czech Republic.

Janetschek, H. 1982. Okologische Feldmethoden. E. Ulmer Verlag, Stuttgart, Germany.

Johnston, C. A. 1991. Sediment and nutrient retention by freshwater wetlands: effects on surface water quality. CRC Critical Reviews in Environmental Control 21:491–565.

Judy, B. M., W. R. Lower, C. D. Miles, M. W. Thomas, and G. F. Krause. 1990. Chlorophyll fluorescence of a higher plant as an assay for toxicity assessment of soil and water. Pages

308–318 *in* W. Wang, J. W. Gorsuch, and W. R. Lower (eds.), Plants for toxicity assessment. American Society for Testing and Materials, Philadelphia.

Kaswadji, R. F., J. G. Gosselink, and E. G. Turner. 1990. Estimation of primary production using five different methods in a *Spartina alterniflora* salt marsh. Wetlands Ecology and Management 1:57–64.

Kershaw, K. A. 1964. Quantitative and dynamic ecology. American Elsevier, New York.

Kirby, C. J., and J. G. Gosselink. 1976. Primary production in a Louisiana Gulf Coast *Spartina alterniflora* marsh. Ecology 57:1052–1059.

Klopatek, J. M. 1975. The role of emergent macrophytes in mineral cycling in a freshwater marsh. Pages 367–383 *in* F. G. Howell, J. B. Gentry, and M. H. Smith (eds.), Mineral cycling in southeastern ecosystems. Symposium Series, CONF 740513. Energy Research and Development Administration, Washington, DC.

Koerselman, W., and A. F. M. Meuleman. 1996. The vegetation N:P ratio: a new tool to detect the nature of nutrient limitation. Journal of Applied Ecology 33:1441–1450.

Kruczynski, W. L., C. B. Subrahmanyam, and S. H. Drake. 1978. Studies on the plant community of a north Florida salt marsh: I. Primary production. Bulletin of Marine Science 28:316–334.

Lewis, M. A., F. L. Mayer, R. L. Powell, M. K. Nelson, S. J. Klaine, M. G. Henry, and G. W. Dickson (eds.). 1999. Ecotoxicology and risk assessment for wetlands. Setac Press, Pensacola, FL.

Lieth, H., and R. H. Whittaker. 1975. Primary productivity of the biosphere. Ecological Studies Vol. 14. Springer-Verlag, Berlin.

Linthurst, R. A., and R. J. Reimold. 1978. Estimated net aerial primary productivity for selected estuarine angiosperms in Maine, Delaware and Georgia. Ecology 59:945–955.

Lomnicki, A., E. Bandola, and K. Jankowska. 1968. Modification of the Wiegert–Evans method for estimation of net primary production. Ecology 49:147–149.

Mcgonigal, J. P., and F. P. Day. 1992. Effects of flooding on root and shoot production in large experimental enclosures. Ecology 74:1182–1193.

Milner, C., and R. E. Hughes, 1968. Methods for the measurement of the primary production of grassland. IBP Handbook 7. Blackwell, Oxford.

Morris, J. T., and B. Haskin. 1990. A 5-yr record of aerial primary production and stand characteristics of *Spartina alterniflora*. Ecology 71:2209–2217.

Mueller-Dombois, D., and H. Ellenberg. 1974. Aims and methods of vegetation ecology. Wiley, New York.

Oosting, H. J. 1956. The study of plant communities: an introduction to plant ecology, 2nd ed. W.H. Freeman, San Francisco.

Page, A .L., R. H. Miller, and D. B. Keeney (eds.). 1982. Methods of soil analysis, Part 2, Chemical and microbiological properties, 2nd ed. American Society of Agronomy and Soil Science Society of American, Madison, WI.

Phillips, E. A. 1959. Methods of vegetation study. Holt, Rinehart & Winston, New York.

Reader, R. J., and J. M. Stewart. 1972. The relationship between net primary production and accumulation for a peatland in southeastern Manitoba. Ecology 53:1024–1037.

Richardson, C. J. 1978. Primary productivity values in fresh water wetlands. Pages 131–145 *in* P. E. Greeson, J. R. Clark, and J. E. Clark (eds.), Wetland functions and values: The state of our understanding. American Water Resources Association, Minneapolis, MN.

Richardson, C. J., and C. W. Cares. 1976. An analysis of elm *(Ulmus americana)* mortality in a second-growth hardwood forest in southeastern Michigan. Canadian Journal of Botany 54(10):1120–1125.

Richardson, C. J., and P. E. Marshall. 1986. Processes controlling the movement, storage, and export of phosphorus in a fen peatland. Ecological Monographs 56:279–302.

Richardson, C. J., and S. Qian. 1999. Long-term phosphorus assimilative capacity in freshwater wetlands: a new paradigm for maintaining ecosystem structure and function. Environmental Science and Technology 33(10):1545–1551.

Richardson, C. J., D. L. Tilton, J. A. Kadlec, J. P. M. Chamie, and W. A. Wentz. 1978. Nutrient dynamics of northern wetland ecosystems. Pages 217–241 *in* R. E. Good, D. F. Whigham, and R. L. Simpson (eds.), Freshwater wetlands: ecological processes and management potential. Academic Press, New York.

Richardson, C. J., G. M. Ferrell, and P. Vaithiyanathan. 1999. Effects of N and P additions on stand structure, nutrient resorption efficiency and secondary compounds of sawgrass (*Cladium jamaicense* Crantz) in the subtropical Everglades. Ecology, 80(7):2182–2192.

Schlesinger, W. H. 1978. Community structure, dynamics, and nutrient ecology in the Okefenokee cypress swamp-forest. Ecological Monographs 48:43–65.

Sculthorpe, C. D. 1967. The biology of aquatic vascular plants. St. Martin's Press, New York.

Shew, D. M., R. A. Linhurst, and E. D. Seneca. 1981. Comparison of production computation methods in a southeastern North Carolina *Spartina alterniflora* salt marsh. Estuaries, 4:97–109.

Simpson, R. L. (ed.). 1978. Freshwater wetlands: ecological processes and management potential. Academic Press, New York, pp. 217–241.

Singh, J. S., W. K. Lauenroth, and R. K. Steinhorst. 1975. Review and assessment of various techniques for estimating net aerial primary production in grasslands from harvest data. Botanical Reviews 41:181–232.

Smalley, E. A. 1959. The role of two invertebrate populations, *Littorina irrorata* and *Orchelimum fidicinium* in the energy flow of a salt marsh ecosystem. Dissertation, University of Georgia, Athens, GA.

Sørensen, T. 1948. A method of establishing groups of equal amplitude in plant sociology based on similarity of species content. Kongelige Danske Videnskabernes Selskab Biologiske Skrifter 5(4):1–34.

Szumigalski, A. R., and S. E. Bayley. 1996. Net above-ground primary production along a bog-rich fen gradient in central Alberta, Canada. Wetlands 16:467–476.

Tiner, R. W. 1999. Wetland indicators: a guide to wetland identification, delineation, classification, and mapping. Lewis Publishers, Boca Raton, FL.

Valiela, I., J. M. Teal, and W. J. Sass. 1975. Production and dynamics of salt marsh vegetation and the effects of experimental treatment with sewage sludge. Journal of Applied Ecology 12:973–981.

Verhoeven, J. T. S., W. Koerselman, and A. F. M. Meuleman. 1996. Nitrogen- or phosphorus-limited growth in herbaceous, wet vegetation: relations with atmospheric inputs and management regimes. Trends in Ecology and Evolution 11:494–497.

Vollenweider, R. A. (ed.). 1969. A manual on methods for measuring primary production in aquatic environments. International Biological Program Handbook 12. Blackwell, Oxford.

Vymazal, J. 1995. Algae and element cycling in wetlands. Lewis Publishers, Boca Raton, FL.

Wang, W., J. W. Gorsuch, and W. R. Lower (eds.). 1990. Plants for toxicity assessment. American Society for Testing and Materials, Philadelphia.

Westlake, D. F. 1963. Comparisons of plant productivity. Biological Reviews 38:385–425.

Westlake, D. F. 1969. Macrophytes. Pages 103–107 *in* R. A. Vollenveider (ed.), A manual on methods for measuring primary production in aquatic environments. International Biological Program Handbook 12. Blackwell, Oxford.

Wetzel, R. G.,1983. Limnology, 2nd ed. CBS College Publishing, Philadelphia.

Whigham, D. F., J. McCormick, R. E. Good, and R. L. Simpson. 1978. Biomass and primary production in freshwater tidal wetlands of the Middle Atlantic coast. Pages 3–20 *in* R. E. Good, D. F. Whigham, R. L. Simpson (eds.), Freshwater wetlands: ecological processes and management potential. Academic Press, New York.

Whittaker, R. H. 1961. Estimation of net primary production of forest and shrub communities. Ecology 42:177–180.

Whittaker, R. H. 1962. Net production relations of shrubs in the Great Smokey Mountains. Ecology 43:357–377.

Whittaker, R. H. 1965. Branch dimensions and estimation of branch production. Ecology 46:365–370.

Whittaker, R. H., and G. M. Woodwell. 1968. Dimensions and production relations of trees and shrubs in the Brookhaven forest, New York. Journal of Ecology 56:1–25.

Wiegert, R. H., and F. C. Evans. 1964. Primary production and the disappearance of dead vegetation on an old field in southeastern Michigan. Ecology 45:49–63.

Williams, R. B., and M. B. Murdoch. 1972. Compartmental analysis of the production of *Juncus roemerianus* in a North Carolina salt marsh. Chesapeake Science 13:69–79.

15 Sampling Invertebrates in Wetlands

DAROLD P. BATZER, AARON S. SHURTLEFF,
and RUSSELL B. RADER

Difficulties in sampling have long hindered research on wetland macroinvertebrates. With the increasing interest in using macroinvertebrate populations to monitor the environmental health of wetlands, sampling of these organisms has become an important research focus. For this chapter we summarized sorting and subsampling procedures and queried many of the prominent researchers who study freshwater wetland macroinvertebrates about their preferences in samplers. For each device we provide a synopsis of their comments, both pro and con, and provide direction on how to use each sampler. Based on the results of this survey as well as published studies that contrast sampler efficacies, we conclude that the sweep net should probably become the sampler of choice for most bioassessment efforts that use wetland macroinvertebrates. We also recommend that most programs sort in the laboratory using either a selective or random technique (depending on the level of taxonomic expertise) and a fixed count of 100 to 300 individuals.

The relatively recent interest in using wetland macroinvertebrates in bioassessment programs has focused attention on the sampling of these organisms (Cheal et al. 1993, Turner and Trexler 1997). For bioassessment programs, sampling wetland macroinvertebrates can be challenging because numerous habitats may need to be sampled accurately while working within a limited budget. In this chapter we critique the various kinds of samplers that are commonly used by wetland invertebrate researchers and make recommendations for their use in bioassessment programs.

WETLAND MACROINVERTEBRATES AND BIOASSESSMENT PROGRAMS

Rosenberg and Resh (1993) outline the rationale for using macroinvertebrates to monitor the environmental health of aquatic habitats (e.g., streams, river, lakes).

Bioassessment and Management of North American Freshwater Wetlands, edited by Russell B. Rader, Darold P. Batzer, and Scott A. Wissinger.
0-471-35234-9

Unfortunately, similar habitat-specific information is not yet available for wetland macroinvertebrates. In Table 15.1 we list many of Rosenberg and Resh's criteria for using aquatic macroinvertebrate biomonitors in aquatic habitats and comment on their applicability to wetland macroinvertebrates and wetland habitats. This review

TABLE 15.1. Criteria for Using Macroinvertebrate Bioassessment in Aquatic versus Wetland Habitats

Rosenberg and Resh (1993) Criteria for Using Macroinvertebrate Biomonitoring in Aquatic Habitats	Applicability of Criteria for Wetlands
Advantages	
1. Macroinvertebrates are ubiquitous.	True for wetlands.
2. A large number of species occur offering a spectrum of responses.	True for wetlands.
3. Macroinvertebrates are usually sedentary, allowing for effective spatial analysis.	Many wetland taxa are highly mobile (e.g., water bugs and water beetles).
4. The somewhat long life cycles of macroinvertebrates facilitate temporal analysis.	Most wetland taxa are relatively short lived (<1–2 months).
5. Sampling macroinvertebrates is a simple procedure.	True for wetlands.
6. The taxonomy is quite well known for many groups.	Only true for a few wetland groups (e.g., odonates, beetles).
7. The responses of many species to different types of pollution are well known.	Tolerances of wetland taxa to pollutants are largely unknown.
8. Macroinvertebrates can be quite easily manipulated in experimental studies (e.g., bioassays).	True for wetlands.
Disadvantages	
1. Macroinvertebrates may not respond to all types of pollution or all types of human impacts.	True for wetland taxa.
2. Large numbers of samples are often needed to achieve desirable precision.	True for wetlands.
3. There are potentially confounding seasonal effects (e.g., emergence).	True for wetlands.
4. Some groups are taxonomically difficult to use (e.g., Chironomidae).	Especially true for wetlands.
5. Biotic indices are just beginning to be developed.	Especially true for wetlands.

suggests that although the disadvantages of using aquatic macroinvertebrates for bioassessment all seem to apply to wetland macroinvertebrates, only some of the advantages apply. Much baseline information on wetland macroinvertebrates is lacking (e.g., their responses to pollutants), largely because the use of wetland macroinvertebrates for bioassessment is only beginning to be explored. However, examples exist where macroinvertebrates have been used successfully in wetland bioassessment programs, and prominent case studies are provided in Chapters 7, 8, 10, and 11.

SAMPLING DEVICES

Several papers have quantitatively evaluated the efficacy of different devices for sampling macroinvertebrates from specific wetlands (Table 15.2). Conclusions among these studies vary, with the device(s) shown to be most effective changing from wetland to wetland. Benke et al. (1999) suggest that accurate quantitative sampling of macroinvertebrate communities in heterogeneous wetland environments will require many types of sampling gear, and they used seven different devices to obtain such a sample from a single wetland. However, the degree of accuracy required for Benke's study may not be necessary for most bioassessment programs, and simpler procedures may yet be effective.

We surveyed 25 authors from a recently published wetland invertebrate text (Batzer et al. 1999) about their preferences in sampling devices, and this survey suggests that in practice most researchers rely on only a few samplers. Below we review the results of that survey, discussing the most widely used samplers first. This treatment is not intended as a thorough review of all possible samplers, only the ones most commonly used in current wetland invertebrate research. [Refer to Murkin et al. (1994), Merritt et al (1996), and the papers cited in Table 15.2 for descriptions of the complete range of samplers available.]

Corers (and Quadrat Box Samplers). Twenty-one of the 25 researchers responding to our survey used some sort of corer in their research. A corer is simply a tube made of durable plastic or metal (Fig. 15.1*a*), with the bottom edge sharpened or serrated to enable easier penetration into sediments. Small corers (4 to 10 cm in diameter) are typically used to extract a standard-sized portion of bottom substrate and sometimes to sample the water column. Survey respondents using small corers typically collect organisms from plants with a second sampler (usually, a sweep net). Large corers (10 to 30 cm in diameter) or quadrat box samplers, which are essentially large, square corers, can enclose entire plants and are effective at collecting epiphytic as well as benthic macroinvertebrates.

To collect samples, corers are plunged through the water column and pushed into the bottom sediments until a watertight seal is made (usually about 10 cm). Most benthic macroinvertebrates live near the sediment surface of wetlands, so penetration deeper than 10 cm is unnecessary. Both sediments and water are removed and inspected for invertebrates. With small corers, water can be sampled by decanting it through a sieve, and then the sediment core can be removed. With large corers, the

TABLE 15.2. Statistical Comparisons Among Devices Used for Sampling Wetland Macroinvertebrates

Reference	Wetland Type	Samplers Compared	Conclusions
Anderson and Smith 1996	Vegetated playa wetlands	Gerking, quadrat, and water-column samplers	*Abundance, biomass, collection time*: Quadrat/water column combination > Gerking *Richness*: all similar
Brinkman and Duffy 1996	Deep marsh zone of a semipermanent wetland	Gerking sampler, core (small diameter), activity trap, plant-mimic artificial substrate	*Abundance:* Gerking > activity trap > core > artificial substrate *Richness:* Gerking > activity trap > artificial substrate > core *Sorting time:* (1) Artificial substrate (least time), (2) activity trap, (3) core, (4) Gerking (most time)
Cheal et al. 1993	Multiple wetlands	Core, plankton net, sweep net	*Richness:* Sweep > plankton > core
Kaminski and Murkin 1981	Marsh channels and ponds	Sweep net, modified Gerking sampler	Sweep net = Gerking
Murkin et al. 1983	Northern marsh	Activity trap, sweep net	Activity trap = sweep net, except activity traps more efficient in dense vegetation
Turner and Trexler 1997	Everglades marshes	Sweep net, stovepipe sampler, throw trap, corer (small dia.), funnel trap, minnow trap, Hester–Dendy sampler, plankton net	*Abundance:* Funnel > Hester–Dendy > sweep > stovepipe > throw trap > plankton > core > minnow trap *Richness:* Sweep > funnel > stovepipe = Hester–Dendy > throw *Equitability:* Funnel > sweep > stovepipe = Hester–Dendy > throw *Species composition:* No clear correlations among samplers *Sorting time:* (1) Funnel (least time), (2) Hester–Dendy, (3) throw, (4) sweep, (5) stovepipe (most time)

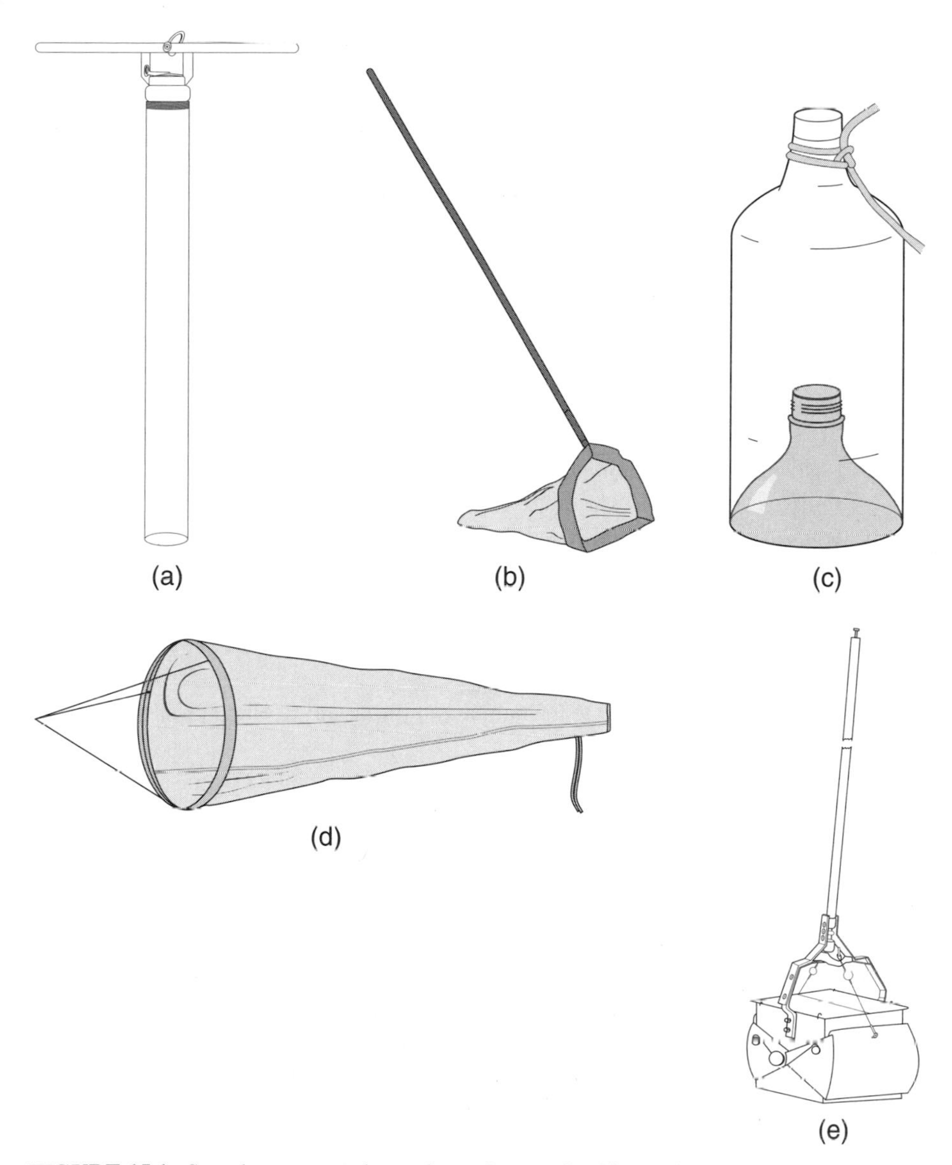

FIGURE 15.1. Samplers commonly used to collect wetland invertebrates: *(A)* small-diameter corer; *(B)* D-frame sweep net; *(C)* bottle activity trap; *(D)* plankton tow net; *(E)* Ekman grab.

water can be pumped through a sieve using a bilge pump, and then plant material and surface sediments can be removed manually. Rather than removing contents from large corers or quadrat boxes, some contributors simply take a standard number of sweeps with a dip net inside enclosures. However, this technique requires an initial calibration study to determine the capture efficiency of sweeping. According to survey respondents, the advantages of corers include:

1. A standardized, quantitative sample is collected.
2. Even small invertebrates are collected.
3. Both a benthic and a water-column sample are collected, and contents from both strata can be separated (see Euliss et al. 1992).
4. For small corers, the ease of use enables inexperienced personnel to sample efficiently.
5. Most corers are inexpensive to make or purchase.
6. The devices are portable.
7. Some corers can be operated from a boat.

Disadvantages of corers include:

1. It is very time consuming to extract invertebrates (particularly small ones) from samples when large volumes of organic sediments and debris are collected.
2. Wood and fibrous plant material can prevent the corer from penetrating into substrates.
3. Some of the more mobile organisms can evade capture (e.g., freshwater shrimps and water bugs), as might individuals living in deep crayfish burrows or old tree-root holes.
4. Sampling is destructive to the habitat (an important consideration for repeated sampling of small plots).
5. Even with small corers, removing a sediment core intact from sandy or loose substrates is difficult.
6. Large corers or quadrat boxes cannot be used in water deeper than the sampler's height, and even in shallow water the devices can be cumbersome to use.

Solutions to some of these problems are being developed. Euliss and Swanson (1989) describes a screening technique that allows easier sorting of macroinvertebrates from mud and plant debris (see also Brinkman and Duffy 1996, Merritt et al. 1996). We should note that the flotation or elutrication techniques developed to separate macroinvertebrates from the largely mineral-based sediments of lakes, rivers, or streams are not as effective for highly organic wetland sediments. The time-consuming process of hand-sorting screened sediments for macroinvertebrates remains the norm for processing core samples (or for that matter any kind of wetland macroinvertebrate samples).

Given their popularity with researchers, it is perhaps surprising that studies evaluating samplers (Table 15.2) generally give corers low rankings, particularly in terms of the numbers of taxa that they collect. However, only the performance of small-diameter corers were included in those tests. Our survey respondents acknowledge that small corers are only efficient at collecting sedentary benthic macroinvertebrates and plankton (few highly motile and epiphytic taxa are captured), and they indicate that small corer samples must be supplemented with other efforts to collect the complete range of macroinvertebrates that might be present.

Sweep Nets. Fourteen of 25 survey respondents used aquatic dip or sweep nets (this does not include their use inside large corers). D-frame sweep nets are the norm (Fig. 15.1*b*), although other shapes are used. However, the specific shape is probably less important to standardization than is the mesh size of the net. Small taxa or early developmental stages of large macroinvertebrates will pass through the large 1- to 2-mm mesh sizes that are most commonly available.

Sweep nets are generally used to sample the water column or submersed plant surfaces but can be used to sample along benthic substrates. The net can be swept vertically up through the water column, horizontally through plants and the water column, or at a 45° angle from the bottom to the water surface. In shallow habitats, a horizontal sweep is the norm. Although the volume of water passing through the net during sampling can be quantified, such a volumetric measure may not be ecologically relevant. Except for a few planktonic forms, most macroinvertebrates in wetlands are associated with substrates, either plant or benthic, and water volume is not meaningful. Sweep-net data probably should not be expressed in terms of area or volume, but rather, in terms of sampling effort (i.e., numbers of organisms/standardized sweep). Thus, sweep nets may not be suitable for some quantitative analyses. However, sweep nets yield results of high precision, making them particularly useful for among-site or among-year comparisons (Cheal et al. 1993, Turner and Trexler 1997, Rader 1999). According to respondents, the advantages of sweep nets include:

1. Virtually all macroinvertebrate taxa are collected.
2. Samples are relatively clean, so extensive sorting time is not required for processing.
3. Nets are commercially available and inexpensive.
4. Nets are easy for personnel to use, either within sites or from boats.
5. Protocols can be developed to stratify sampling across a range of subhabitats within individual wetlands.

Disadvantages of nets include:

1. It is difficult to extrapolate results into truly quantitative measures of macroinvertebrate density (i.e., numbers/m^2).
2. Small organisms can pass through the net and be missed when large mesh sizes are used.
3. Nets may clog if small mesh sizes are used, particularly where filamentous algae are abundant.
4. Dense emergent vegetation or woody debris inhibits operation.
5. Highly mobile macroinvertebrates can evade capture.

Emergence Traps. Eight of the 25 survey respondents reported that they had used emergence traps in their research. An emergence trap collects flying adult insects that developed from aquatic immatures. Traps are commonly floating pyramid- or cone-shaped tents, with an open bottom that allows emerging insects to enter. The sloping mesh-covered walls direct insects to a collection bottle (with preservative) that is fitted on the trap top. However, a small aspirator is still needed to gather those individuals that do not enter the collection bottle. According to respondents, the advantages of emergence traps include:

1. They provide a quantitative density measure of flying insect emergence.
2. Minimal time is required to process samples.
3. They capture adult insects, which are easier to identify than immatures because better species-level taxonomic keys are available.

Disadvantages of emergence traps include:

1. Sampling is biased because traps do not collect nonflying macroinvertebrates and inefficiently collect those flying insects that primarily swim (hemipterans and beetles).
2. Emergence rates can be overestimated for insects that use trap surfaces for habitat or as emergence substrate, or they can be underestimated for those insects that avoid traps.
3. Spiders or other predators living inside traps can consume emerged insects.
4. Traps are easily damaged by muskrats, birds, or inclement weather.

Although emergence traps are commonly used for taxonomic studies of wetland insects (e.g., Wrubleski and Rosenberg 1990, Leeper and Taylor 1998), it remains difficult to classify many taxa (especially chironomid midges) to species with only a single life stage. Optimally, adult males and females, pupae, and last instar larvae should all be provided to taxonomists for species determinations, which usually requires laboratory rearing of specimens (Merritt et al. 1996).

Activity Traps. Seven of 25 respondents used activity traps in their research. Activity traps capture wetland macroinvertebrates that swim or crawl. They are constructed by removing the top of a glass or plastic bottle and replacing it with a funnel opening. A simple trap fashioned out of plastic beverage bottles is described by Riley and Bookhout (1990) (Fig. 15.1*c*). Macroinvertebrates will enter the trap through the funnel and are retained because they have difficulty reexiting. Sampling is standardized by using a set-trapping period, usually 24 hours. Traps can be "fished" vertically or horizontally and can be placed to sample depth strata at various depths. According to contributors, the advantages of activity traps include:

1. Samples are extremely clean, so sample sorting is easy.
2. They are one of the few samplers whose efficiency is not impaired by dense emergent vegetation.
3. Traps can be used to monitor horizontal or vertical movements of invertebrates.
4. Sampling is not destructive to the habitat.

Disadvantages of activity traps include:

1. Sampling is biased because particularly active swimmers may dominate collections even when they are not the most abundant.
2. Only a qualitative index of macroinvertebrate abundance can be obtained (even though the traps enclose a standard amount of water, trap volume should not be used to quantify density).
3. Sampling requires two visits, one to set the trap and a second to pick it up, which may be a hardship in difficult-to-access locations.
4. Predators that are captured may consume other organisms in the trap.
5. Nontarget small fish and amphibians readily enter traps and may inadvertently be killed.

Tow Nets. Five of 25 contributors had used small plankton tow nets (Fig. 15.1d) in their research. Because these samplers are typically used to collect microinvertebrates, nets are made of fine mesh (<100 μm). The net terminates in a small collecting vessel that can be detached to remove invertebrates. Nets can be pulled either vertically or horizontal through the water column. The limited use of these samplers may reflect a bias among survey respondents toward the study of macro- rather than microinvertebrates. According to contributors, the advantages of plankton tow nets include:

1. The fine mesh retains even the smallest invertebrates (e.g., small crustaceans, rotifers).
2. A quantitative measure of water volume can be obtained, which is ecologically relevant for planktonic taxa.

Disadvantages include:

1. The fine-meshed nets clog easily where algae are abundant.
2. Macrophytes inhibit towing efficiency.
3. Nonplanktonic invertebrates are rarely collected.

Grab Samplers. Three of 25 respondents had used a grab sampler in their research. The most popular of these was a pole-mounted Ekman-type grab, which is a rectangular box with paired scoop jaws on the bottom (Fig. 15.1e). With its jaws set open, the device is shoved into the bottom substrate, and then the jaws are closed to gather a sample of sediment. According to our contributors, the advantages of grabs include:

1. A large, quantitative sample is collected.
2. In deep water, grabs can be used effectively from a boat
3. The sampler collects the near-surface sediments only where most benthic macroinvertebrates of wetlands occur.

Disadvantages of grabs include:

1. Woody debris or coarse plants prevent the jaws from closing, so many attempts may be required before a sample is collected.
2. Processing the copious amount of sediment and debris in samples is time consuming.
3. Even after closing, the scoop box may still not retain loose sediment samples.
4. Highly mobile organisms may evade capture.

Artificial Substrates. Three of 25 respondents used artificial substrates in their research, including Hester–Dendy multiplate samplers (see Merritt et al. 1996) and artificial mimics of macrophytes. Clean artificial substrates are placed into habitats and biota begin to colonize them. After a set period of time (usually, 2 to 6 weeks), the samplers are retrieved by transferring them into mesh bags, and colonizing macroinvertebrates are removed in the laboratory. According to contributors, the advantages of artificial substrates include:

1. A standardized sample is collected, and the standard relates to colonizable substrate surface.
2. Human bias in sampling is minimized.
3. Samples can be sorted quickly.

Disadvantages of artificial substrates include:

1. Nonnatural attributes of samplers introduce bias (multiplate samplers do not mimic natural wetland substrates, and the texture and surface epiphytes of plant-mimic samplers may differ from those of natural plant surfaces).
2. Calibration studies are required to determine an appropriate incubation period (i.e., optimally substrates should remain in sites until community structures equilibrate).

SAMPLER SELECTION FOR BIOASSESSMENT PROGRAMS

In Figure 15.2 we provide a simple flowchart for sampler selection in bioassessment programs. Until we have a better understanding of the response of specific taxa to impacts (i.e., indicator taxa), the entire macroinvertebrate community probably needs to be sampled, either to calculate biotic indices or to develop ratio metrics. Only two samplers commonly used in wetland macroinvertebrate research will collect whole community samples, large corers and sweep nets. Most other samplers collect only a subset of the species present, and for wetland bioassessment purposes these samplers should be used only if specific organisms are being targeted (e.g., threatened species, indicator taxa).

Large corers (or quadrat boxes) yield a truly quantitative measure of macroinvertebrate numbers (i.e., individuals per unit area), and this kind of measure has obvious appeal (hence the common use of these devices for research projects). However, as discussed above, large corers have numerous logistic limitations. For purposes of a bioassessment program the most important limitation is that large amounts of time are required to collect and process samples, probably making large corers impractical to use in most cases.

Despite (or perhaps because of) the simplicity of sweepnets, they seem to have become the sampler of choice for wetland bioassessment (Cheal et al. 1993, Turner and Trexler 1997). Even researchers use them extensively, although the corer remains the most popular sampler for research. Quantitative testing (Table 15.2) indicates that sweep nets effectively collect most macroinvertebrates and the numerical measure generated (albeit semiquantitative) has high statistical precision (see Chapter 3 for a discussion of precision). Sweep-net data are suitable for analyses with either multimetric or multivariate approaches to bioassessment (see Chapter 8). Processing samples remains a logistic obstacle, but sweep-net samples are usually easier to process than corer samples. The development of more innovative sample processing techniques remains an important need.

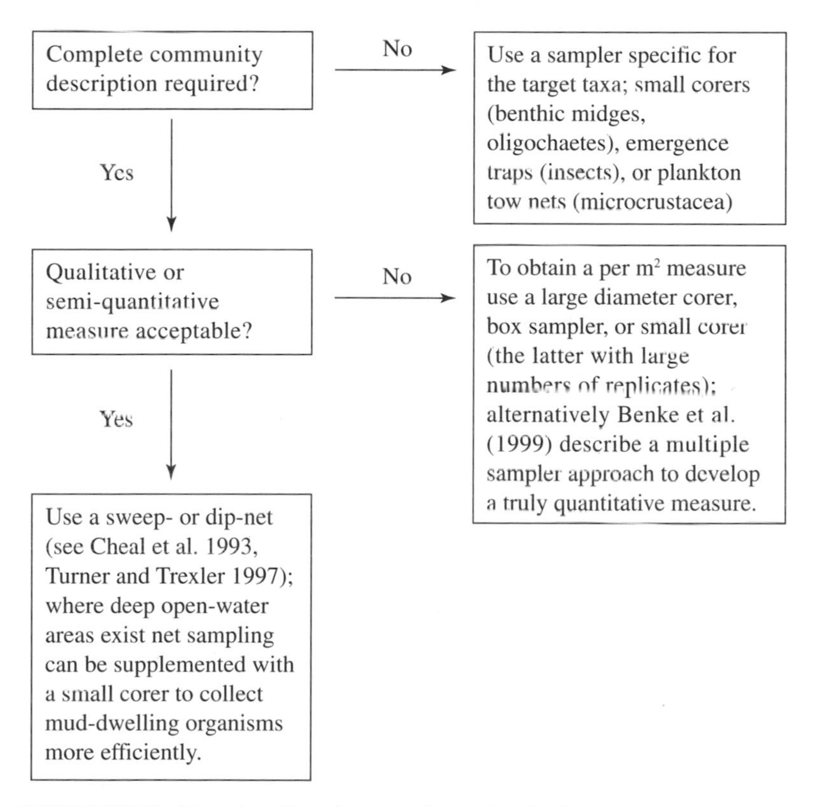

FIGURE 15.2. Flowchart for selection of samplers for bioassessment programs.

Some sort of standardization in sampling is required before wetland macroinvertebrates can be used successfully as biomonitors. The ability to compare results from potentially impacted sites to reference standards is required, and this will be difficult or impossible if a variety of sampling procedures are being used. One device should probably be singled out for use, and we believe that sufficient evidence exists that the sweep net should be that sampler.

SORTING AND SUBSAMPLING PROCEDURES

One of the fundamental objectives of bioassessment is to ensure that the data obtained are a good representation of the sample(s) collected. Sorting/subsampling procedures are often designed to minimize sorting time while maintaining the ability to detect differences among reference conditions and test sites. Decisions concerning quantitative versus qualitative sampling, type of sampling devise, taxonomic resolution, number of samples, number of habitats, and many separate versus a single composite sample have been discussed above and in Chapter 2. The purpose of this section is to discuss the advantages and disadvantages of various sorting and subsampling procedures. Although we emphasize minimizing time and costs, each of these procedures, or some combination thereof, can be applied to more quantitative designs that may utilize genus or species taxonomic resolution to detect early signs of impairment. There are three interrelated sorting/subsampling decisions: (1) live sorting in the field versus sorting of preserved samples in the laboratory, (2) selective subsampling versus random subsampling, and (3) timed subsamples versus fixed-count subsamples.

Field Sorting Versus Laboratory Sorting

Samples can be collected and sorted in the field for a specified length of time (e.g., 30 minutes) or until a specified number of individuals (e.g., 100) have been obtained (e.g., Growns et al. 1997). The advantages of live-sorting samples in the field include:

1. Easier detection of small cryptic taxa by their movements
2. Eliminates transporting preservative and large samples to and from remote locations

Disadvantages of this procedure include difficulties in:

1. Detecting taxa because of associated vegetation, sediment, and so on
2. Seeing very small taxa (e.g., ostracods) without the aid of a microscope
3. Processing samples in the field during foul-weather conditions

Unless the transport of preservative and samples into and out of remote areas is of overriding importance, it seems that laboratory processing has several advantages that may increase sampling accuracy. For instance, washing samples through sieves in the laboratory to remove large pieces of vegetation, sticks, and so on, and sorting samples under a microscope can facilitate the detection of small cryptic taxa. Also, various dyes (e.g., Rhodamine B or Congo Red) can be added to the preservative (e.g., ethyl alcohol) that will stain invertebrates and enhance their visibility (e.g., Brinkman and Duffy 1996). Also, various elutriation and floating techniques (supersaturated solution of table sugar or salt) can be used to separate organic material, including invertebrates, from inorganic particles (e.g., Fairchild et al. 1987). However, the relative merits of field processing of live organisms versus laboratory procedures have rarely been examined, especially in wetlands.

Selective Subsampling versus Random Subsampling

During selective subsampling a technician decides which individuals will be selected to represent a sample. That is, the technician is instructed to remove no more than 10 individuals of any "type" for a specified length of time or until a specified number of individuals have been collected. This method is selective in that technicians can maximize the number of taxa sorted while limiting the number of individuals per taxon. In a random procedure, the sample is placed in a white sorting tray that is divided into square sections of identical size (e.g., 6 × 6 cm). All organisms are removed from randomly selected sections for a specified length of time or until a specified number of individuals have been collected (e.g., Barbour et al. 1999). Although the experience of the technician is important in both procedures, there is a greater chance of producing a biased representation using the selective procedure because the technician must decide which forms represent different types. An experienced taxonomist will see many more taxa in a sample than will an inexperienced technician. The random technique can avoid this bias, when untrained taxonomists are used to sort or subsample. However, if an experienced taxonomist is available to process all the samples, the selective technique may maximize the number of taxa sorted per time expended. Growns et al. (1997) suggested that the selective process is better suited to detecting early signs of degradation, or small differences between reference sites and test sites, because a larger proportion of the taxa will be identified. Because some taxa are more sensitive to impairment than others, the greater the number of taxa used for bioassessment comparisons (reference versus test), the greater our ability to detect small differences.

Timed versus Fixed-Count Subsamples

As mentioned above, individuals can be removed from a sample for a specified length of time or until a specific number of individuals are sorted. This decision depends on the importance of providing unbiased estimates of richness. The number of species collected will increase with the size of the area or volume of water sampled (e.g., number of sweep samples) and the number of individuals collected (e.g., Vinson and Hawkins 1996). Therefore, the fixed-count method will provide a better estimate of richness (unbiased) compared to a timed procedure where different numbers of individuals may be sorted from different samples. Richness is one of the most important indices used in bioassessment because it provides information on both the number of taxa and the community composition. Some multivariate methods rely solely on the similarity between reference conditions and test sites using richness and community composition to detect impairment (see Chapter 4). Multimetrics also rely on unbiased estimates of invertebrate richness (e.g., Chapters 7 and 8). Therefore, the fixed-count method should be used to standardize sorting/subsampling techniques because it provides an unbiased estimate of richness.

The next question has been the topic of several investigations: How many individuals should be counted per sample when using a fixed-count method? A count of 100 individuals appears sufficient for metrics that do not rely directly on richness or programs that do not require the detection of rare taxa (e.g., Sovell and Vondracek 1999). Depending on the severity of impairment, richness measures may require 200 or 300 individuals per sample to detect differences (reference versus test).

Missing from most studies comparing various sampling devices, sorting procedures, and subsampling techniques is an estimate of the trade-off between timed saved versus lost information to detect small differences between reference sites and test sites. To detect early signs of degradation, it may be necessary to compare more taxa among sites (reference versus test), including rare species. The search for rare taxa includes two sorting/subsampling techniques: (1) a two-phase large-rare search, or (2) serial processing. The two-phase large-rare search involves searching the entire sample for large-rare taxa prior to using a standard subsampling procedure (e.g., fixed count). Larger individuals are typically rare and could easily be missed without such preliminary searching efforts. However, some rare taxa are small and may best be detected using a serial processing approach. The serial process involves increasing the number of individuals removed beyond the number in a fixed count or until a richness–effort curve becomes asymptotic. That is, the investigator stops sorting when the addition of more individuals does not appreciably increase the number of species encountered (Vinson and Hawkins 1996). A large-rare search combined with serial processing may provide the best opportunity of detecting early signs of impairment, especially if a taxonomic expert is used for subsampling and identification.

In summary, most bioassessment programs will sort in the laboratory using either a selective or a random technique (depending on the level of taxonomic expertise) and a fixed count of 100 to 300 individuals. More quantitative techniques designed to detect early signs of degradation may be inclined to sort selectively using a large-rare search, serial processing, and identifying taxa to the genus or species level. It is important to note that larval, pupal, adult, and even egg stages can be helpful in identifying wetland invertebrates to finer taxonomic levels (genus or species).

ACKNOWLEDGMENTS

We thank Erica Chiao for providing drawings of samplers. This work was sponsored by funds from the Hatch program and the Brigham Young University Zoology Department.

REFERENCES

Anderson, J. T., and L. M. Smith. 1996. A comparison of methods for sampling epiphytic and nektonic invertebrates in playa wetlands. Journal of Freshwater Ecology 11:219–224.

Barbour, M. T., J. Gerritsen, B. D. Snyder, and J. B. Stribling. 1999. Rapid bioassessment protocols for use in streams and wadeable rivers: periphyton, benthic macroinvertebrates, and fish, 2nd edition. Report EPA 841-B-99-002. U.S. Environmental Protection Agency, Office of Water, Washington, DC.

Batzer, D. P., R. B. Rader, and S. A. Wissinger (eds.). 1999. Invertebrates in freshwater wetlands of North America: ecology and management. Wiley, New York.

Benke, A. C., G. M. Ward, and T. D. Richardson. 1999. Beaver-impounded wetlands of the southeastern coastal plain: habitat-specific composition and dynamics of invertebrates. Pages 217–246 *in* D. P. Batzer, R. B. Rader, and S. A. Wissinger (eds.), Invertebrates in freshwater wetlands of North America: ecology and Management. Wiley, New York.

Brinkman, M. A., and W. G. Duffy. 1996. Evaluation of four wetland aquatic invertebrate samplers and four sample sorting methods. Journal of Freshwater Ecology 11:193–200.

Cheal, F., J. A. Davis, J. E. Growns, J. S. Bradley, and F. H. Whittles. 1993. The influence of sampling method on the classification of wetland macroinvertebrate communities. Hydrobiologia 257:47–56.

Euliss, N. H., Jr. and G. A. Swanson. 1989. Improved self-cleaning screen for processing benthic samples. Californai Fish and Game 75:126–128.

Euliss , N. H., Jr., G. A. Swanson, and J. McKay. 1992. Multiple tube sampler for benthic and pelagic invertebrates in shallow wetlands. Journal of Wildlife Management 56:186–191.

Fairchild, W. L., M. C. O'Neil, and D. M. Rosenberg. 1987. Quantitative evaluation of the behavioral extraction of aquatic invertebrates from samples of sphagnum moss. Journal of the North American Benthological Society 6:281–287.

Growns, J. E., B. C. Chessman, J. E. Jackson, and D. G. Ross. 1997. Rapid assessment of Australian rivers using macroinvertebrates: cost and efficiency of 6 methods of sample processing. Journal of the North American Benthological Society 16:682–693.

Kaminski, R. M., and H. R. Murkin. 1981. Evaluation of two devices for sampling nektonic invertebrates. Journal of Wildlife Management 45:493–496.

Leeper, D. A., and B. E. Taylor. 1998. Insect emergence from a South Carolina (USA) temporary wetland pond, with emphasis on the Chironomidae (Diptera). Journal of the North American Benthological Society 17:54–72.

Merritt, R. W., V. H. Resh, and K. W. Cummins. 1996. Design of aquatic insect studies: collecting, sampling and rearing procedures. Pages 12–28 *in* R. W. Merrit and K. W. Cummins (eds.), An introduction to the aquatic insects of North America. Kendall/Hunt, Dubuque, IA.

Murkin, H. R., P. G. Abbott, and J. A. Kadlec. 1983. A comparison of activity traps and sweep nets for sampling nektonic invertebrates in wetlands. Freshwater Invertebrate Biology 2:99–106.

Murkin, H. R., D. A. Wrubleski, and F. A. Reid. 1994. Sampling invertebrates in aquatic and terrestrial habitats. Pages 349–369 *in* T. A. Bookhout (ed.), Research and management techniques for wildlife and habitats, 5th ed. Wildlife Society, Bethesda, MD.

Rader, R. B. 1999. The Florida Everglades: natural variability, invertebrate diversity, and foodweb stability. Pages 25–54 *in* D. P. Batzer, R. B. Rader, and S. A. Wissnger (eds.),

Invertebrates in freshwater wetlands of North America: ecology and management. Wiley, New York.

Riley, T. Z., and T. A. Bookhout. 1990. Responses of aquatic macroinvertebrates to the early-spring drawdown in nodding smartweed marshes. Wetlands 10:173–185.

Rosenberg, D. M., and V. H. Resh. 1993. Introduction to freshwater biomonitoring and benthic macroinvertebrates. Pages 1–9 *in* D. M. Rosenberg and V. H. Resh (eds.), Freshwater biomonitoring and benthic macroinvertebrates. Chapman & Hall, New York.

Sovell, L. A., and B. Vondracek. 1999. Evaluation of the fixed-count method for Rapid Bioassessment Protocol III with benthic macroinvertebrate metrics. Journal of the North American Benthological Society 18:420–426.

Turner, A. M., and J. C. Trexler. 1997. Sampling aquatic invertebrates from marshes: evaluating the options. Journal of the North American Benthological Society 16:694–709.

Vinson, M. R., and C. P. Hawkins. 1996. Effects of sampling area and subsampling procedure on comparisons of taxa richness among streams. Journal of the North American Benthological Society 15:392–399.

Wrubleski, D. A., and D. M. Rosenberg. 1990. The Chironomidae (Diptera) of Bone Pile Pond, Delta Marsh, Manitoba, Canada. Wetlands 10:243–275.

PART 2
Managing Freshwater Wetlands

16 Management of Wetland Fish Populations: Population Maintenance and Control

JOEL W. SNODGRASS and JOANNA BURGER

Despite the harsh environmental conditions that occur in many North American freshwater wetlands, fish are often an important component of wetland faunas. Fishes occurring in freshwater wetlands often possess physiological, morphological, or behavioral adaptations that allow them to persist or move to adjacent, more benign aquatic habitats during periods of harsh environmental conditions within wetlands. Wetland fish populations may be managed for the protection of threatened or endangered species, to provide forage for other threatened or endangered species, to protect indigenous faunas from introduced species, or to provide harvestable fish for recreational use. In this chapter we define different groups of wetland habitats based on factors that influence fish use and production. We provide a brief review of factors influencing fish occurrence, assemblage structure and production, and practices used to manage fish populations in each wetland type.

When one thinks of a wetland, an ephemeral aquatic habitat comes to mind, one in which water is present for only a portion of the year. Intuitively, it would seem that fish would not be a significant component of the wetland biota because they cannot survive long out of water. However, this is not the case in many systems. For example, in southern freshwater wetlands, from 12 to 62 species have been reported from individual habitats (Hoover and Killgore 1998).

Fishes are integral parts of many wetland systems for two reasons. First, although wetlands are often environmentally severe habitats (e.g., low levels of dissolved oxygen and high temperatures are common), many fishes have adapted to wetland life. Many fishes can tolerate high temperatures (Matthews 1987), or are physiologically, morphologically, or behaviorally adapted to survive low-dissolved-oxygen conditions (Kramer 1987, Matthews 1987) or temporary drying (Hoover and Killgore 1998). Second, many wetlands are integrated with deeper aquatic habitats and fishes

Bioassessment and Management of North American Freshwater Wetlands, edited by Russell B. Rader, Darold P. Batzer, and Scott A. Wissinger.
0-471-35234-9

can move from wetlands to aquatic refuges when environmental conditions in wetlands deteriorate.

In this chapter we (1) outline reasons for managing fish populations in wetlands; (2) define different groups of wetland habitats based on factors that influence fish use and production in each; (3) review briefly the factors that influence fish occurrence, assemblage structure, and production in wetland systems; (4) review practices used to manage wetland fish populations in each wetland type; and (5) recommend priorities for future research that will be of particular use in managing wetland fish populations. We have made a specific effort to include management practices that view wetlands as parts of larger, integrated aquatic systems, and that include population control as well as fish production. We include population control because fishes can have negative impacts on populations of other organisms. Intentional and inadvertent releases of fishes into wetlands have occurred in the past and are likely to increase in the future, and many managers are now charged with conserving overall biological diversity in addition to maintaining fish populations of sport and commercial interest. The chapter is not intended to be an exhaustive review of all literature pertaining to management of wetland fish populations. Rather, we provide an introduction to pressing issues and current methods, and list representative references that are useful starting points if more in-depth coverage of these topics is desired.

REASONS FOR MANAGING FRESHWATER FISH POPULATIONS

Protection of Fish Populations to Maintain Biodiversity

Fishes endemic to wetlands are most thoroughly documented in the desert southwest (Naiman and Soltz 1981) and Lake Waccamaw in North Carolina (Frey 1951). Ciénegas, marshy wetlands associated with springs and streams in the desert southwest, harbor 24 endemic species of fish (Meffe 1989). Two species, the Waccamaw darter (*Etheostoma prolongum*) and the Waccamaw glassminnow (*Mendia extensa*), are endemic to Lake Waccamaw, North Carolina, a large permanently flooded depression wetland. The Waccamaw killifish (*Fundulus waccamensis*) is found in Lake Waccamaw and proximate wetlands. Because ranges of endemics are so limited, protection of wetlands harboring endemics should be a priority in conserving these species.

Less understood is the role of wetlands in supporting fish species considered rare because of lack of scientific collections in wetland systems. Hoover and Killgore (1998) provide several examples of wetland fish species initially considered rare but later found in large numbers. This trend is likely to continue as ichthyologists more thoroughly explore wetland systems. For example, the mud sunfish (*Acantharchus pomotis*) is ubiquitous in sluggish, blackwater streams of the Atlantic and Gulf coastal plains from north Florida to southern New York, although it never reaches high densities and is considered rare. Recently, however, investigations of isolated depression wetlands in South Carolina found that mud sunfish were the fourth most abundant species (Snodgrass et al. 1996). In one wetland, densities were 2.5- to 10-fold higher than in blackwater streams of North Carolina (Pardue 1993; J. W. Snodgrass, unpublished data). Thus, to assure the continued existence of many fishes presently considered rare, preservation and management of many wetland types may be required.

Wetlands can provide refuges for native fish from introduced predatory fish. In Africa, the introduction of Nile perch (*Lates niloticus*) caused the extirpation of several native fishes from the open water areas of Lake Victoria (Ogutu-Ohwayo 1990). However, some of these species have subsequently been found in the fringing wetlands of the lake (Chapman et al. 1996). Apparently, low-dissolved-oxygen conditions and complexity of the fringing wetlands inhibit the efficiency of Nile perch foraging, and native species found refuge there.

Protection of Fish Prey Populations to Maintain Predator Populations

Many species of wildlife depend on wetland fishes as their primary source of forage. Therefore, protection and management of fish production in many wetland systems will be a critical component in managing wildlife. Some wildlife species are listed by state or federal agencies as "endangered," "threatened," or "species of special concern." For example, the wood stork (*Mycteria americana*) is listed by the U.S. Fish and Wildlife serves as "threatened." Wood storks are tactile feeders that do not use vision to capture their prey (Kushlan 1978). The concentration of fishes in shallow drying wetlands provides wood storks with ample food during their nesting season (Coulter and Bryan 1993). Some wetland habitats have been created, and forage fish populations stocked and managed, for the express purpose of providing forage for endangered species (e.g., Coulter et al. 1987, Batzer et al. 1999).

Control of Fish Populations to Protect Biodiversity of Other Groups of Organisms

Fish are often the top predators in wetlands systems, and fish predation can have direct and indirect effects on other groups of organisms, such as invertebrates and amphibians (Batzer and Wissinger 1996, Wellborn et al. 1996). The intentional or unintentional introduction of fishes into wetland habitats previously lacking fishes has been identified as a cause of declines of some North America amphibians (Alford and Richards 1999). Besides direct predation, fishes may compete with other organisms for prey, or fish introductions may spread diseases to native biota (Blaustein et al. 1994). Population control and eradication of introduced fishes may be needed to protect wetland biota from these threats.

As pointed out in the introduction, native fishes are integral parts of many wetland systems. Many organisms are well adapted to coexist with fish and actually do poorly in wetlands lacking native fish populations (Wellborn et al. 1996). Therefore, maintenance of native fish populations will be essential to conserving the full range of biota found in many types of wetland systems (Snodgrass et al. 2000a).

Maintenance of Fish Populations for Harvest

Few fish populations are exploited commercially within wetland systems of North America, although commercially important fish species from adjacent systems such as rivers and lakes may use wetlands during portions of their life cycle. These species may be managed using traditional methods of limiting harvest. Harvest limits can be derived from optimal or sustainable yield population models for single species. However, because these models assume that limiting environmental factors are constant, a condition often not met in highly variable systems such as riverine wetlands, the applicability of these approaches to wetland dependent fisheries is limited at present (Bayley and Li 1992, Hoover and Killgore 1998).

Sunfish (*Lepomis* spp.), largemouth bass (*Micropterus salmoides*), and crappie (*Pomoxis* spp.) in river systems, and such species as walleye (*Stizostedion vitreum*) in lake systems, provide extensive recreational fisheries. As with species of commercial interest, those species of recreational interest depend on wetland systems to complete a portion of their life cycle. These species are often managed using traditional catch limits and stocking. Including considerations of wetland systems in their management could provide increased yield and enjoyment from these fisheries.

RIVERINE FLOODPLAIN WETLANDS

Riverine floodplain wetlands occur along many large North American rivers and consist of a diversity of fish habitats that are maintained by the spatial and temporal dynamics of river discharge (Junk et al. 1989, Johnson et al. 1995, Ward 1998). Two types of large rivers occur in North America: (1) constrained rivers where geological controls limit channel meandering and wetland development is minimal, and (2) alluvial rivers where broad floodplains allow channel meandering and wetland development is extensive. Here we concentrate on the latter.

The energy of waters flowing through the floodplains of rivers results in erosion and deposition of sediments and channel migration back and forth across the floodplain. As a result, the occurrences of specific habitat types on floodplains are dynamic in nature, shifting with each flood event. Wetland habitats occurring in floodplains of rivers include oxbow lakes in abandoned meanders, backwater marshes and swamps, dead-arms, drainage sloughs, side channels, spring seeps, fringing marshes of the main channel, and inundated floodplain forests (Fig. 16.1). Hydrology varies seasonally among wetland habitats of river floodplains. The floodplain forest is inundated only during high-water periods of the wet season. Backwater marshes and swamps, fringing main channel marshes, and some side channels hold water after floods have receded but dry out during low-water periods. Dead-arm channels, oxbow lakes, and spring seeps may hold water throughout the dry season, due to their connections with groundwaters.

Many riverine fishes use floodplain wetlands during various portions of their life cycles and two major patterns exist (Welcomme 1979). First, some riverine fishes require access to floodplains to complete their life cycle. They migrate laterally from the main channel to floodplains to use the inundated floodplain as feeding, spawning, and nursery areas. During the dry season, these fishes move back to the main channel. Second, some species may use floodplain wetlands throughout their life cycle. These fishes move onto the fully inundated floodplain during the wet season and seek

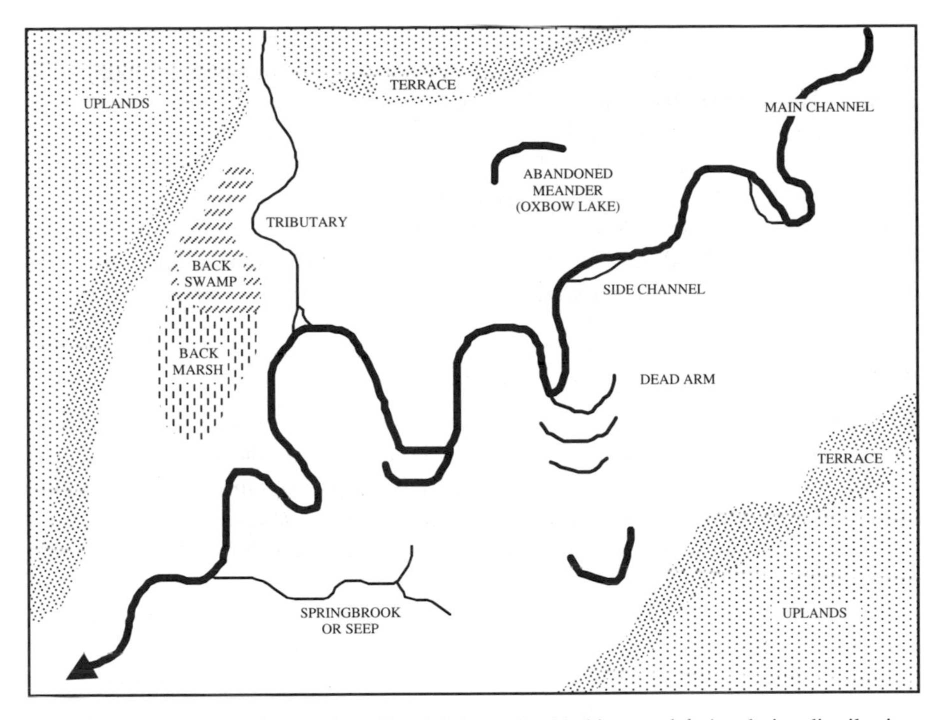

FIGURE 16.1. Some of the various floodplain wetland habitats and their relative distribution. (Adapted from Ward 1998.)

refuge in permanent floodplain wetlands (e.g., oxbow lakes) during the dry season. We refer to both groups that use floodplain wetlands extensively as *floodplain-dependent fishes*.

Threats to Floodplain-Dependent Fishes

Two factors have greatly impacted links between main channels and floodplains: (1) isolation of floodplain wetlands from main channel habitats by dikes, levees, and sedimentation; and (2) alteration of natural flow regimes (Ligon et al. 1995, Poff et al. 1997). Hydrologic factors important for floodplain-dependent fishes include the timing, duration, and magnitude of inundation, and the rate of water level change at the beginning and ending of inundation events (Poff et al. 1997). Fishes that spawn and feed on floodplains have reproductive patterns that are adapted to the natural seasonal flooding of river floodplains (Junk et al. 1989). If the seasonal occurrence of floods is shifted, fishes will become ripe during seasons when the floodplain is not available for spawning. Additionally, prolonged inundation and elevated flows in floodplain habitats can displace fish eggs to unfavorable habitats resulting in reproductive failure (Poff et al. 1997).

The floodplain is a mosaic of habitats with varying current velocities ranging from lentic to lotic conditions. Spawning and growth of larval and juvenile fish of different species occur at different locations along the lentic–lotic gradient (Cope 1989, Turner et al. 1994). Thus, maintenance of the complete nursery function of the floodplain requires maintenance of the complete range of habitats. Rivers or sections of rivers with more diverse and expansive habitats are likely to be more productive (Schiemer and Zalewski 1992), and fishery yields from tropical floodplain rivers are correlated with floodplain extent (Welcomme 1979). Along altered river channels, dikes, levees, sedimentation, and channeled flow prevent the movement of water and fishes between the main channel and floodplain wetlands. In some cases, reductions in sport and commercial fisheries have been correlated with levee construction (Lambou 1963).

Managing Floodplain-Dependent Fishes

Because floodplain wetlands are functionally linked to main channels of rivers, management of fish populations in these systems should consider the processes responsible for maintaining links between main channels and floodplains as well as habitat management along the main channel and on the floodplain (Ward 1998). Because of the economical and social demands placed on larger river systems in North America, efforts to manage fish populations at the whole river scale are limited, with most projects concentrating on selected river reaches (Gore and Shields 1995). Here we concentrate on management methods that are applicable at the individual reach or smaller scales. For an overview of a project directed at restoring and managing an entire watershed, see Trexler (1995).

Methods to reestablish aquatic connections for fish to floodplain wetland habitats include dredging of sediments from entrances to side channels and constructing flow diversion structures in the main channel (Sparks et al. 1990, Gore and Shields 1995). To reestablish natural hydrology to floodplain wetlands, weirs, levees, and diversion structures can be placed so that channelized flow is directed into side channels and wetlands (Fig. 16.2*A*). Additionally, weirs and levees can be placed so that sediment-laden flows are prevented from entering backwaters where slower current velocities will allow deposition of sediments (Fig. 16.2*B*). Excessive sedimentation will alter local hydrology and ultimately reduce habitat diversity on the floodplain. In cases where existing levees isolate floodplains from the main channel, connections between some or all of the floodplain and the main channel can be reestablished by shutting down pumps used to remove floodwaters outside levees (Sparks et al. 1990), excavating breaches in levees (Kern 1992), setting back levees from the main channel (Welcomme 1989), or removing levees all together (Bayley 1991).

Individual floodplain wetlands that hold water during all or portions of low-water periods can be managed to enhance fish habitat. Artificial habitats can be constructed by excavating depressions in the floodplain. Borrow pits excavated to provide soil for roads, dams, and other elevated structures in floodplains may also provide fish habitat on floodplains (Nunnally et al. 1987). Reforestation of floodplains using a mix of native hardwood species may restore spawning habitat of seasonally flooded portions of floodplains (Hoover and Killgore 1998). Although it has not been attempted, managing the geomorphology of seasonally inundated portions of floodplains also may hold potential for managing fish habitats during floodplain inundation.

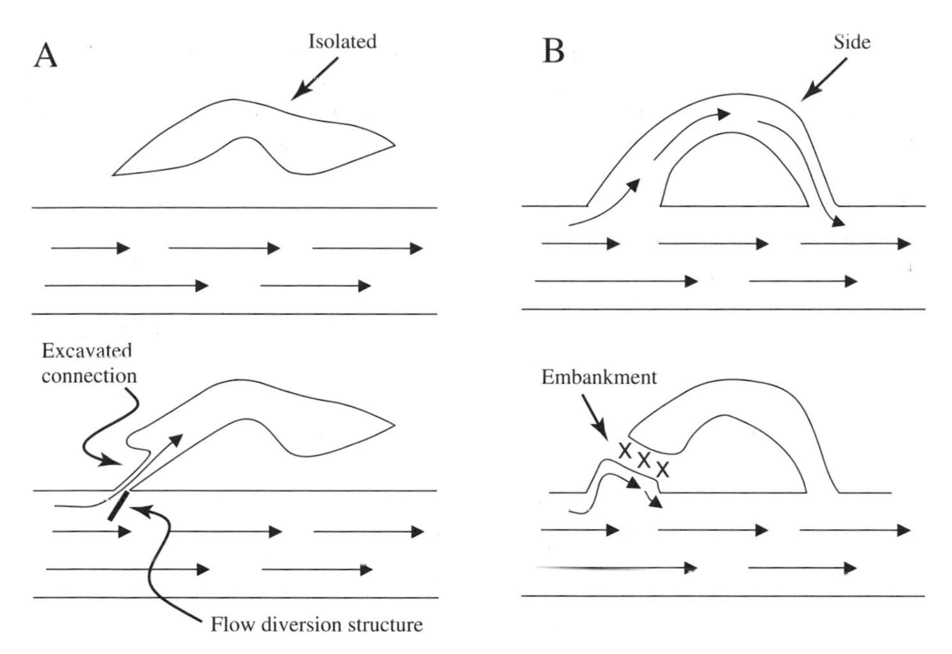

FIGURE 16.2. Methods of *(A)* reestablishing floodplain wetland hydrological connection and *(B)* preventing sedimentation in floodplain habitats. Conditions before restoration are presented above and conditions after restoration are presented below. In *(A)*, a small channel is excavated and a flow diversion structure is placed to direct flow into an isolated backwater. In *(B)*, an embankment is constructed to prevent sediment-laden waters from entering a side channel.

Methods for measuring and determining potential or existing effects of hydrologic alterations to riverine systems are reviewed by Gordon et al. (1992) and Richter et al. (1997) and include the instream flow incremental method (IFIM, Bovee 1982) and its modifications, and the indicators of hydrological alteration method (IHA; Richter et al. 1996). Here we provide a brief introduction to these methods as they relate to management of floodplain-dependent fishes. The IFIM uses data on specific species–habitat relationships and habitat availability under different discharge scenarios to estimate the minimum flow needed to maintain sufficient main channel habitat for persistence of target species (Bovee 1982). In contrast, the IHA method uses hydrologic data from pre- and postdisturbance or impacted and control rivers to assess the effects of human alterations on five general types of metrics that describe characteristics of river flow regimes. Metrics describe the timing, duration, magnitude, and rate of water-level change at the beginning and ending of inundation events rather than the habitat characteristics of the channel. The IHA approach is most applicable for maintaining floodplain-dependent fishes because it considers the factors that control the use of floodplain wetlands by floodplain-dependent fishes.

Quantitative relationships among physical and chemical habitat variables of floodplain wetlands, and larval, juvenile and adult fish densities have been developed (Cope 1989, Sabo and Kelso 1991, Sabo et al. 1991, Turner et al. 1994). These relationships may be useful in directing management of floodplain fish habitat. Current velocity and associated differences in water temperature and substrata characteristics appear to be the most important determinants of differences in fish assemblage structure among habitats (Cope 1989, Turner et al. 1994). For example, sunfish prefer warmer lentic conditions, and many minnow species prefer cooler more lotic conditions. In artificial floodplain ponds, high fish production is correlated with relatively high dissolved oxygen levels, lower total organic carbon concentrations, and the relatively high conductivity and turbidity that occur in larger ponds with more sinuous shorelines (Sabo and Kelso 1991, Sabo et al. 1991).

Models linking hydrological fluctuations, floodplain access, and production of fishes also have been developed (Welcomme and Hagborg 1977, Power et al. 1995), and may prove useful in managing fish harvest and assessing the effects of human alterations to floodplains. Welcomme and Hagborg's (1977) model was based on fishes that migrate between the main channel and floodplains and included components of fish recruitment, growth, and mortality as they are influenced by low- and high-water conditions during the annual hydrologic cycle. Relationships among hydrologic conditions, recruitment, growth, and mortality were based on data from African rivers and fish species, and the model assumed a two-phase hydrological cycle involving a wet season when the floodplain is inundated and a dry season when fishes are confined to the main channel. Although Welcomme and Hagborg's model was developed for African rivers, their general approach may be used in developing models for North American rivers.

Hydraulic models connecting hydrology and floodplain access to trophic dynamics (Power et al. 1995) may be particularly useful to fisheries managers because they include the effects of isolating floodplain wetlands from the main river channel. Hydraulic models incorporate the direct effects of floodplain isolation on fish populations as well as the indirect effects of floodplain isolation and alteration of hydrology on resources consumed by fishes. Hydraulic models have yet to be incorporated into river fisheries management strategies, but hold promise for use in stock assessment and setting harvest limits and seasons.

LAKE MARGIN WETLANDS

The open waters of many natural and man-made lake systems of North America are separated from uplands by narrow to wide, broken, or continuous bands of wetlands along the shoreline, which we refer to as *lake margin wetlands*. Lake margin wetlands encompass the littoral zone of lakes as well as seasonally flooded shoals and shorelines. Lake margin wetlands are used by many fishes, with some requiring access to these wetlands to complete their life cycle. For example, Trautman (1981) considered 28 species of Ohio fishes to be dependent on Lake Erie wetlands for at least part of their life cycle. In an analysis of fish assemblage information from open waters and wetlands of the Great Lakes, Jude and Pappas (1992) identified a gradient in fish species use of wetland habitats from species that occurred almost exclusively in wetlands to species that occurred almost exclusively in open waters of the lakes. Forty-seven species (41% of the species examined) were identified as a wetland taxocene that included permanent residents of wetlands (completing their life cycle within the wetlands), species that used wetlands as spawning or nursery areas, or both. Recent food-web studies using stable isotopes of carbon and nitrogen have demonstrated the importance of wetland habitats as nursery areas for young-of-the-year species of sport and commercial interest (Keough et al. 1996). For example, walleye feed in wetlands as young-of-the-year and shift to feeding in open waters as juveniles and adults. Additionally, positive relationships between aquatic macrophyte densities and sport fish production in small lakes (Wiley et al. 1984) suggest the importance of lake margin wetlands for fish production.

For small species that spend their entire life within lake margin wetlands, and larger species that spend their first year of life in these wetlands, protection from larger predators is probably as important as the rich food resources base. Experimental studies in small ponds have demonstrated complex interactions between predation and habitat use of lake margin wetlands and open waters. For example, when predators are absent, small bluegill sunfish (*Lepomis macrochirus*) forage on benthic organisms in the littoral zone of ponds, but in the presence of largemouth bass they shift to foraging in the emergent marsh along the pond edge, resulting in reduced growth (Werner et al. 1983a,b).

Lake margin wetlands also present an array of habitat types that may be viewed as a mosaic of patches (Lodge et al. 1988), among which fishes partition habitat (Chick and McIvor 1994, Weaver et al. 1997). Important factors influencing use of different types of habitat patches by fishes include vegetation density, structure, diversity and patchiness at smaller spatial scales (ca. 1 m^2), and differences in physicochemical water characteristics among patches (Frodge et al. 1990).

In large lake systems, geomorphology and adjacent characteristics of open water habitats also influence fish use of lake margin wetlands (Randall et al. 1996). As the slope from nearshore to offshore in a wetland increases, species richness and abundance of fishes decreases while mean fish size increases. Additionally, fish abundance and species richness are highest in wetlands with effective fetches (i.e., degree of exposure to lake-induced winds from the prevailing wind direction) of 0.5 to 0.7; at lower and higher fetches, fish species richness and abundance are lower and mean fish sizes are larger. Further investigations of these patterns are needed to determine their generality and the mechanisms responsible.

Threats to Lake Margin Wetlands

Threats to lake margin wetlands include destruction due to filling, isolation from the open waters of the lake by dikes, development of shorelines adjacent to wetlands, pollution, and habitat disturbance by exotic fishes. Diked wetlands and wetlands adjacent to developed shorelines are used by fewer fish and fewer species than are undeveloped wetlands (Brazner and Beals 1997). Additionally, fish condition can be poorer in diked wetlands than in adjacent undiked areas (Johnson et al. 1997). The presumed causes of these degradations are limited exchange of fishes between wetlands and open water areas, damped water-level fluctuations in diked systems, and runoff of polluted waters from developed uplands. Input of fine sediments and nutrients from developed upland areas can interact with other abiotic (e.g., wave action) and biotic (e.g., foraging of introduced species) disturbances to accelerate degradation processes (Whillans 1996).

Introduced species may disturb lake-margin wetlands through their foraging behavior. The common carp (*Cyprinus carpio*) is an example of an exotic species that has been particularly detrimental to lake-margin wetlands. At high densities, the foraging activities of common carp result in sediment resuspension, increased turbidity and nutrient levels, and increased phytoplankton production (Richardson et al. 1990, Breukelaar et al. 1994). Uprooting of aquatic macrophytes and changes in food webs induced by common carp reduce macrophyte and benthic algae production (Hanson and Butler 1990, Meijer et al. 1990), resulting in reduced wetland habitat quality for some native fishes.

Managing Lake Margin Wetlands

The first step in managing fishes associated with lake margin wetlands in any lake system is preservation of existing relatively unaffected systems. In identifying and prioritizing habitats for protection, connections to open waters of the lake, upland land use, vegetation structure and geomorphology within the wetland, and existing fish use should be considered. Methods of assessing the ecological integrity of littoral zone fish assemblages of the Great Lakes have been developed (Minns et al. 1994) and could be adapted for use in assessment and monitoring of other systems. It should be kept in mind that fish densities and diversity often increase with vegetation density and diversity; however, areas of very dense vegetation, where vegetation nearly fills the water column, will not be used by most fishes.

Restoring connections between lake margin wetlands and open water areas can be accomplished by breaching or removing dikes, which should also restore natural water-level fluctuations. In large systems, where common carp are abundant, dikes may isolate wetlands so that carp access can be controlled. Carp access may also be controlled using electrical fences. Preliminary investigations of carp exclusion from a degraded system in Lake Ontario suggest that exclusion will improve water quality by reducing turbidity and nutrient levels. However, turbidity may not be reduced to the point that native aquatic vegetation will recover (Lougheed et al. 1998). Because of the trade-offs between controlling access of carp and allowing more free exchange of fishes between wetlands and open waters, we recommend an adaptive management approach to removing dikes from wetlands. First, breaches in the dike could be established and movement of fishes into and out of wetlands through the breaches could be monitored as well as conditions within the wetland. If the initial breaches are found to increase the habitat value of the wetland for native fishes, the degree of breaching could be increased. Additionally, establishment of control structure in breaches could be used to limit wetland access to native fishes only (although this would be labor intensive and costly), or to exclude fish altogether while still allowing natural water-level fluctuations.

Restoration of lake margin wetlands may involve reestablishment of vegetation through planting. Techniques for restoring wetland vegetation can be found in Chapter 9 and in Levine and Willard (1990). In lake margin wetlands, native plants should be used and protection from wave active in the open water portions of the lake may be required in the initial stages of the planting project. Planning should consider the factors outlined above for identifying lake margin wetlands for protection.

Quantitative approaches for assessing and managing fish habitat over larger scales are emerging (Minns 1997). Quantitative approaches concentrate on the principle of "no net loss" of fish production due to habitat destruction, including lake margin wetlands. Again, these methods were developed using the Great Lakes as examples, but may be transferred to other large lake systems and used as guides for managing wetland habitats at the whole-lake scale.

Fish habitat quality in wetland portions of smaller lakes and reservoirs may decline as conditions in the entire system decline from nutrient pollution, sedimentation, and introduction of exotic species (both plants and animals). Cooke et al. (1993) provide an introduction to methods of controlling sediment and nutrient inputs and nuisance growths of algae and introduced aquatic plants in whole-lake and whole-reservoir systems. If the goal of management is fish production and preservation in smaller lake systems, care must be taken to maintain or reestablish natural lake margin vegetation while controlling nuisance species.

Finally, in lake systems where water-level fluctuations can be controlled, water-level manipulations can be used to enhance fish production. Production of sport fishes is often enhanced when higher water levels are maintained during the breeding season (Beam 1983, Miranda et al. 1984). It is hypothesized that new habitat and plentiful resources in flooded terrestrial vegetation contribute to enhanced fish production when higher water levels are maintained during the spawning season. However, water-level fluctuation may also influence habitat quality for species spawning in habitats other than wetlands, so planned water-level manipulations should consider potential effects on both target and nontarget species.

DEPRESSION WETLANDS

Depression wetlands share two distinct characteristics: (1) they are not a part of a larger surface water drainage, so they have no permanent surface inlet or outlet; and (2) their hydrology is driven by the trapping of precipitation runoff or temporary connections to underlying groundwater. Types of fish-bearing depression wetlands that are particularly abundant include Carolina bays and pocosins of the Atlantic coastal plain, prairie pothole wetlands of the Midwest, limestone sink wetlands of regions of karst topography, and late succession bog lakes (Fig. 16.3). We include smaller pocosins here because they occur in shallow basins on divides between ancient rivers and sounds (Richardson and Gibbons 1993). Larger pocosins are discussed in the section "Large Contiguous Wetlands." We also include late-succession bog lakes because the physical characteristics of these lakes (e.g., maximum depth (5.0 m, wetland area/lake area ratios > 10, and no permanent inlet or outlet) are similar to other types of depression wetlands and they support a unique fish assemblage among seepage lakes (Rahel 1984). While the formative factors responsible for creating basins may vary among depression wetland types, humate-impregnated sands and

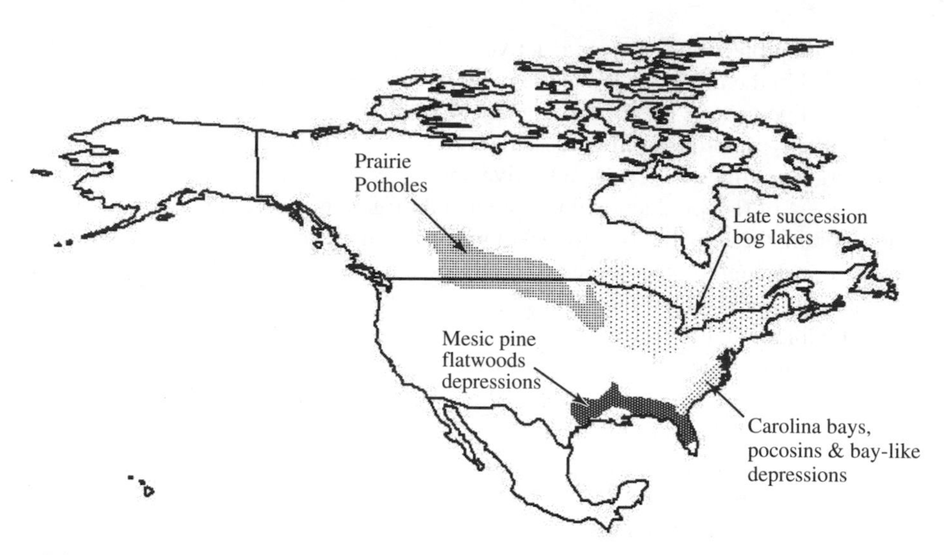

FIGURE 16.3. Distribution of some common depression wetland types in North America.

clays are responsible for holding water in most types of depression wetlands (Kantrud et al. 1989, Richardson and Gibbons 1993).

Fishes may occur in only a relatively minor portion of the depression wetlands in a landscape (Sharitz and Gibbons 1982, Kantrud et al. 1989). However, where high densities of depression wetlands exist, a large absolute number of individual wetlands may support fish populations. In a survey of depression wetlands at the Savannah River site in South Carolina, Snodgrass et al. (1996) reported that 20% of the wetlands contained at least one fish species. If the percentages from the Savannah River site survey are multiplied by the total number and density of depression wetlands (ca. 400 and 0.52 km^{-2}, respectively, Savannah River Ecology Laboratory, unpublished data), there are about 80 wetlands that contain fish, or 0.10 wetland km^{-2} with fishes on the site. These estimates are conservative because the Savannah River site is located on the western edge of the Atlantic coastal plain, where depression wetland density is low (Richardson and Gibbons 1993). In a survey of wetlands in the prairie region of Manitoba, Lawler et al. (1974) found that 10 to 20% of the wetlands contained fish populations. If the estimated number of individual depression wetlands in North Dakota (2.7 million, Stewart and Kantrud 1973) is multiplied by 10 to 20%, an estimated 270,000 to 540,000 depression wetlands contain fish in North Dakota alone. Although these estimates of wetlands with fish populations are rough, they clearly demonstrate that a significant number of depression wetlands in many regions contain fish populations.

Surveys of fish assemblages in depression wetlands are limited. Sixty species have been reported as occurring in at least one type of depression wetland (Table 16.1), further pointing to the potential significance of these habitats for fishes. Although extensive surveys of fish use of depression wetlands are lacking for most types, information from pocosins is particularly incomplete. In prairie pothole wetlands and late-succession bog lakes, harsh environmental conditions including low

TABLE 16.1. Fishes Documented as Occurring in Different Types of Depression Wetlands[a]

Species (Common Name, Scientific Name)	MF	SE	PP	BL
1. American eel, *Anguilla rostrata*		×[b]		
2. Longnose gar, *Lepisosteus osseus*		×[b]		
3. Florida gar, *Lepisosteus platyrhincus*	×			
4. Bowfin, *Amia calva*		×[c]		
5. Gizzard shad, *Dorosoma cepedianum*		×[b]		
6. Pirate perch, *Aphredoderus sayanus*		×		
7. Eastern mudminnow, *Umbra pygmaea*		×		
8. Central mudminnow, *Umbra limi*			×	×
9. Redfin pickerel, *Esox americanus*		×		
10. Chain pickerel, *Esox niger*		×		
11. Northern pike, *Esox lucius*			×[d]	
12. Golden shiner, *Notemigonus crysoleucas*	×	×		×
13. Fathead minnow, *Pimephales promelas*		×[e]	×	×
14. Blacknose shiner, *Notropis heterolepis*				×
15. Ironcolor shiner, *Notropis chalybaeus*		×		
16. Coastal shiner, *Notropis petersoni*		×[b]		
17. Pearl dace, *Margariscus margarita*				×
18. Northern redbelly dace, *Phoxinus eos*				×
19. Finescale dace, *Phoxinus neogaeus*				×
20. Common shiner, *Luxilus cornutus*				×
21. White sucker, *Catostomus commersoni*				×
22. Lake chubsucker, *Erimyzon sucetta*		×		
23. Creek chubsucker, *Erimyzon oblongus*		×[b]		
24. White catfish, *Ameiurus catus*		×[b]		
25. Yellow bullhead, *Ameiurus natalis*		×		
26. Black bullhead, *Ameiurus melas*			×[d]	×
27. Brown bullhead, *Ameiurus nebulosus*	×	×		
28. Tadpole madtom, *Noturus gyrinus*		×[c]		
29. Golden topminnow, *Fundulus chrysotus*	×			
30. Lined topminnow, *Fundulus lineolatus*		×		
31. Waccamaw killifish, *Fundulus waccamensis*		×[b]		
32. Bluefin killifish, *Lucania goodei*	×			
33. Flagfish, *Jordanella floridae*	×			

(*continues*)

TABLE 16.1. (*continued*)

Species (Common Name, Scientific Name)	MF	SE	PP	BL
34. Sailfin molly, *Poecilia latipinna*	×			
35. Mosquitofish, *Gambusia holbrooki*	×	×		
36. Least killifish, *Heterandria formosa*	×			
37. Brook silverside, *Labidesthes sicculus*	×			
38. Waccamaw silverside, *Menidia extensa*		×[b]		
39. Brook stickleback, *Culaea inconstans*			×	×
40. White perch, *Morone americana*		×[b]		
41. Largemouth bass, *Micropterus salmoides*	×	×[d]		
42. Flier, *Centrarchus macropterus*		×		
43. Warmouth, *Lepomis gulosus*	×	×		
44. Green sunfish, *Lepomis cyanellus*		×[d]		
45. Bluegill, *Lepomis macrochirus*	×	×[d]		
46. Spotted sunfish, *Lepomis punctatus*		×[b]		
47. Redear sunfish, *Lepomis microlophus*	×	×		
48. Pumpkinseed sunfish, *Lepomis gibbosus*		×[b]		
49. Dollar sunfish, *Lepomis marginatus*	×	×		
50. Redbreast sunfish, *Lepomis auritus*		×		
51. Mud sunfish, *Acantharchus pomotis*		×		
52. Bluespotted sunfish, *Enneacanthus gloriosus*	×	×		
53. Blackbanded sunfish, *Enneacanthus chaetodon*		×		
54. Banded pygmy sunfish, *Elassoma zonatum*		×		
55. Everglades pygmy sunfish, *Elassoma evergladei*	×			
56. Yellow perch, *Perca flavescens*		×[b]	×[d]	×
57. Walleye, *Stizostedion vitreum*			×[d]	
58. Swamp darter, *Etheostoma fusiformes*	×	×		
59. Iowa darter, *Etheostoma exile*				×
60. Waccamaw darter, *Etheostoma perlongum*		×[b]		

[a]MF, mesic flatwoods; data from Dunson et al. (1997); SE, southeastern Carolina bays, baylike depressions and pocosins; data from Frey (1951), Snodgrass et al. (1996), Snodgrass and Bryan (unpublished data); PP, prairie pothole wetlands; data from Peterka (1989), Hanson and Riggs (1995), Duffy (1998); BL, late succession bog lakes; data from Rahel (1984), Mallory et al. (1994).

[b]Species reported from Lake Waccamaw of North Carolina (Frey 1951).

[c]Species reported from permanent Carolina bays of North Carolina (Frey 1951).

[d]Species that may naturally occur in permanent wetlands but are probably present in semipermanent wetlands as a result of human introduction.

[e]Species introduced to systems outside their natural range.

dissolved oxygen, low pH, high sulfate and bicarbonate levels, and salinity, limit fish assemblage diversity. In winter, shallow potholes (< 0.7 to 0.8 m depth) probably freeze to the bottom (Kantrud et al. 1989), eliminating fishes from these systems. Low dissolved-oxygen levels under winter ice cover or during warm summer months in deeper wetlands also limit fish assemblage structure to two native species, fathead minnows (*Pimephales promelas*) and brook sticklebacks (*Culaea inconstans*). Both are capable of surviving under these conditions. However, groundwater seepage areas may provide refugia for other species in some wetlands (Peterka 1989). In saline depression wetlands, total dissolved solid concentrations limit the distribution of fathead minnows (McCarraher 1971, Held and Peterka 1974, Burnham and Peterka 1975). In late-succession bog lakes, low dissolved-oxygen levels under winter ice cover limit assemblages to a group of cyprinids, central mudminnows (*Umbra limi*), and yellow perch (*Perca flavescens*); low pH may further limit these assemblages to only the latter two species (Rahel 1984, Magnuson et al. 1989)

Harsh environmental conditions appear to impose fewer limits on fish occurrence in southeastern U.S. wetlands, where winters are milder and geological conditions do not produce saline water chemistry. Although pH is low (often, 4.5 to 6.0) and fluctuates in southeastern systems, pH in individual wetlands is not necessarily related to fish occurrence (Snodgrass et al. 1996). Limited duration and extent of ice cover and patchy distribution of anoxic conditions probably allow fishes to avoid these conditions by moving among habitat patches within wetlands.

Because hydroperiod varies among individual isolated wetlands, and even wetlands with long hydroperiods may dry completely during severe droughts, hydroperiod and location of depression wetlands in relation to other aquatic habitats are important determinants of fish assemblage structure. In southeastern depression wetlands the occurrence of fish is restricted to wetlands that dry infrequently (about every 2 to 5 years) and are "close" to sources of colonists from permanent aquatic habitats (e.g., lakes and perennial streams; Snodgrass et al. 1996). It should be noted that what determines if a wetland is "close" to a source of colonists is not the absolute distance of a wetland from the nearest permanent aquatic habitat, but the characteristics of the landscape between the wetland and nearest permanent aquatic habitat. For example, distances to the nearest intermittent aquatic habitat (e.g., another wetland) and elevation differences between the wetland and the nearest permanent aquatic habitat determine fish occurrence in southeastern depression wetlands (Snodgrass et al. 1996; Fig. 16.4).

Stable water levels and more permanent status of fishes are unique to late-succession bog lakes. Because bog lakes are undergoing succession involving gradual filling of lake basins over geological time scales, extinction of fishes, rather than colonization dynamics, are most important in structuring fish assemblages in these systems (Tonn et al. 1995). Therefore, loss of fish populations from bog lakes is a natural process.

Threats to Depression Wetlands

Threats to depression wetland fishes mostly involve habitat destruction by filling or drainage. Most depression wetlands are small and thus are particularly vulnerable to development. In fact, the U.S. Army Corps of Engineers currently allows the

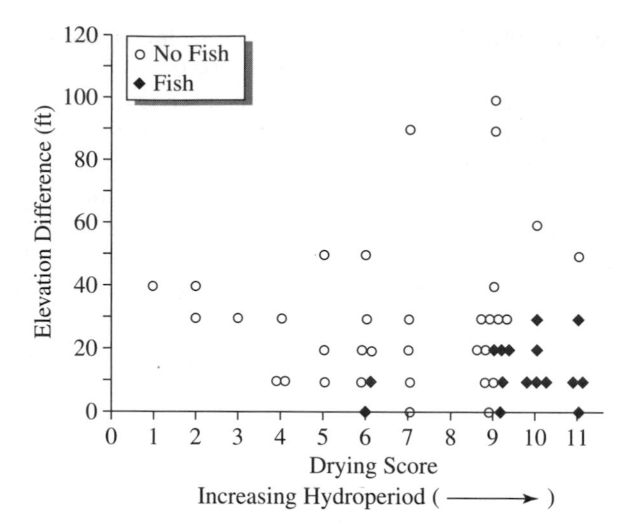

FIGURE 16.4. Relationship among hydroperiod length, elevation difference between wetlands and their nearest permanent aquatic habitat, and the occurrence of fishes in wetlands at the Savannah River Site, South Carolina. Wetlands were sampled from 1994 through 1998 using the methods of Snodgrass et al. (1996). Drying score is a relative measure of hydroperiod length and is calculated as the total number of times out of 11 visits that a wetland held water.

destruction of most wetlands < 4.0 ha in size under a blanket permit (Nationwide Permit 26). In many cases, ditching did not accomplish the goal of complete wetland drainage, so agricultural use was limited. However, because ditching alters a wetland's hydroperiod and the landscape characteristics between the wetland and the nearest permanent aquatic habitat (e.g., decreases elevation difference), ditching alone probably affects fish use of depression wetlands. Excavations within depression wetlands may compromise the integrity of the relatively impermeable soils under lying depression wetlands, allowing faster percolation of water into the groundwater table and reducing hydroperiod lengths. Reducing hydroperiod length will reduce fish use of depression wetlands because fish will be eliminated more often.

Introduction of fishes into depression wetlands may occur in some areas, posing a risk to other wetland organisms. In the southeast, occurrence of largemouth bass, golden shiners (*Notemigonas crysoleucas*), and fathead minnows in wetlands surrounded by residential development suggests bait bucket releases (J. W. Snodgrass, personal observation). Occurrence of young-of-the-year of all three species in one of these wetlands suggests that these populations were established, at least on a short-term basis. Fathead minnows, which are common in prairie pothole wetlands, are popular baitfish because of their tolerance of harsh environmental conditions, and many prairie pothole wetlands are used as culture ponds for this species (Carlson and Berry 1990). Additionally, species of sports interest may be stocked to prairie pothole wetlands to produce fingerlings for stocking to more permanent water bodies or catchable-sized fish (Peterka 1989). While use of depression wetlands as fish culture sites may provide some economic benefits (e.g., the bait fish industry in North America is estimated at more than $1 billion annually; Litvak and Mandrak 1993), establishment of self-sustaining sport fisheries in depression wetlands is unlikely, due to harsh environmental conditions.

Accumulation of pollutants in depression wetlands surrounded by agricultural or urban development may threaten fish populations and pose a risk to wildlife that feed on wetland fishes (e.g., Lemly et al. 1993, Gariboldi et al. 1998). Important factors that influence accumulation of pollutants by wetland fishes and risk to wildlife feeding on fish are wetland morphology and hydroperiod (Snodgrass et al. 2000b). Additionally, many small wetlands have been created to intercept polluted groundwater and surface water runoff. The purpose of these wetlands is to remove and sequester pollutants to protect downstream biological communities. However, if fishes colonize these systems, they may act as a trophic link, transferring pollutants to wildlife that feed on them (Helfield and Diamond 1997).

Managing Depression Wetland Fish Populations

Presently, it is difficult to make recommendations for managing naturally occurring fish populations in depression wetlands because the extent of use by fishes and degree of fish movement between depression wetlands and other aquatic habitats are poorly known. Clearly, some depression wetlands are productive fish habitats (e.g., Duffy 1998; Table 16.1). While fish production in depression wetlands is available for human and wildlife consumption, it is not clear how this production affects fish population dynamics in nearby permanent aquatic habitats or over larger spatial scales. If young-of-the-year produced in depression wetlands do not disperse from their wetland of origin, production of fishes in depression wetlands will not affect fish population dynamics over larger spatial scales. However, if these fish disperse to other aquatic habitats, depression wetlands may be critical habitats in managing fishes that are capable of successful reproduction in them. We believe more study of fish production in, and movement of fish among, depression wetlands and other aquatic habitats is needed.

From a management standpoint it is important to maintain a number of wetlands with and without fishes across larger landscapes. Fishes in depression wetlands are important food resources for many wildlife species, and some birds and animals have life history strategies adapted to wetlands with fish populations. On the other hand, fishes affect other wetland species negatively. Restoring natural hydrological conditions to as many wetlands in a landscape as possible should be a good first step in establishing a diversity of habitats with and without fish. Management options include plugging ditches and filling excavations within wetlands. Additionally, alterations to the landscapes between depression wetlands and permanent aquatic habitats may facilitate or retard fish recolonization of ephemeral wetlands. Reestablishing lost wetlands and removing barriers to fish dispersal, such as raised roads, should facilitate natural recolonization processes. Alternatively, reestablishing natural overland water flow patterns by filling ditches may prevent colonization of historically isolated wetlands.

Introduced fish populations in wetlands that dry frequently will be eliminated relatively rapidly. However, in permanently flooded depression wetlands, introduced fish populations that become established will have to be actively eradicated to restore the ecological integrity of these systems. Traditional ichthyocides such as rotenone may be used to eradicate fish populations but have the disadvantage of affecting nontarget organisms as well. Gill nets have proven effective at eradicating trout populations from high mountain lakes of the western United States with fewer nontarget effects (Knapp and Matthews 1998) and may be effective at eradicating larger fishes from depression wetlands. However, it should be noted that gill nets would have their own nontarget effects in lower-elevation depression wetlands where larger vertebrates (e.g., turtles and muskrats) are common. In semipermenant wetlands where fish are stocked on an annual bases for production purposes, rotation of stocking among a number of wetlands may allow these activities while reducing the impacts to other native biota and deserves further attention from managers and researchers.

Use of polluted depression wetlands by fishes should be restricted as much as possible to reduce the risk of poisoning to wildlife using these habitats. Barriers to colonization can be used in both natural and constructed systems. If fish use cannot be prevented, efforts should be made to limit the bioavailability of pollutants to fish. Because accumulation of pollutants such as mercury in fish tissues is correlated with alternating oxidative (during dry periods) and anoxic (during flooded periods) conditions, minimizing water-level fluctuations may reduce pollutant accumulation in fishes (Snodgrass et al. 2000). In constructed systems, steep sides and flat bottoms should reduce the amount of substrata exposed during normal water-level fluctuations, further restricting the occurrence of alternating oxidative and anoxic conditions to a narrow band around the wetland edge.

Efforts to manage depression wetlands for fish production have been concentrated in prairie pothole systems (Peterka 1989). In wetlands that are stocked and harvested at the end of the growing seasons, empirical relationships among the likelihood of summer fish kills, phytoplankton blooms, and ammonia concentration during winter stagnation can be used to evaluate the production potential of individual wetlands. To reduce the chance of summer kills following stocking, algacides and inorganic nitrogen treatments can be used to reduce blue-green algae blooms and encourage the growth of green algae. If long-term establishment of fish populations (more than one growing season) is desired, empirical relationships among oxygen depletion rates, mean depth, and dissolved oxygen storage at freeze-up can be used to evaluate the potential of winter fish kills in wetlands. Bioenergetic models for fathead minnows have been developed (Duffy 1998) and can be used to assess the effects of harvest on minnow populations and the effects of stocked minnows on other wetland-associated organisms.

LARGE CONTIGUOUS WETLANDS

Large contiguous wetlands are recognized here by their diversity of wetland types (e.g., marshes, forested wetlands, submerged aquatic beds, and open water areas) and regional significance. Although large contiguous wetlands have areas of deeper, permanent, open water, these areas are relatively small compared to the area of wetlands found in the system. Examples include the Everglades of south Florida, the Okefenokee Swamp of southern Georgia, the Great Dismal Swamp of Virginia and North Carolina, and the Great Kankakee Marsh of Indiana and Illinois. Although many of the midwestern systems have been almost completely destroyed, the eastern and southeastern systems remain, although their hydrology has been altered extensively and their overall size diminished (Mitsch and Gosselink 1993).

Large contiguous wetlands support diverse fish assemblages. For example, Laerm and Freeman (1986) list 36 species of fishes from the Okefenokee Swamp, and at least 30 species are common in the Everglades (Loftus and Kushlan 1987). Fish assemblages in large contiguous wetlands are dominated by small rapidly maturing species such as livebearers, but populations of larger fishes of recreational interest are common. As in lake margin wetlands, habitat use is governed by vegetation structure and water depth, with smaller species and the young-of-the-year of larger species occupying shallower vegetated areas and larger adults occupying deeper more open water areas.

Hydrologic fluctuations control the density, abundance, and production of fish in large contiguous wetlands (Loftus and Kushlan 1987, Loftus and Eklund 1994). Periodic drying and refilling of wetland habitats maintains vigorous plant and invertebrate communities, which in turn can benefit fish. Secondary production of small fishes is high in southeastern systems (Freeman and Freeman 1985) and is dependent on water-level fluctuations (Freeman 1989). In areas that are dry on a semiannual basis, fish production is highest following drying. During low-water periods, fishes are concentrated into the remaining deep, open water areas, where harsh environmental conditions and predation result in high mortality (Kushlan 1974, 1976). During high-water periods, reduced predation risks in more structurally complex reflooded marsh habitats and relatively benign environmental conditions allow populations of small and large fishes to grow (Loftus and Eklund 1994).

Threats to Large Contiguous Wetlands

Alteration of hydrology and nutrient enrichment are the main threats to large contiguous wetland systems. In the Middlewest, systems have been ditched, drained, and converted to agricultural use with only remnants of the original area remaining. In the east and southeast, ditching, draining, and installation of water control structures have altered the hydrology of the remaining large contiguous wetlands. Nutrient enrichment in the Everglades has increased standing stocks of fishes but threatens the unique trophic structure of the Everglades ecosystem (Turner et al. 1999). Additionally, ditches and stabilized water levels may increase fish standing stocks by moderating severe environmental conditions that otherwise would control fish abundance, again compromising biotic integrity.

Shortening the duration and frequency of high-water periods by diverting waters away from, or increasing the rate of movement of waters through, large contiguous wetlands may also affect fishes. Small fishes that depend on inundated emergent vegetation are particularly at risk, because decreasing the duration of high-water periods will limit their access to refuges and food in marshes. In turn, abundances of larger predatory fishes may decline with the loss of forage fish (Rader 1999).

Managing Large Contiguous Wetlands

Because large contiguous wetlands cross numerous political boundaries, multiple local, state, and federal agencies are often involved in managing these systems. Restoration of natural water-level fluctuations and treatment of polluted runoff before it enters wetlands are presently used to manage these wetlands. Such practices should benefit native fish populations by maintaining the diversity of habitat types within large contiguous wetlands. Methods for determining natural water-level fluctuations have not been developed but may follow the strategies used in large river systems (see above).

BEAVER POND WETLANDS

Beavers (*Castor canadensis*) were historically, and are again becoming, significant wetland creating agents in North America. Since their near extirpation at the beginning of the nineteenth century, beavers have recovered in many areas because of decreased natural predator abundance, laws regulating hunting and trapping, and reintroductions throughout their range (Naiman et al. 1988). Historically, the range of beavers included areas of North America from the Arctic tundra to northern Florida and the Mexican deserts. Beavers are capable of converting 0.5 to 15% of the landscape of first- to third-order drainages into semipermanent and permanent wetland impoundments, or beaver ponds (Johnston and Naiman 1990a, Snodgrass 1997). Because of their shallow depth (often < 2 m maximum depth), small size, and abundance of aquatic vegetation, beaver ponds are clearly wetlands. However, beaver ponds differ from other wetland systems because many are temporally dynamic over ecological time scales, with individual ponds persisting for less than a year to over 100 years (Remillard et al. 1987, Johnston and Naiman 1990a).

The benefits of beaver ponds to fish and wildlife have long been recognized. Impoundment of small streams by beavers increases habitat diversity in headwater streams, which influences the diversity and productivity of fishes at the reach and drainage basin scales. At the reach scale, the density and diversity of fishes in beaver ponds is greater than that found in adjacent stream reaches and pools (Hanson and Campbell 1963, Snodgrass and Meffe 1998). At the basin scale, a pattern of increasing species richness with increasing stream size (i.e., drainage basin size) has been described for many North American streams; however, when beaver ponds are considered there is little change in species richness across drainage basins (Snodgrass and Meffe 1998). The increase in fish diversity in headwater streams associated with beaver ponds is a result of species turnover among ponds of different age and location within the drainage basin. As ponds in headwater streams age, there is a shift from dominance of the fish assemblage by small minnows and the young-of-the-year of larger species to dominance by larger adults of piscivorous species. In older ponds (more than 20 years old), sport fisheries may develop (Pullen 1971).

Dispersal of fishes from beaver ponds and the physical structure of ponds along stream corridors influence biological communities in adjacent stream reaches. Young-of-the-year fishes dispersing from ponds can reduce invertebrate colonization in adjacent stream reaches (Schlosser 1995). Ponds may also act as barriers to movement, resulting in concentrations of lotic fishes near ponds (Snodgrass and Meffe 1999). Additionally, impoundment of stream reaches can result in the local extirpation of lotic species (Rohde and Arndt 1991).

Threats to Beaver Pond Fishes

Because beavers are the creative agents of beaver pond wetlands, impacts to beaver populations are the main threats to fishes that use these habitats. Historically, beavers were extirpated from many areas of North America by fur trappers. Presently, there is little or no market for beaver pelts, although some beavers are still trapped to reduce nuisance interaction with humans (e.g., flooding of property and timber or destruction of road and railroad structures). Although it is difficult to assess the impacts of past beaver population reductions on fishes, anecdotal observations suggest that some fish populations declined with beaver populations. Many fishes that prefer lentic conditions in southeastern streams are listed by southeastern states as "endangered," "threatened," or of "special concern" (Hoover and Killgore 1998). Other species that are endemic to specific physiographic regions of the southeast, such as the pinewoods darter (*Etheostoma mariae*), prefer small, cool, fast-flowing first- and second-order streams with abundant aquatic vegetation (Rohde and Arndt 1991), conditions that develop after beaver ponds are abandoned. On the other hand, beaver impoundments may threaten lotic species that prefer small headwater stream channels (Rohde and Arndt 1991).

Managing Beaver Populations

To maintain or enhance the value of beaver ponds for fish production and diversity in headwater streams, beaver populations should be managed to maintain the temporal and spatial dynamics of beaver ponds across the drainage basin landscape. Pond conditions can be monitored using geographic information systems (GISs) and aerial photographs (Remillard et al. 1987, Johnston and Naiman 1990a). Pond dynamics can be linked to beaver population trends (Johnston and Naiman 1990b) and then rates of harvest (Snodgrass 1997) to balance conflicting management goals of protecting human property and natural resources, and maintaining benefits to fishes.

At the individual pond scale, sport fisheries can be enhanced by managing water levels and using ichthyocide treatments. Following pond drawdown (by partially breaching the dam) and applying rotenone, game fish populations increased in some southeastern ponds (Pullen 1971). Water-level control devices also can be used to protect property while maintaining some of the wetland area created by beavers (Fig. 16.5). Currently, several state and local agencies, as well as the forestry industry, use such devices to control damage by beavers.

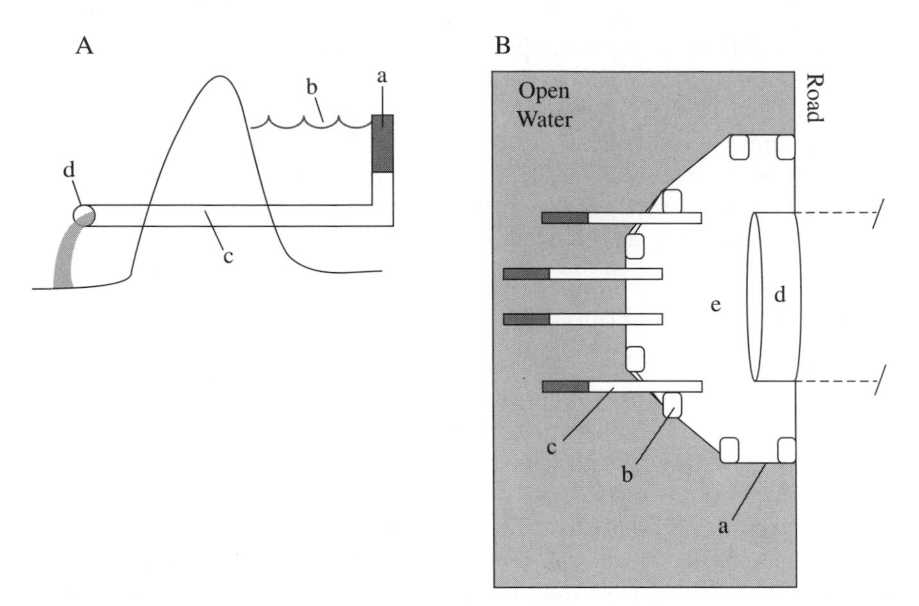

FIGURE 16.5. Methods used to control water levels in beaver ponds to prevent flooding of *(A)* private property and timber, and *(B)* roads. In *(A)* a PVC pipe with a slotted and screened riser is placed through the dam to maintain water levels below the top of the dam. Because the riser is in open water, away from the dam, beavers will not be able to locate or prevent water loss through the pipe. In *(B)*, a wire mesh fence is placed around a culvert entrance to prevent damming of the entrance and flooding of the adjacent road. If beavers place a dam along the fence, a series of PVC pipes can be placed across the top of the fence with slotted and screened ends extending into the open waters of the pond to maintain water levels at the top of the fence. *(A)* a, slotted and screened riser; b, water surface; c, PVC pipe; d, water outflow below the pond; *(B)* a, wire mesh fence; b, fence poles or stakes; c, PVC pipe for water level control; d, culvert; e, protected area.

CONCLUSIONS

From our review of wetland fish ecology and management it is clear that more research is needed, and a lack of information limits our ability to draw conclusions regarding management of fish populations in wetlands. However, our review does suggest a set of common management considerations that arranged in a hierarchical manner may serve as a template for developing management strategies for individual wetlands or groups of wetlands. These considerations can be viewed as a set of overlapping circles (Fig. 16.6) that decrease in size with the spatial extent of the management effort that is required. It is also expected that the degree of organization and cost of management efforts will increase as you move up this hierarchy.

At the local habitat scale, management activities that enhance or maintain fish habitat such as planting aquatic vegetation may take place at small spatial scales (tens to hundreds of square meters). In such systems as lake margin and large con-

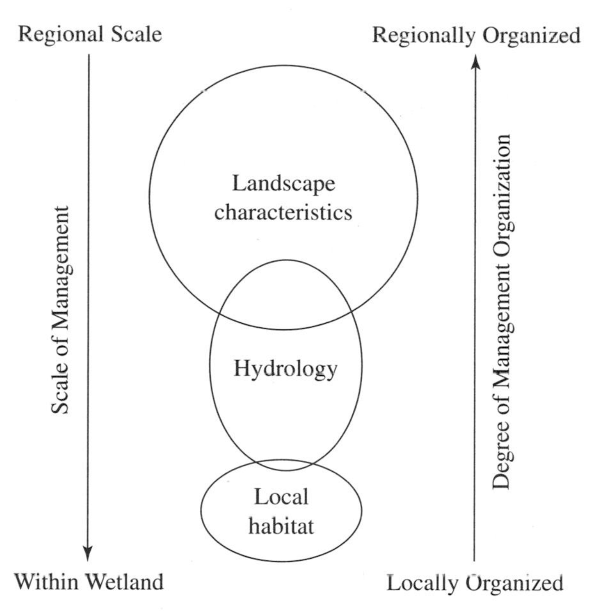

FIGURE 16.6. Hierarchy of wetland fish management considerations. The size of circles represents the spatial scale of management activities, ranging from regional to within individual wetlands. The degree of organization among regulatory and management agencies is expected to increase with higher levels on the hierarchy.

tiguous wetlands, maintenance or enhancement of fish habitat may also involve managing wetland topography to increase habitat diversity. The hydrology circle in Fig. 16.6 is shown as overlapping with the local habitat circle because management activities at the local habitat scale, such as filling ditches, may affect hydroperiods. In turn, the hydrology circle overlaps with landscape characteristics because landscape features will influence hydrology. For example, hydrology of floodplain wetlands will be affected by upstream dams or dikes on the floodplain. Landscape characteristics and hydrology also influence wetland access and emigration of fishes from wetlands. Because some fishes spend only a portion of their life in wetland systems or depend on open water refuges during harsh environmental conditions, it is important to maintain the natural connections of wetlands to open water systems.

We believe that these considerations represent a true hierarchy because processes at higher levels affect processes at lower levels and constrain the degree of success of management efforts at lower levels. For example, planting of floating aquatic vegetation will not be successful if hydrological conditions required to maintain the plantings are not restored. Therefore, while management efforts at the local habitat scale may meet with some degree of success, maintenance and restoration of wetland fish populations, as well as other wetland associated organisms, will require cooperation among governmental agencies and management at all hierarchical levels.

Finally, because our knowledge of wetland fish ecology is limited, restoration and management projects that incorporate an "adaptive management" approach are most likely to succeed and will provide much needed information for planning future projects. Adaptive management involves monitoring components of interests in wetland ecosystem (in this case, fish abundance and productivity), periodically assessing progress toward stated goals, and making decisions on future actions based on the data gathered (Walters and Holling 1990). If systems are not maintaining, or recovering toward desired conditions, management and restoration plans can be altered based on the data gathered. The results of adaptive management projects should yield new insights into wetland fish ecology and the role of fish in wetland ecosystem function.

REFERENCES

Alford, R. A., and S. J. Richards. 1999. Global amphibian declines: a problem in applied ecology. Annual Review of Ecology and Systematics 30:133–165.

Batzer, D. P., and S. A. Wissinger. 1996. Ecology of insect communities in nontidal wetlands. Annual Review of Entomology 41:75–100.

Batzer, D. P., A. S. Shurtleff, and J. R. Robinette. 1999. Managing fish and invertebrate resources in a wood stork feeding pond. Journal of Freshwater Ecology 14:159–165.

Bayley, P. B. 1991. The flood pulse advantage and the restoration of river-floodplain systems. Regulated Rivers Research and Management 6:75–86.

Bayley, P. B., and H. W. Li. 1992. Riverine fishes. Pages 251–281 *in* P. Calow and G. E. Petts (eds.), The rivers handbook: hydrological and ecological principles. Blackwell, Oxford.

Beam, J. H. 1983. The effects of annual water level management on population trends of white crappie in Elk City Reservoir, Kansas. North American Journal of Fisheries Management 3:34–40.

Blaustein, A. R., D. G. Hokit, R. K. O'Hara, and R. A. Holt. 1994. Pathogenetic fungus contributes to amphibian losses in the Pacific Northwest. Biolological Conservation 67:251–254.

Bovee, K. D. 1982. A guide to stream habitat analysis using the instream flow incremental methodology. Instream Flow Information Paper 12. FWS/OBSERVATION/82/86. U.S. Fish and Wildlife Service, Washington, DC.

Brazner, J. C., and E. W. Beals. 1997. Patterns in fish assemblages from coastal wetland and beach habitats in Green Bay, Lake Michigan: a multivariate analysis of abiotic and biotic forcing factors. Canadian Journal of Fisheries and Aquatic Sciences 54:1743–1761.

Breukelaar, A. W., E. H. R. R. Lammens, J. G. P. K. Breteler, and I. Tatria. 1994. Effects of benthivorous bream (*Abramis brama*) and carp (*Cyprinus carpio*) on sediment resuspension and concentrations of nutrients and chlorophyll-*a*. Freshwater Biology 32:113–121.

Burnham, B. L., and J. J. Peterka. 1975. Effects of saline water from North Dakota lakes on survival of fathead minnow (*Pimephales pomelas*) embryos and sac fry. Fisheries Research Board of Canada 32:809–812.

Carlson, B. N., and C. R. Berry. 1990. Population size and economic value of aquatic bait species in palustrine wetlands of eastern South Dakota. Prairie Naturalist 22:119–128.

Chapman, L. J., C. A. Chapman, R. Ogutu-Ohwayo, M. Chandler, L. Kaufman, and A. E. Keiter. 1996. Refugia for endangered fishes from an introduced predator in Lake Nabugabo, Uganda. Conservation Biology 10:554–561.

Chick, J. H., and C. C. McIvor. 1994. Patterns in the abundance and composition of fishes among beds of different macrophytes: viewing a littoral zone as a landscape. Canadian Journal of Fisheries and Aquatic Sciences 51:2873–2882.

Cooke, G. D., E. B. Welch, S. A. Peterson, and P. A. Neworth. 1993. Restoration and management of lakes and reservoirs, 2nd ed. CRC Press, Boca Raton, FLA.

Cope, G. H. 1989. The habitat diversity and fish reproductive function of floodplain ecosystems. Environmental Biology of Fishes 26:1–27.

Coulter, M. C., and A. L. Bryan, Jr. 1993. Foraging ecology of wood storks (*Mycteria americana*) in east-central Georgia. I. Characteristics of foraging sites. Colonial Waterbirds 16: 59–70.

Coulter, M. C., W. D. McCort, and A. L. Bryan, Jr. 1987. Creation of artificial foraging habitat for wood storks. Colonial Waterbirds 10:203–210.

Duffy, W. G. 1998. Population dynamics, production, and prey consumption of fathead minnows (*Pimephales promelas*) in prairie wetlands: a bioenergetic approach. Canadian Journal of Fisheries and Aquatic Sciences 54:15–27.

Dunson, W. A., C. J. Paradise, and R. L. Van Fleet. 1997. Patterns of water chemistry and fish occurrence in wetlands of hydric pine flatwoods. Journal of Freshwater Ecology 12:553–565.

Freeman, B. J. 1989. Okefenokee Swamp fishes: abundance and production dynamics in an aquatic macrophyte prairie. Pages 529–540 *in* R. R. Sharitz and J. W. Gibbons (eds.), Freshwater wetlands and wildlife. U.S. Department of Energy, Office of Scientific and Technical Information, Oak Ridge, TN.

Freeman, B. J., and M. C. Freeman. 1985. Production of fishes in a subtropical blackwater ecosystem: the Okefenokee Swamp. Limnology and Oceanography 30:686–692.

Frey, D. G. 1951. The fishes of North Carolina's bay lakes and their intraspecific variation. Journal of the Elisha Mitchell Science Society 67:1–44.

Frodge, J. D., A. L. Thomas, and G. B. Pauley. 1990. Effects of canopy formation by floating and submerged aquatic macrophytes on the water quality of two shallow Pacific Northwest lakes. Aquatic Botany 38:231–248.

Gariboldi, J. C., C. H. Jagoe, and A. L. Bryan, Jr. 1998. Dietary exposure to mercury in nestling wood storks (*Mycteria americana*) in Georgia. Archives of Environmental Contamination and Toxicology 34:398–405.

Gordon, N. D., T. A. McMahon, and B. L. Finlayson. 1992. Stream hydrology: an introduction for ecologists. Wiley, New York.

Gore, J., and F. D. Shields. 1995. Can large rivers be restored? BioScience 45:142–152.

Hanson, M. A., and M. G. Butler. 1990. Early responses of plankton and turbidity to biomanipulation in a shallow prairie lake. Hydrobiologia 200/201:317–327.

Hanson, W. D., and R. S. Campbell. 1963. The effects of pool size and beaver activity on distribution and abundance of warm-water fishes in a Missouri stream. American Midland Naturalist 69:136–149.

Hanson, M. A., and M. R. Riggs. 1995. Potential effects of fish predation on wetland invertebrates: a comparison of wetlands with and without fathead minnows. Wetlands 15:167–175.

Held, J. W., and J. J. Peterka. 1974. Age, growth, and food habitats of fathead minnow, *Pimephales promelas*, in North Dakota saline lakes. Transactions of the American Fisheries Society 103:743–756.

Helfield, J. M., and M. L. Diamond. 1997. Use of constructed wetlands for urban stream restoration: a critical analysis. Environmental Management 21:329–341.

Hoover, J. J., and K. J. Killgore. 1998. Fish communities. Pages 237–260 *in* M. G. Messina and W. H. Conner (eds.), Southern forested wetlands: ecology and management. Lewis Publishers, Chelsea, MI.

Johnson, B. L., W. B. Richardson, and T. J. Naimo. 1995. Past, present, and future concepts in large river ecology: how rivers function and how human activities influence river processes. BioScience 45:134–141.

Johnson, D. L., W. E. Lynch, Jr., and T. W. Morrison. 1997. Fish communities in a diked Lake Erie wetland and an undiked adjacent area. Wetlands 17:43–54.

Johnston, C. A., and R. J. Naiman. 1990a. The use of geographical information systems to analyze long-term landscape alteration by beaver. Landscape Ecology 1:47–57.

Johnston, C. A., and R. J. Naiman. 1990b. Aquatic patch creation in relation to beaver population trends. Ecology 71:1617–1621.

Jude, D. J., and J. Pappas. 1992. Fish utilization of Great Lakes coastal wetlands. Journal of Great Lakes Research 18:651–672.

Junk, W. J., P. B. Bayley, and R. E. Sparks. 1989. The flood-pulse concept in river–floodplain systems. Canadian Journal of Fisheries and Aquatic Sciences 106:110–127.

Kantrud, H. A., G. L. Krapu, and G. A. Swanson. 1989. Prairie basin wetlands of the Dakotas: a community profile. Biological Report 85(7.28). U.S. Fish and Wildlife Service, Washington, DC.

Keough, J. R., M. E. Sierszen, and C. A. Hagley. 1996. Analysis of Lake Superior coastal food web with stable isotope techniques. Limnology and Oceanography 41:136–146.

Kern, K. 1992. Restoration of lowland rivers: the German experience. Pages 279–297 *in* P. A. Carling and G. E. Petts (eds.), Lowland floodplain rivers. Wiley, Chichester, West Sussex, England.

Knapp, R. A., and K. R. Matthews. 1998. Eradication of nonnative fish by gill netting from a small mountain lake in California. Restoration Ecology 6:207–213.

Kramer, D. L. 1987. Dissolved oxygen and fish behaviour. Environmental Biology of Fishes 18:81–92.

Kushlan, J. A. 1974. Effects of a natural fish kill on water quality, plankton, and fish population of a pond in Big Cypress Swamp, Florida. Transactions of the American Fisheries Society 103:235–243.

Kushlan, J. A. 1976. Environmental stability and fish community diversity. Ecology 57:821–825.

Kushlan, J. A. 1978. Feeding ecology of wading birds. Pages 249–297 *in* A. Sprunt IV, J. C. Ogden, and S. Winkler (eds.), Wading birds. Research Report 7. National Audubon Society, New York.

Laerm, J., and B. J. Freeman. 1986. Fishes of the Okefenokee Swamp. University of Georgia Press, Athens, GA.

Lambou, V. W. 1963. The commercial and sports fisheries of the Atchafalaya Basin floodway. Proceedings of the Annual Conference of the Southeastern Association of Game and Fish Commissions 17:256–281.

Lawler, G. H., L. A. Sunde, and J. Whitaker. 1974. Trout production in prairie ponds. Journal of the Fisheries Research Board of Canada 31:929–936.

Lemly, A. D., and S. E. Finger, and M. K. Nelson. 1993. Sources and impacts of irrigation drainwater contaminants in arid wetlands. Environmental Toxicology and Chemistry 12:2265–2279.

Levine, D. A., and D. E. Willard. 1990. Regional analysis of fringe wetlands in the Midwest: creation and restoration. Pages 299–321 *in* J. A. Kusler and M. E. Kentula (eds.), Wetland creation and restoration: the state of the science. Island Press, Washington, DC.

Ligon, F. K., W. E. Dietrich, and W. J. Trush. 1995. Downstream ecological effects of dams: a geomorphic perspective. BioScience 45:183–192.

Litvak, M. K., and N. E. Mandrak. 1993. Ecology of freshwater baitfish use in Canada and the United States. Fisheries 18:6–13.

Lodge, D. M., J. W. Barko, D. Strayer, J. M. Melack, G. G. Mittelbach, R. W. Howarth, B. Menge, and J. E. Titus. 1988. Spatial heterogeneity and habitat interactions in lake communities. Pages 181–208 *in* S. R. Carpenter (ed.), Complex interactions in lake communities. Spring-Verlag, New York.

Loftus, W. F., and A.-M. Eklund. 1994. Long-term dynamics of an Everglades small-fish assemblage. Pages 461–483 *in* S. M. Davis and J. C. Ogden (eds.), Everglades: the ecosystem and its restoration. St. Lucie Press, Delray Beach, FL.

Loftus, W. F., and J. A. Kushlan. 1987. Freshwater fishes of south Florida. Bulletin of the Florida State Museum, Biological Sciences 31:147–344.

Lougheed, V. L., B. Crosbie, and P. Chow-Fraser. 1998. Predictions on the effects of common carp (*Cyprinus carpio*) exclusion on water quality, zooplankton, and submergent macrophytes in a Great Lakes wetland. Canadian Journal of Fisheries and Aquatic Sciences 55:1189–1197.

Magnuson, J. J., C. A. Paszkowski, F. J. Rahel, and W. M. Tonn. 1989. Fish ecology in severe environments of small isolated lakes in north Wisconsin. Pages 487–515 *in* R. R. Sharitz and J. W. Gibbons (eds.), Freshwater wetlands and wildlife. U.S. Department of Energy, Office of Scientific and Technical Information, Oak Ridge, TN.

Mallory, M. L., P. J. Blancher, P. J. Weatherhead, and D. K. McNicol. 1994. Presence or absence of fish as a cue to macroinvertebrate abundance in boreal wetlands. Hydrobiologia 279/280:345–351.

Matthews, W. J. 1987. Physicochemical tolerance and selectivity of stream fishes as related to their geographic ranges and local distribution. Pages 111–120 *in* W. J. Matthews and D. C. Heins (eds.), Community and evolutionary ecology of North American stream fishes. University of Oklahoma Press, Norman, OK.

McCarraher, D. B. 1971. Survival of some freshwater fishes in the alkaline eutrophic waters of Nebraska. Journal of the Fisheries Research Board of Canada 28:1811–1814.

Meffe, G. K. 1989. Fish utilization of springs and Ciénegas in the arid southwest. Pages 475–485 *in* R. R. Sharitz and J. W. Gibbons (eds.), Freshwater wetlands and wildlife. U.S. Department of Energy, Office of Scientific and Technical Information, Oak Ridge, TN.

Meijer, M.-L., M. W. de Haan, A. W. Breukelaar, and H. Buiteveld. 1990. Is reduction of the benthivorous fish an important cause of high transparency following biomanipulation in shallow lakes? Hydrobiologia 200/201:303–315.

Minns, C. K. 1997. Quantifying "no net loss" of productivity of fish habitat. Canadian Journal of Fisheries and Aquatic Sciences 54:2463–2473.

Minns, C. K., V. W. Cairns, R. G. Randall, and J. E. Moore. 1994. An index of biological integrity (IBI) for fish assemblages in the littoral zone of Great Lakes' areas of concern. Canadian Journal of Fisheries and Aquatic Sciences 51:1804–1822.

Miranda, L. E., W. L. Shelton, and T. D. Bryce. 1984. Effects of water level manipulation on abundance, mortality and growth of young-of-the-year large mouth bass in West Point Reservoir, Alabama–Georgia. North American Journal of Fisheries Management 4:314–320.

Mitsch, W. J., and J. G. Gosselink. 1993. Wetlands, 2nd ed. Van Nostrand Reinhold, New York.

Naiman, R. J. and D. L. Soltz (eds.), 1981. Fishes in North American deserts. Wiley, New York.

Naiman, R. J., C. A. Johnston, and J. C. Kelley. 1988. Alteration of North American streams by beaver. BioScience 38:753–762.

Nunnally, N. R., J. R. Hynson, and F. D. Shields, Jr. 1987. Environmental considerations for levees and floodwalls. Environmental Management 11:183–191.

Ogutu-Ohwayo, R. 1990. The decline of the native fishes of Lake Victoria and Kyoga (East Africa) and the impact of introduced species, especially Nile perch, *Lates niloticus*, and Nile tilapia, *Oreochomis niloticus*. Environmental Biology of Fishes 27:81–96.

Pardue, G. B. 1993. Life history and ecology of the mud sunfish (*Acantharchus pomotis*). Copeia 1993:533–540.

Peterka, J. J. 1989. Fishes in northern prairie wetlands. Pages 302–315 *in* A. G. van der Valk (ed.), Northern prairie wetlands. Iowa State University Press, Ames, IA.

Poff, N. L., J. D. Allan, M. B. Bain, J. R. Karr, K. L. Prestegaard, B. D. Richter, R. E. Sparks, and J. C. Stronmberg. 1997. The natural flow regime: a paradigm for river conservation and restoration. BioScience 47:769–784.

Power, M. E., A. Sun, G. Parker, W. E. Dietrich, and J. T. Wooton. 1995. Hydraulic food-chain models: an approach to the study of food-chain dynamics in large rivers. BioScience 45: 159–167.

Pullen, T. M., Jr. 1971. Some effects of beaver *(Castor canadensis)* and beaver pond management on the ecology and utilization of fish populations along warm-water streams in Georgia and South Carolina. Ph.D. dissertation, University of Georgia, Athens, GA.

Rader, R. B. 1999. The Florida Everglades: natural variability, invertebrate diversity, and foodweb stability. Pages 25–54 *in* D. P. Batzer, R. B Rader, and S. A. Wissinger (eds.), Invertebrates in freshwater wetlands of North America: ecology and management. Wiley, New York.

Rahel, F. J. 1984. Factors structuring fish assemblages along a bog lake successional gradient. Ecology 65:1276–1289.

Randall, R. G., C. K. Minns, V. W. Cairns, and J. E. Moore. 1996. The relationship between an index of fish production and submerged macrophytes and other habitat features at three littoral areas in the Great Lakes. Canadian Journal of Fisheries and Aquatic Sciences 53(Suppl. 1):35–44.

Remillard, M. M., G. K. Gruendling, and D. G. Bogucki. 1987. Disturbance by beaver (*Castor canadensis* Kuhl) and increased landscape heterogeneity. Pages 103–122 *in* M. G. Turner (ed.), Landscape heterogeneity and disturbance. Springer-Verlag, New York.

Richardson, C. J., and J. W. Gibbons. 1993. Pocosins, Carolina bays, and mountain bogs. Pages 257–310 *in* W. H. Martin, S. G. Boyce, and A. C. Echternacht (eds.), Biodiversity of the southeastern United States/lowland terrestrial communities. Wiley, New York.

Richardson, W. B., S. A. Wickman, and S. T. Threlkeld. 1990. Foodweb responses to the experimental manipulation of a benthivore (*Cyrinus carpio*), zooplanktivore (*Mendia beryllina*) and benthic insects. Archiv fuer Hydrobiologia 119:143–165.

Richter, B. D., J. V. Baumgartner, J. Powell, and D. P. Braun. 1996. A method for assessing hydrological alteration within ecosystems. Conservation Biology 10:1163–1174.

Richter, B. D., J. V. Baumgartner, R. Wigington, and D. P. Braum. 1997. How much water does a river need? Freshwater Biology 37:231–249.

Rohde, F. C., and R. G. Arndt. 1991. Distribution and status of the sandhills chub, *Semotilus lumbee*, and the pinewoods dart, *Etheostoma mariae*. Journal of the Elisha Mitchell Scientific Society 107:61–70.

Sabo, M. J., and W. E. Kelso. 1991. Relationship between morphometry of excavated floodplain ponds along the Mississippi River and their use as fish nurseries. Transactions of the American Fisheries Society 120:552–561.

Sabo, M. J., W. E. Kelso, C. F. Bryan, and D. A. Rutherford. 1991. Physicochemical factors affecting larval fish densities in Mississippi River floodplain ponds, Louisiana USA. Regulated Rivers Research and Management 6:109–116.

Schiemer, F., and M. Zaleweski. 1992. The importance of riparian ecotones for diversity and productivity of riverine fish communities. Netherlands Journal of Zoology 42:323–335.

Schlosser, I. J. 1995. Dispersal, boundary processes, and trophic-level interactions in streams adjacent to beaver ponds. Ecology 76:908–925.

Sharitz, R. R., and J. W. Gibbons. 1982. The ecology of southeastern shrub bogs (pocosins) and Carolina bays: a community profile. Report FWS/OBS-82/04. U.S. Fish and Wildlife Service Biological Services Program, Slidell, LA.

Snodgrass, J. W. 1997. Temporal and spatial dynamics of beaver-created patches as influenced by management practices in a south-eastern North American landscape. Journal of Applied Ecology 34:1043–1056.

Snodgrass, J. W., and G. K. Meffe. 1998. Influence of beavers on stream fish assemblages: effects of pond age and watershed position. Ecology 79:928–942.

Snodgrass, J. W., and G. K. Meffe. 1999. Habitat use and temporal dynamics of blackwater stream fishes in and adjacent to beaver ponds. Copeia 1999:628–639.

Snodgrass, J. W., A. L. Bryan, Jr., R. F. Lide, and G. W. Smith. 1996. Factors affecting the occurrence and structure of fish assemblages in isolated wetlands of the upper coastal plain, U.S.A. Canadian Journal of Fisheries and Aquatic Sciences 53:443–454.

Snodgrass, J. W., M. J. Komoroski, A. L. Bryan, Jr., and J. Burger. 2000a. Relationships among isolated wetland size, hydroperiod, and amphibian species richness: implications for wetland regulations. Conservation Biology 14:414–419.

Snodgrass, J. S., C. H. Jagoe, A. L. Bryan, Jr., H. A. Brant, and J. Burger. 2000b. Effects of trophic status and wetland morphology, hydroperiod, and water chemistry on mercury concentrations in fish. Canadian Journal of Fisheries and Aquatic Sciences 57:171–180.

Sparks, R. E., P. B. Bayley, S. L. Kohler, and L. L. Osborne. 1990. Disturbance and ecology of large floodplain rivers. Environmental Management 14:699–709.

Stewart, R. E., and H. A. Kantrud. 1973. Ecological distribution of breeding waterfowl populations in North Dakota. Journal of Wildlife Management 37:39–50.

Tonn, W. M., and R. E. Vandenbos, and C. A. Paszkowski. 1995. Habitat on a broad scale: relative importance of immigration and extinction for small lake fish assemblages. Bulletin Francais de la Pêche et da la Pisciculture 337/338/339:47–61.

Trautman, M. B. 1981. The fishes of Ohio. Ohio State University Press, Columbus, OH.

Trexler, J. C. 1995. Restoration of the Kissimmee River: a conceptual model of past and present fish communities and its consequences for evaluating restoration. Restoration Ecology 3:195–210.

Turner, T. F., J. C. Trexler, G. L. Miller, and K. E. Toyer. 1994. Temporal and spatial dynamics of larval and juvenile fish abundance in a temporate floodplain river. Copeia 1994: 174–183.

Turner, A. M., J. C. Trexler, C. F. Jordan, S. J. Slack, P. Geddes, J. H. Chick, and W. F. Loftus. 1999. Targeting ecosystem features for conservation: standing crops in the Florida Everglades. Conservation Biology 13:898–911.

Walters, C. J., and C. S. Holling. 1990. Large-scale management experiments and learning by doing. Ecology 71:2060–2068.

Ward, J. V. 1998. Riverine landscapes: biodiversity patterns, disturbance regimes, and aquatic conservation. Biological Conservation 83:269–278.

Weaver, M. J., J. J. Magnuson, and M. K. Clayton. 1997. Distribution of litteral fishes in structurally complex macrophytes. Canadian Journal of Fisheries and Aquatic Sciences 54: 2277–2289.

Welcomme, R. L. 1979 Fisheries ecology of floodplain rivers. Longman, London.

Welcomme, R. L. 1989. Floodplain fisheries management. Pages 209-234 *in* J. A. Gore and G. E. Petts (eds.), Alternatives in regulated rivers management. CRC Press, Boca Raton, FL.

Welcomme, R. L., and D. Hagborg. 1977. Toward a model of a floodplain fish population and its fishery. Environmental Biology of Fishes 2:7–24.

Wellborn, G. A., D. K. Skelly, and E. E. Werner. 1996. Mechanisms creating community structure across a freshwater habitat gradient. Annual Review of Ecology and Systematics 27:337–363.

Werner, E. E., G. G. Mittlebach, D. J. Hall, and F. J. Gilliam. 1983a. Experimental tests of optimal habitat use in fish: the role of relative habitat profitability. Ecology 64:1525–1539.

Werner, E. E., F. J. Gilliam, D. J. Hall, and G. G. Mittlebach. 1983b. An experimental test of the effects of predation risk on habitat use in fish. Ecology 64:1540–1548.

Whillans, T. H. 1996. Historic and comparative perspectives on rehabilitation of marshes as habitat for fish in the lower Great Lakes basin. Canadian Journal of Fisheries and Aquatic Sciences 53(Suppl. 1):58–66.

Wiley, M. J., R. W. Gorden, S. W. Waite, and T. Powless. 1984. The relationship between aquatic macrophytes and sportfish production in Illinois ponds: a simple model. North American Journal of Fisheries Management 4:111–119.

17 Managing Wetlands for Waterbirds

MURRAY K. LAUBHAN and JAMES E. ROELLE

Over 70 species of waterbirds (excluding waterfowl) occur in North America. Although the habitat requirements of many species have been documented, there remains a paucity of information on selecting management strategies to benefit a diversity of waterbirds. Further, attempts to manage specific aspects of a habitat without ensuring that processes important to wetland functions also remain intact probably will result in reduced long-term wetland values. Thus, we focus on management directed toward maintaining wetland productivity, and in that context, manipulations that benefit waterbirds in six taxonomic groups (cranes, grebes, ibises, herons, rails, and shorebirds). The timing, frequency, and intensity of five primary management strategies (water manipulation, burning, grazing, mechanical disturbances, and herbicides) are discussed in relation to the composition and structure of plant and invertebrate communities. In addition, the potential effects of these manipulations on the habitat requirements of each waterbird group are presented for each annual cycle event.

Over 70 species of waterbirds (excluding waterfowl) either breed, migrate, winter, or are resident in North America. There is substantial published information on the biology and ecology of these birds, but relatively little regarding active management of their habitats. Further, management strategies that do exist often are directed toward a single species or group rather than entire waterbird communities, and potential impacts to nontarget species are rarely discussed (Heitmeyer et al. 1996). Our understanding of factors affecting wetland productivity has improved greatly (Weller 1979, Mitsch and Gosselink 1993). As a result, we better understand wetland functions as they relate to resource requirements of waterbirds. However, evaluating management techniques to benefit waterbirds relative to disruption of wetland processes has received much less attention (see Williams et al. 1999). Although both species autecology and wetland ecology are important to improving management, there has been little effort to integrate species requirements with general wetland ecology principles.

Bioassessment and Management of North American Freshwater Wetlands, edited by Russell B. Rader, Darold P. Batzer, and Scott A. Wissinger.
0-471-35234-9

Consequently, our ability to provide long-term benefits for multiple waterbird species is constrained even though much information necessary to make sound decisions is available (Fredrickson and Laubhan 1994a).

Managing for waterbirds is a two-part process: (1) producing required resources, and (2) making resources available. Resource production is clearly paramount and requires an understanding of wetland functions and processes. Because resource production is intricately linked to dynamic wetland processes, managers should not attempt to provide resources for the same species or group on the same wetland each year. Static management ultimately disrupts the short- and long-term dynamics necessary to ensure wetland productivity (Weller 1979). Rather, management should attempt to maintain the inherent productivity of wetlands because multiple abiotic and biotic factors often contribute to an observed waterbird response. Attempts to manage specific aspects of a habitat (e.g., vegetation structure) without ensuring that processes important to wetland functions (e.g., hydrology, nutrient cycling, detrital processing, vegetation and invertebrate productivity) also remain intact probably will result in reduced benefits to waterbirds.

The habitats used by waterbirds are diverse, ranging from southern forested swamps to arctic tundra, and coastal marshes to temporary wetlands in the arid west. In addition, some species are migratory and exploit habitats at continental scales, whereas others are residents that use much smaller geographic areas. It is not possible to cover the entire habitat spectrum in detail; thus, we concentrate on palustrine wetlands in the contiguous United States. Further, we focus on management directed toward maintaining long-term wetland productivity and, in that context, manipulations that benefit waterbirds. Recommendations are conceptual and general in nature; managers are encouraged to refine techniques as new information becomes available.

WATERBIRD REQUIREMENTS

We provide information on six taxonomic groups: cranes (family Gruidae), grebes (Podicipedidae), ibises (Threskiornithidae), herons (Ardeidae), rails (Rallidae), and shorebirds (Recurvirostridae, Charadriidae, and Scolopacidae). Many members of the heron, ibis, and rail families complete all annual cycle events in the contiguous states, although some populations also breed and winter elsewhere (Ehrlich et al. 1988). In contrast, only 13 shorebird species breed in temperate North America; most shorebirds use the contiguous states only during migration (Helmers 1992). Numerous species occur in some families, but many have habitat requirements sufficiently similar to allow discussion of general management actions. Although waterfowl are not discussed, the management techniques described also benefit waterfowl that use palustrine wetland complexes.

Detailed information on diet is available for many waterbird species. In contrast, literature pertaining to habitat requirements often uses vague terminology (e.g., marsh, swamp) that lacks detail regarding parameters (e.g., vegetation height, horizontal and vertical structure of vegetation, water depth) that determine habitat suitability. Collectively, information on vegetation and water depths is necessary to (1) identify similarities and differences in habitat requirements of various waterbirds, (2) evaluate the benefits of various wetland conditions to different waterbirds, and (3) make decisions regarding appropriate management strategies.

In general, foods can be broadly grouped as plant (browse, seeds, tubers) and animal (invertebrate, vertebrate), whereas vegetation structure can be characterized based on density (vertical and horizontal) and height. Although simplistic, these categories are consistent with our current management capabilities.

Foods

Food selection often varies seasonally depending on location, availability, and dietary requirements of annual cycle events. Grebes and herons eat mostly fish, but also consume some invertebrates, reptiles, and amphibians (Table 17.1). Shorebirds consume primarily invertebrates during all phases of the annual cycle (Helmers 1992), although seeds contribute to the diet, particularly during fall migration (Skagen and Oman 1996, Davis and Smith 1998). Some of the larger shorebirds also consume fish and amphibians (Table 17.1). Rails consume varied amounts of seeds and invertebrates, depending on the annual cycle event and time of year. In general, seeds constitute a greater proportion of the diet during fall and winter (Rundle and Sayre 1983, Meanley 1992). For some waterbirds, foods are not located in wetlands. Cranes consume mostly small grains (>90%) produced in agricultural fields and lesser amounts of tubers, amphibians, and terrestrial and aquatic invertebrates (Tacha et al. 1994; Table 17.1).

TABLE 17.1. Foods Consumed by Selected Waterbird Species Occurring in North America

		Foods[a]						
Group/Common Name	Scientific Name	A	F	I	M	R	S	V
Cranes								
Sandhill	*Grus canadensis*	P		2[b]	P	P	3[c]	
Whooping	*Grus americana*	1	2	7	1	1	7[c]	
Grebes								
Pied-billed	*Podilymbus podiceps*	1	7	7			P	P
Western	*Aechmophorus occidentalis*	1	P	3+				
Herons								
American bittern	*Botaurus lentiginosus*	2	8	5	P	1		
Black-crowned night heron	*Nycticorax nycticorax*	P	16	4	P	1		
Great blue heron	*Ardea herodias*	P	P	P	P	P		
Green heron	*Butorides striatus*	2	10	10	P	1		
Least bittern	*Ixobrychus exilis*	2	3	5	P	1		
Little blue heron	*Egretta caerulea*	1	11	10				
Tricolored heron	*Egretta tricolor*	P	7	9		1		

(continues)

TABLE 17.1. (*continued*)

		Foods[a]						
Group/Common Name	Scientific Name	A	F	I	M	R	S	V
Ibis								
White	*Eudocimus albus*	1	P	6		1		
White-faced	*Plegadis chihi*	1	1	13	P		P	P
Rails								
Clapper	*Rallus longirostris*	1	P	14		1	4	2
King	*Rallus elegans*	1	2	6			P	
Sora	*Porazana carolina*			7			4	2
Virginia	*Rallus limicola*		P	13		1	11	
Yellow	*Coturnicops noveboracensis*			9			3	
Shorebirds								
American avocet	*Recurvirostra americana*	1	P	16			4	
Greater yellowlegs	*Tringa melanoleuca*	1	4	11			P	
Pectoral sandpiper	*Calidris melanotos*		1	10			P	P
Semipalmated sandpiper	*Calidris pusilla*			18			P	
Snowy plover	*Charadrius alexandrinus*			12			P	
Spotted sandpiper	*Actitis macularia*	1	P	10				
Western sandpiper	*Calidris mauri*			13				
Whimbrel	*Numenius phaeopus*		P	9			3	
White-rumped sandpiper	*Calidris fuscicollis*			9			3	
Wilson's phalarope	*Phalaropus tricolor*			7			P	

[a]Information on foods obtained from Poole and Gill (various dates) and Poole et al. (various dates). A, amphibians; F, fish; I, invertebrates; M, mammals; R, reptiles; S, seeds; V, vegetation. Numbers in columns refer to number of orders (invertebrates, amphibians, reptiles) or families (fish, seeds, vegetation); P indicates documented consumption but no information on specific items.

[b]Two classes (Insecta, Gastropoda).

[c]Includes cultivated grains in Graminae family (barley corn, oats, rice, sorghum, timothy, wheat).

Although some waterbirds tend to consume specific foods during some annual cycle events, most species exhibit considerable dietary breadth (Table 17.1). Over 400 genera of invertebrate prey are consumed by 43 species of shorebirds in the western hemisphere (Skagen and Oman 1996). Similarly, numerous fishes in several families are eaten by herons, and the diet of rails includes numerous genera of invertebrates and seeds (Table 17.1). Size of food items relative to bill morphology; profitability relative to search, capture, and handling time; and foraging mode often determine which specific foods are consumed (Stephens and Krebs 1986, Stephens et al. 1986).

VEGETATION STRUCTURE AND WATER DEPTH

Breeding Habitat

The primary requisites of breeding habitat include appropriate interspersion of nesting, brood rearing, and foraging areas. Shorebirds that breed in the contiguous states nest on the ground, typically on elevated areas near water (Helmers 1992). Substrate type varies from exposed sand/gravel beaches [e.g., piping plover (*Charadrius melodus*), snowy plover (*C. alexandrinus*)] to vegetation of moderate height and density [e.g., Wilson's phalarope (*Phalaropus tricolor*), willet (*Catoptrophorus semipalmatus*), marbled godwit (*Limosa fedoa*)]. Species that nest in open habitats often select sites near clumps of vegetation or large objects (e.g., logs) that provide partial nest cover (Helmers 1992). Nesting habitat often varies depending on geographic area and type and density of predators. In cases of high predation risk, some species [e.g., spotted sandpiper (*Actitis macularia*)] nest in denser vegetation (Oring et al. 1997). Because nesting habitats are semiopen or open and are located near water, vegetation structure in areas surrounding nest sites also provides suitable brood habitat if foods are available.

Breeding habitat of rails and bitterns is characterized by tall, dense, emergent cover for nesting interspersed with sparsely vegetated, shallowly flooded areas that permit movement of broods (rails) to foraging sites (Johnson and Dinsmore 1985, 1986, Davidson 1992, Gibbs et al. 1992a, Eddleman and Conway 1998). Most species nest in elevated areas with vegetation of sufficient density and height to support nests and provide concealment. This structure is typically provided by robust emergents [e.g., cattail (*Typha* spp.), bulrush (*Scirpus* spp.), sedges (*Carex* spp.)], but clapper rails (*Rallus longirostris*; Eddleman and Conway 1998) and least bitterns (*Ixobrychus exilis;* Weller 1961) also nest in shrubs. Nests are located over or adjacent to water. Depth of water at nests varies depending on preferred foraging depths, risks associated with flooding, and vegetation structure. Rails typically nest in habitats that range from moist soil to 46 cm of water (Meanley 1992, Bookhout 1995), whereas American bitterns (*Botaurus lentiginosus*) and least bitterns nest in habitats flooded from 5 to 20 cm and 8 to 96 cm (Gibbs et al. 1992a,b), respectively.

Remaining members of the heron family occurring in temperate North America typically nest in trees and shrubs or occasionally in emergent vegetation or on the ground, depending on available substrate and geographic area (Butler 1992, Davis and Kushlan 1994, Watts 1995, Frederick 1997). Nests often occur in single- or mixed-species colonies near or over water. In mixed colonies, separation of nesting locations among species is often related to vegetation height and structure. For example, tricolored herons (*Egretta tricolor*) and green herons (*Butorides striatus*) typically nest in denser vegetation near the ground (tricolored heron < 4 m, green herons < 10 m) relative to other species (Davis and Kushlan 1994, Frederick 1997). Colony location is often influenced by distribution of foraging habitats and predators (Davis 1993, Davis and Kushlan 1994).

Grebes nest in stands of flooded emergent vegetation, but occasionally on bare ground or mats of submerged aquatic vegetation (Terres 1980). Suitable nest sites have emergent vegetation surrounded by large expanses of open water that provide protection from predators (Storer and Nuechterlein 1992). Emergent vegetation must be of sufficient density and height to support nests and provide concealment. Most nests are placed over water > 25 cm deep (Terres 1980). Breeding habitat of ibis is similar, although nests may also be placed in low shrubs and trees as well as in emergent vegetation (Kushlan and Bildstein 1992, Ryder and Manry 1994). Height of nests above water varies depending on water-level, vegetation structure, and magnitude of water level fluctuations. Location of ibis colonies is influenced by proximity to foraging sites. Cranes also nest over water in emergent vegetation, but vegetation typically is shorter and less dense (Tacha et al. 1992, Lewis 1995). Typical nesting areas include open marshes, wet meadows, and open, low shrubs. Parents feed young until they can forage with adults; thus, there are no differences in the structural features or water depths at foraging sites on breeding grounds.

Migration Habitat

Migration habitats primarily allow birds to acquire the foods necessary to complete movements between breeding and wintering sites. Although foraging habitats for waterbirds span a broad range of water depths and vegetation structure, there is considerable overlap among species (Table 17.2, Fig. 17.1). Most shorebirds require water depths of less than 10 cm and less than 25% cover of short vegetation (Helmers 1992). Cranes, herons, and ibis also forage in such habitats, but herons and ibis are also capable of foraging in taller, more densely vegetated habitats with deeper water (Kushlan and Bildstein 1992, Ryder and Manry 1994, Fig. 17.1). Bitterns often pre-

TABLE 17.2. Comparison of Relative Vegetation Structure at Preferred Nesting and Foraging Sites of Selected Waterbird Groups That Breed in North America

	Nesting			Foraging	
Group	Type	Height	Density	Height	Density
Bitterns	Emergent	High	High	Medium–high	Medium–high
Cranes	Emergent/upland	Low	Medium–high	Low–medium	Low–high
Grebes	Emergent	High	Medium–high	Low	Low
Herons	Forested/emergent	High	Medium–high	Low–medium	Low–high
Ibis	Emergent	High	Medium–high	Low–medium	Low–high
Rails	Emergent	High	High	Medium–high	Medium–high
Shorebirds	Beach/upland	Low–medium	Low–medium	None–low	None–low

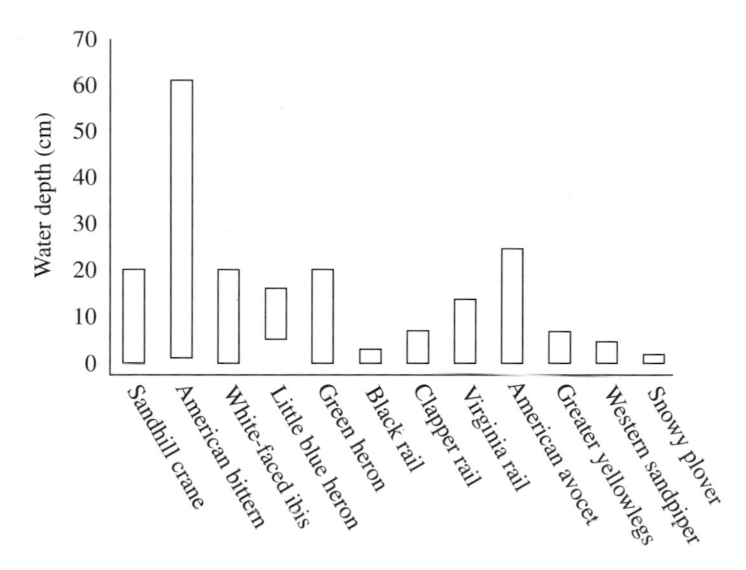

FIGURE 17.1. Foraging depths of selected waterbird species in North America. (Data from Poole and Gill various dates, Poole et al. various dates.)

fer to forage in more densely vegetated habitats that provide concealment (Gibbs et al. 1992a,b). Rails also require shallow flooding for optimum foraging, but prefer tall, dense, emergent cover interspersed with openings (Meanley 1992). In contrast, grebes forage at depths up to 6 m in waters where fish populations are reliable. Because grebes forage visually and pursue fish under water, the most preferred habitats are also characterized by sparse vegetation (Storer and Nuechterlein 1992).

Within these general groups, subtle differences in foraging habitat exist among species. Among shorebirds, noted exceptions include common snipe (*Gallinago gallinago*), which forage in short, dense vegetation (Helmers 1992). Pectoral sandpipers (*Calidris melanotus*) tend to forage in wetter sites and in taller vegetation than co-occurring dunlin (*C. alpina*) and Baird's sandpipers (*C. bairdii*) on breeding grounds, and during winter frequently forage in dense grasses (Holmes and Pitelka 1998). Some of the larger shorebirds [e.g., American avocet (*Recurvirostra americana*), greater yellowlegs (*Tringa melanoleuca*)] and phalaropes can forage in open, more deeply flooded habitats than other species (Fig. 17.1). Similar differences exist among members of the rail and heron families. Black rails (*Laterallus jamaicensis*) use shallower water than do other North American rails (Eddleman et al. 1988), and soras (*Porzano carolina*) exploit a wider range of water depths than do Virginia rails (*Rallus limicola*, Melvin and Gibbs 1996). Compared to least bitterns, American bitterns use a wider variety of cover types, less densely vegetated sites, and exclusively freshwater wetlands. Regardless of these differences, however, species exhibiting greater flexibility in choice of habitats also tend to forage in habitats used by other members of the same family. Thus, from a management perspective, such differences should be recognized, but they should not necessarily alter management strategies significantly.

Wintering Habitat

Most members of the heron, ibis, crane, and rail families that breed in temperate North America either reside in breeding areas year round or migrate to the southern United States during winter (Ehrlich et al. 1988). In contrast, only a few shorebirds are residents (e.g., spotted sandpiper, snowy plover, piping plover) or winter in this geographic area [e.g., whimbrel (*Numenius phaeopus*), greater yellowlegs, western sandpiper (*Calidris mauri*), least sandpiper (*C. minutilla*), stilt sandpiper (*C. himantopus*)] (Hayman et al. 1986). In general, wintering habitats are similar to breeding and migration habitats. However, conditions that restrict habitat use during breeding (e.g., location of nest sites relative to foraging sites) do not influence use of wintering habitats. Consequently, many species exhibit greater flexibility in winter habitat preferences. Important parameters include vertical and horizontal vegetation structure that provides roosting areas and protection from extremes in climatic conditions, as well as foraging habitats.

WATERBIRD MANAGEMENT

Management for waterbird diversity requires determining the (1) chronology of use by individual species or guilds, (2) annual cycle requirements of species at a particular geographic location (resident, breeding, migration, wintering), and (3) resource requirements, primarily foods and vegetation structure, necessary to complete annual cycle requirements. Understanding which life cycle events occur in a geographic area of interest is essential for determining dietary and structural requirements. Foraging and nesting often require different vegetation structure. Further, timing is critical because some events are of short duration (e.g., 1 month). Consequently, understanding resource requirements and chronology of species use often allows managers to maintain dynamic hydroperiods necessary for sustained wetland productivity while simultaneously providing critical resources for waterbirds.

A single wetland basin can often provide resources to multiple species (Helmers 1992), but not all life requisites for all species (Fredrickson and Laubhan 1994a). Therefore, decisions must be made regarding what management strategies to implement. Appropriate strategies will vary depending on goals and objectives, which are influenced by geographic location, number and type of wetlands composing a complex, management capability, ownership, and wetland processes. Identifying disrupted wetland processes often helps determine the cause of reduced productivity and refine the number of plausible management options. In some cases, deficiencies in management capability (e.g., degree of water-level control) can limit success. Early detection of disrupted processes reduce the need for costly and time-consuming techniques. In cases where management is feasible, this approach will help identify appropriate actions (e.g., water level manipulation, burn), as well as the time (e.g., growing/dormant season, early/late drawdown), extent (e.g., portion of wetland or entire wetland, depth of disking), and duration (e.g., entire growing season, entire year) of implementation.

Spatial Considerations

Waterbirds use habitats at scales varying from local (e.g., resident herons) to continental (e.g., migratory shorebirds), depending on annual cycle event and species (Soule 1991). However, even at the scale of a single wetland complex, numerous factors influence waterbird use, including the size, topographic complexity, and dominant vegetation of individual wetlands. Compared to small, relatively flat wetlands, large wetlands with diverse topography often exhibit greater structural diversity and variable water depths that can be exploited by different species of waterbirds (Laubhan and Fredrickson 1993). For example, breeding birds are most abundant and diverse when emergent vegetation and open water occupy about equal proportions in a basin (Kaminski and Prince 1981). However, some species (e.g., herons) use herbaceous wetlands as foraging sites and forested wetlands or uplands for nesting. In such cases, distances between different wetland types can influence waterbird use. Issues of scale are complex, and a detailed discussion is beyond the scope of this chapter. However, the topic is extremely relevant to management and development, and managers are encouraged to consider interspersion and configuration of all wetland types comprising a complex when choosing management strategies.

Water Management

Water-level manipulation is one of the most effective methods of influencing plant community composition, vegetation structure, and production of plant and invertebrate foods (Fredrickson 1991). In general, annual plants are more prolific seed producers and decompose at faster rates, whereas perennial species have underground structures (e.g., rhizomes, tubers, stolons) that are important foods for some species, decompose more slowly, are more tolerant of flooding, and often form dense stands capable of supporting and concealing nests. The timing and rate of water-level fluctuations influence the distribution and proportion of these two plant types (Fredrickson and Taylor 1982, Merendino and Smith 1991). Among the most important parameters controlling germination are light and soil temperature, salinity, oxygen, and moisture (Simpson et al. 1989). These conditions change throughout the growing season; thus, time of soil exposure is critical. From a management perspective, the growing season can be defined as the period between last spring frost and first fall frost. Two (early, late) or three (early, middle, late) periods can be defined depending on whether the length of the growing season is less than or greater than 160 days (Fredrickson 1991). Drawdowns that expose soil during the first 45 days of the growing season are defined as early, whereas those exposing soil during the last 90 days are considered late. In areas with growing seasons greater than 160 days, midseason drawdowns are variable in length but occur at least 45 days after the start of the growing season and at least 90 days before the end of the growing season. For growing seasons 135 to 160 days in length, there is a midseason (up to 25 days), but it is of insufficient length to be considered for separate management actions.

These periods are useful because they are related to many abiotic factors (e.g., day length, temperature, rate of soil drying) that control germination (Table 17.3).

TABLE 17.3. Relative Comparison of the Influence of Drawdown Dates and Rates on the Germination, Growth, and Seed Production of Plants, Invertebrate Availability, and Selected Abiotic Factors

	Drawdown Rate	
	Fast[a]	Slow[b]
Plants		
Germination		
Length of ideal conditions	Short	Long
Root development		
Wet year	Good	Excellent
Dry year	Poor	Excellent
Seed production		
Early season drawdown	Good	Excellent
Midseason and late drawdown	Poor	Good
Wet year	Good	Good
Dry year	Poor	Good
Invertebrates		
Availability	Good	Excellent
Early season	Good	Excellent
Midseason and late drawdown	Poor	Good
Length of availability	Short	Long
Potential for nutrient export	High	Low
Potential for reducing soil salinity	Good	Poor

[a] < 7 days.
[b] > 14 days.

Obviously, water-level manipulations conducted at different times of the growing season influence soil moisture, but timing influences other factors as well. For example, late drawdowns can increase soil salinity if saline water is allowed to evaporate, whereas early drawdowns expose substrates during shorter day lengths and cooler soil temperatures. Rate of drawdown is also important because the rate of soil drying influences the type and density of seeds that germinate (Fredrickson and Taylor 1982). In general, slow drawdowns create a broader range of environmental conditions that resulted in germination of more plant species. In addition, slow drawdowns maximize the time available for root growth, which increases seedling survival during drought (Fredrickson 1991). In contrast, fast drawdowns often cause rapid drying of the soil, which may result in sparse germination of fewer species or poor survival because roots do not develop sufficiently prior to decreases in available water near the soil surface. However, fast drawdowns may periodically be necessary in some geographic areas to reduce accumulation of soil salts that can prevent germination of many plant species. Often, similar drawdowns (time and rate) conducted in different wetlands in the same geographic area can result in germination of different plant species or, in some cases, complete lack of germination. Many factors may affect such differences; among the most common are soil properties (pH, salinity, texture) and water quality (e.g., salinity, chemical) (Smith and Kadlec 1983).

Although germination requirements of most species have not been documented, some general information exists to help select appropriate times to conduct manipulations (Table 17.4). Some species germinate under a wide range of environmental conditions; thus, controlling plant community establishment can be difficult. However, seeds of many perennials such as cattail and bulrush, and common reed germinate under moist to shallowly flooded conditions when surface soil temperatures are warm (about 30°C). In contrast, species such as annual smartweeds (*Polygonum* spp.) tend to germinate in cooler soils that are moist to saturated. Therefore, early season drawdowns tend to stimulate greater production of smartweed, whereas late season drawdowns are more likely to stimulate germination of some perennials. Consequently, highest seed production typically occurs following early and midseason drawdowns, although seed production can also be high following late drawdowns in wet years (Fredrickson 1991; Table 17.4).

TABLE 17.4. Relative Response of Selected Plant Species to Drawdown Date

			Drawdown date[a]		
Family	Common Name	Scientific Name	Early	Midseason	Late
Buckwheat	Pennsylvania smartweed	*Polygonum pennsylvanicum*	+++	+	
	Curlytop ladysthumb	*Polygonum lapathifolium*	+++	+	
Composite	Aster	*Aster* spp.	+++	++	+
	Beggarticks	*Bidens* spp.	+	+++	+++
	Cocklebur	*Xanthium strumarium*	++	+++	++
Grass	Rice cutgrass	*Leersia oryzoides*	+++	+	
	Sprangletop	*Leptochloa* spp.		++	+++
	Panic grass	*Panicum dichotomoflorum*		+++	++
	Barnyardgrass	*Echinochloa crusgalli*	+++	++	+
Pea	Sesbania	*Sesbania exalta*	+	++	
	Sweet clover	*Melilotus* spp.	+++		
Sedge	Chufa	*Cyperus esculentus*	+++	+	
	Red-rooted sedge	*Cyperus erythrorhizos*	+	+++	
	Spikerush	*Eleocharis* spp.	+++	+	+

[a] Drawdown dates defined as time of soil exposure: early, first 45 days of growing season; late, last 90 days of growing season.

Although seed germination of most annuals requires exposure of the substrate, seeds of some perennials are capable of germinating in shallow water and can also initiate growth from remnant underground structures. In addition, most herbaceous perennials have physical and/or physiological adaptations that enable survival during unfavorable environmental conditions. Thus, prolonged shallow flooding through consecutive growing seasons tends to favor growth and expansion of perennial species and hinders germination of annuals. In contrast, periodic drawdowns to stimulate annuals rarely cause complete mortality of perennial species. However, periodic drawdowns often are an effective means of reducing encroachment of many perennial species, thereby promoting plant diversity (Fredrickson 1991).

Relatively little is known about the direct response of invertebrates to drawdowns (Kadlec and Smith 1992). Many invertebrates exhibit adaptations to long-term hydrologic cycles that permit survival during drought, but short-term cycles may determine the actual occurrence and abundance of invertebrates (Murkin and Kadlec 1986, Fredrickson and Reid 1988). Thus, entirely different invertebrate communities can occur in basins with differing hydroperiods. The initial, or short-term, response of most aquatic invertebrates to drawdowns may be negative (Kadlec and Smith 1992). However, prolonged flooding (>1 year) of uniform depth can also reduce wetland invertebrate numbers (Fredrickson and Reid 1988). In many cases, this phenomenon can be attributed to decreased wetland productivity, degradation of habitat structure for invertebrates, decreased nutrient availability, or increased predation (by fish). Therefore, the long-term impacts of drawdowns may be beneficial (e.g., higher invertebrate biomass) because water manipulations can alter plant species composition, release nutrients, and influence fish populations.

In addition to influencing vegetation structure and food (seeds, invertebrates, fish) production, water-level manipulations can also increase resource availability to multiple waterbird species. Although nutritional differences exist among plant species, managers need not attempt to produce a community of one or two "desirable" species. Rather, management should be directed toward stimulating plant diversity, which provides some protection against failure in the seed crop of a particular species and also tends to increase structural diversity as well as invertebrate diversity. Further, because the life history strategies of fishes and invertebrates vary, there is much flexibility in managing wetlands to satisfy the requirements of numerous waterbird species. For example, slow drawdowns timed to coincide with spring shorebird migration can concentrate invertebrates in small, shallowly flooded areas where they are readily available and, simultaneously, stimulate germination of annual plants. Because drawdowns conducted during the growing season promote vegetation development, flooding these same wetlands in fall provides habitat for species (e.g., ibis, rails) that forage or roost in medium to dense cover. However, these wetlands probably will not receive extensive use by shorebirds that prefer sparse vegetation; thus, different wetlands should be selected to provide appropriate habitat conditions and water depths for these species.

Although fish tend to be most abundant in permanent and semipermanent wetlands, seasonal wetlands can also harbor fish if they are not completely dewatered.

This is particularly true of seasonal wetlands that can be managed intensively (e.g., moist-soil impoundments). Partial drawdowns often concentrate fish in remaining water. For example, many fish occur in the borrow ditches of moist-soil impoundments following drawdowns. Further, many fish use wetlands associated with lotic systems (rivers, streams, ditches connecting wetlands) only as brood and nursery habitat; thus, permanent water in the basin is not necessary. In fact, drawdowns are sometimes necessary to stimulate the movement of young fish to habitats required as adults. Partial drawdowns not only allow better access to fish by herons, but also promote plant diversity. Annuals can colonize areas that are dewatered, whereas areas that remain flooded can continue to support aquatic vegetation or emergent perennials that provide structure for nesting.

In summary, water-level manipulations enhance energy and nutrient flow in wetlands (Magee 1993). The impacts of manipulations vary depending on the extent time, and rate of drawdown and flooding, but can include oxidation of soils, consolidation of suspended sediments, release of bound nutrients, improved water clarity, and increased dissolved oxygen in the water column. Ultimately, such changes can result in increased plant productivity and biomass of invertebrates. Because plant germination requirements cover a broad spectrum of conditions, and because vertebrates and invertebrates inhabiting wetlands are diverse, much flexibility exists in selecting water management strategies that maintain wetland productivity and also provide for the needs of waterbirds. Of primary importance is that diverse structure and numerous food types can be produced and made available if traditional water management strategies are altered slightly. Partial rather than complete drawdowns, timing drawdown or flooding to coincide with migration or periods of high food requirements (e.g., brood rearing), and leaving borrow ditches flooded year-round are just a few possible techniques to consider.

Burning

Periodic fire was a normal phenomenon in presettlement wetlands, and prescribed burning is used in an attempt to restore this component of wetland ecology. The effects of burning depend on a number of interrelated factors (Kantrud 1986, Kadlec and Smith 1992), the most important of which include:

1. *Wetland type*. Plant species occupying different wetland types vary greatly in their response to fire. Most of our management experience involves permanent or semipermanent wetlands dominated by emergents, whereas relatively little information is available on the effects of fire in seasonal (e.g., moist-soil impoundments) and temporary wetlands. Generalizations among wetland types are difficult due to basic differences in their ecology.
2. *Fire intensity*. Fires can be characterized according to their intensity as surface burns, root burns, or peat burns. Fire intensity is an important determinant of the resulting pattern of vegetation response in terms of height, density, and species composition.

3. *Timing*. The most obvious aspect of timing of burns is that the nesting season should be avoided if at all possible. In addition, timing with respect to plant phenology can influence vegetation response.

Perhaps the primary purpose of prescribed fire in wetlands is to alter vegetation structure and stimulate primary productivity by removing dead vegetation and litter, releasing minerals and nutrients, and depending on timing, promoting increased soil temperatures earlier in the growing season (Kantrud 1986). Surface burns (including burns occurring over ice), usually conducted during the dormant period, are sufficient to accomplish these purposes. The results of surface burns are rarely uniform over an entire marsh, often producing a desirable interspersion of vegetative cover and open water (after reflooding). However, vegetation response following surface burns is usually rapid, the effects seldom lasting more than a year (Kadlec and Smith 1992).

Although surface burns during the dormant season temporarily set back succession, they do not normally alter species composition. However, timing of surface burns during the growing season can sometimes be used to favor more desirable species. For example, late summer burning in southeastern coastal marshes favored Olney bulrush (*Scirpus olneyi*) over marshhay cordgrass (*Spartina patens*) (Lay 1945). In southeastern Missouri, spring fires reduced percent cover of marsh elder (*Iva ciliata*), a species considered undesirable in moist-soil impoundments (Laubhan 1995). Summer burns were more effective than spring burns in reducing the combined percent cover of all undesirable species, but summer fires also increased the frequency of occurrence and percent cover of two undesirable species, sesbania (*Sesbania exalta*) and morning glory (*Ipomoea* sp.).

More intense fires (root burns and peat burns) can be used to alter species composition. For example, if a wetland can be completely dried, root burns during the growing season can destroy root crowns, rhizomes, and tubers at or below ground level, in addition to removing aboveground biomass. Such fires can be used to control some species (e.g., cattail), particularly if followed by complete inundation of the residual stubble for at least one growing season. Variations in drying often lead to spotty burns, again producing a desirable interspersion of vegetation and open water following reflooding. Vegetation response on exposed substrates depends on a variety of factors, including the soil seedbank and water management (see the preceding section). In addition, peat burns often reduce the surface elevation of parts of a marsh, which may result in pockets of open, perhaps deeper water after reflooding. Extreme caution should be exercised with peat fires because they can be difficult to extinguish.

Most of the existing information on burning focuses on vegetation and waterfowl. As a result, likely impacts of burning on other waterbirds must often be extrapolated from knowledge about the physical and chemical effects of fire. In relation to waterbird requirements, we know most about the impacts of fire on the height and density of vegetation, which are important characteristics of foraging, nesting, and brood-rearing habitat. We know relatively less about how burning affects production of plant and invertebrate foods. The most obvious short-term effect of burning is a significant reduction in both the height and density of vegetation. Thus, burns initially tend to favor those species that prefer more open conditions for foraging or nesting (Table 17.2). How long such conditions persist depends on complex interactions of all the factors discussed above. For example, shallow reflooding following a winter or early spring surface burn may produce good foraging conditions for early-migrating shorebirds. However, shorebirds migrating later in spring may not benefit if vegetation regrowth is rapid. In contrast, Laubhan (1995) found that vegetation regrowth was relatively slow following July burns on seasonal wetlands in Missouri. He suggested that such sites might be shallowly flooded for fall-migrating shorebirds. Interspersion of water and vegetation caused by patchy fires may also benefit some waterbirds. Rails prefer to forage at the vegetation–water interface, and deeper pockets of water created by peat fires may benefit herons. Obvious negative effects include loss of vegetation structure that can support nests of some waterbird species.

In addition to altering vegetation structure, fire can influence production and availability of seeds and other edible plant parts. Results may again depend on timing of the burn. For example, Linde (1985) suggested that winter burns conducted over ice can expose seeds produced in the previous growing season and make them more available to foraging birds. Laubhan (1995) reported that spring burning of seasonal wetlands in Missouri significantly increased total seed production and production of beggarticks (*Bidens* spp.). Summer burning, on the other hand, may have decreased seed production. Of course, water conditions must also be appropriate to make increased production available to waterbirds.

Although information regarding the effects of wetland fires on invertebrates is lacking, one of the principal benefits of burning is to stimulate primary production through release of minerals and nutrients bound in plant material. Therefore, it is likely that these benefits extend up the food chain to invertebrates. However, invertebrate response may be delayed because fires also remove much of the vegetative substrate necessary for production of many invertebrate species.

Grazing

Grazing is a naturally occurring process in wetlands (Heitmeyer 1994). Grazing systems that emulate natural timing and intensity of consumption and trampling will probably be most compatible with maintaining the structure and function of wetlands (Heitmeyer 1994). Although more complex than other commonly used management techniques, grazing can result in diverse vegetation structure (height, density) at multiple scales that cannot be achieved with mechanical disturbances. Grazing has also been used with varying degrees of success to control certain plant species (Lacey 1987). The effects of grazing on invertebrate abundance in wetlands are little known but may not be detrimental unless livestock consume large portions of plant biomass or excessively trample detrital litter (Logan 1975). Herbivores can also influence numerous factors that affect wetland processes, including the moisture retention capacity of soils (Gifford 1986) and vegetation decomposition (Brown and Allen 1989).

The effects of grazing tend be highly variable (Maschinski and Whitham 1989), depending on complex plant–herbivore interactions that operate at various spatial and temporal scales (Belsky 1987, Brown and Allen 1989). Herbivores can affect the distribution, abundance, and diversity of plant species through either selective defoliation or disproportionate trampling of vegetation in specific areas. Plant community composition, vegetation structure, and the distribution and growth phenology of individual species often influence the extent and location of these disturbance types. For example, trampling tends to be most extensive in areas of water availability, and cattle often selectively forage on different plants throughout the growing season, depending on palatability, growth form, stage of maturity, and nutrient content (Brown and Allen 1989). Further, because availability is constrained by plant species composition, selection by herbivores often varies within and among years on the same area.

Among the most important factors to consider when developing a grazing strategy are the type of herbivore, intensity (unit area/animal unit months), and time (frequency, season, duration) of application relative to plant community composition and phenology of individual species. In addition, geographic location as it relates to environmental conditions (e.g., time and amount of precipitation, soil type, growing season length) is extremely important (Milchunas and Lauenroth 1993). For example, intensive grazing during the growing season can create openings that stimulate plant germination. The plant species favored will be determined largely by the time of grazing relative to the abiotic factors (soil moisture, light penetration, nutrient availability) that control germination. In contrast, less intense grazing typically affects only vegetation structure. In general, the vertical structure of vegetation decreases the longer grazing occurs on a site (Kirby et al. 1992). However, the season and frequency of grazing must also be considered because many effects of grazing change over time (Mundinger 1976). For example, moderate winter grazing in Utah improved waterfowl habitat the following spring (Kadlec and Smith 1992), whereas continuous grazing or grazing during the growing season can preclude waterfowl nesting in that season due to significant decreases in vegetation height and density (Gilbert et al. 1996).

The impacts of a particular grazing strategy vary from short to long term depending on plant type, environmental conditions, and subsequent management actions. Of primary importance in wetland systems is the hydrology of the treated area. For example, the presence or absence of water will determine the rate of plant regrowth. In many cases, the values of short- and long-term effects differ. The short-term effects of intensive grazing during the growing season might be a significant reduction in vegetation height, density, and biomass of residual and living plant matter, whereas the long-term effects might include increased primary productivity.

The impacts of grazing in relation to waterbird habitats vary by species. Much of the existing information tends to focus on the detrimental impacts of grazing on waterfowl nesting (Kirsch 1969, Jarvis and Harris 1971); however, other studies indicate that waterfowl response varies by species and grazing system (Flake et al. 1977, Kantrud 1986). In relatively few cases have the effects of grazing on other waterbirds, or on waterfowl during other annual cycle events, been examined specifically. However, high-intensity grazing can create short, sparse vegetation structure that benefits shorebird nesting and foraging, whereas less intensive grazing can result in areas of low or moderate vegetation height and density interspersed with more open areas that are used by foraging cranes and ibis. In addition, herbivores often create openings in dense stands of robust emergents. Although most reports indicate that the resulting conditions are favorable for waterfowl (Weller 1978, Schultz et al. 1994), similar conditions would also benefit nesting and foraging rails and bitterns. Although these examples suggest that many benefits can be achieved through the use of herbivores, successful implementation requires clearly defined objectives, a sound grazing plan, and careful monitoring to determine when objectives have been met.

Mechanical Disturbance

Various forms of mechanical disturbance are often used to emulate the effects of natural disturbances that have become less common. Mechanical disturbance can be used to alter plant structure, density, and species composition, and to stimulate invertebrate productivity. As with other techniques, the type, timing, and intensity of implementation relative to plant community composition and waterbird chronology significantly affect the results achieved. Some mechanical actions disturb the soil profile (hereafter termed belowground), whereas others do not (aboveground). Depending on location, time, and type of equipment, some mechanical disturbances can be conducted in flooded wetlands or on ice. More generally, however, mechanical disturbance is restricted to periods when the substrate is sufficiently dry to support equipment. Intensity can be defined as the number of times a given action is implemented within a growing season, or in the case of belowground disturbances, the depth of treatment.

Aboveground Treatments. Mowing is the principal aboveground mechanical treatment used in wetland management. Impacts of mowing and surface burns are similar in that both affect aboveground vegetation by reducing vegetation height but do not significantly affect soil structure or belowground reproductive structures. As with surface burns, mowing can be used in certain situations to favor desirable plant species. However, an additional benefit of mowing is more precise control; mowing can be used to spot-treat specific areas, and height can be adjusted to achieve specific objectives. For example, mowing above the meristematic tissue of grasses (which grow from the base) but below the apical buds of dicots (which grow from the tips) can improve seed production of grasses and other low-growing annuals by increasing light penetration and, in come cases, increasing availability of soil moisture and nutrients. In general, this technique tends to be most useful if taller, undesirable plants can be mowed without affecting shorter, desirable species. Depending on height and pattern, mowing can improve habitats for species (e.g., rails) that consume seeds and prefer interspersion of tall, dense vegetation with more open areas. However, timing is critical. For example, mowing conducted too late in the growing season may preclude seed production or adequate plant regrowth.

Mowing also differs from surface burns with respect to vegetation structure. Although both disturbances reduce vegetation height, surface burns tend to reduce vegetation density to a greater extent than mowing. Therefore, mowing might be used to produce desirable foraging (e.g., ibis, cranes, herons, common snipe) or nesting (e.g., cranes, phalaropes) conditions for waterbirds that prefer short, dense vegetation, whereas burning might be more favorable for species (e.g., many shorebirds) that prefer short, sparse vegetation for foraging and nesting. As with surface fires, vegetation response following mowing is likely to be rapid if conducted during the growing season; thus, timing of treatment relative to the period of use by target species must be considered.

In combination with other treatments such as flooding, mowing can sometimes be used to alter species composition. Mowing followed by flooding has been used to control emergent species such as cattail and bulrush (Kadlec and Smith 1992). For best results, flooding should occur immediately after mowing, and cut stalks should remain overtopped for the remainder of the growing season to prevent oxygen transport to roots and rhizomes. In some situations, mowing extensive areas of dense vegetation may result in large amounts of dead plant biomass. Upon reflooding, the initial release of nutrients from this newly created detrital base can be rapid and excessive, which may stimulate extensive algae blooms or reduce the dissolved oxygen content of water and ultimately result in lower invertebrate productivity. Thus, in some cases, other techniques (e.g., haying) that remove some litter may be required.

Specialized underwater mowing equipment (e.g., cookie cutter) has also been used to increase cover–water interspersion in stands of dense emergents. The perceived benefit of this technique is that more desirable interspersion can be created without draining the wetland. Although valuable in situations where water levels cannot be manipulated, this technique often does not address other problems associated with extensive stands of robust vegetation, including decreased nutrient availability and lowered invertebrate productivity (Murkin and Kadlec 1986, Fredrickson and Reid 1988). In such cases, mowing without dewatering the basin may not improve invertebrate productivity because nutrient availability is not increased and, in some cases, may be further decreased (see above).

Belowground Treatments. Disking and plowing are the primary belowground disturbances in wetland management. Although these techniques are potentially applicable to all wetland types, they are used most extensively in seasonally flooded wetlands managed for food production (e.g., moist-soil impoundments). In addition to altering the structure and density of vegetation, such treatments can change species composition because they disturb root systems of existing vegetation, alter the distribution of seeds in the soil seedbank, and increase light penetration (Simpson et al. 1989). In addition, they can affect soil texture, moisture retention, and salinity. Therefore, they are often used to reduce the extent of perennial vegetation and improve conditions for germination of annuals.

Timing of belowground disturbance during the growing season, in combination with hydroperiod following disturbance, has a significant impact on the types and densities of seeds that germinate (see the section on hydrology). In addition to improving germination conditions for annuals, disking can also increase the density of some desirable perennial species. For example, shallow (<10 cm) disking of established chufa (*Cyperus esculentus*) can separate tubers from the parent plant, which stimulates additional buds to break dormancy and grow (Kelley 1990). However, the same mechanism (e.g., severing rhizomes into smaller but still viable fragments) can stimulate increased production of perennials (e.g., cattail, common reed) that are the target of control. More intense mechanical treatments (deep disking, multiple disking, plowing) are sometimes required to achieve adequate control of some robust emergent plants (Cross and Fleming 1989, Sojda and Solberg 1993), especially stands having extensive, deep root systems. In addition, control of undesirable perennials often requires an extended period of drought (at least one and often two complete growing seasons) following disking, whereas improved chufa production requires that soils be moist following treatment.

Disking can also increase the rate of invertebrate production by fragmenting coarse organic matter into finer material (Magee 1993). This technique is most beneficial if vegetation [e.g., American lotus (*Nelumbo lutea*)] has a thick, waxy cuticle that is difficult for microbes to colonize. When reflooded, decomposition begins quickly, and invertebrate response can be rapid if water temperatures are warm. The best response typically follows shallow disking (<10 cm) because fragmented material remains on or near the surface of the soil, whereas deep disking or plowing tends to bury fragmented plant parts. Depending on time of implementation, this technique can benefit species that forage for invertebrates in sparsely vegetated habitats. For example, shallow disking in late summer can result in appropriate structure and good invertebrate production for fall migrant shorebirds (Fredrickson and Reid 1988).

As with mowing, disking and plowing are precise techniques that can be applied on specific areas within a wetland basin. Therefore, belowground disturbances do not necessarily result in the loss of all vegetation structure. Rather, these techniques can be used to alter vegetation communities and simultaneously create a mosaic of habitat types in close proximity that can benefit a diversity of waterbirds, including shorebirds, herons, and rails. Specific waterbird values will depend on the area disturbed, the structure of undisturbed vegetation, and the ability to manipulate water levels following disturbance.

Herbicides

In situations where other management actions cannot be used, chemicals sometimes represent the most practical approach for addressing certain vegetation problems. Herbicides often are used to control dense, monotypic stands of vegetation. In many cases the desired objective is to improve the cover–water interspersion ratio or increase seed production by removing less desirable vegetation. Applications of various herbicides have been reported to reduce undesirable vegetation, including dense stands of common reed, bulrush, sedge, American lotus, white-water lily (*Nymphaea advena*), and black willow (*Salix nigra*). However, control is often incomplete or only temporary. Although exact causes for not achieving desired results often go unresolved, type of herbicide, rate and timing of application, environmental factors, and plant characteristics are important considerations. Thus, knowledge of active ingredients, types of surfactants, and modes of action are critical. For example, different forms of the same herbicide can affect similar vegetation communities differentially.

Another consideration in the use of herbicides is preventing reestablishment of the undesirable species following treatment (Malecki and Rawinski 1985). This is often difficult in the context of natural wetland processes because herbicides do not modify the ecological conditions that originally caused the vegetation problem (Fredrickson and Laubhan 1994b). Therefore, herbicides are often used in combination with other techniques, including planting agricultural crops (Malecki and Rawinski 1985) or flooding for extended periods of time (Steenis and Warren 1969). Unfortunately, such strategies tend to increase the total cost of using herbicides, often result in only temporary relief of the problem, and may provide fewer benefits to waterbirds (Fredrickson and Laubhan 1994b).

In addition to altering vegetation structure and density, herbicides can indirectly affect chemical and physical factors important in regulating wetland processes (Huckins et al. 1986, Johnson 1986). Suppression of macrophyte growth by herbicides can increase the oxidation–reduction potential of bottom sediments, which may result in decreased availability of nitrogen and phosphorus and, ultimately, reduced invertebrate biomass (Boyle 1984). There is also evidence suggesting that some herbicides may negatively affect invertebrate populations by reducing phytoplankton (Hamilton et al. 1988). Finally, some herbicides can cause direct mortality of invertebrates (Sanders 1980, Dewey 1986, Johnson 1986, Buhl and Faerber 1989). Although these studies indicate potential negative impacts to invertebrates, results are not conclusive because numerous factors affect toxicity (e.g., concentration, sediment type, temperature, residence time). Further, many invertebrate species recover partially or completely following exposure (Hill 1989, but see Tome et al. 1990).

Based on the information above, the primary value of chemicals in relation to waterbird habitat is to improve cover–water interspersion. This may be most beneficial for breeding species (e.g., rails, bitterns) that nest in tall, dense vegetation but also prefer small open areas for foraging, or species (e.g., shorebirds, ibis, herons) that prefer more open foraging habitats. However, benefits may be of short duration (Fredrickson and Laubhan 1994b), so time of application relative to the chronology and type of waterbird use is critical.

REFERENCES

Belsky, A. J. 1987. The effects of grazing: confounding of ecosystem, community, and organism scales. American Naturalist 129:777–783.

Bookhout, T. A. 1995. Yellow rail. Number 139 *in* A. Poole, P. Stettenheim, and F. Gill (eds.), The birds of North America. Academy of National Sciences, Philadelphia, and American Ornithologists' Union, Washington, DC.

Boyle, T. P. 1984. Effect of environmental contaminants on aquatic algae. Pages 237–256 *in* L. E. Shubert, (ed.), Algae as ecological indicators. Academic Press, London.

Brown, B. J., and T. S. H. Allen. 1989. The importance of scale in evaluating herbivory impacts. Oikos 54:189–194.

Buhl, K. J., and N. L. Faerber. 1989. Acute toxicity of selected herbicides and surfactants to larvae of the midge *Chironomus riparius*. Archives of Environmental Contaminant Toxicology 18:530–536.

Butler, R. W. 1992. Great blue heron. Number 25 *in* A. Poole, P. Stettenheim, and F. Gill (eds.), The birds of North America. Academy of Natural Sciences, Philadelphia, and American Ornithologists' Union, Washington, DC.

Cross, D. H., and K. L. Fleming. 1989. Control of phragmites or common reed. Waterfowl management handbook. Leaflet 13.4.12. U.S. Fish and Wildlife Service. Washington, DC.

Davidson, L. M. 1992. Black rail, *Laterallus jamaicensis*. Pages 119–134 *in* K. J. Schneider and D. M. Pence (eds.), Migratory nongame birds of management concern in the northeast. U.S. Fish and Wildlife Service, Newton Corner, MA.

Davis, W. E., Jr. 1993. Black-crowned night-heron. Number 74 *in* A. Poole and F. Gill (eds.), The birds of North America. Academy of Natural Sciences, Philadelphia, and American Ornithologists' Union, Washington, DC.

Davis, W. E., Jr., and J. A. Kushlan. 1994. Green heron. Number 129 *in* A. Poole and F. Gill (eds.), The birds of North America. Academy of Natural Sciences, Philadelphia, and American Ornithologists' Union, Washington, DC.

Davis, C. G., and L. M. Smith. 1998. Ecology and management of migrant shorebirds in the playa lakes region of Texas. Wildlife Monographs 140.

Dewey, S. L. 1986. Effects of the herbicide atrazine on aquatic insect community structure and emergence. Ecology 67:148–162.

Eddleman, W. R., and C. J. Conway. 1998. Clapper rail. Number 340 *in* A. Poole and F. Gill (eds.), The birds of North America. Academy of Natural Sciences, Philadelphia, and American Ornithologists' Union, Washington, DC.

Eddleman, W. R., F. L. Knopf, B. Meanley, F. A. Reid, and R. Zembal. 1988. Conservation of North American rallids. Wilson Bulletin 100:458–475.

Ehrlich, P. R., D. S. Dobkin, and D. Wheye. 1988. The birder's handbook. Simon & Schuster, New York.

Flake, L. D., G. L. Peterson, and W. L. Tucker. 1977. Habitat relationships of breeding waterfowl on stock ponds in northwestern South Dakota. Proceedings of the South Dakota Academy of Science 56:135–151.

Frederick, P. C. 1997. Tricolored heron. Number 306 *in* A. Poole and F. Gill (eds.), The birds of North America. Academy of Natural Sciences, Philadelphia, and American Ornithologists' Union, Washington, DC.

Fredrickson, L. H. 1991. Strategies for water level manipulations in moist-soil systems. Waterfowl management handbook. Leaflet 13.4.6. U.S. Fish and Wildlife Service, Washington, DC.

Fredrickson, L. H., and M. K. Laubhan. 1994a. Intensive wetland management: a key to biodiversity. Transactions of the North American Wildlife and Natural Resources Conference 59:555–565.

Fredrickson, L. H., and M. K. Laubhan. 1994b. Managing wetlands for wildlife. Pages 623–647 *in* T. A. Bookhout (ed.), Research and management techniques for wildlife and habitats, 5th ed. Wildlife Society, Bethesda, MD.

Fredrickson, L. H., and F. A. Reid. 1988. Invertebrate response to wetland management. Waterfowl management handbook. Leaflet 13.3.1. U.S. Fish and Wildlife Service, Washington, DC.

Fredrickson, L. H., and T. S. Taylor. 1982. Management of seasonally flooded impoundments for wildlife. Resource Publication 148. U.S. Fish and Wildlife Service, Washington, DC.

Gibbs, J. P., S. Melvin, and F. A. Reid. 1992a. American bittern. Number 18 *in* A. Poole, P. Stettenheim, and F. Gill (eds.), The birds of North America. Academy of Natural Sciences, Philadelphia, and American Ornithologists' Union, Washington, DC.

Gibbs, J. P., F. A. Reid, and S. Melvin. 1992b. Least bittern. Number 17 *in* A. Poole, P. Stettenheim, and F. Gill (eds.), The birds of North America. Academy of Natural Sciences, Philadelphia, and American Ornithologists' Union, Washington, DC.

Gifford, G. F. 1986. Watershed response to short duration grazing. Pages 145–150 *in* J. A. Tiedeman (ed.), Short duration grazing. Proceedings of the short duration grazing and current issues in grazing management shortcourse, Kennewick, WA. Washington State University, Pullman, WA.

Gilbert, D. W., D. R. Anderson, J. K. Ringelman, and M. R. Szymczak. 1996. Response of nesting ducks to habitat and management on the Monte Vista National Wildlife Refuge, Colorado. Wildlife Monographs 131.

Hamilton, P. B., G. S. Jackson, N. K. Kaushik, K. R. Solomon, and G. L. Stephenson. 1988. Impact of two applications of atrazine on the plankton communities of in situ enclosures. Aquatic Toxicology 13:123–140.

Hayman, P., J. Marchant, and T. Prater. 1986. Shorebirds: an identification guide to the waders of the world. Houghton Mifflin, Boston.

Heitmeyer, M. E. 1994. Wetlands and cattle grazing: impacts and management considerations for waterfowl. *In* F. M. Byers (ed.), Cattle on the land: a conference. Texas A&M University Press, College Station, TX.

Heitmeyer, M. E., P. J. Caldwell, B. D. J. Batt, and J. W. Nelson. 1996. Integration of landscape principles into waterfowl conservation. Pages 125–138 *in* J. T. Ratti (ed.), 7th International Waterfowl Symposium, Memphis, TN.

Helmers, D. L. 1992. Shorebird management manual. Western Hemisphere Shorebird Reserve Network, Manomet, MA.

Hill, I. R. 1989. Aquatic organisms and pyrethroids. Pesticide Science 27:429–465.

Holmes, R. T., and F. A. Pitelka. 1998. Pectoral sandpiper. Number 348 *in* A. Poole and F. Gill (eds.), The birds of North America. Academy of Natural Sciences, Philadelphia, and American Ornithologists' Union, Washington, DC.

Huckins, J. N., J. D. Petty, and D. C. England. 1986. Distribution and impact of trifluralin, atrazine, and fonofos residues in microcosms simulating a northern prairie wetland. Chemosphere 15:563–588.

Jarvis, R. L., and S. W. Harris. 1971. Land-use patterns and duck production at Malheur National Wildlife Refuge. Journal of Wildlife Management 35:767–773.

Johnson, B. T. 1986. Potential impact of selected agricultural chemical contaminants on a northern prairie wetland: a microcosm evaluation. Environmental Toxicology and Chemistry 5:473–485.

Johnson, R. R., and J. J. Dinsmore. 1985. Brood-rearing and postbreeding habitat use by Virginia rails and soras. Wilson Bulletin 97:551–554.

Johnson, R. R., and J. J. Dinsmore. 1986. Habitat use by breeding Virginia rails and soras. Journal of Wildlife Management 50:387–392.

Kadlec, J. A., and L. M. Smith. 1992. Habitat management for breeding areas. Pages 590–610 *in* B. D. J. Batt, A. D. Afton, M. G. Anderson, C. D. Ankney, D. H. Johnson, J. A. Kadlec, and G. L. Krapu (eds.), Ecology and management of breeding waterfowl. University of Minnesota Press, Minneapolis, MN.

Kaminski, R. M., and H. H. Prince. 1981. Dabbling duck activity and foraging responses to aquatic macroinvertebrates. Auk 98:115–126.

Kantrud, H. A. 1986. Effects of vegetation manipulations on breeding waterfowl in prairie wetlands: a literature review. Fish and Wildlife Technical Report 3. U. S. Fish and Wildlife Service, Washington, DC.

Kelley, J. R., Jr. 1990. Biomass production of chufa (*Cyperus esculentus*) in a seasonally flooded wetland. Wetlands 10:61–67.

Kirby, R. E., J. K. Ringelman, D. R. Anderson, and R. S. Sojda. 1992. Grazing on national wildlife refuges: do the needs outweigh the problems? Transactions of the North American Wildlife and Natural Resources Conference 57:611–626.

Kirsch, L. M. 1969. Waterfowl production in relation to grazing. Journal of Wildlife Management 33:821–828.

Kushlan, J. A., and K. L. Bildstein. 1992. White ibis. Number 9 *in* A. Poole, P. Stettenheim, and F. Gill (eds.), The birds of North America. Academy of Natural Sciences, Philadelphia, and American Ornithologists' Union, Washington, DC.

Lacey, J. R. 1987. The influence of livestock grazing on weed establishment and spread. Proceedings of the Montana Academy of Science 47:131–146.

Laubhan, M. K. 1995. Effects of prescribed fire on moist-soil vegetation and soil macronutrients. Wetlands 15:159–166.

Laubhan, M. K., and L. H. Fredrickson. 1993. Integrated wetland management: concepts and opportunities. Transactions of the North American Wildlife and Natural Resources Conference 58:323–334.

Lay, D. W. 1945. Muskrat investigations in Texas. Journal of Wildlife Management 9: 56–76.

Lewis, J. C. 1995. Whooping crane. Number 153 *in* A. Poole and F. Gill (eds.) The birds of North America. Academy of Natural Sciences, Philadelphia, and American Ornithologists' Union, Washington, DC.

Linde, A. F. 1985. Vegetation management in water impoundments: alternatives and supplements to water-level control. Pages 51–60 *in* M. D. Knighton (compiler), Water impoundments for wildlife: a habitat management workshop, General Technical Report NC-100. U.S. Forest Service, St. Paul, MN.

Logan, T. H. 1975. Characteristics of small impoundments in western Oklahoma, their value as waterfowl habitat and potential for management. M.S. thesis, Oklahoma State University, Stillwater, Oklahoma, OK.

Magee, P. A. 1993. Detrital accumulation and processing in wetlands. Waterfowl management handbook. Leaflet 13.3.14. U.S. Fish and Wildlife Service, Washington, DC.

Malecki, R. A., and T. J. Rawinski. 1985. New methods for controlling purple loosestrife (*Lythrum salicaria*). New York Fish and Game Journal 32:9–19.

Maschinski, J., and T. G. Whitham. 1989. The continuum of plant responses to herbivory: the influence of plant association, nutrient availability, and timing. American Naturalist 1134:1–19.

Meanley, B. M. 1992. King rail. Number 3 *in* A. Poole, P. Stettenheim, and F. Gill (eds.), The birds of North America. Academy of Natural Sciences, Philadelphia, and American Ornithologists' Union, Washington, DC.

Melvin, S. M., and J. P. Gibbs. 1996. Sora. Number 250 *in* A. Poole, P. Stettenheim, and F. Gill (eds.), The birds of North America. Academy of Natural Sciences, Philadelphia, and American Ornithologists' Union, Washington, DC.

Merendino, M. T., and L. M. Smith. 1991. Influence of drawdown date and reflood depth on wetland vegetation establishment. Wildlife Society Bulletin 19:143–150.

Milchunas, D. G., and W. K. Lauenroth. 1993. Quantitative effects of grazing on vegetation and soils over a global range of environments. Ecological Monographs 63:327–366.

Mitsch, W. J., and J. G. Gosselink. 1993. Wetlands, 2nd ed. Van Nostrand Reinhold, New York.

Mundinger, J. G. 1976. Waterfowl response to rest-rotation grazing. Journal of Wildlife Management 40:60–68.

Murkin, H. R., and J. A. Kadlec. 1986. Responses by benthic macroinvertebrates to prolonged flooding of marsh habitat. Canadian Journal of Zoology 64:65–72.

Oring, L. W., E. M. Gray, and J. M. Reed. 1997. Spotted sandpiper. Number 289 *in* A. Poole and F. Gill (eds.), The birds of North America. Academy of Natural Sciences, Philadelphia, and American Ornithologists' Union, Washington, DC.

Poole, A., and F. Gill (eds.). Various dates. The birds of North America. Academy of Natural Sciences, Philadelphia, and American Ornithologists' Union, Washington, DC.

Poole, A., P. Stettenheim, and F. Gill (eds.). Various dates. The birds of North America. Academy of Natural Sciences, Philadelphia, and American Ornithologists' Union, Washington, DC.

Rundle, W. D., and M. W. Sayre. 1983. Feeding ecology of migrant soras in southeastern Missouri. Journal of Wildlife Management 47:1153–1159.

Ryder, R. R., and D. E. Manry. 1994. White-faced ibis. Number 130 *in* A. Poole and F. Gill (eds.), The birds of North America. Academy of Natural Sciences, Philadelphia, and American Ornithologists' Union, Washington, DC.

Sanders, H. O. 1980. Sublethal effects of toxaphene on daphnids, scuds and midges. Report 600/3-80-006. U.S. Environmental Protection Agency, Washington, DC.

Schultz, B. D., D. E. Hubbard, J. A. Jenks, and K. F. Higgins. 1994. Plant and waterfowl responses to cattle grazing in two South Dakota semipermanent wetlands. Proceedings of the South Dakota Academy of Science 73:121–134.

Simpson, R. L., M. A. Leck, and V. T. Parker. 1989. Seed banks: general concepts and methodological issues. Pages 3–21 *in* M. A. Leck, V. T. Parker, and R. L. Simpson (eds.), Ecology of soil seed banks. Academic Press, San Diego, CA.

Skagen, S. K., and H. D. Oman. 1996. Dietary flexibility of shorebirds in the western hemisphere. Canadian Field Naturalist 10:419–444.

Smith, L. M., and J. A. Kadlec. 1983. Seed banks and their role during drawdown of a North American marsh. Journal of Applied Ecology 20:673–684.

Sojda, R. S., and K. L. Solberg. 1993. Management and control of cattails. Waterfowl management handbook. Leaflet 13.4.6. U.S. Fish and Wildlife Service, Washington, DC.

Soule, M. E. 1991. Theory and strategy. Pages 91–104 *in* W. E. Hudson (ed.), Landscape linkages and biodiversity. Island Press, Washington, DC.

Steenis, J. H., and J. Warren. 1969. Management of needlerush for improving waterfowl habitat in Maryland. Proceedings of the Annual Conference of the Southeastern Association of Game and Fish Commissioners 13:296–298.

Stephens, D. W., and J. R. Krebs. 1986. Foraging theory. Princeton University Press, Princeton, NJ.

Stephens, D. W., J. F. Lynch, A. E. Sorensen, and C. Gordon. 1986. Preference and profitability: theory and experiment. American Naturalist 127:533–553.

Storer, R. W., and G. L. Nuechterlein. 1992. Western and Clark's grebe. Number 26 *in* A. Poole, P. Stettenheim, and F. Gill (eds.), The birds of North America. Academy of Natural Sciences, Philadelphia, and American Ornithologists' Union, Washington, DC.

Tacha, T. C., S. A. Nesbitt, and P. A. Vohs. 1992. Sandhill crane. Number 31 *in* A. Poole, P. Stettenheim, and F. Gill (eds.), The birds of North America. Academy of Natural Sciences, Philadelphia, and American Ornithologists' Union, Washington, DC.

Tacha, T. C., S. A. Nesbitt, and P. A. Vohs. 1994. Sandhill crane. Pages 77–94 *in* T. C. Tacha and C. E. Braun (eds.), Migratory shore and upland game bird management in North America. International Association of Fish and Wildlife Agencies, Allen Press, Lawrence, KS.

Terres, J. K. 1980. The Audubon Society encyclopedia of North American birds. Alfred A. Knopf, New York.

Tome, M. W., C. E. Grue, S. Borthwick, and G. A. Swanson. 1990. Effects of an aerial application of ethyl parathion on selected aquatic invertebrate populations inhabiting prairie pothole wetlands. Pages 31–33 *in* Proceedings of a symposium: environmental contaminants and their effects on biota of the Northern Great Plains. North Dakota Chapter, The Wildlife Society, Bismarck, ND.

Watts, B. D. 1995. Yellow-crowned night-heron. Number 161 *in* A. Poole and F. Gill (eds.), The birds of North America. Academy of Natural Sciences, Philadelphia, and American Ornithologists' Union, Washington, DC.

Weller, M. W. 1961. Breeding biology of the least bittern. Wilson Bulletin 73:11–35.

Weller, M. W. 1978. Management of freshwater marshes for wildlife. Pages 267–284 *in* R. E. Good, D. F. Whigham, and R. L. Simpson (eds.), Freshwater wetlands. Academy Press, New York.

Weller, M. W. 1979. Wetland habitats. Pages 210–234 *in* P. E. Greeson, J. R. Clark, and J. E. Clark (eds.), Wetland functions and values: the state of our understanding. American Water Resources Association, Minneapolis, MN.

Williams, B. K., M. D. Koneff, and D. A. Smith. 1999. Evaluation of waterfowl conservation under the North American Waterfowl Management Plan. Journal of Wildlife Management 63:417–440.

18 Mosquito Control and Habitat Modification: Case History Studies of San Francisco Bay Wetlands

VINCENT H. RESH

Case histories of four types of habitat manipulations and their effects on wetland biota and physical features that were conducted in San Francisco Bay (California) marshes from 1976 through 1999 are described. These studies evaluated (1) the effect of the addition of mosquito control recirculation ditches to tidal marshes, (2) the management of freshwater marshes created from wastewater, (3) whether water-level manipulations and vegetation modifications that reduce mosquitoes can enhance wildlife habitat, and (4) the effect that all-terrain vehicles used in marsh management have on marsh vegetation. The adoption and implementation of research into management was strongly influenced by a variety of factors besides the outcomes of the research itself. Research that involved collaboration among agencies with apparently conflicting mandates often lead to implementation of effective wetland management that enhance wildlife habitat and provide concurrent control of mosquitoes.

The association of wetland habitats with pestiferous and disease-vectoring insects has long been known. In fact, until recent times, control of these nuisance organisms was probably the main interest that most regulatory agencies had in these habitats, and throughout the world, wetlands have been filled in or drained for this purpose. However, with the rise of environmental awareness in the 1970s, an understanding of the need for preserving wetland habitats has increased dramatically. Since then there has also been concern that approaches used for pest management in wetland habitats, such as pesticide applications or site drainage, may damage these important ecosystems.

Recently, there have been attempts to develop mosquito control techniques that are, at minimum, benign in their effects on wetlands and, ideally, that serve to benefit marshland habitat. These management approaches typically involve habitat manipulations. Physical modification of wetlands for mosquito management is not

Bioassessment and Management of North American Freshwater Wetlands, edited by Russell B. Rader, Darold P. Batzer, and Scott A. Wissinger.
0-471-35234-9

a new approach—in 1831 the Pontine Marshes in Italy were drained to reduce the incidence of malaria! In North America, marshes have been manipulated for insect pest management since the turn of this century.

In this chapter I review four case histories of habitat manipulations related to control of mosquitoes conducted in wetlands of the San Francisco Bay area, California. The mosquito abatement agencies of this region have tended to have forward-looking administrators; thus, their concerns and efforts usually preceded national U.S. trends. The first case history involves a series of studies conducted from 1976 though 1983 to evaluate the effect of the addition of mosquito control recirculation ditches on marsh flora, fauna, and essential physical processes in these habitats. In this case, a technique had been developed that controlled mosquitoes successfully, but its effect on the wetland environment was unknown and presumed to be deleterious. The second case history involved examination of the management of urban freshwater marshes created from treated wastewater, and the development of mosquito control techniques that would not adversely affect the biota occurring there. This is based on research that was conducted from 1982 through 1989. The third case history involved attempts to develop marsh management approaches that not only controlled mosquitoes but also concurrently enhanced the habitat for wildlife. Started in 1988, it was conducted through 1997. Finally, the fourth case history involved an evaluation of the effects of a common mosquito control practice, the use of all-terrain vehicles (ATVs) to travel though marshes, on wetland biota. This research was conducted from 1997 to 1999 and in essence was a return to both the site and object of the original case history—what effect does a specific control practice have on marsh biota? Citations to the scores of published research articles that resulted from the research conducted on these projects are described in Table 18.1.

For each case history I outline the problem as related to mosquito control and environmental considerations, how it was addressed from a research perspective, a summary of the major findings, and management implications of the research. Because the science is often a small part of what ultimate leads to policy decisions, I also comment on some of the "human" aspects of the interactions between the research(ers) and the government agencies.

Throughout California, most wetlands are managed as overwintering habitat for waterfowl (de Szalay et al. 1999), but in the highly urbanized San Francisco Bay area, recreational activities (e.g., birdwatching) and the conservation of remaining habitat are important management goals. Salt marsh mosquitoes and their control have been a continuous problem in the San Francisco Bay area; in fact, the first Mosquito Abatement District in California was formed in response to problems caused by salt marsh mosquitoes (Kramer et al. 1995). Today, the mandates of mosquito control and public health agencies—control of nuisance and disease-vectoring species—are often perceived as being in conflict with mandates of conservation and wildlife concerns, and this conflict is certainly perceived elsewhere as well (Batzer and Resh 1992c). Through these case histories, we show how collaboration led to the development of wetland management techniques that would be endorsed by agencies with apparently conflicting mandates and, ultimately, led to enhancement of wetland habitat and concurrent control of mosquitoes.

TABLE 18.1. Citations to Published Research Examining Specific Features of the Four Case History Studies[a]

Feature	Effects of Recirculation Ditches	Wastewater Marshes	Water and Vegetation Management	Effect of All-Terrain Vehicles
Terrestrial arthropods	Balling and Resh 1982, 1991	—	de Szalay and Rest 2000	RFT
Nontarget aquatic invertebrate populations	Balling and Resh 1984a; Barnby and Resh 1980, 1984	Bergey et al. 1992; Collins and Resh 1985a; Feminella and Resh 1986, 1989; Lamberti and Resh 1984; Orr and Resh 1991, 1992	Batzer and Resh 1991; Batzer et al. 1993, 1997; de Szalay and Resh 1996, 1997, 2000	RFT
Nontarget aquatic invertebrate communities	Barnby et al 1985	Collins et al. 1983	de Szalay and Resh 1996, 1997, 2000; de Szalsy et al. 1996	RFT
Mosquitoes	Balling and Resh 1983b	Balling and Resh 1984b; Collins and Resh 1984, 1985b; Collins et al. 1983, 1985, 1989; Orr and Resh 1986, 1988, 1989, 1991, 1992; Resh and Grodhaus 1983	Batzer and Resh 1992a,c, 1994; de Szalay and Resh 1996, 2000; Schlossberg and Resh 1997 and Resh 1997	—
Plant species composition	Balling and Resh 1983c	—	Schlossberg and Resh 1997	—
Plant biomass	Balling and Resh 1983c	Balling and Resh 1984b; Collins et al. 1983; Feminella and Resh 1986, 1989; Orr and Resh 1992	—	Hannaford and Resh 1999

(continues)

TABLE 18.1. *(continued)*

Feature	Effects of Recirculation Ditches	Wastewater Marshes	Water and Vegetation Management	Effect of All-Terrain Vehicles
Plant productivity	Balling and Resh 1983c	—	—	Hannaford and Resh 1999
Fish	Balling et al. 1979, 1980	Gall et al. 1980; Orr and Resh 1988	McGee et al. 1991	RFT
Birds	Collins and Resh 1985b	Collins et al. 1983; Collins and Resh 1984, 1985b	Batzer et al. 1993	RFT
Physical characteristics	Barnby et al. 1985; Collins et al. 1986	Collins et al. 1985	—	RFT
Experimental design	Resh and Balling, 1979, 1983b; Resh et al. 1980	Collins and Resh 1989; Jackson and Resh 1988, 1989; Orr and Resh 1986	Batzer and Resh 1989; de Szalay et al. 1995, 1996; Resh et al. 1991	—
Techniques	Balling and Resh 1983a; Resh et al. 1990	Resh et al. 1985, 1990	Batzer and Resh 1989; de Szalay et al. 1995, 1996	Balling and Resh 1983c
Management implications	Balling and Resh 1983b; Resh and Balling 1983a	Collins and Resh 1989	Batzer and Resh 1992a,b,c, 1994; 1994; de Szalay et al. 1995	—
Reviews	Resh and Balling 1983b	—	Batzer and Resh 1992c	—

[a] RFT, research funding was terminated before proposed research was completed; —, topic was not examined in that case history.

CASE STUDY 1: ECOLOGICAL EFFECTS OF THE ADDITION OF MOSQUITO CONTROL RECIRCULATION DITCHES TO TIDAL SALT MARSHES

The main alternative to chemical control of salt marsh mosquitoes has been a habitat manipulation approach that involves removing standing water from marsh surfaces. In tidal marshes, this consists of using ditches to connect depressions where mosquito larvae occur to the open water; tidal flushing and/or increased accessibility of predators results in control of mosquitoes. In some marshes along the east coast of North America, a system of deep, parallel ditches were installed in the 1930s to lower the water table purposefully; this also altered salt marsh plant competition and, consequently, food webs (reviewed by Resh and Balling 1979, 1983a). More recently, ditches have been designed to be shallow connectors between natural tidal channels and ponds where mosquito breeding occurs, and to function as mechanisms to increase tidal flushing and improve access for predatory fish.

The success of shallow, connecting ditches as mosquito control measures is well known (Resh and Balling 1983a,b) but perhaps as a consequence of the original water-table-lowering effects of Atlantic coast ditching, the environmental consequences of adding ditches in Pacific coast marshes was perceived by wildlife and conservation agencies as being deleterious. In this first case study, a 5-year program to evaluate ditching effects on aquatic biota is described and summarized.

Research was conducted in two types of San Francisco Bay salt marshes: a salt marsh where the plant community is dominated by a single species, pickleweed (*Salicornia virginica*), and in a marsh where the plant community is more diverse and the inundating waters are less saline. The first marsh type was represented by Petaluma Marsh in northern San Francisco Bay and Albrae Slough in southern San Francisco Bay; the second by Suisun Marsh in northern San Francisco Bay.

Early in these investigations it was apparent that the response of the physical environment to the addition of ditches is localized. For example, drainage of the water table by ditches at low tide is limited to 3 to 4 m from the ditch banks; during high tides, water is replenished 2 to 3 m from the banks, resulting in only the soil adjacent to the ditches being alternatively drained and recharged during each tidal cycle. This flushing also ameliorates the normally harsh salt marsh conditions (e.g., groundwater salinity is significantly lower near a ditch than in the open marsh), although the pattern is evident only during the dry season. The localized effect of physical changes was regularly incorporated into the experimental design of studies; thus, measurements at ditch margins could be compared with those at various distances from these manipulations.

In Petaluma Marsh, which is essentially a monoculture of pickleweed, the aboveground plant production was 40% higher near a ditch than in the open marsh and over 200% higher than along natural channels. This resulted from the lower woody biomass near ditches and the higher plant growth adjacent to ditches. In Suisun Marsh, which is a mixed-plant-species habitat, plant richness was highest near the ditch, probably because the soil flushing near the ditches lowers groundwater salinity.

Just as plants respond to the physical environment, terrestrial arthropods respond to both vegetation and salinity changes. Areas alongside ditches in Petaluma Marsh had lower wet-season species diversity than in the open marsh or alongside natural channels, probably because of the reduction in heavy, woody pickleweed that provides shelter from high winter tides. In contrast, terrestrial invertebrate diversity was higher alongside ditches during the dry season.

Because the addition of recirculating ditches alters the hydrology of salt marsh temporary ponds, which is why they are effective in eliminating mosquitoes, other aquatic invertebrates would be expected to be affected by this activity as well. Although biomass does not differ significantly between ditched and natural ponds, species diversity is reduced in the former habitat. This is the result of differences in species richness, which also reflects frequency of tidal inundation. In terms of specific populations, most common species show similar distributions along ditches and natural channels, and in the case of the numerically dominant arthropods, densities in ditched ponds exceeded those in natural temporary ponds and even approach density levels found in the larger permanent ponds!

Responses of the fish communities were studied in the Albrae Slough pickleweed marsh and that of birds in Petaluma marsh. Twice as many fish species occurred in ditched than natural channels, and densities of three resident fish species was 300% higher (Balling et al. 1980). This probably results from the improved access for fish to travel from the bay to the salt marsh ponds. Ditches provide additional habitat for song sparrow (*Melospiza melodia samuelis*), the most abundant and visible of the birds in the northern San Francisco Bay marshlands. However, sites along ditches were the last to be chosen for nesting and do not mimic conditions created by natural channels (Collins and Resh 1985b).

The addition of recirculation ditches is effective in controlling mosquitoes, but as indicated above, they do result in changes in some features of marsh ecosystems. From an environmental assessment viewpoint (Rosenberg et al. 1981), only the decrease in aquatic insect diversity would be considered as an adverse impact (Table 18.2). This is in contrast to the studies reported from Atlantic coast marshlands; Resh and Balling (1983b) concluded that this was the result of substantial differences in the marsh types and in how the ditches are installed.

Kramer et al. (1995) followed up this study and demonstrated how carefully planned designs that are based on marsh hydrology could provide even longer-term solutions to mosquito problems in the San Francisco Bay marshlands and even further minimize the need for repeated mosquito larvicide applications. They concluded that their techniques for increasing tidal circulation not only reduced mosquito densities but also improved habitat for many types of wildlife. Early in the research process described in this case history a positive effect of the addition of ditches was evident an—increase in fish species richness and abundance. Given the deleterious effects from ditching expected by some conservation agency personnel familiar with the Atlantic coastal marsh studies, the finding changed the mindset of the many of the concerned agencies. This new perception continued through the project even after functional alterations and localized, deleterious effects were found (Table 18.2). When the research was completed, a workshop was held to explain the findings and to answer questions related to recirculation ditches from agency personnel involved in the issues. This approach was highly successful and was used in the next two case studies described below; it was equally successful in each.

TABLE 18.2. Summary of Studies Designed to Evaluate the Effects of Mosquito Control Recirculation Ditches on Selected Physical and Biological Features of San Francisco Bay Marshlands

Feature	Localized Effect
Water table height	Decreases (tidal marsh); no change (seasonal marsh)
Groundwater salinity	Decreases
Soil surface salinity	Decreases
Pickleweed (*Salicornia virginica*) production	Increases
Plant diversity	Increases
Terrestrial arthropod diversity	
Wet season	Decreases
Dry season	Increases
Year round	No change
Terrestrial arthropod biomass	No change
Terrestrial arthropod population densities	No change (in >50% species)
Aquatic invertebrate diversity	Decreases
Aquatic invertebrate biomass	No change
Fish diversity	Increases
Fish density	Increases
Salt marsh song sparrow (*Melospiza melodia samuelis*) density	Increases

Source: Data from Balling and Resh (1983b).

Although initially, this research settled concerns about the environmental effects of this manipulation, every 5 years or so, often corresponding with periods of permit renewal, the same concerns about environmental effects are raised. This appears to be the result of new personnel coming into regulatory agencies with different approaches about how impact should be assessed. Sometimes, explanations of existing results suffice.

Tidal marsh habitats are highly politicized systems, and various mitigation measures to the actual practice of adding recirculation ditches have been proposed and attempted, ranging from the ridiculous (substituting peanut oil for crank case oil in vehicles used to construct ditches) to the appropriate (having ditch contours follow natural marsh contours). More comprehensive studies, such as those that use recently developed experimental designs (e.g., B-A-C-I; Cooper and Barmuta 1993), will probably be conducted, perhaps even within the next decade, to reevaluate the effects of adding recirculation ditches to San Francisco Bay wetlands.

CASE STUDY 2: WASTEWATER MARSHES AND MOSQUITO CONTROL

In the mid-1970s, numerous wetland restoration projects were begun in the San Francisco Bay area to enhance ecological, aesthetic, and recreation values of the bay waterfront. Initially focusing on restoration of salt marshes, emphasis shifted toward creation and restoration of both seasonal and permanent freshwater wetlands, often using treated wastewater as part of this process. This restoration movement also created habitats that often resulted in mosquito production. Efforts to restore and create wetlands near human populations (along with urbanization of rural areas) brings people and mosquitoes closer together at many locations. There is also a general decrease in the public's tolerance of mosquitoes that accompanies urbanization. This combination creates highly sensitive issues for marsh-management agencies. The mosquito abatement districts wanted assistance in the development of effective wetland management programs, and this was the impetus for the research described in this case history: the development of a mathematical model that could provide management guidelines for the ecological control of mosquito populations in nontidal, palustrine wetlands of the San Francisco Bay area. The model was to be developed from studies conducted in marshes at Coyote Hills Regional Park in Fremont, Alameda County, California.

A key to understanding the control of nuisance wetland mosquitoes is an understanding of the relationship between vegetation and the habitat requirements of the mosquito species to be controlled, and this was a major focus of this research. The vegetation of a wetland can impede ecological control of mosquitoes by providing them a refuge from predation and physical disturbance of the water surface, and by increasing the development rate of mosquito larvae by raising habitat temperatures and enhancing food resources. In the case of *Anopheles* mosquitoes, which are both a nuisance and a public health concern, vegetation is important primarily in providing the third element necessary to develop the meniscus that characterizes the *intersection line*. This meniscus is formed at the point where the air, the plant surface, and the water surface are in contact and provides a refuge from predation, sources of food, and increased development times (Collins and Resh 1989).

Research conducted through this project demonstrated that the density of anopheline larvae is related positively to the amount of intersection line that is available as microhabitat (Collins and Resh 1989). The density of intersection line depends on plant architecture; plants with the greatest ratio of wetted perimeter to total tissue surface area will provide the most intersection line. Research demonstrated that females choose aquatic vegetation with high amounts of intersection line as sites to lay their eggs, and that high amounts of intersection line increase the maturation rate of larvae. The reduction in amounts of intersection line depends largely on control of vegetation, which can be done by surface water regulation, controlled burning, harvesting, or by poisoning plants. Research conducted during this project evaluated surface water regulation by both manipulation of hydrologic regime and by enhancing conditions for herbivory of plants, such as by waterfowl and crayfish, which are common herbivores in San Francisco Bay marshlands.

Predation of mosquitoes was also examined in this project. Consumption of mosquitoes in wetlands depends on a variety of factors, including dietary preference of predators for mosquitoes, abundance of alternative prey, the degree of congruity between the habitat of the predator and the mosquito, predator and prey density, and the quality of the mosquito habitat as a refuge from predators. A variety of natural enemies of mosquitoes, especially predaceous aquatic insects and insectivorous fish, were examined in research described in this case history. Results indicate that the habitats of some predaceous insects were enhanced by hydrological and vegetation management. For example, in the absence of insectivorous fish, the relative abundance of predaceous Notonectidae (Insecta : Hempitera) can be increased by removal of emergent and submergent vegetation; in contrast, the addition of short emergent and submergent vegetation will increase the relative abundance of predaceous Odonata. Research demonstrated that efforts to enhance predation of mosquitoes will usually be more effective when vegetation is removed rather than enhanced.

Insectivorous fish in wastewater marshes are usually more important than invertebrates as predators of mosquitoes; at least in part, this is because they are larger and consume more prey. An important outcome of this research is that in both the field and laboratory, rates of predation by mosquito fish (and notonectid) predators can be predicted based on fish density, larval size, and intersection line value (Collins et al. 1989, Orr and Resh 1992).

The final outcome of this research was the creation of a monograph (Collins and Resh 1989) describing guidelines for the control of mosquitoes in wastewater marshes. Besides describing how historical, land use, hydrological, and biological factors influence the production of mosquitoes (Table 18.3), this compendium described research and sampling methods for studying marshes, including techniques for cartography, water balance estimation, water quality assessment, enumeration of wildlife, and enumeration and assessment of mosquitoes and their habitats.

The creation of the monograph described above was not the intended outcome of this project. The original goal was to produce a model that could be used to predict

TABLE 18.3. Summary of Guidelines for the Ecological Control of Mosquitoes

Historical and lab use factors	Develop soils map and map of local historical wetlands; avoid intermittent inundation of fine-grained sediments; avoid freshwater inundation of historic salt marshes; avoid unnecessary excavations that retain water.
Hydrologic factors	Regulate runoff from non-point sources; maintain constant water quality; develop flow-through hydrological systems; develop a water balance; monitor water quality.
Biological factors	Develop a plan for hydrological control of wetland vegetation; optimize herbivory by crayfish; optimize herbivory by waterfowl; optimize predation by aquatic invertebrates; optimize predation by insectivorous fish; optimize predation of mosquito adults.
Field research methods	Develop a schedule of field research; stratify the control site according to ecological factors; calibrate and standardize field methods.

Source: Data from Collins and Resh (1989).

how physical condition and biological features of wasteland marshes resulted in (or did not result in) mosquito production. The research described alone was conducted from 1982 through 1986 and the model was scheduled for completion in 1987. However, the model was never done because the investigator responsible moved from California and abandoned the project. The dilemma faced was that a variety of basic research studies were conducted and published, but that the impetus for the funding was an applied product (e.g., the model). The "guidelines" monograph was an afterthought and a substitute but one that satisfied the needs of the wetland managers.

CASE STUDY 3: INFLUENCE OF WATER DEPTH AND VEGETATION MANIPULATION

In this case history, research was expanded from documenting the effects of mosquito control activities to developing techniques for concurrent control of mosquitoes and enhancement of wildlife habitat. To do this, emphasis was placed on evaluating the effect of manipulating water depths and plant cover. In 1988, a series of 12 experimental ponds were built in Suisun Marsh and flooded to different water depths: 20, 40, and 60 cm (Table 18.4). As water depth increased, the densities of invertebrates that are important in waterfowl diets increased (such as hydrophilid beetles, water boatmen, midge larvae, and amphipods). Mosquito densities (e.g., of *Culiseta inornata*) were lowest in the 60-cm treatment but high at both 20- and 40-cm depths. The conclusion of this research was that flooding seasonal wetlands to water depths that cover most of the thick emergent vegetation is a management option that may provide high-quality waterfowl habitat and low numbers of mosquitoes, and may be a priority option for mosquito control in some areas (e.g., near human population centers).

Removal of vegetation by mowing prior to flooding is another way to increase the amount of open water available in densely vegetated wetlands (which typically support larger numbers of mosquitoes). Using experimental ponds and then large pools, studies indicated that food items used by waterfowl were higher in mowed pickleweed areas, whereas mosquitoes were significantly higher in unmowed areas. Studies also demonstrated that manipulations of plant cover can affect food webs in complex ways. For example, predaceous beetle larvae readily colonize mowed habitats and can reduce densities of chironomid midge prey. After beetle densities decline, however, midge larval densities rebound rapidly because their periphyton food supply has increased. In

TABLE 18.4. Results of Manipulation Studies to Concurrently Enhance Waterfowl and Reduce Mosquitoes

Manipulation of water depths	20 cm: low open water; low density of waterfowl food items; highest densities of mosquitoes
Manipulation of plant cover	50% reduction: more waterfowl food items; lower densities of mosquitoes; more food items for waterfowl during critical periods
Manipulation of flooding date	Increase food items for waterfowl and decrease mosquitoes

Source: Data from Batzer and Resh (1992b).

study ponds, this increase in midge numbers occurred in winter, which is when invertebrates are particularly important as waterfowl food. Therefore, manipulations that resulted in numerous midges being available for waterfowl consumption also resulted in lower mosquito densities. Similar results were obtained in larger-scale studies.

Why does mowing vegetation improve waterfowl habitat and provide mosquito control? Mowing can create a ratio of open water to plant cover preferred by ducks, and increase densities of invertebrates important in waterfowl diets. Mowing also either reduces mosquito densities or causes mosquitoes to accumulate along upland edges, which spatially reduces the area required for mosquito control.

In other research conducted in this project, the responses of the plants (in this case, saltgrass) and invertebrates important in the waterfowl diet were examined using prescribed burning and mowing. These treatments in saltgrass had somewhat different results than in pickleweed. Similar to pickleweed, reducing saltgrass cover through burning resulted in many numerically dominant macro- and microinvertebrates having higher densities in burned treatment areas than in unmanipulated control areas. Similarly, plant species richness and percent cover of some taxa were higher in burned than in control areas. In contrast, mowing saltgrass had no detectable affect on densities of most taxa.

The techniques for vegetation manipulation examined—mowing, disking, and burning—alter the physical structure of the plant stands differently, and this has corresponding influences on the invertebrate communities colonizing these habitats. In terms of implications for managers of wildlife habitats, densities of food items important in waterfowl diets are higher, and densities of mosquitoes lower, in areas where thick vegetation is removed effectively by any of these management practices.

A significant outcome of this project was the publication of an article with the provocative title "Wetland management strategies that enhanced waterfowl habitats can also control mosquitoes" (Batzer and Resh 1992c) that appeared in a respected, peer-reviewed journal. This publication provided legitimacy to counter a commonly perceived conflict: the incompatibility of mosquito control and wildlife management. This research has led to proposed legislation in California for the creation of a Wetland Habitat Mosquito Abatement Fund to provide funding for water and vegetation management for private landowners (especially duck-hunting clubs) to control their mosquito problems. This, plus the implementation of these control measures in government-owned lands, will result in major improvements in both wildlife habitat quality and effective mosquito control.

CASE STUDY 4: EFFECT OF ALL-TERRAIN VEHICLES ON WETLAND FLORA AND FAUNA

All-terrain vehicles (ATVs) are widely used in a variety of wetland management activities (including mosquito control); the effects of their use in tundra and desert ecosystems have raised concern about their effects on wetland habitats (Hannaford and Resh 1999). ATVs are viewed as essential by wetland management agencies for the transport of personnel and heavy equipment throughout the marshes. Different types of ATVs are used for transversing marshes, and this study examined the effects of two commonly used amphibious tracked vehicles, the Argo and Lightfoot, in

Petaluma Marsh. The vehicles differ in that the Lightfoot is heavier (18.5 versus 7.5 kN), has a larger track "footprint" (25.5 versus 1.8 m^2), and exerts more ground pressure (7.3 versus 4.3 kN/m^2) on marsh vegetation.

Studies were conducted in Petaluma Marsh (as was Case Study 1), which is dominated by pickleweed. ATV impact was evaluated in terms of immediate mechanical effects (reductions in pickleweed stem height and breakage of stems) and in terms of long-term effects (changes in biomass and growth of pickleweed). Stem height was reduced significantly by both types of vehicles immediately following use and was similar whether either vehicle passed two or 20 times over the vegetation. However, biomass of broken stems was significantly higher for the Lightfoot than for the Argo, and (not unexpectedly) was higher with 20 than with two passes over vegetation. Although even limited ATV use caused immediate damage to pickleweed, 1 year after treatment the pickleweed crossed by the Argo was not different from that of the untreated areas 1 year after treatment.

Several recommendations were made about the operation of vehicles in marsh management from this study: use of heavy, steel-tracked vehicles should be greatly reduced or eliminated in Pacific coastal marshes; a path should not be transversed more than twice; and routine monitoring should be done by foot whenever possible or by using the lightest vehicles available.

This research project was proposed and funded as a 3-year study and endorsed by both mosquito control and state and federal wildlife agencies concerned with management of marsh flora and fauna. However, when results were released at the end of the first year's research, funding for continuation of this project was withdrawn; this was probably because deleterious effects of ATVs were discovered. Only some of the planned subprojects could be completed without the expected financial support, and this was done by using funds drawn from other sources. The faunal component (including a study on the effects of vehicles on the salt marsh harvest mouse, a federally listed endangered species) was never conducted.

CONCLUSIONS

The research described in these four case histories of management practices in San Francisco Bay area wetlands demonstrate that manipulations of wetland habitats for mosquito control can have mixed effects on marsh flora and fauna (e.g., the addition of mosquito control recirculation ditches), can have negative effects (e.g., use of certain all-terrain vehicles), or can clearly enhance waterfowl habitat and reduce mosquitoes (e.g., vegetation and water management in wastewater and seasonal marshes). It is likely that studies of this kind could only have been done in the San Francisco Bay area, because in this region there has been a long tradition of innovative mosquito control and generally enlightened managers concerned with minimizing the environmental impact of their practices.

A lesson learned from these case histories is that basic research can lead to a great improvement in management practices in wetlands. However, it is also apparent that research conducted in politically sensitive habitats such as wetlands is more subject to political pressures than that experienced by basic researchers in general, or by applied researchers on less sensitive topics. Perhaps a major but unforeseen result of these four projects was that the collaboration between mosquitoe control agencies and conservation agencies in facilitating and supporting this research led to a lessening of perceived conflicts and better long-term cooperation between them.

ACKNOWLEDGMENTS

I thank the mosquito abatement districts of the California coastal region and the Mosquito Research Technical Committee of the University of California for continuous funding on these projects from 1976 through 1998. I also thank the score of excellent graduate and undergraduate students that I have been able to conduct research with on these projects.

REFERENCES

Balling, S. S., and V. H. Resh. 1982. Arthropod community responses to mosquito control recirculation ditches in San Francisco Bay salt marshes. Environmental Entomology 11:801–808.

Balling, S. S., and V. H. Resh. 1983a. A method of determining pond inundation height for use in salt marsh mosquito control. Mosquito News 43:239–240.

Balling, S. S., and V. H. Resh. 1983b. Mosquito control and salt marsh management: factors influencing the presence of *Aedes* larvae. Mosquito News 43:212–218.

Balling, S. S., and V. H. Resh. 1983c. The influence of mosquito control recirculation ditches on plant biomass, production, and composition in two San Francisco Bay salt marshes. Estuarine, Coastal and Shelf Science 16:151–161.

Balling, S. S., and V. H. Resh. 1984a. Life history variability in the water boatman *Trichocorixa reticulata* (Hemiptera: Corixidae) in San Francisco Bay salt marsh ponds. Annals of the Entomological Society of America 77:14–19.

Balling, S. S., and V. H. Resh. 1984b. Seasonal patterns of pondweed standing crop and *Anopheles occidentalis* densities in Coyote Hills Marsh. Proceedings of the California Mosquito and Vector Control Association 52:122–125.

Balling, S. S., and V. H. Resh. 1991. Seasonal patterns in a San Francisco Bay salt marsh arthropod community. Pan-Pacific Entomologist 67:138–144.

Balling, S. S., T. Stoehr, and V. H. Resh. 1979. Species composition and abundance of fishes in ditched and unditched areas of a San Francisco Bay salt marsh. Proceedings of the California Mosquito and Vector Control Association 47:88–89.

Balling, S. S., T. Stoehr, and V. H. Resh. 1980. The effects of mosquito control recirculation ditches on the fish community of a San Francisco Bay salt marsh. California Fish and Game 66:25–34.

Barnby, M. A., and V. H. Resh, 1980. Distribution of arthropod populations in relation to mosquito control recirculation ditches and natural channels in the Petaluma salt marsh of San Francisco Bay. Proceedings of the California Mosquito and Vector Control Association 48:100–102.

Barnby, M. A., and V. H. Resh 1984. Distribution and seasonal abundance of brineflies (Diptera: Ephydridae) in a San Francisco Bay salt marsh. Pan-Pacific Entomologist 60:37–46.

Barnby, M. A., J. N. Collins, and V. H. Resh. 1985. Aquatic macroinvertebrate communities of natural and ditched potholes in a San Francisco Bay salt marsh. Estuarine, Coastal and Shelf Science 20:331–347.

Batzer, D. P., and V. H. Resh. 1989. Waterfowl management and mosquito production in diked salt marshes: preliminary considerations and mesocosm design. Proceedings of the California Mosquito and Vector Control Association 56:153–157.

Batzer, D. P., and V. H. Resh. 1991. Trophic interactions among a beetle predator, a chironomid grazer, and periphyton in a seasonal wetland. Oikos 60:251–257.

Batzer, D. P., and V. H. Resh. 1992a. Macroinvertebrates of a California seasonal wetland and responses to experimental habitat manipulation. Wetlands 12:1–7.

Batzer, D. P., and V. H. Resh. 1992b. Recommendations for managing wetlands to concurrently achieve waterfowl enhancement and mosquito control. Proceedings of the California Mosquito and Vector Control Association 60:202–206.

Batzer, D. P., and V. H. Resh, 1992c. Wetland management strategies that enhance waterfowl habitats can also control mosquitoes. Journal of the American Mosquito Control Association 8:117–125.

Batzer, D. P., and V. H. Resh. 1994. Wetland management strategies, waterfowl habitat management, and mosquito control. Pages 825–832 *in* W. J. Mitsch (ed.), Global wetlands: old world and new. Elsevier, Amsterdam.

Batzer, D. P., M. McGee, V. H. Resh, and R. R. Smith. 1993. Characteristics of invertebrates consumed by mallards and prey response to wetland flooding schedules. Wetlands 13:41–49.

Batzer, D. P., F. A. de Szalay, and V. H. Resh. 1997. Opportunistic response of a benthic midge (Diptera: Chironomidae) to management of California seasonal wetlands. Environmental Entomology 26: 215–222.

Bergey, E. A., S. F. Balling, J. N. Collins, G. A. Lamberti, and V. H. Resh. 1992. Bionomics of invertebrates within an extensive *Potamogeton pectinatus* bed of a California marsh. Hydrobiologia 234:15–24.

Collins, J. N., and V. H. Resh. 1984. Do waterfowl affect mosquitoes in Coyote Hills Marsh? Proceedings of the California Mosquito and Vector Control Association 52:129–133.

Collins, J. N., and V. H. Resh. 1985a. Factors that limit the role of immature damselflies as natural mosquito control agents at Coyote Hills Marsh. Proceedings of the California Mosquito and Vector Control Association 53:87–92.

Collins, J. N., and V. H. Resh. 1985b. Utilization of natural and man-made habitats by the salt marsh song sparrow, *Melospiza melodia samuelis* (Baird). California Fish and Game 71:40–52.

Collins, J. N., and V. H. Resh. 1989. Guidelines for the ecological control of mosquitoes in nontidal wetlands of the San Francisco Bay area. Special Publication of the California Mosquito and Vector Control Association and the University of California Mosquito Research Program.

Collins, J. N., S. S. Balling, and V. H. Resh. 1983. The Coyote Hills Marsh model: calibration of interactions among floating vegetation, waterfowl, invertebrate predators, alternate prey, and *Anopheles* mosquitoes. Proceedings of the California Mosquito and Vector Control Association 51:69–73.

Collins, J. N., K. D. Gallagher, and V. H. Resh. 1985. Thermal characteristics of aquatic habitats at Coyote Hills Marsh: implication for simulation and control of *Anopheles* populations. Proceedings of the California Mosquito and Vector Control Association 53:83–86.

Collins, J. N., L. M. Collins, L. B. Leopold, and V. H. Resh. 1986. The influence of mosquito control ditches on the geomorphology of tidal marshes in the San Francisco Bay area: evolution of salt marsh mosquito habitat. Proceedings of the California Mosquito and Vector Control Association 54:91–95.

Collins, J. N., E. P. McElravy, B. K. Orr, and V. H. Resh. 1989. Preliminary observations on the effects of the intersection line upon predation of *Anopheles* mosquito larvae. Bicovas (Proceedings of the International Conference on Biological Control of Vectors with Predaceous Arthropods) 1:1–12.

Cooper, S. D., and L. A. Barmuta. 1993. Field experiments in biomonitoring. Pages 399–441 *in* V. H. Resh, and D. M. Rosenberg (eds.), Freshwater biomonitoring and benthic macroinvertebrates. Chapman & Hall, New York.

de Szalay, F. A., and V. H. Resh. 1996. Spatial and temporal variability of trophic relationships among aquatic macroinvertebrates in a seasonal marsh. Wetlands 16:458–466.

de Szalay, F. A., and V. H. Resh. 1997. Responses of wetland invertebrates and plants important in waterfowl diets to burning and mowing of emergent vegetation. Wetlands 17:149–156.

de Szalay, F. A., D. P. Batzer, E. B. Schlossberg, and V. H. Resh. 1995. A comparison of small and large scale experiments examining the effects of wetland management and practices on mosquito densities. Proceedings of the California Mosquito and Vector Control Association 63:86–90.

de Szalay, F. A., D. P. Batzer, and V. H. Resh. 1996. Mesocosm and macrocosm experiments to examine the effects of mowing emergent vegetation on wetland invertebrates. Environmental Entomology 25:303–309.

de Szalay, F. A., and V. H. Resh. 2000. Factors influencing macroinvertebrate colonization of seasonal wetlands: response to emergent plant cover. Freshwater Biology 45:295–308.

de Szalay, F. A., N. H. Euliss, Jr., and D. P. Batzer. 1999. Seasonal and semi-permanent wetlands of California. Invertebrate community ecology and responses to management methods. Pages 829–555 *in* D. P. Batzer et al. (eds.), Invertebrates in freshwater wetlands of North America: Ecology and management. Wiley, New York.

Feminella, J. W., and V. H. Resh. 1986. Effects of crayfish grazing on mosquito habitat at Coyote Hills Marsh. Proceedings of the California Mosquito and Vector Control Association 54:101–104.

Feminella, J. W., and V. H. Resh. 1989. Submersed macrophytes and grazing crayfish: an experimental study of herbivory in a California marsh. Holarctic Ecology 12:1–8.

Gall, G. A., E. J. Cech, R. Garcia, V. Resh, and R. Washino. 1980. Mosquito fish: an established predator. California Agriculture 34:21–22.

Hannaford, M. J., and V. H. Resh. 1999. Impact of all-terrain vehicles (ATVs) on pickleweed (*Salicornia virginica* L.) in a San Francisco Bay wetland. Wetlands Ecology and Management 7:225–233.

Jackson, J. K., and V. H. Resh. 1988. Sequential decision plans in monitoring benthic macroinvertebrates: cost savings, classification accuracy, and development of plans. Canadian Journal of Fisheries and Aquatic Science 45:280–286.

Jackson, J. K., and V. H. Resh. 1989. Sequential decision plans, benthic macroinvertebrates, and biological monitoring programs. Environmental Management 12:455–468.

Kramer, V. L., J. N. Collins, K. Malamud-Roam, and C. Beesley. 1995. Reduction of *Aedes dorsalis* by enhancing tidal action in a northern California marsh. Journal of the American Mosquito Control Association 11:389–395.

Lamberti, G. A., and V. H. Resh. 1984. Seasonal patterns of invertebrate predators and prey in Coyote Hills Marsh. Proceedings of the California Mosquito and Vector Control Association 52:126–128.

McGee, M., D. P. Batzer, and V. H. Resh. 1991. The influence of flooding date density and age structure of the three-spined stickleback, *Gasterosteus aculeatus,* in Suisun Marsh, California. Proceedings of the California Mosquito and Vector Control Association 58:127–128.

Orr, B. K., and V. H. Resh. 1986. Spatial-scale considerations in predator–prey experiments. Proceedings of the California Mosquito and Vector Control Association 54:105–109.

Orr, B. K., and V. H. Resh. 1988. Interactions among mosquito fish (*Gambusia affinis*), Sago pondweed (*Potamogeton pectinatus*), and the survivorship of *Anopheles* mosquito larvae. Proceedings of the California Mosquito and Vector Control Association 55:94–97.

Orr, B. K., and V. H. Resh. 1989. Experimental test of the influence of aquatic macrophyte cover on the survival of *Anopheles* larvae. Journal of the American Mosquito Control Association 5:579–585.

Orr, B. K., and V. H. Resh. 1991. Interactions among aquatic vegetation, predators, and mosquitoes: implications for management of *Anopheles* mosquitoes in a freshwater marsh. Proceedings of the California Mosquito and Vector Control Association 58:214–220.

Orr, B. K., and V. H. Resh. 1992. Influence of *Myriophyllum aquaticum* cover on *Anopheles* mosquito abundance, oviposition, and larval microhabitat. Oecologia 90:474–482.

Resh, V. H., and S. S. Balling. 1979. Ecological impact of mosquito control recirculation ditches on San Francisco Bay marshlands: preliminary considerations and experimental design. Proceedings of the California Mosquito and Vector Control Association 47:72–78.

Resh, V. H., and S. S. Balling. 1983a. Ecological impact of mosquito control recirculation ditches on San Francisco Bay marshlands: study conclusions and management recommendations. Proceedings of the California Mosquito and Vector Control Association 51:49–53.

Resh, V. H., and S. S. Balling. 1983b. Tidal circulation alteration for salt marsh mosquito control. Environmental Management 7:79–84.

Resh, V. H., and G. Grodhaus. 1983. Aquatic insects in urban environments. Pages 247–276 *in* G. Frankie and C. Koehler (eds.), Urban entomology: interdisciplinary perspectives, Praeger, New York.

Resh, V. H., S. S. Balling, M. A. Barnby, and J. N. Collins. 1980. What is the ecological impact of mosquito control recirculation ditches on San Francisco Bay marshlands? California Agriculture 34:38–39.

Resh, V. H., D. M. Rosenberg, and J. W. Feminella. 1985. The processing benthic samples: responses to the 1983 NABS questionnaire. Bulletin of the North American Benthological Society 2:5–11.

Resh, V. H., J. W. Feminella, and E. P. McElravy. 1990. Sampling aquatic insects. Videotape. Office of Media Services, University of California, Berkeley, CA.

Resh, V. H., N. G. Kobzina, F. de Szalay, and D. P. Batzer. 1991. Where is mosquito research published? Journal of the American Mosquito Control Association 7:123–125.

Rosenberg, D. M., V. H. Resh, S. S. Balling, M. A. Barnby, J. N. Collins, D. V. Durbin, T. S. Flynn, D. D. Hart, G. A. Lamberti, E. P. McElravy, J. R. Wood, T. E. Blank, D. M. Schultz, D. L. Marrin, and D. G. Price. 1981. Recent trends in environmental impact assessment. Canadian Journal of Fisheries and Aquatic Sciences 38:591–624.

Schlossberg, E. B., and V. H. Resh. 1997. Mosquito control and waterfowl habitat enhancements by vegetation manipulation and water management: a two year study. Proceedings of the California Mosquito and Vector Control Association 65:11–15.

19 Timber Harvest in Wetlands: Strategies and Impact Assessment

JOSEPH P. PRENGER and THOMAS L. CRISMAN

Forested wetlands are important features of the North American landscape, contributing to flood abatement and improved water quality, among other services. Although relatively young geologically, these systems exhibit a wide variety of plant communities and provide important habitat for herpetofauna, migratory birds, and other animals. The extent of these wetlands has been reduced drastically from precolonial levels, with recent losses occurring mainly in the southeastern United States, where agriculture and expanding urbanization are a continuing impact. Although many forestry practices are aimed at maintaining and renewing the resource, tree removal ultimately affects the hydrology, soil, plant community composition, and faunal utilization of these areas. This situation points out the need for increased understanding of and development of tools to assess these impacts. In this chapter we review the ecology of common forested wetland types, current research on forestry practices and their effects, and possible assessment. Some implications for management are presented.

Forested wetlands are important features of the landscape, historically comprising most of the wetland area in temperate North America. These systems include riverside swampland, bog forests, bottomland forests, and alder swamps, and in the southeastern coastal plain such diverse systems as bottomland hardwoods, deepwater cypress and tupelo swamps, and pocosins (Messina and Conner 1998). Over 6% (12 million acres) of the 191 million acres under Forest Service management have been identified as wetlands and riparian areas (Bartuska 1993). These wetlands provide critical habitat for avifauna and herpetofauna and are essential elements for the maintenance of biological diversity. In addition, forested wetlands contribute services such as flood control and improved water quality. Economic values of forested wetlands are primarily in the areas of timber production, fisheries, and recreation.

Bioassessnent and Management of North American Freshwater Wetlands, edited by Russell B. Rader, Darold P. Batzer, and Scott A. Wissinger.
0-471-35234-9

The current community structure of forested wetlands in North America is young geologically. Within the glaciated portion of the continent, the maximum age of forested wetlands can be no older than approximately 12,000 years before present (ybp), the beginning of the postglacial period in midcontinent. Wetland maximum age decreases toward the pole, reflecting patterns of deglaciation, with some Alaskan muskegs dated at approximately 2000 ybp (Kangas 1990). The age of forested wetlands in the southeastern United States often lacks the latitudinal synchroneity demonstrated by systems of glaciated areas and is strongly controlled by regional oscillations in rainfall patterns and linkages with the water table. Although focusing heavily on ponds and lakes, the rich palynological record from the southeastern United States can provide valuable insight into the developmental history of swamp forests in the region.

Although it appears that the community structure of forested wetlands in the glaciated zone has remained relatively intact for millennia since basin formation and species establishment, that of forested wetlands in the south has been extremely dynamic. This dynamic quality is associated with pronounced hydrological fluctuations during both the glacial and postglacial periods and the ability of individual species to tolerate prolonged drought. Delcourt et al. (1983) noted that swamp taxa were limited to the margin of Cahaba Pond, a 0.2-ha pond in the Ridge and Valley physiologic provence of north-central Alabama, from the time of its formation 12,000 ybp until 10,000 ypb, when they were eliminated as the pond dried. It was not until 8400 ybp that the current plant community, including black gum (*Nyssa sylvatica*) and buttonbush (*Cephalanthus occidentalis*), became established, when the pond reflooded due to climatic change.

Similar community oscillations in plant community structure for swamps have been reported for the coastal plain and Carolina Bays of the southeast (Whitehead 1972, 1973; Watts 1980), south of the Dismal Swamp in Virginia. At White Pond, a small 19,000-ybp site near Columbia, South Carolina, the current swamp forest community of *Liquidambar*, *Magnolia*, *Myrica*, *Cephalanthus,* and *Acer* did not develop until 9550 ybp, associated with regional changes in climate and hydrology (Watts 1980).

Both the longest record and greatest fluctuations in swamp forest communities have been recorded from Florida. Despite extremely long records of deposition at Lakes Tulane (55,000 ybp), Camel (33,000 ybp), and Sheelar (20,000 ybp), the current community of swamp tree species did not appear until between 5000 and 8000 ybp, as rainfall increased and sea level rose associated with deglaciation patterns in North America (Watts and Hansen 1994). Swamp taxa appeared sporadically in the pollen record at each site prior to that, but it was not until regional hydrology stabilized temporally that the forest community characteristic of present-day Florida developed.

Forested wetland area in the United States was reduced by about 3 million hectares from 1960 to 1990. The rate of loss during the 1980s was 0.3% per year, and the percentage of forested wetlands lost (13%) exceeded that for forestland in general (7%) during the three decades prior to 1990 (Tulloch 1994). The largest reductions occurred in those states that had the most bottomland hardwoods, which during the 1970s and 1980s were the southeastern states, including Arkansas, Florida, Georgia, Louisiana, Mississippi, North Carolina, and South Carolina (Fig. 19.1; Table 19.1).

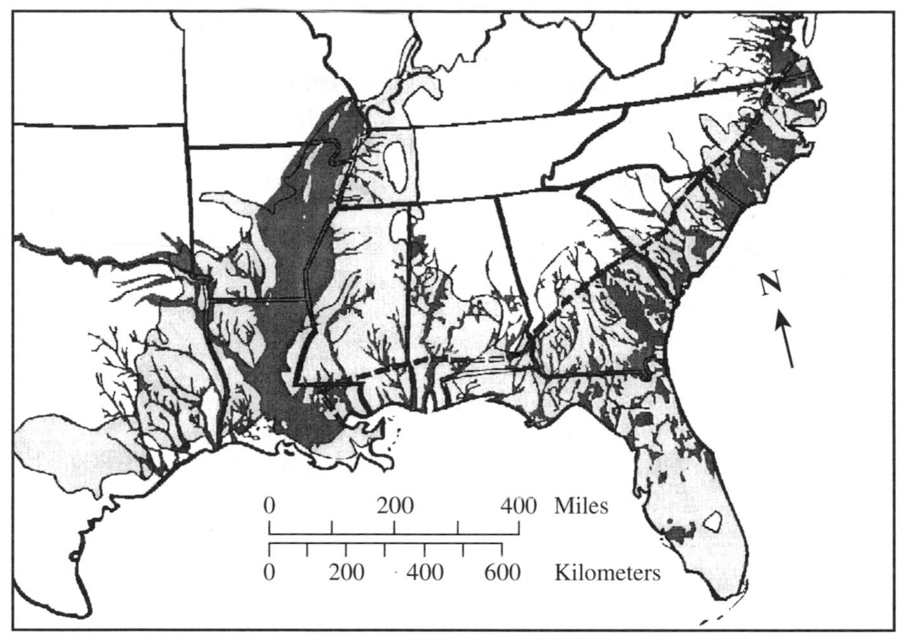

Distribution of bold cypress (*Taxodium distichum*) in SE United States. Dotted line indicates N extent of pond cypress (*Taxodium distichum* var. *nutans*).

Extent of bottomland hardwood forests of SE United States.

FIGURE 19.1. Distribution of bald cypress (*Taxodium distichum*) and extent of bottomland hardwood forests of the southern United States. Dashed line indicates the northern extent of pond cypress (*Taxodium distichum* var. *nutans*).

TABLE 19.1. Estimated Forested Wetlands Area (Presettlement and Present) and Rate of Loss for the late 1970s and Early 1980s

Region	Presettlement Forested Wetlands (ha)	1997 Forested Wetlands (ha)	Rate of Loss, 1974–1983 (ha/yr)
Alaska	5,400,000	5,393,643[a]	Negligible
Great Lakes and New England	11,102,973[a]	3,289,720[a]	22,700 (for all northern conterminous)[b]
Southeastern	21,000,000 (riparian only)[c]	14,541,000[d]	355,500 (for all southern conterminous)[b]

[a]Data from Dahl and Zoltai (1997).

[b]Data from Frayer (1997).

[c]Data from Nature Conservancy (1992).

[d]Data from Shepard et al. (1998).

Losses were due primarily to agriculture and urban development, but cumulative impacts have led to a decrease in the average area of individual wetlands, a shift in proportion of wetland types, changes in spatial configuration of wetlands, and loss of cumulative wetland function at the landscape scale (Johnston 1994, Dahl and Zoltai 1997).

The northern states represent approximately 22% of total forested wetland area of the contiguous United States, but these wetlands have been reduced in area by only 6% since the 1970s (Frayer 1997). The annual loss of wetlands for these states (22,700 ha/yr) for the period 1974–1983 was an order of magnitude less than that of the combined southern states (355,500 ha/yr) for the same period (Table 19.1), reflecting not only the larger total area of wetlands for the latter but also a shift in drainage for agriculture and urban development. The importance of forest management practices and selective harvest of swamp tree species in wetland loss of the southeastern states is less clear. Recent growth rates for bays and magnolia in all southern states exceed harvest, except for Virginia, where the rates are approximately equal (Table 19.2). With few exceptions, annual growth rates of live timber (merchantable portion only) for sweetgum and tupelo/black gum in the region currently exceed the annual removal rate of these species (exceptions are Louisiana and South Carolina for sweetgum and Georgia, North Carolina, and South Carolina for tupelo and black gum combined). In seven of the nine southern states for which data were available (Table 19.2), harvest of cypress (*Taxodium* spp.) either equals or exceeds growth. Cypress is an important component of commercial mulch for residential gardens.

Given the rapid destruction of wetlands in the southern United States, it is important to assess the importance of forestry management practices relative to urbanization and agriculture as a contributive factor. The current chapter focuses, therefore, on understanding impacts to and assessment methods for forested wetlands in the southeastern United States, and we compare these with studies from northern systems.

HYDROGEOLOGY

Wetlands develop in areas where the water table intersects the soil surface or where water losses are retarded by impermeable soils or by lack of hydraulic head. In general, these conditions are met either on flatlands or on slopes where flow is impeded, and provide the necessary conditions for development of characteristic vegetation. Since wetland communities develop through distinctive successional mechanisms, they are geologically relatively young.

Flooding in forested wetlands is highly variable, depending on seasonal rainfall, topography, position relative to water inputs, and other factors. Floodplain wetlands are considered open systems due to the periodic inflow and outflow of sediments and organic matter (Winger 1986), whereas depressional wetlands like pocosins, bays, and pondcypress domes have no or infrequent outflow. The limited outflow of depressional wetlands results in highly organic, often acidic soils, in contrast to the relatively fertile mineral soils of flatwoods and floodplains (Ewel 1998, Sharitz and Gresham 1998).

TABLE 19.2. Average Net Annual Growth and Removals (× 1000 Cubic Feet) of Live Timber on Timberland for Key Wetland Tree Species in Southern U.S. States

		Bays and Magnolias		Cypress		Sweetgum		Tupelo and Black Gum	
	Reporting Date	Growth	Removal	Growth	Removal	Growth	Removal	Growth	Removal
Alabama	1982–1989	NA	NA	NA	NA	89.6	67.9	35.9	18.3
Florida	1987–1994	25.4	8.8	40.2	41.7	15.3	12.1	27.4	15.3
Georgia	1989–1997	10.8	9.9	14.7	10.3	80	71.3	41.5	49.2
Louisiana	1984–1991	NA	NA	NA	NA	59	64.1	14.4	13.4
Mississippi	1987–1994	NA	NA	6.7	2.9	85	72.2	NA	NA
North Carolina	1984–1989	8.8	2.5	9.3	8.1	70.8	63.9	32.9	36
South Carolina	1986–1992	1	0.3	5.4	6.6	35.3	55.3	19.7	24.1
Tennessee	1989–1996	0.2	NA	2.6	7.1	17.1	6.7	3	0.4
Virginia	1986–1991	0.4	0.4	0.1	0.1	4.2	3.6	2.3	1.2

Source: Data compiled from the most recent U.S. Forest Service forest statistical reports for individual states.

[a]NA, not available.

Timing, frequency, intensity, and length of inundation directly influences such abiotic variables as sediment, soils, nutrients, and oxygen availability (Winger 1986), and indirectly affects factors such as organic sediment accumulation and oxidation rates. The length of hydroperiod will determine many biotic factors as well, such as organic matter inputs, productivity and life cycles of wetland plants, fish reproduction and feeding times, and the life cycles of benthic macroinvertebrates. Since the high degree of variability and unpredictability in hydrology is a dominant characteristic of forested wetlands (Fig. 19.2), this feature constrains our ability to assess variation due to human impacts such as forestry management practices. However, this also means that management practices that occur within the historic range of this variability will affect wetland communities in a manner similar to natural disturbances.

TYPES OF FORESTED WETLANDS

Greater than 75% of the wetlands in the south are forested (Shepard et al. 1998). The majority of these (67%) are found along narrow stream margins and small drainages, as compared to 11% in floodplain forests along major rivers and 8% in deepwater swamps (Walbridge 1993). Many of the remaining types of systems are more limited geographically: for example, southern mountain fens and mangrove wetlands.

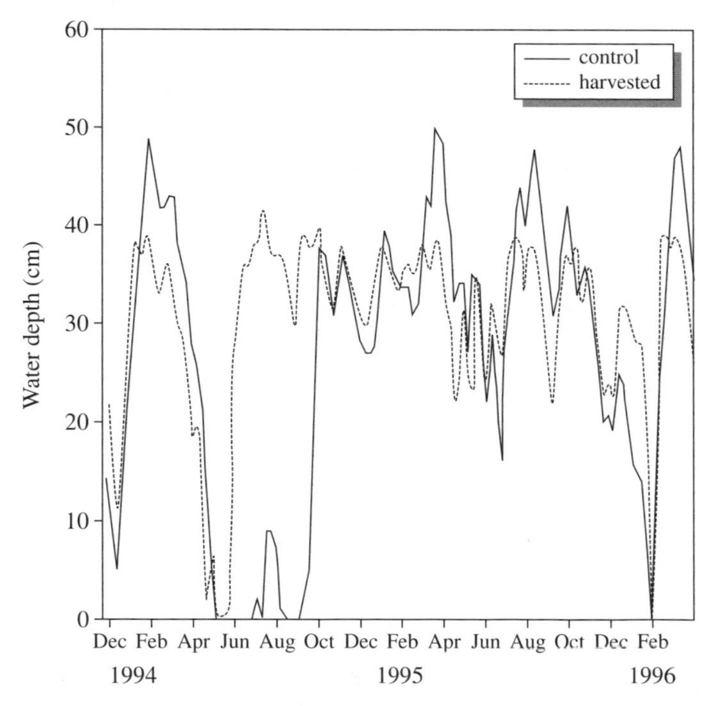

FIGURE 19.2. Water depth, continuously recorded in two 0.6-ha pondcypress swamps near Gainesville, Florida. (From H. Riekirk and L. Korhnak, unpublished data.)

About 90% of the wet timberland in the southeast is occupied by oak–gum–cypress forest (Shepard et al. 1998). Major types of wetlands with respect to timber harvest are deepwater swamps, bottomland hardwoods, wet flatwoods, pocosins and bays, and pondcypress swamps. Important characteristics of these systems are summarized in Table 19.3. There is some overlap between deepwater cypress and pondcypress swamps; however, since pondcypress is considered either a subspecies or a separate species from baldcypress, these two categories are considered separately.

FLORA AND FAUNA

Forested wetlands provide critical habitat for mammals and herpetofauna and are used by breeding and migratory avifauna (Bartuska 1993). Fish may be present in swamps when flooded; reptiles and amphibians are the most common vertebrates year round. Diversity of reptiles in the Okefenokee is quite high (Porter et al. 1999), and alligators may be particularly important ecologically through the creation of "gator holes," which are important refugia for many aquatic animals (Laerm et al. 1980). Amphibian and reptile communities have been the subject of several studies in wet flatwoods (Enge 1984, Domingue-O'Neill 1995) and coastal plain hardwood forests (Harms et al. 1998). Faunal studies of pocosin habitats are limited, but they are considered critical since they are often the only extensive natural areas remaining for wildlife (Sharitz and Gresham 1998).

Although large mammals such as deer and black bear may use swamps, most mammals that actually live in the swamps are squirrels and other arboreal species (Harris and Vickers 1984). Upland birds use mature pondcypress systems, particularly during migration, but water birds seldom use the swamps in northern Florida unless there is open water feeding habitat nearby (Ewel 1998) or after clearcutting. In southern Florida, wading bird rookeries are occasionally found in pondcypress swamps when flooded (Frederick and Spalding 1994). In contrast, aquatic bird diversity is high in the Okefenokee Swamp (Meyers 1982), with sandhill cranes, herons, and storks common. Aquatic avifauna can have an impact on water chemistry and nutrient cycling, particularly near active rookeries or bird colonies (Porter et al. 1999).

IMPACTS OF FORESTRY PRACTICES ON WETLANDS

Impacts of Clear-Cutting

Hydrology. Clear-cutting can change wetland hydroperiods by reducing interception and transpiration and increasing evaporation and runoff concomitantly. Dubé and Plamondon (1995) described how reduction in interception and transpiration rates after clear-cutting of four forested wetland sites caused the habitats to become wetter. Interception ranged from 35 to 41% before and from 8 to 15% after clear-cutting. The reduction of interception on a seasonal basis was responsible for more than 50%

TABLE 19.3. Characteristics of Some Major Forested Wetlands of the Southeastern United States.

Wetland Type	Dominant Species	Subdominant Species	Geomorphology and Soils	Hydrology
Deepwater swamps	baldcypress (*Taxodium distichum*), water tupelo (*Nyssa aquatica*), swamp tupelo (*Nyssa sylvatica* var. *biflora*), Atlantic white cedar (*Chamaecyparis thyoides*) (Conner and Buford 1998)	Red maple (*Acer rubrum*), black willow (*Salix nigra*), swamp cottonwood (*Populus heterophylla*), green and pumpkin ash (*Fraxinus pennsylvanica* and *F. profunda*), pondcypress, pond pine (*Pinus serotina*), loblolly pine (*Pinus taeda*)	Range from broad, flat floodplains to isolated basins, peat deposition; soils range from mucks and clays to silts and sands (Conner and Buford 1998)	Inflows dominated by runoff and overflow from rivers, water levels vary greatly, and may be completely dry for extended periods
Bottomland hardwoods (major and minor alluvial floodplains)	Oak (*Quercus* spp.), red maple (*Acer rubrum*), green ash (*Fraxinus pennsylvanica*), sweetgum (*Liquidambar styraciflua*), American elm (*Ulmus americana*), and baldcypress (Kellison et al. 1998)	Loblolly pine, spruce pine (*Pinus glabra*), baldcypress, pondcypress (*Taxodium distichum* var. *nutans*), Atlantic white-cedar, and sometimes eastern redcedar (*Juniperus virginiana*), cabbage palm (*Sabal palmetto*), saw-palmetto (*Serenoa repens*), and dwarf palmetto (*Sabal minor*) (Kellison et al. 1998)	Strongly infuenced by erosion and depositional processes, moderate levels of organic matter, red rivers—rock dominated (Kellison et al. 1998).	Flooding greatly influences soil structure, species diversity is strongly related to differences in hydrology (Hodges 1998), blackwater rivers—precipitation dominated

Wet flatwoods	Ranges from longleaf pine (*Pinus palustris*) in the drier areas to water-tolerant species such as swamp tupelo and pondcypress	Hardwood tree species are similar to those found in bottomland areas, but may show differences in species associations (Hodges and Switzer 1979)	Poorly drained soils with little or no peat accumulation, occupy interstream areas, including flats, basins or depressions, and drainage ways (Harms et al. 1998)	Connected to stream or river systems only at times of high water
Pocosins and Carolina Bays	Pond pine, loblolly-bay (*Gordonia lasianthus*), pondcypress, and swamp tupelo (Sharitz and Gresham 1998)	Vegetation is characterized by a dense shrub pocosins) or (herbaceous layer (bays)	Nutrient-poor, peat or highly organic soils, not associated with alluvial systems	Largely ombrotrophic, geology and location distinguish them from pondcypress swamps
Pondcypress swamps (cypress strands, dwarf cypress savannas, cypress domes, cypress ponds, and sinkhole ponds or limesinks)	Pondcypress (*Taxodium distichum* var. *nutans*), may include swamp tupelo as codominant	Slash pine (*Pinus elliottii*), swampbay (*Persea palustris*), sweetbay (*Magnolia virginiana*), red maple, and loblolly bay, and in southern Florida, Carolina ash (*Fraxinuscaroliniana*) and willow (*Salix caroliniana*) (Ewel 1998)	Thick peat or organic soils, low nutrient levels	Rainfall is the major hydrologic input, cypress domes, cypress ponds, and sinkhole ponds or limesinks are isolated depressional wetlands, cypress strands drain into rivers or lakes

of the increase in hydroperiod. In some Minnesota peatlands, Verry (1986) described how clear-cutting also caused wetland water tables to fluctuate; ranging from increases of 9 cm to decreases of 19 cm in mature forests. Clear-cutting upland hardwoods or conifers may increase annual streamflow by 30 to 80%, thereby increasing the hydroperiod of riparian wetlands. Streamflow may return to preharvest levels after more than a decade, while annual peak flows are more than doubled and flood-peak increases may persist for 15 years.

Although clear-cutting can drastically alter the hydrology of forested depressional wetlands, in some cases the establishment of emergent vegetation can restore evapotranspiration (ET) to a level that stabilizes surface water at nearly normal levels. In a study of the effects of clear-cutting on pondcypress swamps in a managed pine plantation in north-central Florida (Comerford et al. 1995), water levels were monitored continuously in two 0.6-ha swamps (Fig. 19.2). Immediately after harvesting, a pronounced increase in surface water levels of the harvested swamp relative to an uncut control was observed. By the following growing season, however, emergent vegetation had become established in the clear-cut and surface hydrology was similar in both swamps.

Although monitoring of groundwater and surface-water characteristics may seem a straightforward assessment criteria for impacts on forested wetlands, hydrology can be extremely unpredictable (Fig. 19.2) and subject to periodic influences such as El Niño. Nevertheless, it may be useful to establish detailed monitoring of hydrologic parameters in order to establish baseline conditions and potential impacts. In the study mentioned above, an extensive groundwater monitoring network was established that showed a transient impact on groundwater levels in the harvested area, with increases persisting for 105 days (Comerford et al. 1995).

Soils. Forested wetlands perform two important functions: nutrient removal from surface, subsurface, and groundwaters, and export of organic carbon and associated nutrients to streams and other aquatic ecosystems. These processes are largely soil-mediated. In addition, organic soil substrate forms the basis of the food web, providing an energy source for benthic invertebrates. Disturbance of the soil matrix disrupts many of these functions and changes the substrate character available for the benthic community (Fig. 19.3).

Harvesting and regeneration practices have an effect on organic decomposition in surface soil. Increased temperature and aeration of the soil combine to raise the decomposition rate (Trettin and Jurgensen 1992). These effects on soil temperature and decomposition also correlate with the level of site disturbance. Trettin et al. (1996) found that decomposition rates were greatest when sites were bedded (i.e., aeration was increased), followed by rates for trenching and whole-tree harvest (lower levels of aeration). Since temperature is highly correlated with the rate of decomposition, higher rates of organic matter decay would be expected until canopy closure reduces soil temperature.

In a similar study of a palustrine water tupelo-baldcypress wetland forest in southwestern Alabama, soil temperature and organic carbon were evaluated following three levels of timber-harvest-related disturbances: helicopter logging, rubber-tired

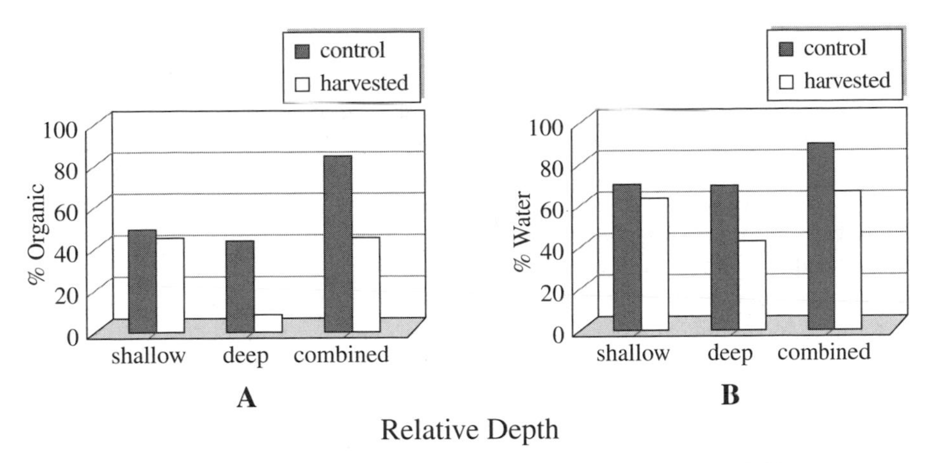

FIGURE 19.3. *(a)* Mean percent organic matter in dried sediment cores and *(b)* mean percent water in wet cores taken in control and harvested pondcypress swamps. Comparisons were made for whole cores (15-cm depth) in shallow and deep-water areas of the swamps and for combined (shallow + deep) surface sediments (0 to 5 cm). (From Leslie 1996.)

skidder logging, and helicopter logging followed by herbicide application (Aust and Lea 1991). In the 2-year postharvest period, soil temperatures in the reference area were lowest, helicopter and skidder logged areas were warmer, and herbicide plots had the highest soil temperatures. Decomposition of organic matter in the soils followed the same trend. These variables were related to amount of vegetative cover and provide an indication of the level of recovery after tree harvest. Most soil parameters of harvested sites had recovered to reference conditions after 7 years (Aust et al. 1997).

Dissolved organic matter (DOM) exported by headwater and riparian wetlands is an important contribution to the energy budget of many streams. Since hydraulic conductivity generally decreases with depth, predominant flow paths often occur in the uppermost organic-rich layers, resulting in higher levels of DOM. In addition, dissolved organic carbon (DOC) varies with degree of interaction with the surrounding upland, distance from the wetland, and internal processes within the wetland. Carbon dating of DOC exported from Precambrian shield wetlands indicates that the organic matter is mostly modern, consistent with shallow flow paths and export of DOM from shallow organic-rich horizons (Schiff et al. 1998).

Organic matter contributions to a stream from the surrounding landscape are particularly important during rain events. Coarse organic matter initially is increased after harvest but declines with time, either through export or conversion to fine particulates and DOC. Transport is affected by leaching and flushing of DOC from the wetland, leading to lower DOC concentrations with successive storms (Hinton et al. 1998). Precipitation, throughfall, and stem flow were minor sources of stream DOC during storms in this study and contributed less than 20% of the total export.

In studies of pondcypress systems affected by clear-cutting, soil redox potential was significantly lower in areas where compaction by large equipment had occurred (Casey and Ewel 1998). Significant changes in soil composition were also observed, from coarse, highly organic soils with high water content to more silty soils with fine organic matter, higher levels of inorganics, and lower water content (Fig. 19.3; Leslie 1996). These changes are likely to have resulted in markedly different food types for benthic invertebrates. Changes in soil characteristics are likely to contribute to shifts in the benthic community of harvested swamps.

Nutrient removal from surface and subsurface water occurs through phosphorus deposition in sediments, phosphorus adsorption, and bacterial denitrification. Biological processes such as plant uptake and microorganism absorption are also important and are closely related to organic matter export. Ground-based harvesting methods have been shown to increase soil bulk density and decrease hydraulic conductivity and redox potential in wetland soils (Aust and Lea 1992). These effects, as well as increases in soil temperature and hydrology, may contribute to the reduced productivity and altered species composition observed following ground-based versus aerial harvesting (Walbridge and Lockaby 1994).

Reductions in plant biomass and uptake, litterfall, and changes in species composition following harvesting can alter both nutrient dynamics and carbon export. When standing water is present, increased insolation may result in higher water column productivity (Leslie 1996), possibly replacing some nutrient uptake functions while changing sedimentation rates and sediment character. Changes in the coarseness of organic sediments will probably induce changes in functional feeding groups among benthic invertebrates.

Algae, Herbaceous Plants, and Fauna. There is often an increase in primary production in surface waters following clearcutting (Leslie 1996). Higher levels of pelagic and periphytic algaes may be due in part to increased nutrients from disturbed soils and to increased insolation from loss of tree canopy. In addition, applications of fertilizer during site preparation often reach surface waters upon reflooding. Increases in water column productivity will probably be transient, as emergent macrophytes and resprouting trees begin to compete for light and nutrients; however, higher algal productivity will result in qualitative changes, such as more silty and flocculent sediments. These changes may in turn result in alterations in benthic macroinvertebrate communities and are discussed in more detail below.

Changes in macrophytic groundcover are correlated with both alterations in hydrology and with the level of soil disturbance. Community composition would reflect the seed bank from preharvest populations and available seed sources in surrounding wetland areas, but density and coverage would depend on factors such as the level of rutting and soil disturbance caused by heavy equipment. Due to the high variance in understory vegetation, this type of impact would be difficult to assess without a large amount of data from reference wetlands of similar type.

The duration of changes in macrophyte densities will to a large degree depend on the rate of resprouting or growth of seedlings. In cases where vegetative communities of forested wetlands are being actively controlled for production purposes, changes in structure and diversity would be artificial in nature and therefore would not be useful. In any case, monitoring of species indicative of wetland hydrology will provide information on long-term hydrologic changes. Regeneration of trees in wet forests will be highly dependent on the timing, level, and duration of flooding, particularly with respect to germination of seeds. Since the seed source is often completely removed or destroyed, active planting may be required. Planting may present some of the same problems as harvesting; for example, compaction and disturbance of soils. In the case of pondcypress, leaving stumps will encourage resprouting and provide seed sources in the future (Ewel 1996). In most cases, if a seed source is desired, 10 to 20 trees per acre are left, or 30 to 50% of trees are left in a shelterwood method. Hydrologic management may be required in this case to promote establishment of seedlings.

Perturbations of existing hydrologic patterns that lead to increased or decreased inundation can result in decreased growth rates or fatality of production species (Conner 1994). An opposing trend results from increased light and nutrient availability when forest canopy is removed. In a study in northern Michigan, the number of species and total cover of all vascular plants were significantly greater in harvested compared to uncut areas. Relative cover of herbaceous vegetation increased with the level of site disturbance (bedded and trenched), due primarily to disturbance of a thick *Sphagnum* mat (Gale et al. 1998).

Changes in sediment organic matter composition and physical-chemical characteristics will probably have a measurable impact on secondary production in the benthic community. Hydrologic alterations directly affect fish and other aquatic fauna, as well as benthic invertebrates. Fish and other macrofauna can be monitored by standard methodology, (e.g., fish traps and seines, herpetofaunal traps, or by simple counts such as bird surveys). Benthic macroinvertebrate community changes are probably more standardized as a method of assessment and may be monitored using dip-net surveys (Plafkin et al. 1989; Chapter 15 this volume), or sediment sampling methods such as cores (Leslie et al. 1997) or sediment grabs.

Impacts of forestry practices on benthic invertebrates of wetlands have not been examined extensively, although several studies have examined benthic macroinvertebrates as an assessment tool for impacts on streams (Campbell and Doeg 1989, Carlson et al. 1990). Studies at the University of Florida have focused on the effects of clear-cutting on benthic communities in six pondcypress swamps located within a 42-ha managed slash pine plantation (Prenger et al. no data). Ranging from 0.4 to 2.5 ha, three swamps were in an unharvested plot and three were in an adjacent plot in which both the swamps and upland were clear-cut in April and May 1994. Benthic macroinvertebrates were sampled bimonthly from December 1993 to April 1996 using a stainless steel corer with a sample area of approximately 40 cm^2. Twenty randomly selected sediment cores were taken per swamp and macroinvertebrates were picked from the sediment and identified to genus or to the lowest taxonomic level practical. Taxon richness and total density were calculated using only aquatic and semiaquatic taxa, and data for the three swamps in each treatment were pooled for each sampling date.

As discussed previously, sediments of harvested swamps were more spatially varied after the clear-cut (Fig. 19.3), due primarily to rutting caused by logging equipment. Sediments had a siltier character in and around the ruts than in undisturbed areas, and soils from deep areas of harvested ponds contained less organic matter and lower percent water than sediments of shallow areas of harvested swamps and sediments of control swamps. There was no difference observed in sediment organic matter and water content between depths within the control ponds.

Three months after harvesting, the total density of benthic macroinvertebrates was significantly greater in harvested swamps than in undisturbed swamps (Fig. 19.4). Maximum average density in the harvested swamps was approximately twice the maximum average density of control ponds. Densities in the control swamps showed no differences before and after harvest, while postharvest density changes were significant in the harvested swamps. Most of the density increase in harvested swamps was due to a single genus, the Dipteran *Chironomus*, the density of which increased approximately tenfold. Although taxon richness generally increased over time, there was no significant difference between the treatments, with 136 and 128 taxa in the three control and three treated swamps, respectively. The most taxonomically rich groups were Diptera and Coleoptera, contributing 20 and 30% of the total, respectively.

Impact of Postharvest Site Preparation and Management

Bedding. Regeneration practices such as bedding affect organic decomposition rates in soils by increasing soil temperature and aeration (Trettin and Jurgensen 1992). Although it is difficult to separate bedding from other treatments (e.g., fertilization), these practices do result in higher productivity in seedlings and the plant community in general (Gale et al. 1998). Although these changes might be expected to enhance

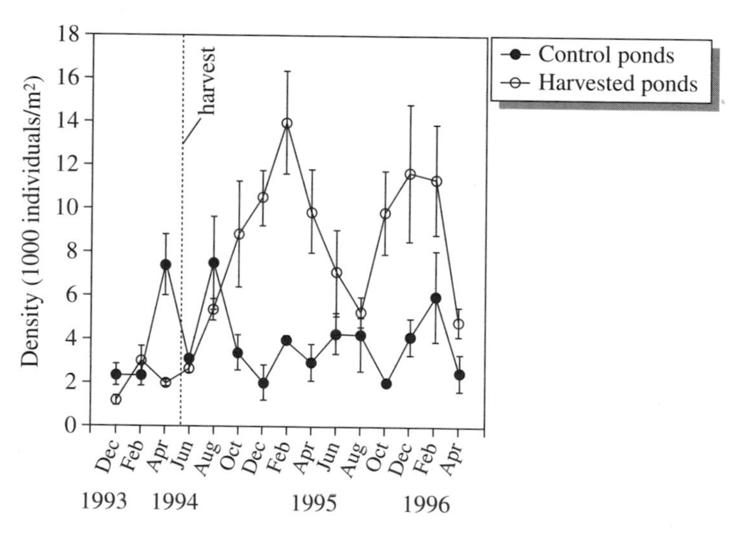

FIGURE 19.4. Mean (±SE) density of benthic macroinvertebrates for three control and three harvested pondcypress swamps.

the export of organic carbon from wet forest areas in the short term, leaching and alteration of hydrology and microtopography will probably cause reductions in DOC over time. In addition, these changes would affect the substrate and energy base for the food web. Soil temperature and aeration are discussed in more detail below with respect to impacts from clear-cutting.

Fertilization. Fertilization of clearcut sites for enhancement of seedling production is a common practice. Moderate increases in nutrients have been shown in streams following fertilization of surrounding forestland (Binkley et al. 1999), but no concomitant decrease in water quality was seen. In the case of depressional wet forestlands with little or no outflow, this may result in eutrophication of surface water and, eventually, changes in sediment character. Deposition from increased algal populations alters the nature of organic sediments, producing a finer, more flocculent surface layer. Fine sediments are fed on by collectors, as opposed to generalists and shredders, which feed on coarse organic material (Anderson and Cummins 1979), and several studies have observed taxonomic changes in benthic communities with increased trophic state (Wiederholm and Eriksson 1979, Devai and Moldovan 1983). The possibility of fertilizer impacts on depressional wetlands should be a focus of future research.

Herbicides. Herbicide treatment of production areas can have both direct and indirect effects on forested wetland sites. As mentioned previously, reduction of the vegetative cover can increase soil temperatures and soil metabolism (decomposition). Additionally, direct reduction of herbaceous vegetation occurs and some herbicides are toxic to the aquatic fauna (Michael and Neary 1990), although toxicity is unlikely at concentrations typically found after operational applications. A recent study in our laboratory showed no effect of the herbicide imazapyr (Arsenal) on benthic invertebrate populations, although there was some indication of increased levels of deformities in chironomids (M. Fowlkes, unpublished data). This increase in deformities was not statistically significant, although chironomid deformities have been linked to other pesticides, such as the organochlorines DDE (Warwick, 1985) and DDT and the herbicides Dacthal (Madden et al. 1993), glyphosate, and atrazine (Norwak et al. 1993).

Prescribed Burns. Fire is a relatively rare occurrence in bottomland and wet flatwoods, but occurs at low frequency (about 13 to 50 years) in pocosin and bay communities. In northern Florida, the fire frequency before settlement was perhaps as high as once every 20 years (Ewel 1990), and in southern Florida the frequency was probably even higher (Wade et al. 1980). Since pondcypress swamps are common in fire climax communities such as pine savannas and marshes, and these swamps are dry 3 to 6 months each year (Marois and Ewel 1983), fire may burn into them often enough to be ecologically significant. Burning removes peat and course detritus and changes the character and availability of carbon and other nutrients. Since substrate is an important determinant of aquatic insect species composition and density (Ward 1992), these changes probably have a significant impact on the benthic communities upon reflooding.

Prescribed burns can be used to reduce undergrowth and the risk of catastrophic fire. Although the use of this management tool is not common in wetland areas, it is becoming more common in state forests and other areas where depressional wetlands are found (Florida Division of Forestry personnel, personal communication). Fire can be an important modifier of sediment and organic matter in wetlands. Changes induced by burning provide a pulse of nutrients upon reflooding of wetlands and is probably an important driving force behind the natural variability and diversity observed in forested wetlands.

Water Management. Secondary changes in water level due to clear-cutting are discussed above, but intentional alteration of water-level for increased production can also have undesirable side effects. In bottomland and cypress/tupelo forests, water level control can have both positive and negative effects; for example, increased flooding or drainage can cause decreased growth rates or even mortality. Bottomland hardwoods respond favorably in the short term to water-level management, but less is known about the long-term response (Conner 1994). Obviously, water-level management has a direct effect on the benthic and aquatic communities of forested wetlands; however, research has shown that benthic invertebrate communities of intermittently flooded wetlands are well adapted to drawdown and are present even during dry periods (Leslie et al. 1997). More research and better assessment methods are needed to understand the ecological consequences of increasing the length of dry periods.

RECOMMENDATIONS FOR MANAGEMENT

Clearcutting of forested wetlands is a common management strategy because it encourages regeneration of commercially desirable species and minimizes the number of times a stand must be entered (Wigley and Roberts 1994). The latter is important in order to avoid changes in soil composition and redox potential that will have impacts on benthic invertebrates and other components of the aquatic food web. Operation of heavy equipment on wet soils often results in rutting, creating deeper water areas upon inundation. In still water wetlands, these areas can be hypoxic or anoxic, making them essentially dead zones until filled in by sedimentation or vegetation. Temporary changes in the benthic community can be expected following clear-cutting, due to an increase in the ratio of coarse-particulate to fine-particulate organic matter (Prenger et al. no data), but these changes appear to become less important after revegetation. Although temporary, the long-term effects of timber harvesting on the benthic communities of wetlands are unknown and more research is needed in this area.

In addition to the aforementioned standard to avoid entering wet sites as much as possible, forested wetlands should be harvested using management techniques designed to avoid sedimentation, nutrient enrichment of surface waters, excessive mixing and alteration of soil, and unnecessary deterioration of drainage patterns. Best management practices that utilize a shelterwood cut and estimation of probability of damage (hazard indices) during the wet season will minimize these types of problems (McKee 1985). Recommendations for management of wet forests for hydrologic considerations include attempts to ensure adequate regeneration and protection of vegetative cover, avoidance of full tree harvesting, and retention of logging debris in the cut area to minimize runoff and maximize interception and evapotranspiration (Dubé and Plamondon 1995). As already stated, increases in coarse particulate organic matter from logging debris will probably affect benthic communities, but these changes are probably transient.

One of the major considerations for management of forested wetlands is the regeneration of the wet forest. Since timing, duration, and depth of inundation is critical to the success of seed germination and seedling establishment, regeneration from this source may require active hydrologic management. Planting of seedlings is one option, but may be impractical in wet conditions or for certain species. Regeneration via sprouting is often depended on for reestablishment of pondcypress, and in this case care must be taken so that the tops of stumps remain above water in order to maintain viability and encourage sprouting.

Invasive exotic species are not discussed in detail here but relate to the overall disturbance regime and regeneration of wet forests. Many exotic tree species compete successfully with natives and may cause secondary alterations to forested ecosystems (Cox 1999). Particularly in Florida, species such as Melaleuca (*Melaleuca quinquenervia*), Brazilian pepper (*Schinus terebinthifolius* Raddi), and Chinese tallow (*Sapium sebiferum*) change forest structure and ecosystem functions (Cameron and Spencer 1989, Harcombe et al. 1993, Simberloff et al. 1997). In many situations, thick growth patterns exclude wildlife, slowly decomposing litter alters detritus and benthic communities, and in the case of Melaleuca, growth actually alters hydrology in the immediate area. Because of the high level of disturbance, areas subject to forestry will require active management to control exotic species.

Although progress continues to be made on assessment techniques for impacts to forested wetlands, no single component of these ecosystems will provide adequate criteria for management decisions. Integration of some of the elements discussed here would provide information on several key wetland functions and perhaps allow overall management of the ecosystem. Although the continued integrity of benthic and aquatic habitat in wet forests is not the only concern, simple measures such as avoidance of excessive rutting may allow these functions to return to preharvesting conditions more quickly.

REFERENCES

Anderson, N. H., and K. W. Cummins. 1979. Influences on the life histories of aquatic insects. Journal of the Fisheries Research Board of Canada 36:335–342.

Aust, W. M., and R. Lea. 1991. Soil temperature and organic matter in a disturbed forested wetland. Soil Science Society of America Journal 55:1741–1746.

Aust, W. M., and R. Lea. 1992. Comparative effects of aerial and ground logging on soil properties in a tupelo–cypress wetland. Forest Ecology and Management 50:57–73.

Aust, W. M., S. H. Schoenholtz, T. W. Zaebst, and B. A. Szabo. 1997. Recovery status of a baldcypress–tupelo wetland seven years after harvesting. Forest Ecology and Management 90:161–170.

Bartuska, A. M. 1993. USDA Forest Service wetlands research. Journal of Forestry 91:25–28.

Binkley, D., H. Burnham, and H. L. Allen. 1999. Water quality impacts of forest fertilization with nitrogen and phosphorus. Forest Ecology and Management 121:191–213.

Brown, M. J. 1996. Forest statistics for Florida, 1995. Resource Bulletin SRS-6. U.S. Department of Agriculture, Forest Service, Asheville, NC.

Cameron, G. N., and S. R. Spencer. 1989. Rapid leaf decay and nutrient release in a Chinese tallow forest. Oecologia 80:222–228.

Campbell, I. C., and T. J. Doeg. 1989. Impact of timber harvesting and production on streams: a review. Australian Journal of Marine and Freshwater Research 40:519–539.

Carlson, J. Y., C. W. Andrus, and H. A. Froehlich. 1990. Woody debris, channel features, and macroinvertebrates of streams with logged and undisturbed riparian timber in northeastern Oregon, U.S.A. Canadian Journal of Fisheries and Aquatic Science 47:1103–1111.

Casey, W. P., and K. C. Ewel. 1998. Soil environment in small pondcypress swamps after harvesting. Forest Ecology and Management 112:281–287.

Comerford, N. B., J. Montgomery, and D. Long. 1995. Water table response to forest harvesting in a N. Florida flatwoods landscape. Pages 4–30 *in* NCASI Wetlands Study. 1995 Annual Report. University of Florida, Gainesville, FL.

Conner, W. H. 1994. Effect of forest management practices on southern forested wetland productivity. Wetlands 14:27–40.

Conner, W. H., and M. A. Buford. 1998. Southern deepwater swamps. Pages 261–287 *in* M. G. Messina and W. H. Conner (eds.), Southern forested wetlands: ecology and management. CRC Press, Boca Raton, FL.

Cox, G. W. 1999. Alien species in North America and Hawaii. Island Press, Washington, DC.

Dahl, T. E., and S. E. Zoltai. 1997. Forested wetlands of North America. Pages 3–17 *in* C. C. Trettin et al. (eds.), Northern forested wetlands: ecology and management. CRC Press, Boca Raton, FL.

Delcourt, H. R., P. A. Delcourt, and E. C. Spiker. 1983. A 12,000-year record of forest history from Cahaba Pond, St. Clair County, Alabama. Ecology 64:874–887.

Devai, G., and J. Moldovan. 1983. An attempt to trace eutrophication in a shallow lake (Balaton, Hungary) using chironomids. Hydrobiologia 103:169–175.

Domingue-O'Neill, E. 1995. Amphibian and reptile communities of temporary ponds in a managed pine flatwoods. M.S. thesis, University of Florida, Gainesville, FL.

Dubé, S., and A. P. Plamondon. 1995. Relative importance of interception and transpiration changes causing watering-up after clearcutting on four wet sites. Pages 113–120 *in* IAHS Publication 230, International Association of Hydrological Sciences, Wallingford, Oxon, England.

Enge, K. M. 1984. Effects of clearcutting and site preparation on the herpetofauna of a north Florida flatwoods. M.S. thesis, University of Florida, Gainesville, FL.

Ewel, K. C. 1990. Swamps. Pages 281–323 *in* R. L. Meyers and J. J. Ewel (eds.), Ecosystems of Florida. University of Central Florida Press, Orlando, FL.

Ewel, K. C. 1996. Sprouting by pondcypress (*Taxodium distichum* var. *nutans*) after logging. Southern Journal of Applied Forestry. 20:209–213.

Ewel, K. C. 1998. Pondcypress swamps. Pages 405–420 *in* M. G. Messina and W. H. Conner (eds.), Southern forested wetlands: ecology and management. CRC Press, Boca Raton, FL.

Frayer, W. E. 1997. Status and trends of forested wetlands in the northern United States. Pages 19–26 *in* C. C. Trettin et al. (eds.), Northern forested wetlands: ecology and management. CRC Press, Boca Raton, FL.

Frederick, P. C., and M. G. Spalding. 1994. Factors affecting reproductive success of wading birds (Ciconiiformes) in the Everglades ecosystem. Pages 659–691 *in* S. M. Davis and J. C. Ogden (eds.), Everglades: the ecosystem and its restoration. St. Lucie Press, Boca Raton, FL.

Gale, M. R., J. W. McLaughlin, M. F. Jurgensen, C. C. Trettin, T. Soelsepp, and P. O. Lydon. 1998. Plant community responses to harvesting and post-harvest manipulations in a *Picea–Larix–Pinus* wetland with a mineral substrate. Wetlands 18:150–159.

Harcombe, P. A., G. N. Cameron, and E. G. Glumac. 1993. Above-ground net primary productivity in adjacent grassland and woodland on the coastal prairie of Texas, USA. Journal of Vegetation Science 4:521–530.

Harms, W. R., W. M. Aust, and J. A. Burger. 1998. Wet flatwoods. Pages 421–444 *in* M. G. Messina and W. H. Conner (eds.), Southern forested wetlands: ecology and management. CRC Press, Boca Raton, FL.

Harris, L. D., and C. R. Vickers. 1984. Some faunal community characteristics of cypress ponds and the changes induced by perturbations. Pages 171–185 *in* K. C. Ewel and H. T. Odum (eds.), Cypress swamps. University Presses of Florida, Gainesville, FL.

Hinton, M. J., S. L. Schiff, and M. C. English. 1998. Sources and flowpaths of dissolved organic carbon during storms in two forested watersheds of the Precambrian Shield. Biogeochemistry 41:175–197.

Hodges, J. D. 1998. Minor alluvial floodplains. Pages 325–341 *in* M. G. Messina and W. H. Conner (eds.), Southern forested wetlands: ecology and management. CRC Press, Boca Raton, FL.

Hodges, J. D., and G. L. Switzer. 1979. Some aspects of the ecology of southern bottomland hardwoods. Pages 360–365 *in* North America's forests: gateway to opportunity, Proceedings of the 1978 Convention of the Society of American Foresters and the Canadian Institute of Forestry, Washington, DC.

Johnston, C. A. 1994. Cumulative impacts to wetlands. Wetlands 14:49–55.

Kangas, P. C. 1990. Long-term development of forested wetlands. Pages 25–51 *in* A. E. Lugo, M. Brinson, and S. Brown (eds.), Ecosystems of the world 15: Forested wetlands. Elsevier, Amsterdam, The Netherlands.

Kellison, R. C., M. J. Young, R. C. Braham, and E. J. Jones. 1998. Major alluvial floodplains. Pages 291–323 *in* M. G. Messina and W. H. Conner (eds.), Southern forested wetlands: ecology and management. CRC Press, Boca Raton, FL.

Laerm, J., B. J. Freeman, L. J. Vitt, J. M. Meyers, and L. Logan. 1980. Vertebrates of the Okefenokee Swamp. Brimleyana 47:47–73.

Leslie, A. J. 1996. Structure of benthic macroinvertebrate communities in natural and clearcut cypress ponds of North Florida. M.S. thesis, University of Florida, Gainesville, FL.

Leslie, A. J., T. L. Crisman, J. P. Prenger, and K. C. Ewel. 1997. Benthic macroinvertebrates of small Florida pondcypress domes and the influence of dry periods. Wetlands 17:447–455.

Madden, C. P., J. Suter, B. C. Nicholson, and A. D. Austin. 1993. Deformities in Chironomid larvae as indicators of pollution (pesticide) stress. Netherlands Journal of Aquatic Ecology 26:551–557.

Marois, K. C., and K. C. Ewel. 1983. Natural and management-related variation in cypress domes. Forest Science 29:627–640.

McKee, W. H. 1985. Forestry and forest management impacts on wetlands. Pages 216–224 *in* Wetlands of the Chesapeake. Proceedings of the conference held Apr. 9–11 1985, Easton, MD.

Messina, M. G., and W. H. Conner (eds.). 1998. Southern forested wetlands: ecology and management. CRC Press, Boca Raton, FL.

Meyers, J. M. 1982. Community structure and habitat associations of breeding birds in the Okefenokee Swamp. Ph.D. dissertation, University of Georgia, Athens, GA.

Michael, J. L., and D. G. Neary. 1990. Fate and transport of forestry herbicides in the South: research knowledge and needs. Proceedings of the 6th Biennial Southern Silvicultural Research Conference. General Technical Report 70, Vol. 2, pp. 641–649. U.S. Department of Agriculture, Forest Service, Asheville, NC.

Nature Conservancy. 1992. The forested wetlands of the Mississippi River: an ecosystem in crisis. Nature Conservancy, Baton Rouge, LA.

Norwak B., G. Clark, V. Pettigrove, J. Hughes and C. Madden. 1993. Biological indicators and biomonitoring of water quality in agricultural areas. Proceedings of the Ecotoxicology Specialist Workshop. Wee Waa, NSW, Australia. Land and Water Resources Research and Development Corporation.

Plafkin, J. L., M. T. Barbour, K. D. Porter, S. K. Gross, and R. M. Hughes. 1989. Rapid bioassessment protocols for use in streams and rivers: benthic macroinvertebrates and fish. Report EPA/440/4-89-001. U.S. Environmental Protection Agency, Office of Water, Washington, DC.

Porter, K. G., A. Bergstedt, and M. C. Freeman. 1999. The Okefenokee Swamp: invertebrate communities and foodwebs. Pages 121–135 *in* D. P. Batzer, R. B. Rader, and S. A. Wissinger (eds.), Invertebrates in freshwater wetlands of North America: ecology and management. Wiley, New York.

Prenger, J. P., A. J. Leslie, T. L. Crisman, and K. C. Ewel. No data. Impacts of clearcutting on macrobenthic invertebrates of pondcypress swamps in north Florida (unpublished manuscript).

Schiff, S., R. Aravena, E. Mewhinney, R. Elgood, B. Warner, P. Dillon, and S. Trumbore. 1998. Precambrian shield wetlands: hydrologic control of the sources and export of dissolved organic matter. Climatic Change 40:167–188.

Sharitz, R. R., and C. A. Gresham. 1998. Pocosins and Carolina Bays. Pages 343–377 *in* M. G. Messina and W. H. Conner (eds.), Southern forested wetlands: ecology and management. CRC Press, Boca Raton, FL.

Shepard, J. P., S. J. Brady, N. D. Cost, and C. G. Storrs. 1998. Classification and inventory. Pages 3–28 *in* M. G. Messina and W. H. Conner (eds.), Southern forested wetlands: ecology and management. CRC Press, Boca Raton, FL.

Simberloff, D., D. C. Schmitz and T. C. Brown (eds.), 1997. Strangers in paradise. Island Press, Washington, DC.

Trettin, C. C, and M. F. Jurgensen. 1992. Organic matter decomposition response following disturbance in a forested wetland in northern Michigan, USA. Report CONF-9206117-1. U.S. Department of Energy, Washington, DC.

Trettin, C. C., M. Davidian, M. F. Jurgensen, and R. Lea. 1996. Organic matter decomposition following harvesting and site preparation of a forested wetland. Soil Science Society of America Journal 60:1994–2003.

Tulloch, D. L. 1994. Recent areal changes in U.S. forested wetlands. Wetlands Ecology and Management 3:49–53.

Verry, E. S. 1986. Forest harvesting and water: the lake states experience. Water Resources Bulletin 226:1039–1047.

Wade, D., J. Ewel, and R. Hofstetter. 1980. Fire in south Florida ecosystems., General Technical Report SE-17. U.S. Department of Agriculture, Forest Service, Washington, DC.

Walbridge, M. R. 1993. Functions and values of forested wetlands in the southern United States. Journal of Forestry 91:15–19.

Walbridge, M. R., and B. G. Lockaby. 1994. Effects of forest management on biogeochemical functions in southern forested wetlands. Wetlands 14:10–17.

Ward, J. V. 1992. Aquatic insect ecology, Vol. I, Biology and habitat. Wiley, New York.

Warwick, W. F. 1985. Morphological abnormalities in Chironomidae (Diptera) larvae as measures of toxic stress in freshwater ecosystems: indexing antennal deformities in *Chironomus* Meigen. Canadian Journal of Fisheries and Aquatic Science 42:1881–1941.

Watts, W. A. 1980. Late-Quaternary vegetation history at White Pond on the inner coastal plain of South Carolina. Quaternary Research 13:187–199.

Watts, W. A., and B. C. S. Hansen. 1994. Pre-Holocene and Holocene pollen records of vegetation history from the Florida peninsula and their climatic interpretations. Palaeogeography, Palaeoclimatology and Palaeoecology 109:163–176.

Whitehead, D. R. 1972. Developmental and environmental history of the Dismal Swamp. Ecological Monographs 42:301–315.

Whitehead, D. R. 1973. Late-Wisconsin vegetational changes in unglaciated eastern North America. Quaternay Research 3:621–631.

Wiederholm, T., and L. Eriksson. 1979. Subfossil chironomids as evidence of eutrophication in Ekoln Bay, Central Sweden. Hydrobiologia 62:195–208.

Wigley, T. B., and T. H. Roberts. 1994. Forest management and wildlife in forested wetlands of the southern Appalachians. Water, Air, and Soil Pollution 77:445–456.

Winger, P. V. 1986. Forested wetlands of the Southeast: review of major characteristics and role in maintaining water quality. RESOURCE PUB-163, NTIS order no. PB87-113155/GAR. National Technical Information Service, Springfield, VA.

20 Biological Control of an Invasive Wetland Plant: Monitoring the Impact of Beetles Introduced to Control Purple Loosestrife

BERND BLOSSEY

*Purple loosestrife (*Lythrum salicaria*) is an invasive, nonindigenous plant species that has negatively affected North American wetlands for decades. Chemical, mechanical, and physical measures have failed to provide long-term control. Current emphasis to control purple loosestrife center around the introduction and distribution of four host-specific insect herbivores from the plant's native range. Mass production methods were developed to increase the availability of these species. By 1999 over 3 million leaf beetles and over 100,000 root-feeding weevil eggs were field released in more than 30 U.S. states and in Canada. These species are now well established and are commercially available. A standardized monitoring protocol was developed that incorporates measures of control agent abundance (number of adults, eggs and larvae) and affect on the host plant (number of stems, plant height, seed output). This protocol will allow long-term evaluations of the impact of control agents on target plant and associated wetland plant communities. At some of the earliest release sites, purple loosestrife was selectively controlled and its biomass reduced to less than 5% of its original level. At many sites, the monotypic stands of purple loosestrife were replaced by a diverse wetland plant community, while at some sites, other invasive species, such as* Phragmites australis *(common reed) or* Phalaris arundinacea *(reed canary grass), expanded as purple loosestrife was controlled. Long-term monitoring will identify local trends, and standardization of the protocol enables meaningful comparisons across North America.*

The invasion of nonindigenous plant species is a major threat to the functioning of the remaining North American wetlands (U.S. Congress 1993). Invading plants displace native plant communities and their associated vertebrate and invertebrate fauna (U.S.

Bioassessment and Management of North American Freshwater Wetlands, edited by Russell B. Rader, Darold P. Batzer, and Scott A. Wissinger.
0-471-35234-9

Congress 1993). The species-poor, cosmopolitan-dominated replacement plant communities prevent the recruitment of native species and accelerate local and global extinction rates (MacDonald et al. 1989). Millions of dollars are spent annually in an effort to curb the spread and reduce the harmful effects of invasive plants, often with little long-term success (U.S. Congress 1993).

One of the most serious invasive plant species in temperate wetlands across North America is purple loosestrife (*Lythrum salicaria* L.), a Eurasian perennial herb introduced in the early nineteenth century (Thompson et al. 1987). Its spread across the continent has degraded many prime wetlands, resulting in large, monotypic stands that lack native plant species (Thompson et al. 1987, Malecki et al. 1993). As for many other invasive plant species, *L. salicaria* invasions alter biogeochemical and hydrological processes in wetlands (Emery and Perry 1996, Grout et al. 1997) and threaten rare and endangered plant and animal species (Thompson et al. 1987, Malecki et al. 1993, Brown 1999). In North America, established *L. salicaria* populations persist for decades, are difficult to control using conventional techniques (chemical, physical, mechanical), and continue to spread into adjacent areas (Thompson et al. 1987). In Europe, *L. salicaria* is a regular but infrequent component of mixed wetland communities (Shamsi and Whitehead 1974) and rarely establishes monotypic stands. When purple loosestrife becomes established, specialized insect herbivores quickly colonize the plants and devastate tissues, both above and below ground (Blossey 1995a). These specialized natural enemies were absent from North America, allowing purple loosestrife once introduced, to grow unchecked (Hight 1990). The dramatic differences in the abundance and persistence of purple loosestrife in Europe compared to North America and the importance of specialized herbivores in suppressing the species in Europe resulted in the development of a classical biological weed control program in 1986 (Malecki et al. 1993). Classical biological weed control is the introduction of host specific natural enemies (herbivores, usually insects, less often pathogens) from a plant's native range in an effort to restore the self-regulatory potential of a particular plant-insect interaction.

HISTORY OF THE BIOLOGICAL CONTROL PROGRAM

There are several essential components to an effective biological weed control program. Natural enemies, in combination with competition from other plants, are an important factor in regulating the abundance of plant species. In turn, the abundance of the plant (acting as a host) influences the abundance of its natural enemies. Biological attributes of insect herbivores (host specificity, fecundity, impact, etc.) influence the effectiveness of biocontrol agents (Harris 1973, Goeden 1983), however, such characteristics are often difficult to observe in the field and therefore require detailed experimental studies. Significant long-term funding is needed for overseas and domestic research. Biological weed control programs often require 5 to 10 years of overseas research at an annual cost of $50,000 to $100,000, followed by studies after the introduction of control agents. In the long term, however, biocontrol is the most cost effective control method available.

In the program targeting purple loosestrife, research in Europe began in 1986 with field surveys for potential control agents. After initial investigations, out of over 100 different species reported from purple loosestrife in Europe (Batra et al. 1986), six species were selected as the most promising for further studies. This selection was based on literature reports on the specificity of species, their distribution and availability in the field, and initial observations on their impact on purple loosestrife performance. The species selected were a root-mining weevil, *Hylobius transversovittatus* Goeze, which attacks the main storage tissue of purple loosestrife; two leaf beetles, *Galerucella calmariensis* L. and *G. pusilla* Duft, which can completely defoliate entire *L. salicaria* populations; a flower-feeding weevil, *Nanophyes marmoratus* Goeze; a seed-feeding weevil, *N. brevis* Boheman; and a gall midge, *Bayeriola salicariae* Gagné, which attacks leaf and flower buds. Detailed investigations followed as to their life history, distribution, impact, and host specificity (Blossey 1993, 1995a–c; Blossey et al. 1994a,b; Blossey and Schroeder 1995). The species finally proposed for introduction into North America were selected based on information about (1) host specificity, (2) impact on the target weed in the field, (3) distribution, and (4) feeding niche on *L. salicaria*.

To ensure that native wetland plant species were not at risk by the introduction of herbivores, all species were tested for their specificity to *L. salicaria.* A list of approximately 50 test plant species was compiled which included species that are taxonomically related to purple loosestrife or considered of ecological or economical importance. Briefly, results from the tests identified two native North American plant species, *Decodon verticillatus* (commonly known as swamp loosestrife or waterwillow) and *Lythrum alatum* (winged loosestrife), as potential hosts under confined laboratory conditions (Blossey et al. 1994a,b). Both plant species are members of the family Lythraceae and closely related to *L. salicaria*. Predictions that at high population densities beetles may nibble at other species (Blossey et al. 1994a,b, Blossey and Schroeder 1995) were confirmed (Corrigan et al. 1998), although such spillover effects will not lead to host shifts and are temporary, as has been observed in many other control programs (McFadyen 1998). Test results were reviewed by an expert panel (the Technical Advisory Group on the Introduction of Biological Control Agents), and final approval was granted to introduce five of the six insect species. Based on results indicating a potential wider host range, the gall midge *B. salicariae* was not proposed for introduction (Blossey and Schroeder 1995). The seed-feeding weevil, *N. brevis*, although approved for introduction, was not brought into North America, as nearly 100% of *N. brevis* females collected in Europe and checked for the presence of entomopathogenic diseases showed a nematode infection. This infection did not increase adult mortality or decrease female fecundity; however, due to the potential for nontarget effects of the nematode after introduction into North America, only disease-free specimens should be introduced, which, at present, effectively eliminates the possibility to introduce *N. brevis*. The two leaf-feeding chrysomelids, *G. calmariensis* and *G. pusilla,* and the root-feeding weevil, *H. transversovittatus,* were introduced in 1992 and the flower-feeding weevil, *N. marmoratus,* in 1994.

This process, as just outlined for purple loosestrife, is standard practice in weed biocontrol programs (the introduction of herbivorous insects or pathogens) in the

United States. Worldwide, more than 1200 programs released 350 species of insects and pathogens targeting 133 plant species (Julien and Griffith 1998), and the track record regarding nontarget effects is excellent (McFadyen 1998). A comprehensive review reported damage to nontarget plants for eight insect species (McFadyen 1998). The attack on nontarget hosts for five of these eight (including *Rhinocyllus conicus,* a flowerhead weevil attacking native North American thistles (Turner et al. 1987, Louda et al. 1997), was known and predicted by the host specificity screening procedures before initial releases were approved (McFadyen 1998). For three species the host-specificity screening did not predict attack of nontarget plants (McFadyen 1998); however, in all cases the damage inflicted was temporary or minor. All agents were released based on risk assessments (see Blossey et al. 1994a,b for a discussion of purple loosestrife) using criteria that were thought important at the time of initial introductions. Since risk assessments and cost/benefit evaluations necessarily involve value judgments and societal values (e.g., the value of endangered species) that may change over time, we may not necessarily agree with all decisions of the past (Simberloff and Stiling 1996).

CONTROL AGENTS

Galerucella calmariensis and *G. pusilla*

Galerucella calmariensis and *G. pusilla* are two extremely similar sympatric species that occur throughout the European range of purple loosestrife (Palmén 1945, Silfverberg 1974). With some experience and the help of a dissecting microscope, the majority of adults can be identified to species; however, eggs and larvae are indistinguishable. The two introduced species can easily be confused with other North American *Galerucella* species (which may result in erroneous reports of nontarget effects), and species identity should first be confirmed through identification and the collection of reference specimens. In a recent paper all species in the genus *Galerucella* (sometimes referred to as *Pyrrhalta*) now known from North America were described and a key for their identification provided (Manguin et al. 1993). With the well-known and widespread waterlily beetle, *G. nymphaeae* L. (feeding on *Nuphar*, *Polygonum*, *Myrica,* and occasionally on *L. salicaria*), *G. quebecensis* Brown (known from Nova Scotia and Quebec to Michigan and Minnesota feeding on *Potentilla palustris*, marsh flower), *G. stefanssoni* Brown (found in the Northwest Territories on its host plant, *Rubus chamaemorus*, cloudberry), and the two species introduced as control agents on purple loosestrife, there are now five species of *Galerucella* in North America.

Besides looking alike, *Galerucella calmariensis* and *G. pusilla* share similar life-history characteristics and occupy the same ecological niche on their host plant (Blossey 1995a,c). Adults overwinter in the leaf litter and emerge in early spring tightly synchronized with host plant phenology (beginning of shoot growth, approximately mid-May in upstate New York). Adults feed on young plant tissue. Females lay eggs in batches of 2 to 10 on leaves and stems from May to July with peak oviposition in June (in upstate New York). First-instar larvae feed concealed within leaf or flower buds; later instars feed openly on all aboveground plant parts. Mature larvae move down the stems and pupate in the litter beneath the host plant. At this time (late June or early July in upstate New York) the damage to purple loosestrife caused by insect feeding becomes most conspicuous. Both species are usually univoltine (i.e., they have one generation a year), although there are reports that a second generation may occur in some parts of North America. Further research is needed to determine if these are true second generations or if early emerging females of the F_1 generation produce a few eggs before overwintering [similar to observations in northern Germany (Blossey 1995b)]. Adults are very mobile and possess good host-finding abilities. Peak dispersal of overwintered beetles is during the first few weeks of spring. New-generation beetles have dispersal flights shortly after emergence and are able to locate patches of host plants as far away as 1 km within a few days (Grevstad and Herzig 1997). Beetles may disperse over much larger distances.

Adult feeding causes a characteristic "shothole" pattern that is particularly obvious in spring. Larval feeding strips the photosynthetic tissue off individual leaves, creating a "window-frass" (generally leaving the upper epidermis intact). At high densities (>2 or 3 larvae/cm shoot), entire purple loosestrife populations can be defoliated. At lower densities, plants retain leaf tissue but show reduced shoot growth, reduced root growth, and fail to produce seeds (Blossey 1995a).

Hylobius transversovittatus

The biology and ecology of *H. transversovittatus* were described in detail by Blossey (1993) and the impact on the host plant by Nötzold et al. (1998). Briefly, in the spring, overwintered adults appear shortly after shoot growth of *L. salicaria* begins (coinciding with the *Galerucella* species). The largely nocturnal adults (10 to 14 mm) consume foliage and stem tissue; oviposition begins approximately 2 weeks after adults emerge from overwintering and lasts into September. Females lay white, oval-shaped eggs into plant stems or into the soil close to the host plant. First-instar larvae mine the root cortex, and older larvae subsequently enter the central part of the rootstock, where they feed for 1 to 2 years. Development time from egg to adult is dependent on environmental conditions (temperature, moisture) and time of oviposition. Pupation chambers are found in the upper part of the root and adults emerge between June and October and can be long-lived (several years).

Adult feeding is of little consequence, however, larval feeding can be very destructive. With increasing attack rates, larval feeding reduces shoot growth, seed output, shoot and root biomass and can ultimately result in plant mortality (Nötzold et al. 1998). Attack rates vary widely with rootstock age and size (up to 1 larva per 10 g of fresh root weight), and up to 40 larvae have been found per rootstock. Large rootstocks can withstand substantial feeding pressure, and several larval generations will be necessary before significant impacts can be expected.

In Europe, the weevil occurs in all purple loosestrife habitats, except permanently flooded sites, from southern Finland to the Mediterranean and from western Europe through Asia. Experiments have shown that adults and larvae can survive extended submergence. However, excessive flooding prevents access to plants by adults and will eventually kill developing larvae. Aside from this restriction, the species appears quite tolerant of a wide range of environmental conditions. Information on movements of *H. transversovittatus* is sparse because of its nocturnal nature and secretive habits during daylight hours. The most likely time to find adults is at night using a flashlight or on overcast days with light rain. Adults move primarily by walking, but dispersal flights of newly emerged adults have been reported (Palmén 1940). The effective host finding ability of the weevil is demonstrated by its occurrence in isolated populations of purple loosestrife in Europe.

Nanophyes marmoratus

Overwintered adults of *N. marmoratus* (1.4 to 2.1 mm) appear on purple loosestrife about the same time as the two *Galerucella* species (middle to late May in upstate New York). The beetles start feeding on the youngest leaves. As soon as flower buds develop, beetles move to upper parts of flower spikes, where they mate and feed on receptacles and ovaries. Oviposition starts soon thereafter and continues into August. Eggs are laid singly into the tips of flower buds before petals are fully developed. Larvae first consume stamens and, in most cases, petals, followed by the ovary. Mature larvae use frass to form pupation chambers at the bottom of the bud. Attacked buds remain closed and are later aborted. The new generation beetles appear mainly in August and feed on the remaining green leaves of purple loosestrife before overwintering in the leaf litter. Complete development from egg to adult takes about 1 month. There is one generation a year. Adult and larval feeding causes flower-bud abortion, thus reducing the seed output of *L. salicaria*. Attack rates can reach over 70%.

CURRENT DISTRIBUTION OF CONTROL AGENTS

Initial releases of the two leaf feeders occurred in seven states and Canada (Hight et al. 1995), but it soon became evident that the demand for these agents far exceeded their availability. Mass rearing procedures (Blossey and Hunt 1999) developed at Cornell University to increase the availability of control agents are now widely used. Both *Galerucella* species were released in >30 states and >1500 wetlands nationwide. The mass production of *H. transversovittatus* was more difficult. Initial techniques focused on the inoculation of eggs into stems, but this method was very labor intensive and did not result in satisfactory establishment rates. Development of an artificial diet in 1998 considerably increased the availability of *H. transversovittatus* (Blossey et al. 2000). *Nanophyes marmoratus* is the least widespread of all control agents and occurs in fewer than 10 states. At present, no mass-rearing techniques have been developed for *N. marmoratus,* and the availability of the species is limited.

PREDICTIONS, MONITORING PROTOCOL, AND EARLY OBSERVATIONS

Much of the emphasis in biological control programs has been on finding, screening, releasing and distributing control organisms, with little efforts on postrelease monitoring (McEvoy and Coombs 1999). Although biocontrol practitioners clearly identified the crucial need for long-term follow-up work decades ago (Schroeder 1983), little progress has been made in collecting quantitative data on the effect of biocontrol agents on target plant performance or in documenting the response of associated plant communities. When insects were first introduced to control purple loosestrife in North America, Malecki et al. (1993) predicted:

1. All species will become established throughout the current range of *L. salicaria* in North America.
2. The root feeder *H. transversovittatus* and the two leaf feeders *G. calmariensis* and *G. pusilla* will be most important in reducing large populations. The flower and seed feeders will stabilize smaller populations, further reducing seed output in such a way that not every disturbance will lead to a new outbreak of *L. salicaria*.
3. Combinations of agents will have greater control effect than will any species alone.
4. Control of *L. salicaria* will be achieved more rapidly in mixed-plant communities, where other plant species compete for space and nutrients.
5. Purple loosestrife abundance will be reduced to 10% of its current level over 90% of its range.

These predictions provided a framework for follow-up studies to evaluate the success of insect introductions. A standardized monitoring protocol incorporating measures of target weed populations, control agent abundance, and wetland plant communities was developed to facilitate comparison of data obtained across North America. Monitoring guidelines and techniques were designed to be simple enough to allow participation by nonspecialists, yet sophisticated enough to allow quantitative scientific evaluation. The procedures suggested represent a minimum effort; more detailed investigations (especially on plant community development and on ecosystem effects) are encouraged. This protocol has been in use across North America since 1996 (instructions and forms are available at *http://www.dnr.cornell.edu/bcontrol/weeds.htm*).

The long-term nature of such investigations (5 to 10 or even 20 years) makes it of overriding importance that changes in personnel do not put the continuation of the monitoring program at risk. Therefore, the protocol asks initially for documentation regarding site location, physical characteristics, global positioning system (GPS) locations, landownership, and so on, as well as insect release history. Because herbivore impact is expected to reduce the number of stems, stem height, and reproductive output dramatically (see Blossey 1995a for European data), our protocol was designed to detect these changes. It was also important to use nondestructive sampling to minimize interference with insect population growth. We decided to collect data in select, randomly placed, permanently marked quadrats (1 m^2). The minimum number of quadrats recommended is five, but higher numbers are encouraged. Because random quadrat placement during each site visit requires more quadrats each year to achieve the same accuracy as permanent quadrats, our monitoring protocol is the minimum effort and we encourage more elaborate studies incorporating more detailed investigations. However, we felt that it was essential to allow nonprofessional staff at nature centers, wildlife management areas, or refuges to take part in postrelease monitoring.

Assessment of insects and plants occurs twice during the season. The first visit coincides with peak insect abundance in spring (early June in upstate New York), the second assesses plant performance at the end of the growing season in the fall (before complete senescence). The exact dates for spring visits vary with latitude and in accordance with local weather conditions; therefore, no specific dates are provided. The fall visit occurs between late August and late September. All insect species introduced to control purple loosestrife overwinter as adults and reappear in the spring and can be counted once the host plant is 20 to 30 cm tall. This allows for easy assessment of individuals of all species at the same time. Since the leaf feeders are readily visible, it is possible to count adults, eggs, and larvae. Root and flower feeders lay concealed eggs; so adults must be censused to assess population levels. To avoid biasing data as a result of variation in search efforts by different investigators, search time is standardized. For each species released and for each visible life stage, search time is 1 minute. For example, at a site where only the *Galerucella* species are present, 3 minutes is spent searching for eggs, larvae, and adults. At a site where only *Hylobius* is present, 1-minute total search time is spent looking for adults. Where both *Galerucella* and *Hylobius* are present, 3 minutes is spent for *Galerucella* and 1 minute for *Hylobius*, and so on. We recommend that three people observe the sample quadrat from different sides. Total search time then has to be divided by the number of observers.

In addition to insect abundance, impact on the host plant (leaf area removed), percent cover, height (five tallest stems), number of stems of purple loosestrife and cattails (*Typha* spp.), and presence and cover of other associated plants are recorded during spring visits. In the fall, recorded data includes percent cover, number of stems and height of purple loosestrife and cattails, cover of other associated plant species, as well as an assessment of the reproductive effort of purple loosestrife (number of inflorescences, number of flower buds per 5 cm of inflorescence, and length of terminal inflorescence). We use wide cover classes to minimize variation among observers, however, this also limits our ability to detect small changes. Observations about the presence and abundance of other herbivores or predators is also recorded during each visit.

We find that teams of two or three trained observers average 10 to 15 minutes per quadrat in spring and fall. These averages vary with stem densities and flower abundances and do not include maneuvering through dense, 2-m-tall stands of purple loosestrife, which can make relocating permanent quadrats a difficult task. In general, time to collect data on purple loosestrife is reduced as the impact of herbivores becomes more pronounced (reduced flowering, shorter and fewer stems).

Predictions put forth by Malecki et al. (1993) will be assessed as sufficient data become available. All insects became established across North America (Hight et al. 1995), confirming prediction 1. Initial observations at a number of sites seem to confirm prediction 4—that control will be better in mixed-plant communities—but more quantitative data are needed. Outbreaks of the *Galerucella* species in monospecific stands of purple loosestrife have been reported across the country (Blossey and others, unpublished data), resulting in widespread defoliation and suppression of purple loosestrife.

In western New York at the Tonawanda Wildlife Management Area the impact of leaf beetles and root feeders released in 1992 has been followed since 1996 using the monitoring protocol. The study area consists of series of impoundments with water control structures managed primarily for waterfowl and other marsh birds. Insect populations at Tonawanda in 1996 were still at very low levels when seven permanent 1-m^2 quadrats were established at 50-m intervals across one impoundment. During the following years the *Galerucella* populations exploded, severely defoliating purple loosestrife by mid-June (Fig. 20.1). Beetles spread through the 25-ha impoundment as a wave, a phenomenon seen at other release sites (B. Blossey, personal observation). A detailed quadrat-by-quadrat comparison revealed this pattern (B. Blossey, unpublished data), another advantage of using permanent sampling quadrats. At present, severe defoliation is restricted to purple loosestrife growing within 2 km of the release area, although individual beetles have spread about 5 km from the initial release site. Stem density in the impoundment declined from a mean of 75 ± 16.86/ m^2 (range 35 to 170) in 1996 to 23 ± 8.4/ m^2 (range 0 to 54) in 1999, and height of the remaining stems declined 65% or more (Fig. 20.2) during the same period. Our measurement of only the five tallest plants in each quadrat is a very conservative assessment of the impact of insect feeding on plant performance. Had we

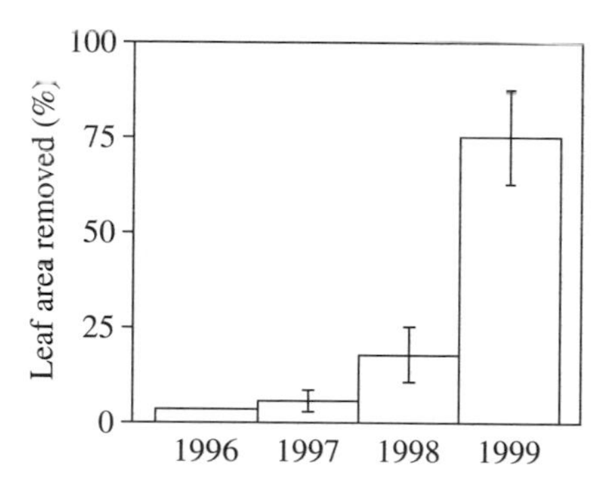

FIGURE 20.1. Leaf area (%) removed by feeding of biological control agents on purple loosestrife at Tonawanda Wildlife Management Area, 1996–1999. Data are means ± SE of seven permanent 1-m^2 quadrats and were collected in early June of each year.

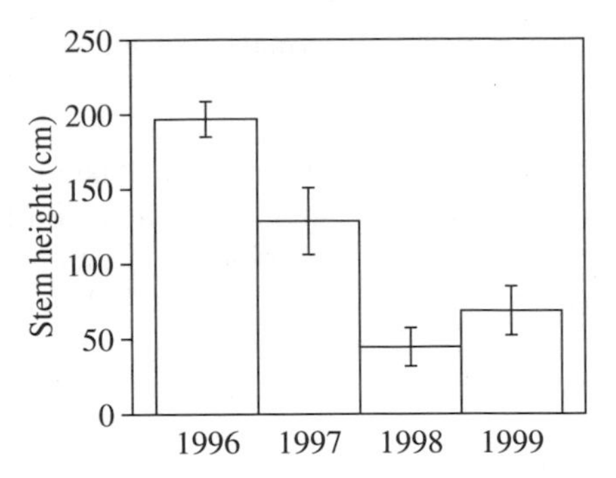

FIGURE 20.2. Stem height (cm) of purple loosestrife at the end of the growing season (mid-September) at Tonawanda Wildlife Management Area from 1996 through 1999. Data are means ± SE of average stem height of five tallest stems in each of seven permanent 1-m_2 quadrats.

measured all stems, mean height would have been lower, but monitoring time would have increased substantially. The persistence of purple loosestrife even under severe feeding pressure (>75% leaf area loss, Fig. 20.2) reflects the compensatory ability of well-established mature plants. The results demonstrate that several years of heavy feeding are necessary to deplete large rootstock reserves and weaken purple loosestrife sufficiently to result in plant death. Contributing to the continued survival of purple loosestrife is the fact that beetles moved from the site to attack purple loosestrife in adjacent areas. Beetle numbers in the defoliated areas declined, allowing plants to somewhat recover. We anticipate that (as seen at other release sites) beetles will return and attack the residual purple loosestrife population.

Ongoing research and monitoring programs are testing the validity of predictions 2, 3, and 4, but it is too early for a judgment. Of particular interest is the assumption of cumulative effects of herbivores (prediction 3). Harris (1981) proposed that biocontrol agents be considered stress factors, the aim being to increase stress load until the balance is tipped against the target weed population. Agent combinations were recently reported to be more destructive to plants than a single species alone (Fowler and Griffin 1995). However, agent combinations may also disadvantage one of the species, as even spatially separated herbivores can compete via their common host plant (Masters et al. 1993). For example, root feeders showed a reduced performance when their host plant was attacked simultaneously by aboveground herbivores. Interestingly, aboveground herbivores showed improved performance on plants simultaneously attacked by root feeders. Whether these interactions have any influence on the success of weed biocontrol in systems where both above- and belowground herbivores are released requires further study. We are currently conducting these experiments for the *L. salicaria–Galerucella–Hylobius* system in upstate New York.

OUTLOOK

Purple loosestrife is an invasive nonindigenous species that has negatively affected temperate North American wetlands for decades. The introduction of four host-specific herbivorous insects in a classical biological weed control program seems likely to achieve long-term control of purple loosestrife. The tremendous seed bank in established purple loosestrife stands (Welling and Becker 1990) presents a long-term management problem because seeds remain viable for many years. Virtually every disturbance in a wetland creates favorable conditions for the germination of purple loosestrife. The continued presence of insect control agents should reduce the abundance of purple loosestrife to low levels, similar to the situation in Europe. However, it is not yet clear what type of replacement communities will develop. At many sites, the once monotypic stands of *L. salicaria* are replaced by a diverse wetland plant community. At several sites, other invasive species, such as *Phragmites australis* (common reed) or *Phalaris arundinacea* (reed canary grass), expand as purple loosestrife is controlled, clearly not a desired result. At yet other sites, dense purple loosestrife litter limits growth of native species. In cooperation with land managers, we are currently investigating means (fire, disking, flooding, mowing, etc.) to accelerate the return of native plant communities (initial test results using flooding and burning have yielded excellent short-term results). Long-term monitoring will determine local trends and help in assessing the need and feasibility of additional restoration work to obtain a diverse wetland flora and prevent the expansion of other invasive species. Together with scientists at the Bureau of Reclamation, we are also evaluating the prospects of aerial photography and remote sensing to monitor landscape-level changes in wetland plant communities as a result of insect feeding on purple loosestrife.

The presence of four new insect species and their feeding on purple loosestrife, a plant that has little value as food or cover for native species will create a new dynamic food web in wetlands. We have already observed birds (catbirds, blackbirds, and marsh wrens) and dragonflies feeding on adult leaf beetles. A suite of invertebrate predators and parasitoids utilizes eggs, larvae, and adults as a new and abundant food source. Attack of native parasitoids on *H. transversovittatus* larvae in the stems and attack of a nematode on adult *Galerucella* remains at 10% (B. Blossey, unpublished data), however, in some instances native predators appear to limit leaf beetle population growth in cages (T. Hunt, unpublished data). In Europe, specialized egg, larval, and adult parasitoids can have dramatic impacts (attack rates of up to 90%) on the leaf beetles and flower-feeding weevils. Although great care was taken to avoid the introduction of these and other natural enemies from Europe, the impact of native predators on the success of purple loosestrife biocontrol and the contribution of biocontrol agents to the wetland food web dynamics need to be assessed.

For those interested in additional information or participation in the long-term monitoring program, pictures of all species and their life stages, a description of their biology and impact on purple loosestrife, as well as descriptions of rearing and monitoring procedures and new developments can be found at *http://www.dnr.cornell.*

edu/bcontrol/weeds.htm. For even more detailed information on biocontrol order the purple loosestrife video, "Restoring the Balance," or a video introducing mass rearing procedures (both available from Cornell Media Services, 607-255-2090). With the exception of *N. marmoratus*, all control agents are currently commercially available. For further information, contact the author.

REFERENCES

Batra, S. W. T., D. Schroeder, P. E. Boldt, and W. Mendl. 1986. Insects associated with purple loosestrife (*Lythrum salicaria*) in Europe. Proceedings of the Entomological Society of Washington 88:748–759.

Blossey, B. 1993. Herbivory below ground and biological weed control: life history of a root-boring weevil on purple loosestrife. Oecologia 94:380–387.

Blossey, B. 1995a. Impact of *Galerucella pusilla* Duft. and *G. calmariensis* L. (Coleoptera: Chrysomelidae) on field populations of purple loosestrife (*Lythrum salicaria* L.). Pages 27–32 *in* E. S. Delfosse and R. R. Scott (eds.), Proceedings of the 8th International Symposium on the Biological Control of Weeds Feb. 2–7, 1992, Canterbury, New Zealand. DSIR/CSIRO, Melbourne, Australia.

Blossey, B. 1995b. Coexistence of two competitors in the same fundamental niche: Distribution, adult phenology and oviposition. Oikos 74:225–234.

Blossey, B. 1995c. A comparison of various approaches for evaluating potential biological control agents using insects on *Lythrum salicaria*. Biological Control 5:113–122.

Blossey, B., and T. Hunt. 1999. Mass rearing methods for *Galerucella calmariensis* and *G. pusilla* (Coleoptera: Chrysomelidae), biological control agents of *Lythrum salicaria* (Lythraceae). Journal of Economic Entomology 92:325–334.

Blossey, B., and L. Skinner. 2000. Design and importance of post release monitoring. Pages 693–706 *in* N. R. Spencer (ed.), Proceedings of the 10th International Symposium on Biological Control of Weeds, July 4–10 1999, Montana State University, Bozeman, Montana, USA.

Blossey, B., and D. Schroeder. 1995. Host specificity of three potential biological weed control agents attacking flowers and seeds of *Lythrum salicaria*. Biological Control 5:47–53.

Blossey, B., D. Schroeder, S. D. Hight, and R. A. Malecki. 1994a. Host specificity and environmental impact of the weevil *Hylobius transversovittatus,* a biological control agent of purple loosestrife (*Lythrum salicaria*). Weed Science 42:128–133.

Blossey, B., D. Schroeder, S. D. Hight, and R. A. Malecki. 1994b. Host specificity and environmental impact of two leaf beetles (*Galerucella calmariensis* and *G. pusilla*) for the biological control of purple loosestrife (*Lythrum salicaria*). Weed Science 42:134–140.

Blossey, B., D. Eberts, E. Morrison, and T. R. Hunt. 2000. Mass rearing the weevil *Hylobius transversovittatus* (Coleoptera: Curculionidae), biological control agent of *Lythrum salicaria,* on semiartificial diet. Journal of Economic Entomology:1644–1656.

Brown, B. 1999. The impact of an invasive species (*Lythrum salicaria*) on pollination and reproduction of a native species (*L. alatum*). Ph.D. dissertation, Department of Biological Sciences, Kent State University, Kent, OH.

Corrigan, J. E., D. L. MacKenzie, and L. Simser. 1998. Field observations of non-target feeding by *Galerucella calmariensis* (Coleoptera: Chrysomelidae), an introduced biological control agent of purple loosestrife, *Lythrum salicaria* (Lythraceae). Proceedings of the Entomological Society of Ontario 129:99–106.

Emery, S. L., and J. A. Perry. 1996. Decomposition rates and phosphorus concentrations of purple loosestrife (*Lythrum salicaria*) and cattail (*Typha* spp.) in fourteen Minnesota wetlands. Hydrobiologia 323:129–138.

Fowler, S. V., and D. Griffin. 1995. The effect of multi-species herbivory on shoot growth in gorse, *Ulex europaeus*. Pages 579–584 *in* E. S. Delfosse and R. R. Scott (eds.), Proceedings of the 8th International Symposium on the Biological Control of Weeds, Feb. 2–7, 1992, Canterbury, New Zealand. DSIR/CSIRO, Melbourne, Australia.

Goeden, R. D. 1983. Critique and revision of Harris scoring system for selection of insect agents in biological control of weeds. Protection Ecology 5:287–301.

Grevstad, F. S., and A. L. Herzig. 1997. Quantifying the effects of distance and conspecifics on colonization: experiments and models using the loosestrife leaf beetle, *Galerucella calmariensis*. Oecologia 110:60–68.

Grout, J. A., C. D. Levings, and J. S. Richardson. 1997. Decomposition rates of purple loosestrife (*Lythrum salicaria*) and Lyngbyei's sedge (*Carex lyngbyei*) in the Fraser River estuary. Estuaries 20:96–102.

Harris, P. 1973. The selection of effective agents for the biological control of weeds. Canadian Entomologist, 105:1495–1503.

Harris, P. 1981. Stress as a strategy in the biological control of weeds. Pages 333–340 *in* G. C. Papavizas (ed.), Biological control in crop production. Allanheld, Osmun, Totowa, NJ.

Hight, S. D. 1990. Available feeding niches in populations of *Lythrum salicaria* L. (purple loosestrife) in the northeastern United States. Pages 269–278 *in* E. S. Delfosse (ed.), Proceedings of the 7th International Symposium on the Biological Control of Weeds, Mar. 6–11, 1988, Rome. Istituto Sperimentale de la Patologià Vegetale (MAF), Rome.

Hight, S. D., B. Blossey, J. Laing, and R. DeClerck-Floate. 1995. Establishment of insect biological control agents from Europe against *Lythrum salicaria* in North America. Environmental Entomology 24:967–977.

Julien, M. H., and M. W. Griffith. 1998. Biological control of weeds: a world catalogue of agents and their target weeds, 4th ed. CAB International, Wallingford, Berkshire, England.

Louda, S. M., D. Kendall, J. Connor, and D. Simberloff. 1997. Ecological effects of an insect introduced for the biological control of weeds. Science 277:1088–1090.

MacDonald, I. A., L. L. Loope, M. B. Usher, and O. Hamann. 1989. Wildlife conservation and the invasion of nature reserves by introduced species: a global perspective. Pages 215–255 *in* J./ A. Drake, H. A. Mooney, F. di Castri, R. H. Groves, F. J. Kruger, M. Rejmanék, and M. Williamson (eds.), Biological invasion: a global perspective. Wiley, Chichester, West Sussex, England.

Malecki, R. A., B. Blossey, S. D. Hight, D. Schroeder, L. T. Kok, and J. R. Coulson. 1993. Biological control of purple loosestrife. Bioscience 43:480–486.

Manguin, S., R. White, B. Blossey, and S. D. Hight. 1993. Genetics, taxonomy, and ecology of certain species of *Galerucella* (Coleoptera: Chrysomelidae). Annals of the Entomological Society of America 86:397–410.

Masters, G. J., V. K. Brown, and A. C. Gange. 1993. Plant mediated interactions between above- and below-ground insect herbivores. Oikos 66:148–151.

McEvoy, P. B., and E. M. Coombs. 1999. Biological control of plant invaders: regional patterns, field experiments, and structured population models. Ecological Applications 9:387–401.

McFadyen, R. E. C. 1998. Biological control of weeds. Annual Review of Entomology 43: 369–393.

Nötzold, R., B. Blossey, and E. Newton. 1998. The influence of below-ground herbivory and plant competition on growth and biomass allocation of purple loosestrife. Oecologia 113: 82–93.

Palmén, E. 1940. Zur Biologie und nordeuropäischen Verbreitung von *Hylobius transverso-vittatus* Steph. (Coleoptera: Curculionidae). Annales Entomologici Fennici 6:129–140.

Palmén, E. 1945. Zur Systematik Finnischer Chrysomeliden. 1. Gattung Galerucella Crotch. Annales Entomologici Fennici 11:140–147.

Schroeder, D. 1983. Biological control of weeds. Pages 41–78 *in* W. E. Fletcher (ed.), Recent advances in weed research. Commonwealth Agricultural Bureau, Farnham Royal, Slough, Buckinghamshire, England.

Shamsi, S. R. A., and F. H. Whitehead. 1974. Comparative ecophysiology of *Epilobium hirsutum* L. and *Lythrum salicaria* L.: I. General biology, distribution and germination. Journal of Ecology 62:279–290.

Silfverberg, H. 1974. The West Palaeartic species of Galerucella Crotch and related genera (Coleoptera, Chrysomelidae). Notulae Entomologicae 54:1–11.

Simberloff, D., and P. Stiling. 1996. How risky is biological control? Ecology 77:1965–1974.

Thompson, D. Q., R. L. Stuckey, and E. B. Thompson. 1987. Spread, impact, and control of purple loosestrife (*Lythrum salicaria*) in North American wetlands. Fish and Wildlife Research 2. U.S. Fish and Wildlife Service, Washington, DC.

Turner, C. E., R. W. Pemberton, and S. S. Rosenthal. 1987. Host utilization of native *Cirsium* thistles (Asteraceae) by the introduced weevil *Rhinocyllus conicus* (Coleoptera: Curculionidae) in California. Environmental Entomology 16:111–115.

U.S. Congress. 1993. Harmful non-indigenous species in the United States. Office of Technology Assessment report OTA-F-565. U.S. Government Printing Office, Washington, DC.

Welling, C. H., and R. L. Becker. 1990. Seed bank dynamics of *Lythrum salicaria* L.: implications for control of this species in North America. Aquatic Botany 38:303–309.

Index